TRANSONIC AERODYNAMICS

Edited by
David Nixon
Nielsen Engineering & Research, Inc.
Mountain View, California

Volume 81
PROGRESS IN
ASTRONAUTICS AND AERONAUTICS

Martin Summerfield, Series Editor-in-Chief
Princeton Combustion Research Laboratories, Inc.
Princeton, New Jersey

Technical papers selected from the Transonic Perspective Symposium, Moffett Field, California, February 18-20, 1981, and subsequently revised for this volume. This Symposium was sponsored by the Aerodynamics Research Branch of NASA/Ames Research Center and the Fluid Dynamics Program of the Office of Naval Research.

Published by the American Institute of Aeronautics and Astronautics, Inc.,
1290 Avenue of the Americas, New York, N.Y. 10104.

American Institute of Aeronautics and Astronautics, Inc.
New York, New York

Library of Congress Cataloging in Publication Data

Transonic Perspective Symposium (1981: Moffett Field, Calif.)
   Transonic aerodynamics.

   (Progress in astronautics and aeronautics; v. 81)
   "Sponsored by the Aerodynamics Research Branch of NASA/Ames Research Center and the Fluid Dynamics Program of the Office of Naval Research."
Includes index.
   1. Aerodynamics, Transonic—Congresses. I. Nixon, David. II. Ames Research Center. Aerodynamics Research Branch. III. United States. Office of Naval Research. Fluid Dynamics Branch. IV. American Institute of Aeronautics and Astronautics. V. Title. V. Series.
TL507.P75   vol. 81 [TL571]   629.1s [629.132'1304]   82-4027
ISBN 0-915928-65-5   AACR2

Copyright © 1982 by
American Institute of Aeronautics and Astronautics, Inc.

All rights reserved. No part of this book may be reproduced in any form or by any means, electronic or mechanical, including photocopying, recording, or by any information storage and retrieval system, without permission in writing from the publisher.

Table of Contents

Authors x

Preface xiii

List of Series Volumes 1-81 xvii

Part I. Review of Transonic Aerodynamics 1

Chapter I. Transonic Aerodynamics—History and Statement of the Problem 3
Introduction 3
Transonic Aerodynamics before 1940 4
Transonic Aerodynamics in 1945 7
Transonic Aerodynamics in 1960 12
Transonic Aerodynamics in 1975 47
Concluding Remarks 66

Chapter II. Commercial Transports—Aerodynamic Design for Cruise Performance Efficiency 81
Introduction 81
Commercial Transports vs Military Aircraft 82
   Transports—Commercial vs Military 82
   Commercial Transports vs High-Performance Fighters 85
Commercial Transports—Cruise Drag Efficiency 90
   Basic Parasite Drag Efficiency 92
   Lift-Dependent Drag Efficiency 96
   Compressibility/Drag Efficiency 98
   Trim Drag 101
Commercial Transport Transonic Wing Design 103
   General Design Considerations 103
      Supercritical Airfoils-Wings 105
   Some Related Computational and Experimental Problems . 113
      Parasite Drag 114
      Lift-Dependent Drag 121
      Compressibility Drag 124
      Buffet Boundary 132
      Post-Buffet Characteristics 136
Interference Drag Problems 139
   Excess Drag at Intersections 139
   Nacelle-Pylon-Wing Interference 140
Recommendations 144

Chapter III. Practical Aerodynamic Problems—Military Aircraft 149
Introduction 149
Performance Environment 150

Transonic Wing Design Problem 153
    Attached Flow Design 155
    Separated Flow Design... 164
    Aeroelastic Tailoring 171
Store Carriage and Separation 174
    Carriage 174
    Separation 179
Concluding Remarks... 182

Chapter IV. Experimental Testing at Transonic Speeds 189
Introduction 190
Experimental Test Process 193
Wind-Tunnel Interference...................................... 194
    Wall Interference—A Significant Problem....................... 194
    Technology for Minimizing Wall Interference Effects Is Growing Rapidly... 196
        Reducing Model Size.. 196
        Utilizing Ventilated Wall Wind Tunnels........................ 196
        Applying Theoretical/Empirical Corrections 198
        Using Self-Correcting or Self-Streamlining Wind-Tunnel Walls........ 203
Wind-Tunnel Models and Support Systems 206
    Wind-Tunnel Models Must Be Tailored To Meet Objective of Experiment .. 206
        Basic Research Models 206
        Concept Development Models 207
        Code Verification Models.................................... 208
        Configuration Development Models........................... 209
        Production Aircraft Models 209
    Model Support System Effects Must Be Minimized.................. 209
        Wall Mount.. 210
        Sting Mount 212
    Power Plant Simulation Must Be Done with Care.................... 213
Reynolds Number Simulation.................................. 215
    Reynolds Number Scale Effects Can Cause Significant Problems.......... 215
    Ways of Quantifying High Reynolds Number Scale Effects
        Are Being Developed 218
        Fixed Transition at Model Leading Edges 218
        Natural Transition 219
        Vortex Generators.. 220
        Fixed Transition Aft on the Model............................. 221
        Panel Model 223
        Component Testing... 224
        High Reynolds Number Facilities 225
Aerodynamic Flow Visualization 227
    Flow Visualization Is Important To Understanding
        the Physics of the Experiment 227
    Flow Visualization Can Be Used to Highlight Aerodynamic
        Flow Phenomena 227
        Boundary-Layer Transition................................... 227
        Surface Streamlines... 228
        Flowfield Visualization 230
Flow Measurement Instrumentation............................ 233
    Standard Instrumentation 233
    Specialized Instrumentation 234
Data Reduction/Presentation 236
    On-Line Data Reduction 236
    Data Presentation.. 236
Concluding Remarks... 236

Chapter V. Potential Equation Methods for Transonic Flow Prediction............................ 239
Introduction 239
Basic Equations for Transonic Flow............................ 241
Some Basic Concepts 248
Weak Solutions 248
Numerical Methods 252
Formulation of the Numerical Method......................... 252
The Murman Difference Scheme 256
Solution of the Exact Potential Flow Equation 259
Approximate Factorization.................................. 264
Finite-Volume Method 265
Multiple Grid Techniques 270
Grid Generation 274
Airfoil Calculations Using a Mapping to a Circle 276
Sheared Parabolic Grid for a Finite Wing......................... 277
Sheared Parabolic Grid for a Wing-Body Combination 279
Body-Oriented Cylindrical Coordinate System..................... 279
Present Computational Methods 282
Two-Dimensional Flows..................................... 282
Three-Dimensional Flows................................... 287

Chapter VI. Reynolds Averaged Navier-Stokes Computations of Transonic Flows—the State-of-the-Art 297
Introduction 297
Governing Equations 300
Navier-Stokes Equations 300
Reynolds Averaged Navier-Stokes Equations with Mass-Weighted Variables 301
Modeling of Turbulence...................................... 303
Conservation Law Forms.................................... 307
Reynolds Averaged Navier-Stokes Equations in Curvilinear Coordinates.... 309
Boundary Conditions....................................... 311
Computational Grids 316
Accuracy Requirements..................................... 317
Methods for Improving Flow Simulation Accuracy.................... 319
Management of Grid Systems 320
Methods for Generating Grid Systems........................... 320
Component Adaptive Grid Systems............................. 323
Numerical Methods... 324
Two Crucial Nonlinear Convective Phenomena..................... 324
Numerical Techniques for Computing Shocks...................... 326
Numerical Techniques for Computing Turbulence Effects 330
Effect of Grid Choice on Numerical Stability 332
The Basic Difference Equations................................ 333
Effect of Grid Topologies on Computational Efficiency................. 335
Comparison between Experiments and Calculations of Turbulent Transonic Flows.. 337
Axisymmetric Steady Flows 339
Two-Dimensional Steady Flows................................ 343
Two-Dimensional Unsteady Flows.............................. 348
Three-Dimensional Steady Flows............................... 358
Concluding Remarks... 361

Chapter VII. Transonic Design Using Computational Aerodynamics 377
  Introduction 377
  Review of Design Methods 379
    Inverse Solution Methods 379
    Evaluation of Inverse Methods 382
    Numerical Optimization 383
  Numerical Optimization Case Study 386
    Configuration Characteristics and Performance 386
    Design Procedure 386
      Design Objective and Constraints 388
      Design Variables 389
      Implementation 390
      Extraction of Solid Wing Geometry 390
      Analysis of Optimized Wing 391
  Configuration Design 391
    Numerical Optimization 392
      Extended Small Disturbance Design 393
      Full Potential Design 393
  Wind-Tunnel Test 394
    Test Facility 394
    Models 394
  Design Evaluation 395
    Analysis of Test Data 395
    Aircraft Performance 396
    Correlations 397
  Conclusions 398

Part II. Prediction Methods—Successes and Failures 403

Chapter VIII. Application of Computational Methods to Transonic Wing Design 405
  Introduction 405
  Analysis Techniques 406
    Bailey-Ballhaus Method 407
    Pandora-Boppe Method 408
    Jameson-Caughey Nonconservative Code (FLO-22) 409
    Jameson-Caughey Finite-Volume Code (FLO-27) 409
    Advanced Jameson Codes (FLO-28/30) 410
    Typical Results 410
  Design Techniques 413
  Design Application 414
    Case I—Trainer Wing Refinement 415
    Case II—Advanced Fighter Flap Design 421
  Conclusions and Recommendations 427

Chapter IX. A-7 Transonic Wing Designs 431
  Introduction 431
  Correlation of FLO-22 Analysis 432
  Airfoil Optimization 433
  Wing Design Procedure 435
  Wing Design Studies 437

Wind-Tunnel Test Results..439
Comparison of Theory with Experiment 441
    FLO-22...441
    FLO-28...442
    FLO-30...444
    WIBCO..447
Concluding Remarks..449

Chapter X. Transonic Computational Experience for Advanced Tactical Aircraft 451
Introduction 452
Discussion..452
    HiMAT Wing Analysis 453
    Forward Swept Wing Analyses 460
    Design Considerations 463
Conclusions...464

Chapter XI. Extension of FLO Codes to Transonic Flow Prediction for Fighter Configurations..............................467
Introduction 467
FLO Code Validation for Low Aspect Ratio Wings 468
FLO Code Application to F-15 Geometry 469
Panel Method Application to F-15 Geometry.....................473
Fuselage Representation by Equivalent Simple Body (ESB)..........476
Applications 485
Conclusions...486

Chapter XII. A Series of Airfoils Designed by Transonic Drag Minimization for Gates Learjet Aircraft 489
Introduction 489
Description of Production Aircraft Wings 489
Wing Design Considerations....................................490
Optimization Technique 492
Single-Point Optimization Results 494
Two-Point Optimization Results................................498
Theory-Experiment Correlation 503
Assessment of the Optimization Technique 506
Conclusions...507

Chapter XIII. Applied Computational Transonics—Capabilities and Limitations 511
Introduction 512
Summary of Computational Methods 512
Flow Analysis Applications 514
    Airfoil Performance Assessment 516
    Wing Performance Assessment 518
    Configuration Performance Improvements 527
    Lift Effects...529
    Reynolds Number Effects 532

Geometry Design Applications 536
    Sample Design Problem 536
    Actual Design Problem 539
Summary 541

Chapter XIV. Evaluation of Full Potential Flow Methods for the Design and Analysis of Transport Wings 545
Introduction 545
Computer Codes Evaluated 546
Prediction of Pressure Distributions 547
Modifications to FLO-22 552
    Viscous Correction 552
    Fuselage Effect Simulation 553
    Consistent Calculation of Velocity 553
Prediction of Pressure Distribution Using the Improved FLO-22 Code 556
Prediction of Drag 557
Conclusions and Recommendations 559

Part III. Alternative Prediction Methods 563

Chapter XV. Nonlinear Green's Function Method for Unsteady Transonic Flows 565
Introduction 567
Review of the Linear Green's Function Method 568
The Green's Function Method for Transonic Flows 571
Shock-Capturing Nature of Method 577
Assessment of Green's Function Method 584
Concluding Remarks 585
Appendix A: Transonic Integral Equation 587
    Green's Theorem for Transonic Potential Aerodynamics 587
    Transonic Integral Equation 589
    Contribution of Wake and Shock Wave 591
    Moving Shock Waves 592
Appendix B: Numerical Solutions of the Integral Equation 593
    Space Discretization 593
    Time Discretization 598

Chapter XVI. Hybrid Approach to Transonic Inviscid Flow with Moderate to Strong Shock Wave 605
Introduction 605
Hybrid Method 606
    Potential Flow Solution (Overall Flowfield) 606
    Solution to Euler Equations (Shock-Wave Region) 608
Solution Procedure 611
    Determination of the Shock Location 612
    Enforcement of Kutta Condition 613
    Shock Geometry 614
Results and Discussion 615
Conclusions and Recommendations 617

Chapter XVII. Application of a Shock-Turbulent Boundary-Layer Interaction Theory in Transonic Flowfield Analysis **621**
  Introduction 621
  Brief Outline of the Local Interaction Theory 622
  Generalized Viscous Ramp Model of the Interaction 624
  Incipient Separation 627
  Application to Global Flowfield Analysis 628
    Downstream Effects from a Shock/Boundary-Layer Interaction 628
    Supercritical Wing Section Flowfields 632
  Concluding Remarks.. 634

Chapter XVIII. Rapid Approximate Determination of Nonlinear Solutions: Application to Aerodynamic Flows and Design/Optimization Problems **637**
  Introduction 637
  Analysis... 638
    Perturbation Concept and Methods 638
    Coordinate Straining 640
    Analytical Formulation 641
    Application to Surface Properties 644
  Results... 648
    Single and Multiple Parameter Perturbations 648
    Combination of Approximation Method with Optimization Procedures..... 652
  Concluding Remarks.. 659

Epilogue 663

Author Index for Volume 81 669

Authors

Chapter I.   Transonic Aerodynamics—History and Statement of the Problem
   John R. Spreiter
   Stanford University, Stanford, Calif.

Chapter II.   Commercial Transports—Aerodynamic Design for Cruise Performance Efficiency
   F.T. Lynch
   Douglas Aircraft Company, McDonnell Douglas Corporation, Long Beach, Calif.

Chapter III.   Practical Aerodynamic Problems—Military Aircraft
   Richard G. Bradley
   General Dynamics Corporation, Fort Worth, Texas

Chapter IV.   Experimental Testing at Transonic Speeds
   James A. Blackwell, Jr.
   Lockheed-Georgia Company, Marietta, Ga

Chapter V.   Potential Equation Methods for Transonic Flow Prediction
   David Nixon and G. David Kerlick
   Nielsen Engineering & Research, Inc., Mountain View, Calif.

Chapter VI.   Reynolds Averaged Navier-Stokes Computations of Transonic Flows—the State-of-the-Art
   Unmeel Mehta and Harvard Lomax
   Nasa/Ames Research Center, Moffett Field, Calif.

Chapter VII. Transonic Design Using Computational Aerodynamics
   M.E. Lores and B.L. Hinson
   Lockheed-Georgia Company, Marietta, Ga.

Chapter VIII. Application of Computational Methods to Transonic Wing Design
   I.C. Bhateley and R.A. Cox
   General Dynamics Corporation, Fort Worth, Texas

Chapter IX. A-7 Transonic Wing Designs
   H.P. Haney
   Lockheed-Georgia Company, Marietta, Ga.

Chapter X. Transonic Computational Experience for Advanced Tactical Aircraft
E. Bonner and P. B. Gingrich
North American Aircraft Division, Rockwell International, Los Angeles, Calif.

Chapter XI. Extension of FLO Codes to Transonic Flow Prediction for Fighter Configuration
A. Verhoff and P.J. O'Neil
McDonnell Aircraft Company, McDonnell Douglas Corporation, St. Louis, Mo.

Chapter XII. A Series of Airfoils Designed by Transonic Drag Minimization for Gates Learjet Aircraft
M.L. Hinson
Gates Learjet Corporation, Wichita, Kans.

Chapter XIII. Applied Computational Transonics—Capabilities and Limitations
P.A. Henne, J.A. Dahlin, and C.C. Peavey
Douglas Aircraft Company, McDonnell Douglas Corporation, Long Beach, Calif.

Chapter XIV. Evaluation of Full Potential Flow Methods for the Design and Analysis of Transport Wings
Luis R. Miranda
Lockheed-California Company, Burbank, Calif.

Chapter XV. Nonlinear Green's Function Method for Unsteady Transonic Flows
Kadin Tseng and Luigi Morino
Boston University, Boston, Mass.

Chapter XVI. Hybrid Approach to Transonic Inviscid Flow with Moderate to Strong Shock Wave
Tsze C. Tai
David Taylor Naval Ship Research and Development Center, Bethesda, Md.

Chapter XVII. Application of a Shock-Turbulent Boundary-Layer Interaction Theory in Transonic Flowfield Analysis
G.R. Inger
University of Colorado, Boulder, Colo.

Chapter XVIII. Rapid Approximate Determination of Nonlinear Solutions: Application to Aerodynamic Flows and Design/Optimization Problems
Stephen S. Stahara
Nielsen Engineering & Research, Inc., Mountain View, Calif.

Preface

As the name implies, transonic aerodynamics is concerned with the flow phenomena that exist where the velocities are in the neighborhood of the local speed of sound. Since it is in this regime that civil aircraft cruise and military aircraft maneuver, flight at transonic speeds is a common occurrence for many aircraft. In order to develop efficient designs it is necessary to understand the changes in the aerodynamic behavior that occur at transonic speeds. This regime is complicated by the formation of shock waves, the location of which depends on a complex interaction of freestream Mach number and body geometry. Transonic flows are very sensitive to small perturbations in various parameters, such as Mach number, and this can make accurate experimentation difficult. Also, since the shock location and strength is a crucial part of a transonic flow, any prediction method must be based on a nonlinear equation or set of equations, as opposed to the linear equations that can adequately model subsonic or supersonic flow. Considerable advances have been made in recent years in the understanding of transonic flow phenomena and, although not all of the problems have been resolved, it is appropriate to examine the present status of transonic aerodynamics.

Bodies, such as artillery shells, have been flying through the transonic regime for many years, but scientific interest in flow phenomena at these speeds dates from only the earlier part of the 20th century. At that time theoretical work was hampered by the nonlinearity of the mathematics in physical variables. This problem was circumvented by interchanging the dependent and independent variables via the hodograph transformation. This transformation linearizes the equations at the expense of complicating the boundary conditions. Experiment was hindered by the choking of the wind tunnel when shock waves appeared on the body.

In the early 1940s the advent of high-performance aircraft that could reach transonic speeds in a dive led to a concentration of research effort in transonic flow. The problem of choking in the wind tunnel was solved by the ingenious idea of making the walls porous so that the streamlines in the tunnel flow could more closely approximate the real flow. Prediction methods evolved via "compressibility" corrections of linear Prandtl-Glauert theory for shock-free flows to the integral equation method that could calculate shock locations. There was also the pioneering numerical solutions of Emmons, but in general the development of accurate prediction methods was delayed by the absence of adequate computational facilities.

During the 1950s the "transonic controversy" was a topic of considerable discussion. The question arose as to whether a transonic flow with an embedded supersonic zone had to contain a shock. A study of the hodograph equations indicated that such a shock-free flow was possible but the experimental data at that time did not show any corroborating evidence. A series of papers by Morawetz examined the problem and concluded that a shock-free flow could not exist since shocks would form under an infinitesimal perturbation.

To a great extent this was the status of the subject in 1962 and an excellent review is contained in the book by Ferrari and Tricomi.[1] Apart from groups in the Netherlands and the United Kingdom, transonic research was abandoned as the more pressing, and interesting, problems connected with hypersonic flight became important. In the early 1960s, however, Nieuwland succeeded in developing the hodograph method to design transonic airfoils that had a substantial zone of supersonic flow but no shock wave. These were the first supercritical airfoil sections. At the same time Pearcy was experimentally developing airfoils with a "peaky" pressure distribution which again had a supersonic shock-free flow. Whitcomb was also experimenting on the development of supercritical airfoils, although these sections had considerably more aft camber than those of Pearcy. The appearance of these supercritical shock-free sections recalled the "transonic controversy" of some 10 years earlier and the various viewpoints were made compatible by recognizing that shock-free flows could exist but that they were bounded on either side by shocked flows. The development of large computers was taking place simultaneously, the availability of which was an important contributing factor in initiating the present resurgence of interest in the topic which began around 1968.

Since 1968 development in prediction methods has been rapid and, together with the impetus given by the cost of fuel to the design of efficient aircraft, has led to great advances in understanding the nature of transonic flow. In spite of this growth in knowledge, no book has appeared that surveys the recent advances in the subject. This book is an attempt to fill that void. For reasons of space only steady transonic flows are considered, although it is unfortunate that the recent advances in both the experimental and theoretical treatment of unsteady transonic flows have had to be neglected. The book is divided into three parts: the first consists of review articles, the second of case histories of the use of prediction methods in a design environment, and the third considers some possible alternatives to existing numerical prediction methods.

The review topics are chosen so that the recent advances in experimental techniques and prediction methods can be placed both in a historical context and in the context of the needs of the aircraft designer. The first three articles concern the history of transonic flow research and the needs of commercial and military aircraft manufacturers. The remaining four articles cover experimental techniques, prediction methods including both the commonly used potential methods and the more complex Navier-Stokes calculation methods, and the use of experiment and computational techniques in present aircraft design.

The second part of the book concerns the behavior of the potential equation prediction methods in a real design environment. The authors have illustrated both the advantages and present disadvantages of these techniques; they are to be congratulated on their candor. Those who develop the prediction methods should take note of the various points raised in these articles.

The third part of the book explores alternatives to the commonly used finite-difference prediction methods. It is not clear if these methods will ever replace the present mainstream techniques but, in any event, it is always a worthwhile exercise to consider different viewpoints of the same problem.

The book concludes with some comments on the present understanding of the nature of transonic flow and on the status of prediction methods. Most of the points contained in this article arose during a panel discussion comprised of most of the authors of the review articles, together with Doris K. Steckel, Paul E. Rubbert, Antony Jameson, and R.W. Barnwell.

The articles in this book, apart from that by David Nixon and G. David Kerlick, were written for the Transonic Perspective Symposium held at NASA/Ames Research Center on Feb. 18-20, 1981, sponsored by the Office of Naval Research and NASA/Ames Research Center. I would like to take this opportunity to thank Robert W. Whitehead of the Fluid Dynamics Program of the Office of Naval Research and Raymond M. Hicks of the Aerodynamics Division of NASA/Ames Research Center for their invaluable help and suggestions, and the authors of the various articles for their excellent work and cooperation. I would also like to express my thanks to my assistant in organizing this Symposium, Bonnie L. Thomas, and to Kathleen A. Woerner for her help in the administration.

David Nixon
August 1981

References

[1] Ferrari, C. and Tricomi, F. *Transonic Aerodynamics,* Academic Press, New York, 1968, pp. 21-104.

**Progress in
Astronautics and Aeronautics**

Martin Summerfield,
Series Editor-in-Chief
*Princeton Combustion Research
Laboratories, Inc.*

Norma J. Brennan,
Director, Editorial Department
AIAA

Brenda J. Hio,
Series Managing Editor
AIAA

VOLUMES

EDITORS

*1. Solid Propellant Rocket
Research. 1960

Martin Summerfield
Princeton University

2. Liquid Rockets and
Propellants. 1960

Loren E. Bollinger
The Ohio State University
Martin Goldsmith
The Rand Corporation
Alexis W. Lemmon Jr.
Battelle Memorial Institute

3. Energy Conversion for
Space Power. 1961

Nathan W. Snyder
Institute for Defense Analyses

*4. Space Power Systems.
1961

Nathan W. Snyder
Institute for Defense Analyses

5. Electrostatic Propulsion.
1961

David B. Langmuir
*Space Technology
Laboratories, Inc.*
Ernst Stuhlinger
*NASA George C. Marshall Space
Flight Center*
J. M. Sellen Jr.
*Space Technology
Laboratories, Inc.*

*Now out of print.

*6. Detonation and Two-Phase S. S. Penner
 Flow. 1962 *California Institute of Technology*
  F. A. Williams
  Harvard University

7. Hypersonic Flow Research. Frederick R. Riddell
 1962 *AVCO Corporation*

8. Guidance and Control. 1962 Robert E. Roberson
  Consultant
  James S. Farrior
  Lockheed Missiles and Space Company

*9. Electric Propulsion Ernst Stuhlinger
 Development. 1963 *NASA George C. Marshall Space Flight Center*

*10. Technology of Lunar Clifford I. Cummings and
 Exploration. 1963 Harold R. Lawrence
  Jet Propulsion Laboratory

11. Power Systems for Space Morris A. Zipkin and
 Flight. 1963 Russell N. Edwards
  General Electric Company

12. Ionization in High- Kurt E. Shuler, Editor
 Temperature Gases. 1963 *National Bureau of Standards*
  John B. Fenn, Associate Editor
  Princeton University

*13. Guidance and Control—II. Robert C. Langford
 1964 *General Precision Inc.*
  Charles J. Mundo
  Institute of Naval Studies

*14. Celestial Mechanics and Victor G. Szebehely
 Astrodynamics. 1964 *Yale University Observatory*

*15. Heterogeneous Combustion. Hans G. Wolfhard
 1964 *Institute for Defense Analyses*
  Irvin Glassman
  Princeton University
  Leon Green Jr.
  Air Force Systems Command

16. Space Power Systems
    Engineering. 1966

    George C. Szego
    Institute for Defense Analyses
    J. Edward Taylor
    TRW Inc.

*17. Methods in Astrodynamics
    and Celestial Mechanics. 1966

    Raynor L. Duncombe
    U. S. Naval Observatory
    Victor G. Szebehely
    Yale University Observatory

18. Thermophysics and
    Temperature Control of
    Spacecraft and Entry
    Vehicles. 1966

    Gerhard B. Heller
    *NASA George C. Marshall Space
    Flight Center*

*19. Communication Satellite
    Systems Technology. 1966

    Richard B. Marsten
    Radio Corporation of America

20. Thermophysics of Spacecraft
    and Planetary Bodies:
    Radiation Properties of Solids
    and the Electromagnetic
    Radiation Environment
    in Space. 1967

    Gerhard B. Heller
    *NASA George C. Marshall Space
    Flight Center*

21. Thermal Design Principles of
    Spacecraft and Entry Bodies.
    1969

    Jerry T. Bevans
    TRW Systems

22. Stratospheric Circulation. 1969

    Willis L. Webb
    *Atmospheric Sciences Laboratory,
    White Sands, and University of
    Texas at El Paso*

23. Thermophysics: Applications
    to Thermal Design of
    Spacecraft. 1970

    Jerry T. Bevans
    TRW Systems

24. Heat Transfer and Spacecraft
    Thermal Control. 1971

    John W. Lucas
    Jet Propulsion Laboratory

25. Communications Satellites for
    the 70's: Technology. 1971

    Nathaniel E. Feldman
    The Rand Corporation
    Charles M. Kelly
    The Aerospace Corporation

26. **Communications Satellites for the 70's: Systems.** 1971

Nathaniel E. Feldman
The Rand Corporation
Charles M. Kelly
The Aerospace Corporation

27. **Thermospheric Circulation.** 1972

Willis L. Webb
Atmospheric Sciences Laboratory, White Sands, and University of Texas at El Paso

28. **Thermal Characteristics of the Moon.** 1972

John W. Lucas
Jet Propulsion Laboratory

29. **Fundamentals of Spacecraft Thermal Design.** 1972

John W. Lucas
Jet Propulsion Laboratory

30. **Solar Activity Observations and Predictions.** 1972

Patrick S. McIntosh and Murray Dryer
Environmental Research Laboratories, National Oceanic and Atmospheric Administration

31. **Thermal Control and Radiation.** 1973

Chang-Lin Tien
University of California, Berkeley

32. **Communications Satellite Systems.** 1974

P. L. Bargellini
COMSAT Laboratories

33. **Communications Satellite Technology.** 1974

P. L. Bargellini
COMSAT Laboratories

34. **Instrumentation for Airbreathing Propulsion.** 1974

Allen E. Fuhs
Naval Postgraduate School
Marshall Kingery
Arnold Engineering Development Center

35. **Thermophysics and Spacecraft Thermal Control.** 1974

Robert G. Hering
University of Iowa

36. **Thermal Pollution Analysis.** 1975

Joseph A. Schetz
Virginia Polytechnic Institute

37. **Aeroacoustics: Jet and Combustion Noise; Duct Acoustics.** 1975

Henry T. Nagamatsu, Editor
General Electric Research and Development Center
Jack V. O'Keefe, Associate Editor
The Boeing Company
Ira R. Schwartz, Associate Editor
NASA Ames Research Center

38. **Aeroacoustics: Fan, STOL, and Boundary Layer Noise; Sonic Boom; Aeroacoustics Instrumentation.** 1975

Henry T. Nagamatsu, Editor
General Electric Research and Development Center
Jack V. O'Keefe, Associate Editor
The Boeing Company
Ira R. Schwartz, Associate Editor
NASA Ames Research Center

39. **Heat Transfer with Thermal Control Applications.** 1975

M. Michael Yovanovich
University of Waterloo

40. **Aerodynamics of Base Combustion.** 1976

S. N. B. Murthy, Editor
Purdue University
J. R. Osborn, Associate Editor
Purdue University
A. W. Barrows and J. R. Ward, Associate Editors
Ballistics Research Laboratories

41. **Communication Satellite Developments: Systems.** 1976

Gilbert E. LaVean
Defense Communications Engineering Center
William G. Schmidt
CML Satellite Corporation

42. **Communication Satellite Developments: Technology.** 1976

William G. Schmidt
CML Satellite Corporation
Gilbert E. LaVean
Defense Communications Engineering Center

43. **Aeroacoustics: Jet Noise, Combustion and Core Engine Noise.** 1976

Ira R. Schwartz, Editor
NASA Ames Research Center
Henry T. Nagamatsu, Associate Editor
General Electric Research and Development Center
Warren C. Strahle, Associate Editor
Georgia Institute of Technology

44. **Aeroacoustics: Fan Noise and Control; Duct Acoustics; Rotor Noise.** 1976

Ira R. Schwartz, Editor
NASA Ames Research Center
Henry T. Nagamatsu, Associate Editor
General Electric Research and Development Center
Warren C. Strahle, Associate Editor
Georgia Institute of Technology

45. **Aeroacoustics: STOL Noise; Airframe and Airfoil Noise.** 1976

Ira R. Schwartz, Editor
NASA Ames Research Center
Henry T. Nagamatsu, Associate Editor
General Electric Research and Development Center
Warren C. Strahle, Associate Editor
Georgia Institute of Technology

46. **Aeroacoustics: Acoustic Wave Propagation; Aircraft Noise Prediction; Aeroacoustic Instrumentation.** 1976

Ira R. Schwartz, Editor
NASA Ames Research Center
Henry T. Nagamatsu, Associate Editor
General Electric Research and Development Center
Warren C. Strahle, Associate Editor
Georgia Institute of Technology

47. **Spacecraft Charging by Magnetospheric Plasmas.** 1976

Alan Rosen
TRW Inc.

48. **Scientific Investigations on the Skylab Satellite.** 1976

Marion I. Kent and
Ernst Stuhlinger
NASA George C. Marshall Space Flight Center
Shi-Tsan Wu
The University of Alabama

49. **Radiative Transfer and Thermal Control.** 1976

Allie M. Smith
ARO Inc.

50. **Exploration of the Outer Solar System.** 1977

Eugene W. Greenstadt
TRW Inc.
Murray Dryer
National Oceanic and Atmospheric Administration
Devrie S. Intriligator
University of Southern California

51. **Rarefied Gas Dynamics, Parts I and II (two volumes).** 1977

J. Leith Potter
ARO Inc.

52. **Materials Sciences in Space with Application to Space Processing.** 1977

Leo Steg
General Electric Company

53. **Experimental Diagnostics in Gas Phase Combustion Systems.** 1977

Ben T. Zinn,
Editor
Georgia Institute of Technology
Craig T. Bowman,
Associate Editor
Stanford University
Daniel L. Hartley,
Associate Editor
Sandia Laboratories
Edward W. Price,
Associate Editor
Georgia Institute of Technology
James G. Skifstad,
Associate Editor
Purdue University

54. **Satellite Communications: Future Systems.** 1977 — David Jarett, *TRW Inc.*

55. **Satellite Communications: Advanced Technologies.** 1977 — David Jarett, *TRW Inc.*

56. **Thermophysics of Spacecraft and Outer Planet Entry Probes.** 1977 — Allie M. Smith, *ARO Inc.*

57. **Space-Based Manufacturing from Nonterrestrial Materials.** 1977 — Gerard K. O'Neill, Editor, *Princeton University*; Brian O'Leary, Assistant Editor, *Princeton University*

58. **Turbulent Combustion.** 1978 — Lawrence A. Kennedy, *State University of New York at Buffalo*

59. **Aerodynamic Heating and Thermal Protection Systems.** 1978 — Leroy S. Fletcher, *University of Virginia*

60. **Heat Transfer and Thermal Control Systems.** 1978 — Leroy S. Fletcher, *University of Virginia*

61. **Radiation Energy Conversion in Space.** 1978 — Kenneth W. Billman, *NASA Ames Research Center*

62. **Alternative Hydrocarbon Fuels: Combustion and Chemical Kinetics.** 1978 — Craig T. Bowman, *Stanford University*; Jørgen Birkeland, *Department of Energy*

63. **Experimental Diagnostics in Combustion of Solids.** 1978 — Thomas L. Boggs, *Naval Weapons Center*; Ben T. Zinn, *Georgia Institute of Technology*

64. **Outer Planet Entry Heating and Thermal Protection.** 1979 — Raymond Viskanta, *Purdue University*

65. **Thermophysics and Thermal Control.** 1979 — Raymond Viskanta, *Purdue University*

66. **Interior Ballistics of Guns.** 1979 — Herman Krier, *University of Illinois at Urbana-Champaign*; Martin Summerfield, *New York University*

67. **Remote Sensing of Earth from Space: Role of "Smart Sensors."** 1979 — Roger A. Breckenridge, *NASA Langley Research Center*

68. **Injection and Mixing in Turbulent Flow.** 1980 — Joseph A. Schetz, *Virginia Polytechnic Institute and State University*

69. **Entry Heating and Thermal Protection.** 1980 — Walter B. Olstad, *NASA Headquarters*

70. **Heat Transfer, Thermal Control, and Heat Pipes.** 1980 — Walter B. Olstad, *NASA Headquarters*

71. **Space Systems and Their Interactions with Earth's Space Environment.** 1980 — Henry B. Garrett and Charles P. Pike, *Hanscom Air Force Base*

72. **Viscous Flow Drag Reduction.** 1980 — Gary R. Hough, *Vought Advanced Technology Center*

73. **Combustion Experiments in a Zero-Gravity Laboratory.** 1981 — Thomas H. Cochran, *NASA Lewis Research Center*

74. **Rarefied Gas Dynamics, Parts I and II** (two volumes). 1981 — Sam S. Fisher, *University of Virginia at Charlottesville*

75. Gasdynamics of
    Detonations and
    Explosions. 1981

    J. R. Bowen
    *University of Wisconsin
    at Madison*
    N. Manson
    Université de Poitiers
    A. K. Oppenheim
    *University of California
    at Berkeley*
    R. I. Soloukhin
    *Institute of Heat and Mass
    Transfer, BSSR Academy
    of Sciences*

76. Combustion in
    Reactive Systems. 1981

    J. R. Bowen
    *University of Wisconsin
    at Madison*
    N. Manson
    Université de Poitiers
    A. K. Oppenheim
    *University of California
    at Berkeley*
    R. I. Soloukhin
    *Institute of Heat and Mass
    Transfer, BSSR Academy
    of Sciences*

77. Aerothermodynamics
    and Planetary Entry. 1981

    A. L. Crosbie
    University of Missouri-Rolla

78. Heat Transfer and
    Thermal Control. 1981

    A. L. Crosbie
    University of Missouri-Rolla

79. Electric Propulsion and
    Its Applications to
    Space Missions. 1981

    Robert C. Finke
    NASA Lewis Research Center

80. Aero-Optical
    Phenomena. 1982

    Keith G. Gilbert
    and Leonard J. Otten
    Air Force Weapons Laboratory

81. Transonic Aerodynamics.
    1982

    David Nixon
    *Nielsen Engineering &
    Research, Inc.*

(Other volumes are planned.)

Part I
Review of Transonic Flow Research

This part of the book is concerned with a review of the present status of the various aspects of transonic flow research. The first article is a historical review of the subject so that the more recent advances can be placed in the context of previous work: the next two articles concern the problems arising in the design of commercial and military aircraft, respectively. These articles provide the environment in which the advances in research can be related to the actual problems encountered in practice. The fourth article describes the present status of experimental techniques and problems for tests at transonic speeds. The next two articles concern the prediction of transonic slow phenomena using potential theory and the Navier-Stokes equations, respectively. The final article in this part of the book discusses the way in which both experimental and prediction methods are used to design aircraft.

Chapter I.

Transonic Aerodynamics—History and Statement of the Problem

John R. Spreiter[*]
Stanford University, Stanford, Calif.

Introduction

It is my assignment to provide a background of the history and development of transonic aerodynamics as a preamble to the remainder of the papers in this book presenting critical reviews of recent advances and suggestions of prospective avenues for further progress. Such assignments are a sure sign of passing years, but I do have the advantage of getting an early start in transonic aerodynamics when I was assigned to the Flight Research Branch at NACA/Ames Aeronautical Laboratory upon arrival as a young engineer in mid-1943. This group, under the direction of Larry Clousing, who was also a leading test pilot, was just beginning research to determine the causes of catastrophic accidents that were occurring when fighter planes were put into steep dives at high altitude, and hopefully to determine ways to cure or avoid the problems. In 1947, I transferred to the Theoretical Aerodynamics Branch under the direction of Dr. Max A. Heaslet, and was soon at work trying to develop theories for dealing with transonic flows. During the following years, it fell my lot to provide summary papers on the state of transonic aerodynamics on a number of occasions. I will draw rather heavily on these papers, inasmuch as they were designed to be comprehensive summaries within imposed limits of length of the state of knowledge of transonic aerodynamics, with somewhat of a bias toward theory and its evaluation in wind-tunnel experiments. While this may seem a bit parochial, since corresponding research was also under way at a number of other places, the work at Ames was both at the forefront and in

---

Presented at the Transonic Perspective Symposium, NASA/Ames Research Center, Moffett Field, Calif., Feb. 18-20, 1981. Copyright © American Institute of Aeronautics and Astronautics, Inc., 1981. All rights reserved.
[*]Professor, Division of Applied Mechanics.

good contact with transonic research elsewhere; and it is these things that I know best.

Rather than develop a continuous chronology of developments, which I found rather difficult when I began to attempt it, I have endeavored to provide a series of accounts of the state of the transonic aerodynamics at selected times between 1940 and 1975, and to provide vignettes of the facilities and supporting services available to those carrying out the research.

Transonic Aerodynamics before 1940

Because of an early awareness of the occurrence of transonic flow in nozzles and around propeller tips, the study of transonic aerodynamics has a long history. Already in 1890, Molenbroeck[1] and in 1904 Chaplygin[2] published mathematical analyses in which the nonlinear equation for steady two-dimensional potential flow of a compressible gas is reduced to a linear equation by use of the hodograph transformation, and applications to nozzle flows were developed. Between those dates and 1940, however, very few theoretical studies were attempted of transonic flow, although there were a number of significant contributions to the theory of purely subsonic or purely supersonic compressible flow.

The first aerodynamic tests at transonic speeds known to me are those of Douglas and Wood[3] in 1924 in which measurements were made in an ordinary wind tunnel of a model airplane propeller driven at supersonic tip speeds. Shortly afterwards in 1925, Briggs, Hull, and Dryden[4] reported a series of airfoil tests conducted in the high-speed jet from an orifice about 12 in. through which air was supplied by a 500 hp air compressor under test by the General Electric Company. In 1927, Briggs and Dryden[5] reported the first airfoil pressure distributions for speeds up to and slightly beyond the speed of sound as measured in a 2 in. diam open jet. The next year, Stanton[6] in England reported the first results obtained under conditions which permitted the determination of airfoil section characteristics. He employed a 3 in. diam closed-throat wind tunnel driven by a 530 hp compressor and capable of Mach numbers up to 1.7. In the same year, Busemann[7] in Germany reported tests in a blow-down wind tunnel in which air from the room was drawn through an entrance section to a small open jet test section, and then through a velocity-controlling restriction into a 10 m^3 tank evacuated by a small compressor.

With the prinicples and results of high-speed testing thus becoming known and the development of more powerful engines tending to push propeller tip speeds toward the speed of sound providing the practical motivation, the design and construction of high-speed wind tunnels began in several of the more advanced technological nations. In the United States, the National Advisory Committee for Aeronautics (NACA) began development in 1927 of more suitable wind tunnels for investigating "compressibility effects" as they were then called. At first, pilot studies were conducted with a small induction jet tunnel with a 1 in. diam test section through which air was driven at high speeds by high-pressure air from the variable density wind tunnel at NACA/Langley Aeronautical Laboratory.[8] Then in 1928, a larger 12 in. diam tunnel was constructed using similar principles.[9] Two years later, the tunnel was made more effective by reducing the test section to 11 in. diam. In 1934, a 24 in. high-speed wind tunnel was completed, and the era of systematic measurements of airfoil characteristics in the high subsonic and lower transonic ranges was begun by Stack and his colleagues.[10-12] Rapid progress was then being made in wind-tunnel design and construction. In 1936, the 8 ft high-speed wind tunnel powered by 8000 hp electric motors was put into operation at Langley; and in 1939, two 16 ft high-speed wind tunnels were designed by NACA personnel. The first was completed in 1941 at Langley and had 16,000 hp. The second was completed the next year at Ames, and had 27,000 hp, a tribute to the great abundance then of hydroelectric power in the West. These and subsequent high-speed wind tunnels built over the next decade in the United States all had closed test sections with solid walls. The flow speed could therefore not exceed the speed of sound in the empty test section, and rarely exceeded a Mach number of 0.9 with a model in place. Parallel developments were taking place in several European nations, although, with two notable exceptions in Britain and Italy, most of the efforts were directed toward supersonic rather than transonic flows.

By 1940, airfoil characteristics, pressure distributions, and Schlieren photographs showing shock waves and boundary layers were being reported for a rapidly increasing number of airfoil shapes. Many new and strange phenomena were being revealed, but theory was of limited usefulness for understanding the observations. In fact, what theory there was consisted mostly of the relations of one-dimensional gasdynamics, together with a few results such as the Prandtl-Glauert rule[13,14] and the Kármán-Tsien rule[15] for two-dimensional subsonic flows and the Ackeret[16] and Busemann[17] first- and second-order formulas for two-dimensional supersonic flows. The Prandtl-Glauert and Kármán-Tsien rules simply relate the pressure coefficient $C_p = (p-p_\infty)/(\rho_\infty v_\infty^2/2)$, where p, ρ, and v refer to the pressure, density, and velocity and subscript ∞ indicates conditions infinitely far upstream, to the corresponding pressure coefficient C_{p_i} for incompressible flow about the same airfoil according to

$$C_p = C_{p_i}/\sqrt{1-M_\infty^2} \qquad \text{Prandtl-Glauert} \quad (1)$$

$$C_p = C_{p_i}/\{\sqrt{1-M_\infty^2} + [M_\infty/(1+\sqrt{1-M_\infty^2})](C_{p_i}/2)\} \quad \text{Kármán-Tsien} \quad (2)$$

where M is the Mach number v/a and a is the speed of sound. The Ackeret and Busemann formulas relate C_p to the local slope δ of the airfoil surface relative to the freestream direction, where δ is taken positive when the flow is on the concave side of the angle between the freestream direction and the tangent to the surface, according to

$$C_p = 2\delta/\sqrt{M_\infty^2-1} \qquad \text{Ackert} \quad (3)$$

$$C_p = 2\delta/\sqrt{M_\infty^2-1} + [(\gamma+1)M_\infty^4 - 4(M_\infty^2-1)]\delta^2/[2(M_\infty^2-1)^2]$$

$$\text{Busemann} \quad (4)$$

These theoretical results were very useful for purely subsonic or supersonic flows, but they provided little guidance for understanding typically transonic phenomena such as strong shock waves and boundary-layer separation and their effects on the pressure distribution at supercritical Mach numbers, and gave obviously erroneous results for $M_\infty = 1$. In actual practice, incompressible flow theory provided the theoretical basis for almost all airplane aerodynamics before 1940. Airplanes did not fly fast enough to encounter significant transonic effects over the main airframe, propellor tip speeds were kept subsonic by the use of gearing and higher solidity propellors to avoid loss of efficiency, and diving speeds were limited by drag or by a warning line on the airspeed indicator. Even the word "transonic" had not been introduced by 1940, as far as I can ascertain, to describe a part of the Mach number range; another four or five years were to elapse before its use was common in the aeronautical world.

Transonic Aerodynamics in 1945

With little warning from the previous research, major problems associated with transonic aerodynamics suddenly revealed themselves about 1941 when fighter planes were pushed over into steep dives at high altitude. As the noise and buffeting rose to levels not experienced before, the airplane developed a powerful tendency to dive even steeper in spite of heroic efforts by the pilot to retain control. Different airplanes exhibited different characteristics. For some, the main problem was that of major shifts in longitudinal trim that caused a strong nose-down effect in the dive, followed by an abrupt pitch-up at lower altitudes as the denser air slowed the aiplane out of the transonic range. For others, severe buffeting, or control surface flutter, ineffectiveness, or even reversal were limiting factors. If the pilot was not knowledgable, alert, strong, and probably lucky as well, the dive could easily culminate in structural failure, destruction of the aircraft, and death for the pilot.

The world of high-speed research that I was plunged into upon arrival at Ames was an exciting one. Ames was a new laboratory, nearly everyone was young, responsibilities and opportunities were great, and many of us happily worked hours far longer than the standard 48 hour week. Within three years of my arrival, I was reporting results of measurements of maximum lift coefficients at the highest Mach numbers attainable in steep dives of the six prominent World War II fighter aircraft illustrated in Fig. 1.[18] Others of my colleagues were reporting other characteristics of high-speed flight as revealed by both flight and wind-tunnel experiments.[19,20] With skill and tremendous courage, our pilots systematically explored the mysteries and dangers of high-speed flight, which never actually exceeded a Mach number of 0.81 in a propeller airplane, using fighter aircraft jammed with every available kind of flight recording instruments. The instruments were bulky, the airplanes were small and compact. Obviously the space normally occupied by guns and ammunition boxes was available for the instruments, but the need for more space was sometimes so great that the radio and all communication equipment had to be removed!

It was risky business for the pilots. During one of Clousing's early dives in a Navy Brewster fighter, the pitch-up was so severe, going off scale on the accelerometer at about 13 g, that the wings were badly bent and the plane had to be retired. On a later flight in a spe-

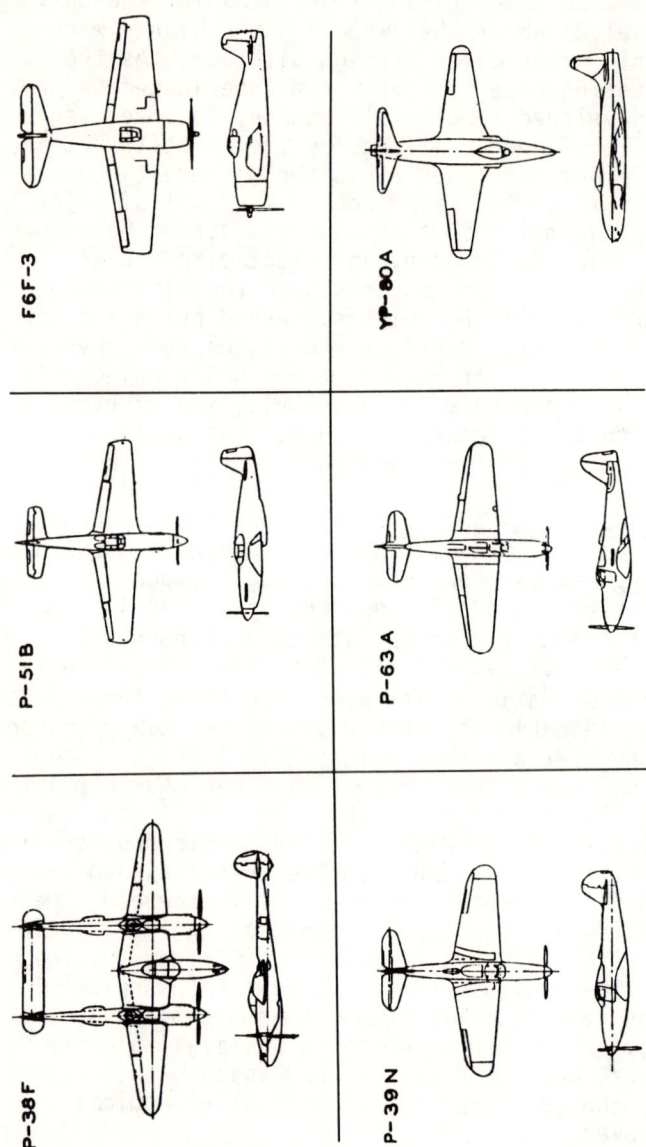

Fig. 1 Airplanes in NACA/Ames dive test program 1943-1946.[18]

cially reinforced Bell P-39, Clousing was alert to the pitch-up, but the combined effects of severe buffeting of the tail (pressure fluctuations of 300 to 400 psf were recorded) and the large up-load on the horizontal tail as he pushed the stick forward with a force of 180 lb to control the pitch-up, whereas a fraction of a second before he had been pulling with a force of 240 lb to control the tuck-under, caused the tail spars and ribs to break in several places and the majority of the hinge brackets joining the elevator to the stabilizer to fracture. In tests with the Bell P-63, which generally resembled the P-39 but had an unusually thick (16%) wing with a cusped low-drag airfoil, the onset of severe aileron buzz limited the dive Mach numbers to about 0.65, barely into the transonic range. In an attempt to hold the aileron more rigidly in place, an irreversible device was installed between the aileron and the control stick.[21] It delayed the Mach number for the onset of buzz by a few hundredths, but then the buffeting set into so severely that a number of aileron hinge brackets were fractured before the plane could be slowed to "safety." Not long after, Clousing dove one of the first Lockheed P-80 jets to Mach numbers as high as 0.87 to explore a rather similar aileron buzz it was experiencing, in spite of having a much thinner wing and a strong irreversible control system, and returned with a buckled aileron. On still another occasion, another Ames pilot, Jack Giosso, found his controls so jammed during a dive of a Ryan RJ-1 that his pull-out extended so low that he cut the rigging of a sunken ship in San Francisco Bay. Not long before a similar plane elsewhere had plunged right into the ground. With the airplane safely retrieved after a very fast landing, the problem was immediately recognized and cured with a simple change. An inspection hole cover plate had buckled inward and jammed against a turnbuckle in the control cables to the elevator! In that case, the solution was simple—just make the cover plate of thicker metal.

Related dive testing of fighter aircraft was, of course, in progress at the time at several aeronautical research and testing centers, and the results were circulated in the usual way among friendly nations. One annonymous memorandum from the Royal Aircraft Establishement in Farnborough, England, created a particular bit of comment because, after detailing nine instructions for the pilot that were in good accordance with the local experiences, it closed with a tenth instruction, "Above all, don't worry!" In every reported case the aircraft flattened out eventually." Indeed! And what about the unreported cases (remember our pilots were flying without radio and never had chase planes); and for that matter where and against what did it eventually flatten out? Years later, after relating this story on opening day of my class in transonic aerodynamics at Stanford, one of the students, older and more experienced than usual, came up afterwards and asked how I knew he wrote that memorandum. I replied that I didn't know, and suggested that he should be telling the story because I felt that he could give the statement its proper English flare—a style impossible for me to emulate. The student was Ronald Smelt, who had been Head of the Flight Test Group at the RAE and is well known in the aeronautical and space communities for both his personal research and his leadership in the AIAA and Lockheed.

Enough reminiscing about flight research in the mid-forties. Now what about the state of knowledge of transonic aerodynamics generally in 1945? From the above it is clear that flight experiences were revealing a desparate problem. Little guidance for its solution was available from theory. In fact, even the counterpart of the Prandtl-Glauert rule for two-dimensional linearized compressible flow theory had not been determined for axisymmetric flow, and Göthert's[22] counterpart for wings of finite span was just being learned in the United States from German documents captured at the end of World War II. Small high-speed wind tunnels had been used for two-dimensional testing for over a decade, and larger tunnels suitable for testing realistic three-dimensional models had been operational for a few years; but the solid walls of their test sections limited the maximum test Mach numbers to less than about 0.9.

Such a Mach number limitation greatly restricts the range of fundamental transonic flow studies, but was not of serious concern in the search for solutions of the high-speed dive problems experienced by existing aircraft since they sometimes became evident at Mach numbers as low as 0.6 and none of the aircraft could attain the maximum wind-tunnel Mach number in flight anyway. The larger high-speed wind tunnels, and particularly the 16 ft wind tunnel at Ames with its greater power and higher speed than its Langley counterpart, were therefore put to work to solve specific high-speed problems exhibited by a number of fighter aircraft.[19] Notable among these was the work at Ames that led to the dive-recovery flap on the Lockheed P-38, one of the first American aircraft to exhibit the notorious nosing-down tendency in a steep dive. Another was solution of a massive rumbling or thumping of the airscoop on the North American P-51; and still another was concerned with the occurrence of high-frequency aileron buzz on the Lockheed P-80.

Throughout this period of intensive wind-tunnel testing, there were nagging questions as to how well the drag and speed of an airplane could be predicted from wind-tunnel measurements. The test models were usually idealized with smoother, truer surfaces than the originals and lacked the gaps, rivet heads, and other irregularities that exist on the actual airplanes; and the influence of wind-tunnel wall and model support struts and airstream turbulence were difficult if not impossible to evaluate. There was, moreover, the matter of the propeller. Since there were no electric motors of the required power that were small enough to fit inside the wind-tunnel model, high-speed wind-tunnel tests were normally performed without a propeller on the model. To resolve these questions, one of the Ames test pilots, James Nissen, proposed a novel solution. He would dive test a P-51 with the propeller off! The data would be compared with measurements from the Ames 16 ft wind tunnel. Nissen was to be towed to 23,000 ft by a two-engine Northrop P-61 night fighter, and upon release from the tow cable go into a vertical dive and land on the extensive flat dry lakebed of Muroc Dry Lake in the California desert. Two flights went successfully, and the data confirmed the usefulness of the wind-tunnel data. The third ended in disaster. The tow cable released inadvertently shortly after takeoff, flew back, and wrapped around the P-51. Nissen quickly got the airplane into a glide. Unfortunately, there were two hazards in all of Muroc Dry Lake, and Nissen encountered them both. He clipped a power line and landed in a quarry! The airplane was finished, but Nissen was spared, probably because of his foresight in wearing one of the new plastic football helmets rather than the soft leather helmets then still in vogue. It was cut right through to his forehead where he struck his head on the instrument panel. The final irony is that the x-ray machine was not working when they took him to the hospital to check for broken bones. He had cut the powerline to the hospital!

The large high-speed wind tunnels were effective tools in solving the problems of existing aircraft, but less satisfactory for supplying data for the design of the next generation of jet-propelled fighter aircraft that would hopefully be capable of flight at higher Mach numbers than could be tested in the wind tunnels. Alternative methods for transonic testing had to be devised to met the challenge. Among those that were employed were 1) in-flight wind flow testing in which semispan models were mounted vertically on a glove-like modified portion of the wing of a fighter aircraft, as the P-51, which was then dived to high speed; 2) rocket launched models; and 3) drop-body models released much like a bomb from a high-flying aircraft. Such testing techniques were both expensive and limited in effectiveness. A better wind tunnel capable of testing all Mach numbers throughout the transonic range was clearly needed, but in 1945 nobody knew how to design such a wind tunnel. One attempt that amused us at the time was that of John D. Akerman of the University of Minnesota who cut up his wife's fur coat and glued it to the test section of his new high-speed wind tunnel to soften the effect of the walls. He was clearly on the right track, but the idea was not pursued to any practical end.

Finally, it should be noted that data analysis and theoretical calcuations in 1945 involved tedious calculations using slide rules and primitive electromechanical desk calculators, few of which could take a square root and some of which could not even divide. Since such a calculator might cost as much as several months salary for a young engineer, even they were in limited supply in most offices. The bulk of the work was done with a slide rule and books of tables, and the more routine or extensive tasks were performed by crews of young women working in crowded offices.

Transonic Aerodynamics in 1960

The period from 1945 to 1960 was one of tremendous progress in all phases of transonic aerodynamics. The development of the jet engine greatly increased the available propulsive power and solved the propeller tip speed problem by removing the propeller from high-speed aircraft. The introduction of Jones' concept of swept wings[23] and low-aspect-ratio wings[24] for efficient flight at higher Mach numbers than possible with the previous high-aspect-ratio straight wings brought a new look to high-speed aircraft. These and independent parallel developments in Germany during World War II enabled the design of aircraft capable of transonic and supersonic speeds, and increased the cruising speed of transport aircraft by 50%. Uncertainties and fears of passing the "sonic barrier" were removed by the successful flights beginning in 1947 of the XS-1 and D-558 experimental aircraft.

The development under Stack at Langley of the modern transonic wind tunnel with ventilated or partially open walls also stems from the beginning of this period,[25] and led directly to the construction of large transonic wind tunnels at Langley, Ames, and several other places in both the United States and abroad. Almost the first result to emerge when the first of these new wind tunnels was put into operation in 1951 at Langley was Whitcomb's reknowned transonic areas rule.[26] It states that the transonic wave drag of a wing-body combination is the same as that of an equivalent body of revolution having the same longitudinal distribution of cross-section area. Whitcomb thereupon proceeded to show that the transonic drag of a wing-body combination could be reduced significantly by indenting the body along the position of the wing so as to make the equivalent body of revolution into a smooth, rather than bumpy, body. This result was so important that the area rule was not revealed publicly for a number of years until it had beem employed to practical advantage on a number of high-speed aircraft.

Aerodynamic theory also experienced an explosive growth during this decade. The first two books on gasdynamics[27,28] in English appeared in 1947. The linearized theory of subsonic and supersonic flow about thin wings and slender bodies was developed extensively during this period,[29-31] and notable progress was made in the solution of the more difficult nonlinear equations of general compressible flow theory.[32-35] Although it is immediately apparent from the simple Prandtl-Glauert rule of Eq. (1) that not all results of linearized theory are valid in the transonic range, it was soon recognized that certain results, notably the aerodynamic loading, lift, and moment on slender wings and bodies[36] and the high-frequency unsteady pressure distribution, forces, and moments generally,[37] were both theoretically consistent and in good accord with experiment.

In the context of this conference, the most outstanding theoretical development of the first few years of this period was the establishment of the nonlinear transonic small disturbance theory[38-43] as a useful mathematical model for the prediction of the aerodynamic properties of airfoils, wings, and bodies throughout the entire Mach number range from 0 to moderate supersonic

values, say 1.5. By 1960, quantitative results had been
determined by a variety of methods, including similitude,
hodograph, separation of variables, function theory, relaxation, integral equations, and local linearization for many
significant cases, and their validity had been established
by comparison with experiments.[44] The latter also revealed
significant shock-wave/boundary-layer interactions, as well
as other more familiar boundary-layer effects, that were
beyond the capability of the inviscid theory to predict.
It was clear that the transonic small disturbance theory
had an important role to play, but the difficulties of
solving the equations were formidable. By inference, it
was also evident that transonic flow could be treated more
accurately by the complete equations for potential flow or
the even more fundamental Euler or Navier-Stokes equations
for compressible inviscid or viscous flow, but that the
mathematical difficulties would be even greater. Alternatively, since the fundamental problems associated with nonlinearity and mixed elliptic-hyperbolic type are both
present in the small disturbance theory, it seemed and has
proven useful to study this theory as a prototype that
retains the fundamental difficulties while avoiding many
of the algebraic and geometrical complexities of the more
complete theories.

We turn now to a more detailed account of some of the
key features of the transonic small disturbance theory and
a sampling of results presented between 1945 and 1960 to
illustrate the state of transonic aerodynamics at that time.
These were the formative years for transonic aerodynamics,
both theoretical and experimental, and progress was substantial and steady. The well-attended and documented
Symposium Transsonicum organized by K. Oswatitsch and held
in Aachen in 1962 provided a comprehensive survey of the
accomplishments of this period with many lengthy review
papers by the principal contributors of the era. There
followed nearly a decade of decline of interest and progress in transonic aerodynamics. The reasons were many.
The existing theories were reaching their limits of
exploitation, the available computers were inadequate for
the extensive numerical calculations that were to characterize the analysis of the seventies, certain key ideas of
numerical analysis remained to be discovered, and the
demands and excitement of new problems arising in space
research were diverting attention and resources elsewhere.

Getting back to specifics, in the original papers on
the transonic small disturbance theory,[38,43] perturbations

were taken about the critical speed of sound a* to obtain a theory for two-dimensional flow with freestream Mach number equal to or near to unity. The governing partial differential equation is

$$\phi_{zz} = [(\gamma+1)/a^*]\phi_x \phi_{xx} \qquad (5)$$

in which ϕ is the perturbation velocity potential related to the x and z perturbation velocity components by $\phi_x = u = U - a^*$ and $\phi_z = w = W$ where U and W are the total velocity components, γ is the ratio of specific heats (usually taken as 1.4 for air), and applications are to be made to flows nearly parallel to thin airfoils that extend along or very near to the x axis. Equation (5) is deceptively simple in appearance, but its solution is complex both because it is nonlinear and because it is of mixed elliptic/hyperbolic type depending on the sign of ϕ_x.

Among the earliest useful results obtained with this theory were the transonic similarity rules of von Kármán.[42,43] They indicate, for approximately sonic Mach numbers, that the flow patterns around an affinely related family of thin airfoils of chord c and ordinates Z given by $Z/c = \tau f(x/c)$ having the same ordinate (upper and lower surface) distribution function $f(x/c)$ but different amplitude ratios τ are similar for equal values of a similarity parameter $K = (1-M_\infty)/[\tau(\gamma+1)]^{2/3}$ and that $(\gamma+1)^{1/3} C_L/\tau^{2/3}$ and $(\gamma+1)^{1/3} C_{D_p}/\tau^{5/3}$, where C_L and C_{D_p} are the lift and pressure drag coefficients defined in the usual way, depend only on K. It follows immediately that the lift and pressure drag coefficients of such a family of airfoils at $M_\infty = 1$ vary as the 2/3 and 5/3 powers of τ, respectively.

In contemporary papers of Guderley and Busemann,[39-41] problems associated with the nonlinearity of Eq. (5) were circumvented without approximation by use of the hodograph transformation in which the dependent and independent variables are interchanged to obtain the Tricomi equation, a linear partial differential equation of mixed elliptic-hyperbolic type

$$\Phi_{uu} = [(\gamma+1)/a^*] u \Phi_{ww} \qquad (6)$$

in which the position coordinate x,z is related to the Legendre potential Φ by

$$x = \Phi_u, \quad a = \Phi_w \qquad (7)$$

where
$$\Phi = xu + zw - \phi \tag{8}$$

Simple solutions of Tricomi's equation can be found in a variety of ways including separation of variables, self-similar, and numerical methods, and these can be superposed to develop solutions of greater generality.

The first example that resembled flow about an airfoil for which solutions were developed for the entire transonic range was that of a nonlifting symmetrical wedge followed by a long straight afterbody of constant thickness.[45-47] At about the same time, Liepmann and Bryson[48,49] completed a wind-tunnel investigation of flow past three such single-wedge sections having semiangles of 4.5 deg, 7.5 deg, and 10 deg. The results indicated reasonably good agreement with both the similarity rules and with the solutions of the small disturbance theory, and also with a newly discovered Mach number freeze principle that states that the local Mach number M at an arbitrary point in the vicinity of an airfoil section or body of revolution is independent of M_∞ for small changes in M near $M_\infty = 1$, that is

$$(dM/dM_\infty)_{M_\infty=1} = 0 \tag{9}$$

Combination with the exact isentropic relation for C_p in terms of M and M_∞ indicates

$$(dC_p/dM)_{M_\infty=1} = 4/(\gamma+1) - [2/(\gamma+1)](C_p)_{M_\infty=1} \tag{10}$$

The approximate relation in the small disturbance transonic theory defined by Eq. (5) is

$$(dC_p/dM_\infty)_{M_\infty=1} = 4/(\gamma+1) \tag{11}$$

from which it follows immediately that $(dC_{D_p}/dM_\infty)_{M_\infty=1} = 4/(\gamma+1)$ for the variation with M_∞ at $M_\infty = 1$ of the pressure drag coefficient for a single-wedge airfoil section. The right-hand side would, of course, be supplemented by an additional term $-[2/(\gamma+1)](C_{D_p})_{M_\infty=1}$ if the exact relation of Eq. (10) were used instead of Eq. (9) of the approximate theory.

At about the same time these results were becoming available, a number of alternatives to Eqs. (5) and (6) were being considered that were equivalent at $M_\infty = 1$, but differed slightly from other M_∞. In an analysis of several of these,[50] it was shown on several theoretical counts, including the accuracy with which the shock relations and critical pressure variation with M_∞ are represented and the ability to blend continuously into the successful linear theory for M_∞ well away from unity, that the following partial differential equation, also nonlinear and of mixed elliptic-hyperbolic type, has considerable merit as an alternative to Eq. (5)

$$(1-M_\infty^2)\phi_{xx} + \phi_{zz} = [M_\infty^2(\gamma+1)/U]\phi_x\phi_{xx} \qquad (12)$$

In deriving this equation, perturbations are taken about the freestream velocity U_∞, as they usually are in linearized compressible flow theory, so that the perturbation velocity components $u = U - U_\infty$ and $w = W - W_\infty$ are related to the perturbation velocity potential ϕ by $u = \phi_x$ and $w = \phi_z$. The other symbols have the same meaning as given above. In place of Eq. (6), the corresponding equation for the hodograph plane is

$$(1-M_\infty^2)\Phi_{ww} + \Phi_{uu} = [M_\infty^2(\gamma+1)/U]u\Phi_{ww} \qquad (13)$$

where $x = \Phi_u$ and $z = \Phi_w$. Part of the merit of this alternative is that all of the results obtained using Eqs. (5) and (6) can be transformed simply to those that would be obtained if the analysis had been carried out with Eqs. (12) and (13). Another merit is that the restriction to freestream Mach numbers near unity is removed. Therefore the results of solutions of the transonic small disturbance theory can be applied throughout the entire Mach number range from 0 to the highest Mach number, say 2, that can be treated with linearized compressible flow theory. But the ultimate test of any of these alternatives is how well it predicts actual results measured in a wind tunnel or in free flight. On this score, it was shown[50] that the alternative described by Eqs. (12) and (13) is substantially superior over a wide range of applications. The search for other alternatives still continues, but this form is probably the most widely used today; although it should be noted that the use of the complete potential equation for compressible flow is becoming increasingly common in current analyses. The latter provides greater accuracy, particularly in regions of large perturbations; and the greatly increased amount of computing required to obtain a numerical solution is being managed through the use of more efficient algorithms and more powerful computers.

With the change to Eq. (12), the von Kármán similarity rules relating the flow about a family of affinely related airfoil sections with ordinates Z given by $Z/c = f(x/c)$ as before alter slightly so that the similarity parameter K is replaced by

$$\xi_\infty = (M_\infty^2 - 1)/[(\gamma+1)M_\infty^2 \tau]^{2/3} \qquad (14)$$

where the sign has been arbitrarily changed from that of von Kármán so that ξ_∞ increases rather than decreases with increasing M_∞. The surface pressure distributions are related in such a way that

$$\bar{C}_p = [(\gamma+1)M_\infty^2]^{1/3} C_p / \tau^{2/3} \qquad (15)$$

depends only on ξ_∞ and x/c; and correspondingly $\bar{C}_{D_p}$ and $\bar{C}_L$ defined by

$$\bar{C}_{D_p} = [(\gamma+1)M_\infty^2]^{1/3} C_D / \tau^{5/3}, \quad \bar{C}_L = [(\gamma+1)M_\infty^2]^{1/3} C_L / \tau^{2/3} \qquad (16)$$

depend only on ξ_∞. The approximate relation that corresponds to Eq. (1)

$$d\xi/d\xi_\infty = 0 \qquad (17)$$

where

$$\xi = [M_\infty^2 - 1 - (\gamma+1)M_\infty^2 u/U_\infty]/[(\gamma+1)M_\infty^2 \tau]^{2/3}$$

leads to the following relations instead of Eq. (11)

$$(d\bar{C}_p/d\xi_\infty)_{\xi_\infty=0} = 2, \quad (dC_p/dM_\infty)_{M_\infty=1} = 4/(\gamma+1) - (2/3)(C_p)_{M_\infty=1} \qquad (18)$$

Application to the single-wedge sections described above indicate that the pressure drag coefficient should vary with M_∞ at $M_\infty = 1$ according to

$$(d\bar{C}_{D_p}/d\xi_\infty)_{\xi_\infty=0} = 2$$

$$(dC_{D_p}/dM_\infty)_{M_\infty=1} = 4/(\gamma+1) - (2/3)(C_{D_p})_{M_\infty=1} \qquad (19)$$

Figure 2 shows the result of plotting the experimental drag data of Liepmann and Bryson[48,49] in this way together

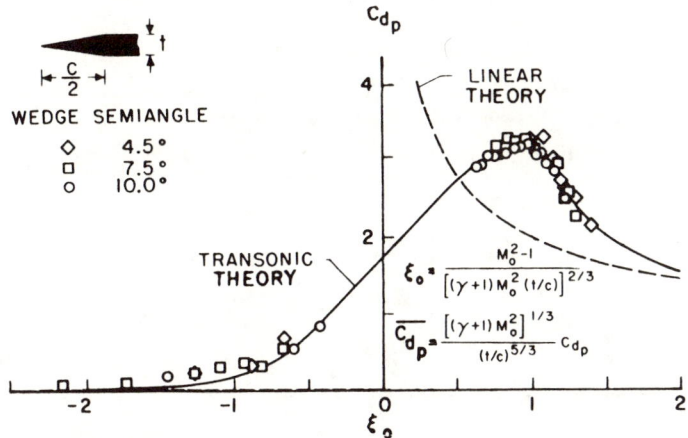

Fig. 2 Transonic pressure drag of nonlifting single-wedge sections. Experimental confirmation of transonic small disturbance theory.[44,50,51]

with the theoretical results of Cole,[45] Guderley and Yoshihara,[46] Vincenti and Wagoner,[47] and the Mach number freeze results of Eq. (19). It is evident that the data from the three experimental wedges fall together to determine a single line in accordance with the prediction of the transonic similarity rule, and that this curve is in excellent agreement with the theoretical curve. Attention is called to the fact that all of the theoretical results shown on this graph have been modified from the form presented in the original papers, in as much as they were derived by solving Eq. (5) rather than Eq. (12). The agreement is not nearly as good when the theoretical results are retained in their original form.[50,51]

The single-wedge section was good for establishing the potential of the transonic small disturbance theory because of the completeness of both the theoretical and experimental results, and because severe viscous phenomena (i.e., separation of shock-wave/boundary-layer interaction) already well known to be important over the rear of a complete airfoil[52-54] were avoided. The variation of drag with Mach number, however, is notably different from that of actual airfoils, which close at the rear. For such airfoils, the relation of Eq. (19) is replaced by

$$(d\bar{C}_{D_p}/d\xi_\infty)_{\xi_\infty=0} = 0, \quad (dC_{D_p}/dM_\infty)_{M_\infty=1} = -(2/3)(C_{D_p})_{M_\infty=1} \quad (20)$$

Figure 3 shows a more typical result, that for the pressure drag of a symmetrical double-wedge section.

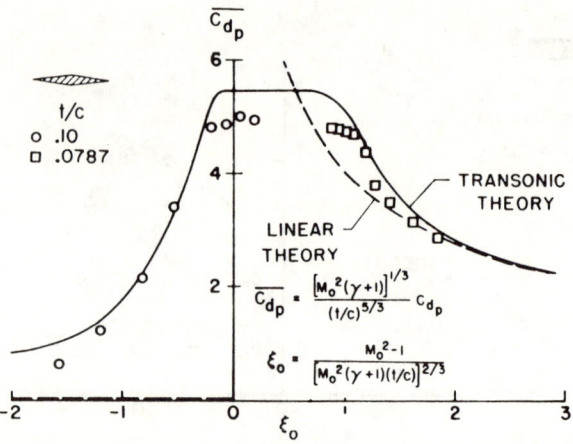

Fig. 3 Transonic pressure drag of nonlifting double-wedge sections. Experimental confirmation of transonic small disturbance theory.[44,50,51]

Again the experimental data[55,56] were shown to be in substantial agreement with the results of transonic small disturbance theory.[46,47,57] Perhaps the principal feature of note is the zero slope of $\bar{C}_{D_p}$ with ξ_∞ at $\xi_\infty = 0$, $M_\infty = 1$, in accordance with Eq. (20) derived from the local Mach number freeze principle. At Mach numbers considerably removed from unity, the predictions of transonic small disturbance theory become substantially the same as those of linear theory, a trend in good accordance with the experimental data. Theoretical results for a number of additional examples of hodograph solutions for transonic flow around wedge and flat-plate airfoils, both nonlifting and lifting, were worked out in the era prior to about 1955.[58-60] Most of these are summarized in comprehensive books by Guderley[61] and Ferrari and Tricomi.[62] Some additional comparisons of theoretical and experimental results similar to those of Figs. 2 and 3 were also published[44] about that time.

All the theoretical results discussed above for wedge and flat-plate sections were obtained with the hodograph transformation. This procedure has the advantage that the differential equation for two-dimensional flow is linearized without approximation, and the disadvantage that the boundary conditions are difficult to specify, except for flat-sided or free-streamline contours. By 1955, there was also a large body of literature in which solutions were sought by iteration of either the approxi-

mate equations of transonic small disturbance theory or the compete equations of compressible flow theory. Of the considerable variety of methods of this type, perhaps the best known and simplest to describe is the method of successive approximation in which the solution is sought in the form of a power series in terms of the thickness ratio.
In this method, the first approximation $\phi^{(1)}$ is determined by solving the Prandtl-Glauert equation, which for two-dimensional flow follows directly from Eq. (12) by replacing the right-hand side by zero

$$(1 - M_\infty^2)\phi_{xx}^{(1)} + \phi_{zz}^{(1)} = 0 \tag{21}$$

The result is precisely that of linear theory, to which the solutions of transonic flow theory converge as the freestream Mach number departs far from unity in either direction. The second approximation $\phi^{(2)}$ is determined by solving the following equation, again linear, or its complete second-order counterpart in which additional terms containing $\phi^{(1)}$ are retained on the right-hand side,

$$(1-M_\infty^2)\phi_{xx}^{(2)} + \phi_{zz}^{(2)} = \frac{M_\infty^2(\gamma+1)}{U_\infty} \phi_x^{(1)}\phi_{xx}^{(1)} \tag{22}$$

in which results of the first approximation are used to evaluate the right-hand side of Eq. (12). Higher approximations for either subsonic or supersonic flows follow in a similar manner.

Although a number of special techniques were developed to aid in the analysis,[63-67] the difficulties of integration are considerable and only a very small number of solutions were ever obtained for approximations higher than the second. Indeed, while the second-order supersonic solution of Busemann[17] indicated by Eq. (4) had been known from the early thirties and the second-order subsonic solution of Hantzsche and Wendt[64] from the early-forties, it was not until 1958, near the arbitrary cutoff date of this section, that the first third-order solution for subsonic flow past an airfoil-like shape was announced in correct form by Asaka.[68] (The problem that Asaka and the others cited immediately above treat explicitly is the determination of higher-order approximate solutions of the complete equations of compressible flow theory, but it is a simple matter to extract the corresponding approximation to the solution of the equations of the small disturbance transonic theory flow from their results.) To my knowledge, this is the only such third-order subsonic solution ever obtained for an airfoil-like shape, although Kaplan determined the fourth-order solution for subsonic flow past a "Kaplan bump",[69] a thin symmetrical shape with cusped leading and trailing edges, and the sixth-order solution for subsonic flow past an infinite sinusoidal wavy wall.[70]

The results of the third-order approximation to the subsonic solution of the transonic small disturbance theory for the variation with freestream Mach number of the pressure coefficient at the midpoint of a nonlifting circular-arc airfoil having a thickness ratio τ are given by[71]

$$C_p = -2\phi_x/U_\infty = -2.5465(\tau/\sqrt{1-M_\infty^2}) - 0.5132(M_\infty^2(\gamma+1)/(1-M_\infty^2)^2]\tau^2$$

$$-0.6339[M_\infty^4(\gamma+1)^2/(1-M_\infty^2)^{7/2}]\tau^3 \qquad (23)$$

and illustrated for the special case of $\tau = 0.06$ in Fig. 4, in which the results indicated by the first three approximations are designated C_{p1}, C_{p2}, C_{p3}. It can be seen that the addition of each term results in an increase in the value of the peak negative pressure coefficient; and that the curves representing each approximation pass without any noticeable effect through the line indicating the critical pressure coefficient at which the local Mach number is unity. It is also a general property of the results obtained by this method of successive approximation that shock-free dragless flow is indicated for all Mach numbers

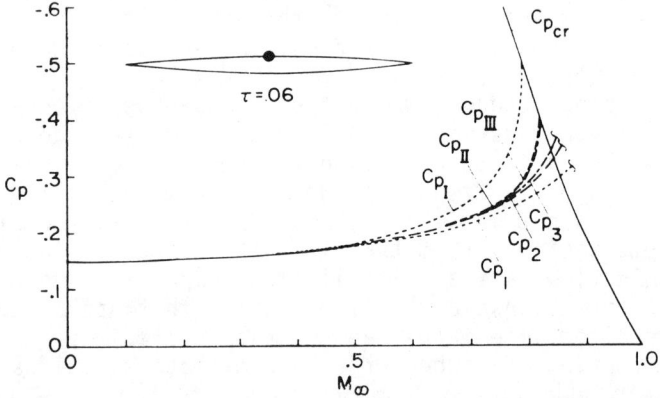

Fig. 4 Successive approximation for the variation with freestream Mach number M_∞ of pressure coefficient C_p at the midpoint of nonlifting 6% thick circular-arc airfoil.[51,71]

up to unity. This is obviously at variance with experience, but the precise limits of convergence of the results of this process have never been determined definitely.

Another, and more successful, method for attempting the solution of the transonic small disturbance equations for flow past thin airfoils developed during the early fifties is the integral equation method[51,72,73] In this method, Eq. (12) and the associated boundary conditions are recast into the form of a nonlinear integral equation by application of Green's theorem to appropriate regions surrounding the airfoil, its wake, and any shock waves that may be present. There are many forms of Green's theorem, and it is possible to derive many different integral equations in this way. The one that was used in the works cited above was

$$\bar{u} - \bar{u}^2/2 = \bar{u}_L - (1/2\pi)\int\int_{-\infty}^{+\infty}(\bar{u}^2/2)\{[(\bar{x}-\bar{\xi})^2$$

$$- (\bar{z}-\bar{\zeta})^2]/[(\bar{x}-\bar{\xi})^2 + (\bar{z}-\bar{\zeta})^2]^2\}d\bar{\xi}d\bar{\zeta} \qquad (24)$$

where

$$\bar{x} = x$$

$$\bar{z} = (1-M_\infty^2)^{1/2}z$$

$$\bar{u} = u/u_{cr} = [M^2(\gamma+1)/(1-M_\infty^2)](u/U_\infty)$$

and $\bar{u}_L$ represents the solution of the linearized theory of compressible flow, i.e., Eq. (21) with $u = \phi_x$. A subtlety not recognized for more than 20 years in the use of this method is that the double integral is semiconvergent and that its value depends on the aspect ratio of the infinitesimal contour used to enclose the field point/singularity where $\bar{u}$ is to be evaluated. In the papers cited above, the singularity was enclosed by an infinitesimal rectangle having a ratio λ of height to width of infinity. In a series of much more recent papers by Nixon and Hancock,[74] a slightly different integral equation in which the term $\bar{u}^2/2$ on the left-hand side of Eq. (24) is replaced by $\bar{u}^2/4$ was derived by enclosing the same point by an infinitesimal circle, and still other values for the coefficient ν of $\bar{u}^2$ have occasionally been put forward. Only recently has this question been clarifed by Ogana and Spreiter[75] who enclosed the field point in an infinitesimal rectangle of arbitrary λ and showed that $\nu = (1/\pi)\arctan\lambda$. This accounts for earlier results, since $\nu = 1/2$ when $\lambda = \infty$ and $\nu = 1/4$ when $\lambda = 1$. These differences are somewhat illusory, however, since the value of the double integral depends on λ in such a way that the differences in $\lambda \bar{u}^2$ are compensated exactly.

Returning to Eq. (24) and the state of knowledge as it existed in the mid-fifties, we proceed by introducing the abbreviation I/2 for the integral and rewriting Eq. (24) in the form[51]

$$\bar{u} = 1 \pm \sqrt{I - L} \tag{25}$$

where $L = 2\bar{u}_L - 1$. It follows immediately that $I \geq L$. The choice of plus or minus sign determines whether the flow is subsonic or supersonic. A change of sign at a point where $I = L$ corresponds to a smooth transition through sonic velocity. A change in sign at a point where $I > L$ results in a discontinuous jump in velocity and corresponds physically to a shock wave. If I and L are continuous at such a point, the discontinuity corresponds to a normal shock wave. Such discontinuities are permissible when they proceed from greater to lesser velocities (compression shock), but are inadmissible when they proceed in the opposite direction.

Equation (25) leads to an alternative method of successive approximation for subsonic flows[51,71] that has the useful property of overestimating, rather than underestimating, the variation of the peak negative pressure coefficient with Mach number, and thereby bracketing the correct answer. This method is closely related to that described above in connection with Eqs. (21) through (23), which with suitable limitations on the airfoil shape leads to a series expansion of C_p in terms τ. An equivalent expansion in terms of the transonic similarity variables $\bar{u}$ and $\bar{\tau} = [M_\infty^2(\gamma+1)/(1-M_\infty^2)^{3/2}]\tau$ is

$$\bar{u}_n = \sum_{n=1}^{n} a_n \bar{\tau}^{-n} \tag{26}$$

where also $\bar{C}_p = \{[M_\infty^2(\gamma+1)]^{1/3}/\tau^{2/3}\}C_p = -(\bar{u}/\bar{\tau}^{2/3})$. The Nth approximation of the alternative method is determined directly from the nth approximation of the series expansion method by use of

$$\bar{u}_N = 1 - \sqrt{1 - 2\bar{u}_n + \bar{u}_{n-1}^2} \tag{27}$$

The result obtained with Eq. (27) for flows that are subsonic everywhere was shown[51] to converge, in the limit of an infinite number of iteration steps, to the same result as would be obtained by using Eq. (26). The result provided by Eq. (27) clearly terminates with the occurrence of sonic velocity somewhere in the flowfield, however, and no results are provided for mixed or transonic flows.

The first three such approximations for the variation with Mach number of the pressure coefficient at the midpoint of a symmetrical circular-arc airfoil with thickness ratio $\tau = 0.06$ are illustrated in Fig. 4, where they are designated C_{pI}, C_{pII}, and C_{pIII}. All of the approximations agree well at low Mach numbers, but substantial differences appear at Mach numbers near the critical. The proper trend is defined within rather narrow limits at all Mach numbers up to the critical, however, since the exact solution of the equations of the transonic small disturbance theory must indicate a variation of C_p with M_∞ between that of C_{p3} and C_{pIII}.

It is clear that the typically asymmetric pressure distribution with embedded shock waves characteristic of supercritical flow past a nonlifting symmetric circular-arc airfoil cannot be calculated using continuous symmetric approximations for u, such as provided by approximating $\bar{u}$ by $\bar{u}_n$, in evaluating I. There is no need to start a successive approximation process with such a distribution for $\bar{u}$, however, and it is of particular importance for the successful successive approximation solution of Eq. (24) for transonic flows that a more appropriate discontinuous initial distribution be used that simulates to some degree the effects of an embedded shock wave. Solutions of Eq. (24) were sought initially by introduction of a velocity profile f of the form $\bar{u}(\bar{x},\bar{z}) = u(\bar{x},0)f(\bar{x},\bar{z},\bar{u}_{z=0},Z)$, where Z represents the ordinates of the airfoil surface and $\bar{u}(\bar{x},0)$ represents the variation of $\bar{u}$ along the airfoil surface and the adjacent portions of the $\bar{x}$ axis. In this way, the double integral I/2 of Eq. (24) was approximated by a single integral, and an initial approximation for I could be obtained with a modest amount of hand computation starting from a reasonable guess for $\bar{u}(\bar{x},0)$. The next approximation for $\bar{u}(\bar{x},0)$ was then determined by Eq. (25) using the resulting values for I, together with the known values for L. The new results were then used in a similar manner to obtain the next approximation, and so on until the process converged to the final solution.

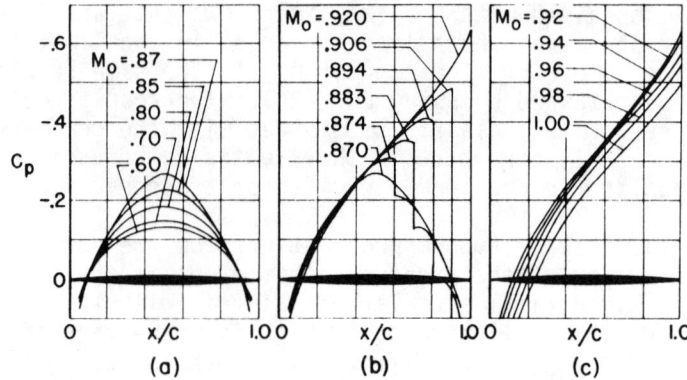

Fig. 5 Integral equation solutions for the pressure distribution of a nonlifting 4% thick circular-arc profile at subsonic and transonic Mach numbers.[44,51]

Figure 5 shows an example of results calculated in this way for a nonlifting 4% thick symmetrical circular-arc airfoil. Part (a) shows that the pressure distributions for subcritical Mach numbers are symmetrical and shock-free, just as the subsonic successive approximation methods indicated by Eqs. (21), (22), (23), (26), and (27) would indicate. Part (b) shows that shock waves make their appearance as the freestream Mach number exceeds the critical, move rapidly rearward with increasing Mach number, and reach the trailing edge at a Mach number of 0.92. Part (c) shows the results for Mach numbers between 0.92 and 1.00. The asymmetric nature of the pressure distributions shown in parts (b) and (c) results in drag at all Mach numbers greater than the critical. Similar satisfactory results for numerous other airfoils have been calculated using this [76] and other methods[77-80] to solve the transonic integral equation. This approach has substantial merit, but its further development[74,81-83] has been rather eclipsed by the spectacular success of finite-difference relaxation methods for solving partial differential equations.

In comparing calculated and measured pressure distribution on airfoils in that part of the supercritical Mach number range for which the shock wave is forward of the trailing edge, there are characteristically found notable discrepancies along the more rearward portions of the airfoil. They were well recognized by 1955 to be associated with strong interaction between the shock wave and boundary layer,[44,52-54] but there was little prospect at the time that such effects could be predicted quantitatively by any solvable theory.

Although many features of transonic flow were revealed by considering two-dimensional flows, either theoretically or experimentally, it was recognized that the effects of finite span are not small and must be determined. Some idea of the magnitudes involved may be had by examining the experimental data[84] shown in Fig. 6, wherein the variations of drag, and lift-curve slope with freestream mach number M_∞ are displayed in the usual coefficient form for a family of rectangular wings of aspect ratio a from 1 to 6. The curves for aspect ratio 6 are much the same as for two-dimensional flow, but the variation with M_∞ diminishes steadily as the aspect ratio is reduced until finally almost no effect of M_∞ is apparent for wings of aspect ratio 1. Another point to observe involves the slope of the drag curves $M_\infty = 1$. The curves for large aspect ratios have a slight negative slope in accordance with Eq. (20) derived from the Mach number freeze principle. For smaller aspect ratio wings, however, C_D increases with $M_\infty = 1$. This is contrary to the prediction, and it would appear that the freeze is no longer operative for these wings.

The first theoretical results to contribute directly to understanding the transonic aerodynamic characteristics of wings of finite span were the transonic similarity rules. These have already been discussed above for two-dimensional flows; their counterparts for the drag and lift-curve slope, both at zero angle of attack, of an affinely related family of uncambered wings are[50]

$$\bar{C}_{D_p} = \mathcal{D}(\xi_\infty, \bar{A}), \quad \bar{C}_{L_\alpha} = L(\xi_\infty, \bar{A}) \qquad (28)$$

where

$$\bar{C}_{D_p} = \{[M_\infty^2(\gamma+1)]^{1/3}/\tau^{5/3}\}C_{D_p}, \quad \xi_\infty = (M_\infty^2-1)/[M_\infty^2(\gamma+1)\tau]^{2/3}$$

$$\bar{A} = [M_\infty^2(\gamma+1)\tau]^{1/3}A, \quad \bar{C}_{L_\alpha} = [M_\infty^2(\gamma+1)\tau]^{1/3}C_{L_\alpha}$$

They indicate that only two paramters, ξ_∞ and $\bar{A}$, are necessary to describe the results, rather than the four parameters (namely, aspect ratio A, thickness ratio τ, freestream Mach number M_∞, and possibly the ratio of specific heats γ if the tests are conducted in more than one gas) that are indicated by traditional considerations of dimensional analysis. Greater simplicity is achieved by considering the situation at $M_\infty = 1$, since $\xi_\infty = 0$ and the results depend only on a single parameter A. If all the tests are conducted in air in the usual manner so that $\gamma = \text{const} = 1.4$, these rules may be simplified further to show that $C_{D_p}/\tau^{5/3}$ and $\tau^{1/3}C_{L_\alpha}$ depend only on $A\tau^{1/3}$ at $M_\infty = 1$. This conclusion is confirmed by the experimental data presented in Fig. 7.[50]

The results of Fig. 7 clearly display one trend for low-aspect-ratio wings and another for high-aspect-ratio wings. The curves approach straight lines through the origin for low-aspect-ratio wings and horizontal lines for high-aspect-ratio wings. The latter behavior indicates that the aspect ratio is dropping out as a parameter and that the lift and drag coefficients are approaching the values they would have for two-dimensional flow. For wings having $A\tau^{1/3}$ less than about unity, the experimental values for the lift practically coincide with the prediction given by linear theory that the lift-curve slope C_{L_α} is equal to $\pi/2$ times the aspect ratio.[36,27,50] Linear theory does not provide a useful approximation for the drag, however, since it predicts infinite values for all wings of finite thickness. It was noted, however, that these results indicate that the drag of the low-aspect-ratio members of this particular family of wings is proportional to the square of the frontal area, provided all of the wings have the same chord and are in flows having the same dynamic pressure.

Additional data of the type illustrated in Fig. 7 were given for the same family of wings by McDevitt[84] and also for a related family of cambered wings of rectangular wings of rectangular planform. Similar results for wings of triangular planform were given by Page.[85]

It was apparent from results such as those of Fig. 7 that a transonic theory for wings of small aspect ratio was needed to supplement that for two-dimensional flow if even a rudimentary predictive capability were to be established. A development commencing with the statement of a transonic equivalence rule by Oswatitsch at the Eighth International Congress of Theoretical and Applied Mechanics in Istanbul[86,87] and the establishment of the transonic area rule by Whitcomb[88] was soon to begin to fill the need. Whitcomb's area rule states that, near the speed of sound, the zero-lift drag rise of a slender wing-body combination

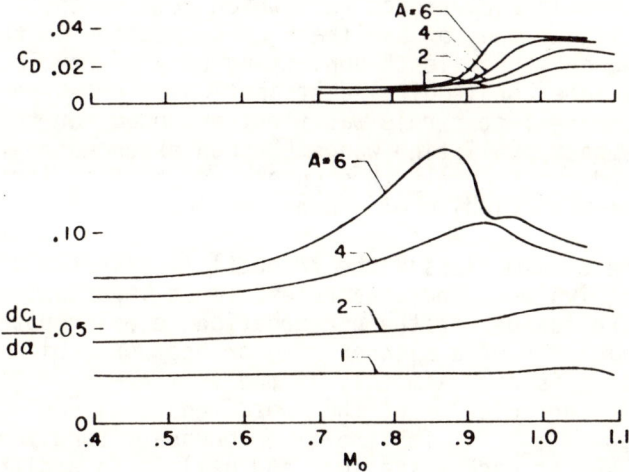

Fig. 6 Experimental variation with freestream Mach number M_∞ of the drag coefficient C_D and lift-curve slope $dC_L/d\alpha$ of a family of rectangular wings of aspect ratio A from 1 to 6 having NACA 65A006 airfoils.[84]

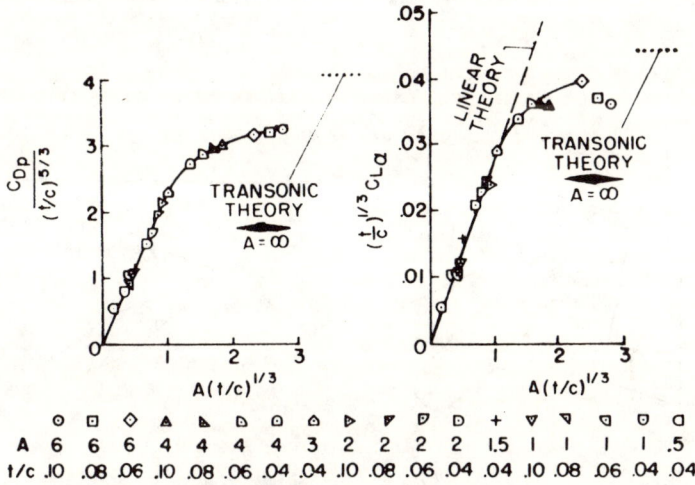

Fig. 7 Lift and drag characteristics at $M_\infty = 1$ of a family of rectangular wings with reduced aspect ratios $A(t/c)^{1/3}$ from 0.5 to 6 with 4% to 10% thick NACA 63A0XX airfoils. Experimental confirmation of the transonic similarity rules.[44,50]

is primarily dependent on the axial distribution of cross-sectional area normal to the airstream. This rule, which was proposed on the basis of certain elementary, yet fundamental, statements regarding the nature of transonic flow-fields and demonstrated experimentally, is closely related to the transonic equivalence rule which relates the flow around a slender body of arbitrary cross section to the flow around an "equivalent" nonlifting body of revolution having the same longitudinal distribution of cross-sectional area $S(x)$. The latter rule was first proposed for transonic flow past nonlifting wings,[86] soon extended to lifting wings,[44] and subsequently to slender bodies of arbitrary cross section, including wing-body combinations.[89]

Figure 8 summarizes[44] the theoretical essentials of both the equivalence and area rules. Most important is that the expression for the perturbation velocity potential ϕ in the vicinity of a slender wing or body or arbitrary cross section is approximately of the form $\phi = \phi_2 + g(x)$ where ϕ_2 is the solution of the two-dimensional Laplace's equation $\phi_{yy} + \phi_{zz} = 0$ for the given boundary conditions in the y,z plane at each x station, and $g(x)$ is an additional contribution dependent upon M_∞ and $S(x)$ but not on the shape of the cross section. It is thus possible to determine $g(x)$ from the solution of the simpler problem of axisymmetric flow past the equivalent body. The equiva-

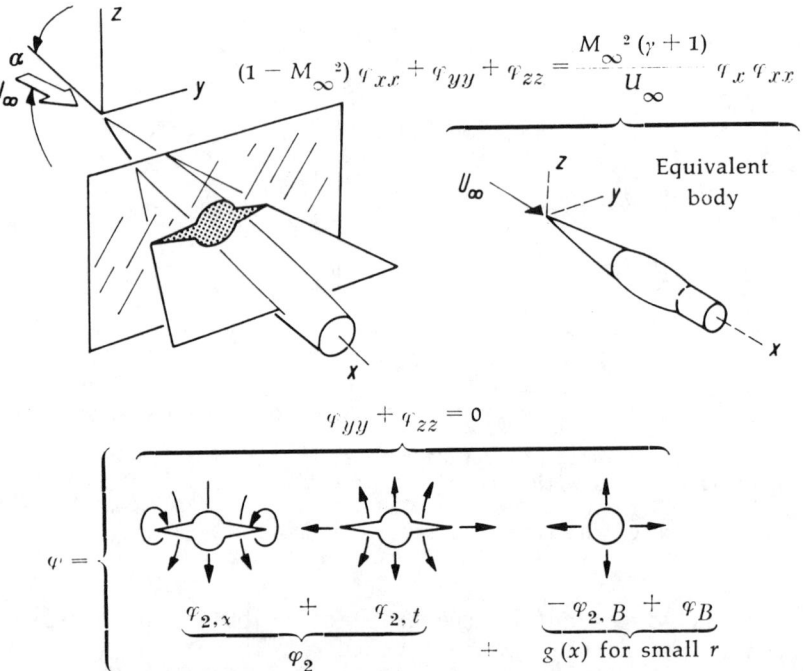

Fig. 8 Sketch illustrating essential ideas and definitions of terms in the transonic equivalence rule.[44,71,89]

lence rule, which is described in mathematical terms by

$$\phi = \phi_{2,\alpha} + \phi_{2,t} - \phi_{2,B} + \phi \qquad (29)$$

in which each component of ϕ has the meaning indicated in Fig. 8, follows immediately by writing $\phi = \phi_2 + g(x)$ for the body of arbitrary cross section and subtracting the corresponding expression for the equivalent body.

While the equivalence rule applies both in transonic small disturbance theory and in the linearized theory of subsonic and supersonic flow, an analysis of the order of error[89] indicates that the equivalence rule should be applicable to wings of greater aspect ratio at $M_\infty = 1$ than at any other Mach number. To be more explicit, it was shown for thin wings of aspect ratio A, chord c, and thickness ratio τ that the magnitude of the quantity $\phi/U_\infty c$ retained in the equivalence rule is $O(A\tau \ln A)$, and that of the quantities discarded in the derivation for $M_\infty = 1$ is $O(A^4\tau^2 \ln A)$; whereas in the linearized theory for either subsonic or supersonic Mach flow the order of magnitude of the discarded quantities is $O(A^3\tau \ln A)$.

Once the appropriate expression has been constructed for ϕ, the pressure distribution on or near the surface of slender bodies may be determined by use of the following expression for C_p:

$$C_p = -\frac{2}{U_\infty}(\phi_x + \alpha\phi_z) - \frac{1}{U_\infty^2}(\phi_y^2 + \phi_z^2) \qquad (30)$$

which differs from the corresponding expression for two-dimensional flow by inclusion of the last three terms. The results may, in turn, be integrated to obtain expressions or values for the total forces, including lift and moments on slender bodies or wing-body combinations of arbitrary cross section. Since the aerodynamic loading, lift, and all lateral forces and moments depend on differences in pressure between pairs of points at the same longitudinal station, these quantities depend solely on ϕ_2, and are therefore independent of M_∞ in this approximation.

At the time the equivalence rule was developed, the only axisymmetric body for which a solution of the transonic small disturbance theory was available was a circular cone-cylinder in flow with a freestream Mach number of 1.[90] That led to the choice of sonic flow past a thin cone-cylinder of elliptic cross section as the first example to be examined in detail, both theoretically and experimentally. The procedure outlined above results in the following relationship between the pressure coefficient C_{p_W} on the surface of the conical part of such a body and C_{p_B} on the surface of a circular cone-cylinder[89]:

$$C_{p_W} + C_{p_B} - (A\tau/8)[1 + \ln(A/8\tau)] \pm [\alpha A/(2\sqrt{1-y^2/s^2})] \quad (31)$$

The results are illustrated in Fig. 9,[44] in which the pressure distribution at zero angle of attack is shown in part (a) and the load distribution due to angle of attack is shown in part (b). Although the pressure distribution at zero angle of attack is not the same as given by linear theory, the load distribution on the lifting wing is contributed solely by $\phi_{2,\alpha}$, and is therefore the same as given by simple slender-wing theory.[24] Thus, the lift of a slender wing at $M_\infty = 1$ is the same as given by linear theory, and all the reciprocal and reverse-flow relations of linear theory for lifting surfaces[91] hold equally for slender wings in transonic flow.

Figure 10 shows a comparison of these results with experimental data for a thin elliptic cone-cylinder having $A = 2$ and $\tau = 0.06$.[71,92] The plot on the left shows the difference between the pressure coefficients at several related points on the nonlifting elliptic and circular

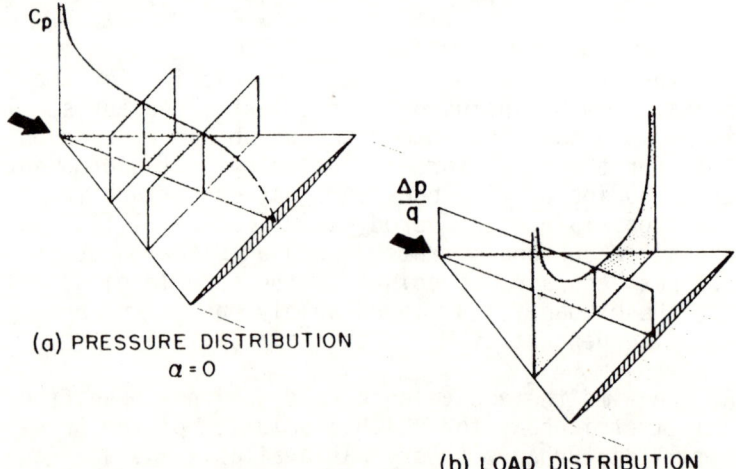

(a) PRESSURE DISTRIBUTION
$\alpha = 0$

(b) LOAD DISTRIBUTION

Fig. 9 Sketches of theoretical pressure and load distributions on a thin elliptic cone-cylinder at $M_\infty = 1$.[44]

cone-cylinders together with a line representing the theoretical prediction of Eq. (31) for $C_{pW} - C_{pB}$. The plot on the right shows a comparison of theoretical and experimental values of the aerodynamic loading, $\Delta p_W/q$, or difference in C_{pW} between the lower and upper surfaces, at a number of points on the elliptic cone-cylinder. The agreement is quite satisfactory except in the small region near the base of the cone.

It is evident that the transonic area rule of Whitcomb[88] is closely related to the transonic equivalence rule. In a detailed examination of the relation between these two rules,[89] the following expression was derived relating the drag D of a slender body of arbitrary cross section and the drag D_B of the equivalent body of revolution,

$$D_W = D_B - (\rho_\infty/2)\left[\oint_{C_W} \phi_{2,\alpha}(\partial\phi_{2,\alpha}/\partial n)ds_C \right.$$
$$\left. + \oint_{C_W} \phi_{2,t}(\partial\phi_{2,t}/\partial n)ds_b - \oint_{C_B} \phi_{2,B}(\partial\phi_{2,B}/\partial n)ds_B \right] \quad (32)$$

Each of the integrals is a line integral along a curve in a plane perpendicular to the x axis that circumscribes the

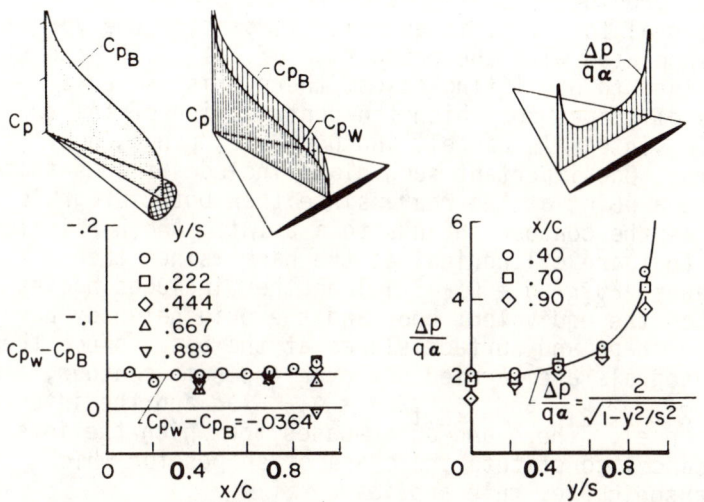

Fig. 10 Pressure and load distributions on a thin elliptic cone-cylinder at $M_\infty = 1$. Experimental confirmation of equivalence rule.[71,89,92]

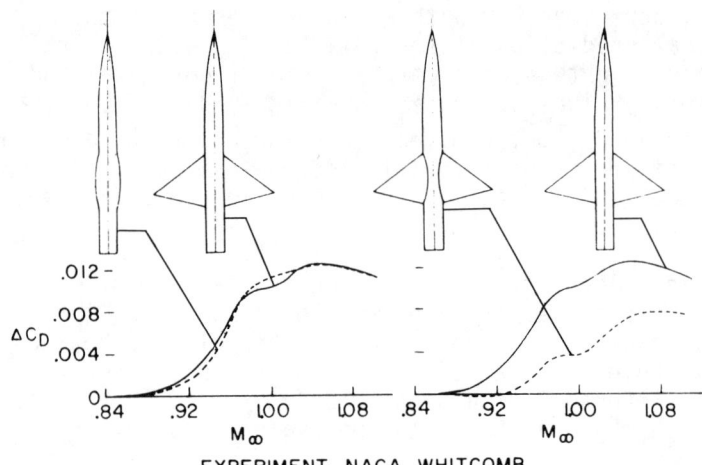

Fig. 11 Experimental pressure drag data[88] demonstrating (left) the transonic area rule that the drag of a wing-body combination is the same as for an equivalent body of revolution and (right) that the drag of a wing-body combination can be reduced substantially by indenting the body so that the equivalent body is smooth rather than bumpy.[71]

base of the body and any vortex wake that may be present. The difference $D - D_B$ is thus independent of M_∞ and is the same as given by linearized slender-body theory. If the arbitrary body is inclined at angle of attack α, the first integral provides a contribution to the drag that is proportional to α^2. This quantity is exactly the vortex drag associated with the production of lift. If attention is confined to nonlifting cases, there exist several classes of shapes for which the contribution of the two remaining integrals cancel, and $D = D_B$ as proposed by Whitcomb. One important such class includes shapes that taper to a point at the rear, since then both integrals vanish as the contour shrinks to a point. Another includes shapes that are cylindrical at the base, since then $\partial \phi_{2,t}/\partial n = \partial \phi_{2,B}/\partial n = 0$. Still another includes bodies for which the equivalent body and the original body have the same shape and surface slopes at the base, since then both integrals are carried out over the same contour, along which $\phi_{2,t} = \phi_{2,B}$ and $\partial \phi_{2,t}/\partial n = \partial \phi_{2,B}/\partial n$ and the integrals again cancel. These and other cases for which the integrals cancel constitute the class of shapes for which the transonic area rule applies.

Some of the data which Whitcomb presented[88] to support his statement of the transonic area rule are shown in

Fig. 11. It can be seen from the plot on the left that the part of the drag that remains after the low-speed friction drag is subtracted is very nearly equal for this particular pair of bodies. Not all of the results, and particularly those for which the equivalent body is indented, are in as good agreement with the area rule as the results shown in this plot, but it is quite possible that some of these discrepancies may be associated with boundary-layer effects.

Whitcomb's main purpose, however, was not to investigate the accuracy of the area rule, but to find shapes for wing-body combinations having low drag through the transonic range. The area rule provides the key to the solution of this problem because it suggests that the shape of a wing-body combination having low drag must be such that the equivalent body has the smooth form of a low-drag body of revolution. Thus, the cross section of the body must be reduced in the vicinity of the wing. The results shown on the right in Fig. 11 illustrate that such wing-body combinations have much lower drag than those without body indentation. This result is of great practical importance in the design of aircraft for efficient flight in the transonic range.

It is important to realize that the integrals in Eq. (32) do not always vanish or cancel, and that the area rule does not hold for all combinations of thin wings and slender equivalent bodies. The cone-cylinders of Fig. 10 are one such case. The drag coefficient C_{D_W}, based on maximum cross-section area of the elliptic cone-cylinder is given by[71,89]

$$C_{D_W} = C_{D_B} - \omega^2 \ln\{(a/4b)[1 + (b/a)]^2\} \qquad (33)$$

in which C_{D_B} represents the drag coefficient of the equivalent body of revolution, the fraction a/b represents the ratio of the major to minor axis of the elliptic cross section, and ω represents the semiapex angle of the equivalent circular cone-cylinder, the drag coefficient for which is[93] given by

$$C_{D_B} = -\omega^2(1.09 + 4 \ln \omega) \qquad (34)$$

This result shows that the drag coefficient is a maximum when the cross section is circular, and diminishes markedly as the elliptic cross section is flattened. This finding is in accordance with a more general conclusion established by Berndt[94,95] in the course of a study of certain minimum and maximum properties of families of equivalent bodies at $M_\infty = 1$.

It is evident that neither the transonic equivalence rule nor the area rule can continue to apply as the lateral dimensions of the wings are extended indefinitely. Insight into the range of applicability of these rules can be gained by examination of the results for the drag of a family of rectangular wings presented in Fig. 7. The values for the drag of various of these wings are related to each other in accordance with the area rule over that part of the range $A_T^{1/3}$ for which the results can be represented by a straight line passing through the origin. It can be seen that the area rule is applicable to all of the wings of this particular family having $A_T^{1/3}$ less than about unity. It should be noted that the equivalent bodies associated with this family of wings tend to be blunt and stubby, rather than pointed and slender. It was apparent that the drag characteristics of the equivalent bodies of revolution might be considerably different from those of the related family of wings, even though the drag characteristics of the low-aspect-ratio members of this family of wings are related to one another in accordance with the transonic area rule. Further discussion of the drag characteristics of this family of wings, as well as a related family of cambered wings, can be found in Ref. 96.

Additional insight into the range of applicability of the area rule was gained at the time, and is still worthy of citation, by examination of the experimental results for the drag of a pair of parallel parabolic-arc bodies of revolution of thickness ratio τ of $\sqrt{2}/12$ in a flow with a freestream Mach number of 1.[97] When the separation distance is less than about a quarter of the body length, the drag coefficient based on maximum cross-section area is very nearly the same as for a single body of revolution of thickness ratio 1/6 as indicated by the area rule. Notable departures from the area rule are apparent as the bodies are moved apart from each other and the drag of each body diminishes until it approaches that of an isolated body of thickness ratio $\sqrt{2}/12$, as it obviously must when the bodies are sufficiently separated. The necessary distance in this case was only somewhat greater than the body length.

An obvious and important need once the area and equivalence rules were known was for a method to determine the solution for axisymmetric transonic flow past a slender body of revolution. A daring proposal by Oswatitsch and Keune[98] that the nonlinear term in the three-dimensional transonic small disturbance equation

$$(1-M_\infty^2)\phi_{xx} + \phi_{yy} + \phi_{zz} = \frac{M_\infty^2(\gamma+1)}{U_\infty} \phi_x \phi_{xx} \qquad (35)$$

could be approximated satisfactorily for sonic flow ($M_\infty=1$) past slender bodies of revolution by replacing $\phi_x\phi_{xx}$ with $K\phi_x$, where K is constant. The name, parabolic method, given to this procedure derives from the parabolic, rather than mixed elliptic-hyperbolic, type of the resulting approximate differential equation for ϕ, which has the form of the equation for heat conduction. This method was subsequently applied to two-dimensional flows past thin airfoils, and extended to other Mach numbers by the simple expedient of retaining the term $(1-M^2)\phi_{xx}$ (see Refs. 99 and 100 for reviews by two of the leading early contributors). Although the results display good accuracy for some applications such as that of sonic flow past a parabolic-arc body of revolution considered originally by Oswatitsch and Keune,[98] they are of such marginal or poor accuracy in other applications that the method is unsuitable for general use.

An effort to improve the general accuracy of the parabolic method led to the local linearization method. This method was developed first for two-dimensional flow past thin airfoils[101] and axisymmetric flow past slender bodies of revolution[93] for the Mach number ranges $M_\infty \approx 1$, $M_\infty < M_{cr,\ell}$ and $M_\infty > M_{cr,u}$, where $M_{cr,\ell}$ and $M_{cr,u}$ refer to the lower and upper critical Mach numbers that bound the transonic range. It was also extended[102] to flows with $M_\infty \approx 1$ past nonlifting wings of finite span having simple planform and airfoil shapes. A comprehensive summary of the method and results obtained with it is given in Ref. 103.

Briefly, the basic idea underlying the local linearization method is that of linearizing the transonic small disturbance equation by replacing either ϕ_x or ϕ_{xx} in the nonlinear term by a constant λ, solving the simplified equation, and then introducing different values for λ for different points in the flow. This procedure might be considered equivalent, in some sense, to replacing the original nonlinear partial differential equation by a different linear partial differential equation at each point. Results obtained by such a procedure depend, of course, on the choice of λ and must be assembled to determine the final results. This step is accomplished by putting the results into such a form that a first-order nonlinear ordinary differential equation is obtained for the streamwise perturbation velocity component $u = \phi_x$ after λ is replaced by the quantity it originally represented. In many cases, this equation is of sufficiently simple form that it can be integrated analytically and the solution expressed in closed form. In other cases, the integration must be performed numerically, but the equation is such a form that standard methods can be applied.

Among the first examples to which the local linearization method was applied was flow with $M_\infty = 1$ past thin nonlifting airfoils having a finite angle at the leading edge. The variation of the pressure coefficient C_p or its transonic similarity counterpart $\bar{C}_p$, along the surface of such an airfoil was shown[101] to be

$$\bar{C}_p = [(\gamma+1)^{1/3}/\tau^{2/3}]C_p$$

$$= -2[3/\pi \int_{x^*}^{x} \left\{ (d/dx_1) \int_0^{x_1} [d(Z/\tau)\sqrt{x_1-\xi}\,]d\xi \right\}^2 dx]^{1/3} \qquad (36)$$

in which Z represents the ordinates of the airfoil surface and x^* represents the x coordinate of the point on the airfoil surface at which the flow accelerates through the speed of sound. Application to a single-wedge profile of chord $c/2$ and semiapex angle τ, for which the sonic point is known from a priori considerations to be at a shoulder, yields the following expression for $\bar{C}_p$[101]

$$\bar{C}_p = [(\gamma+1)^{1/3}/\tau^{2/3}]C_p = -2[(3/\pi)\ln(x/(c/2))]^{1/3}$$

A plot of the results is shown in Fig. 12 together with the corresponding theoretical results obtained with the hodograph method.[46] Although some approximations are introduced in the course of the latter analysis, the results are generally regarded as virtually an exact solution of the equations of the transonic small disturbance theory. This comparison helped establish the accuracy of the local linearization approximation.

Application to smoothly curved airfoils requires the determination of the location of the sonic point x^*.

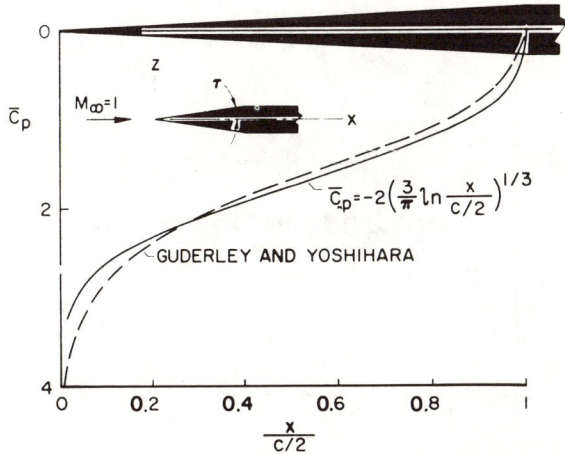

Fig. 12 Transonic small disturbance theory pressure distributions for $M_\infty = 1$ on a nonlifting finite single-wedge section; approximate local linearization[101] and exact hodograph solutions.[46,71]

Specification that du/dx be finite at the sonic point on a smoothly curved airfoil leads to the requirement that x^* is the value for x for which

$$(d/dx)\int_0^x \{[d(Z/\tau)/d\xi]/\sqrt{x-\xi}\}d\xi = 0 \qquad (38)$$

is satisfied.

Application of Eqs. (36) and (38) to a circular-arc airfoil of maximum thickness t and chord c leads to the following expression for the pressure distribution at Mach number 1:

$$\bar{C}_p = [(\gamma+1)^{1/3}\tau^{2/3}]C_p$$
$$= -2\{(12/\pi)[\ln(4x/c)-8(x/c)+8(x/c)^2+3/2\}^{1/3} \qquad (39)$$

where τ represents the thickness ratio t/c. A plot of this result is shown in Fig. 13 together with experimental pressure distributions[104] for four such airfoils having thickness ratios of 6%, 8%, 10%, and 12%. These results are presented in terms of $\bar{C}_p$ rather than C_p because the transonic small disturbance theory indicates that the pressure distributions for all four airfoils should then define a single curve independent of the thickness ratio. It can be seen that the theoretical and experimental

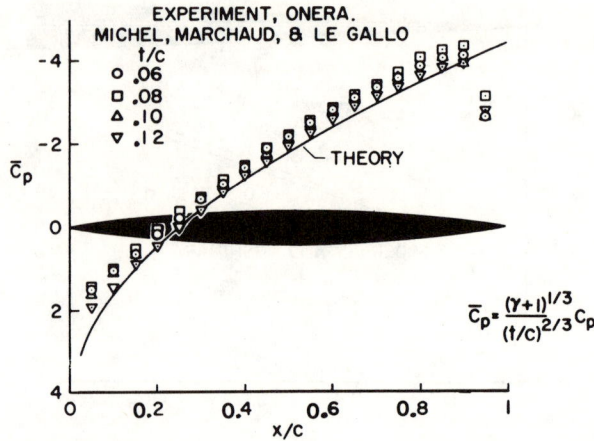

Fig. 13 Local linearization theory[101] and experimental pressure distributions[104] for $M_\infty = 1$ on four nonlifting 6% to 12% thick circular-arc airfoils.[71]

results are in substantial agreements. The most notable discrepancy is that found near the trailing edge, and can be attributed to flow separation induced by boundary-layer/shock-wave interaction. Although part of the remaining discrepancy may be the result of the experimental technique in which the airfoil is simulated by a bump on the tunnel wall and is hence embedded in the wall boundary layer, further analysis[76] indicated that it is plausible to attribute a substantial amount of the discrepancies in this and numerous related examples[71,102,103,105] to wind-tunnel wall interference.

An analysis similar to that outlined above for two-dimensional flows was given for axisymmetric flow in Ref. 93 and reviewed and applied to additional cases in Refs. 71, 103, and 105. Figure 14 shows the pressure distribution calculated with the local linearization method for axisymmetric flow with freestream Mach number 1 past a parabolic-arc body of revolution having ratio τ of maximum diameter to length of 1/6 together with experimental results[97] from tests of the forward 5/6 of a parabolic-arc body mounted on a cylindrical support extending downstream from the rear of the body as indicated in the figure. The theoretical and experimental results are in satisfactory agreement except over a small part of the body immediately forward of the point where the experimental body, in effect, becomes cylindrical. Part of this discrepancy is undoubtedly associated with the occurrence of a shock wave that must be detached from the corner because the local Mach number is too small, but it was soon evident that significant other causes must also be involved.

Close inspection of these and related comparisons of calculated and measured pressure distributions on slender bodies of revolution[71,92,93,103,105] revealed the existence of substantially larger effects of wind-tunnel wall interference near Mach number 1 than had been anticipated. Because of the basic nonlinearity of the theory and the lack of general methods for solving these equations, it was not possible to develop a complete predictive theory at the time. Even so, the transonic small disturbance theory provided considerable guidance for the interpretation of the experimental results[71,105,106] with different size models in several wind tunnels, and the results were examined critically in terms of the available theoretical knowledge. Two of the more striking conclusions that emerged from those investigations were 1) that flows with freestream Mach number 1 could be simulated quite successfully in a conventional solid wall wind tunnel by running it in the choked condition[71,105] and 2) that wall interference effects tended to be more severe for thin bodies than for fatter bodies of the same length. The former is illustrated in Fig. 15, which shows the pressure distributions measured on the same model, a 6 in. diam

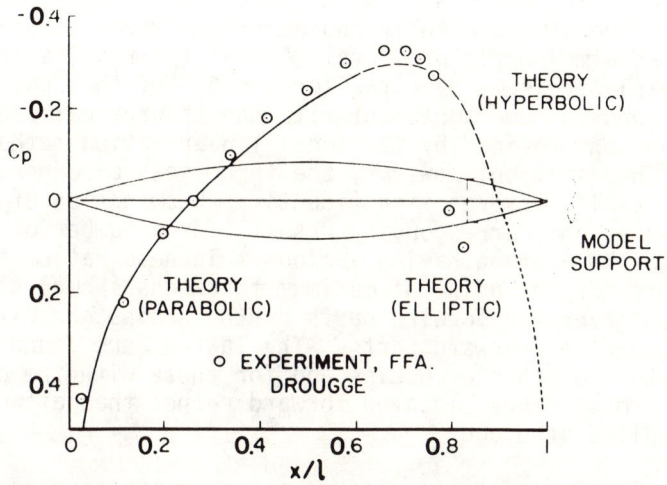

Fig. 14 Local linearization theory[93] and experimental[97] pressure distributions for $M_\infty = 1$ on a nonlifting parabolic-arc body of revolution having a thickness ratio of 1/6.

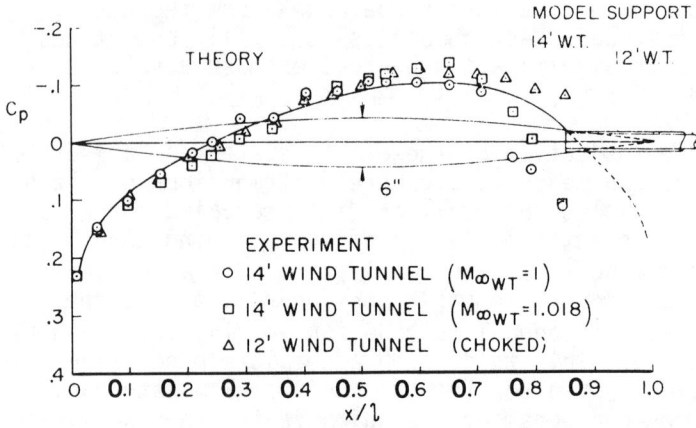

Fig. 15 Local linearization theory[93] and experimental pressure distributions for $M_\infty = 1$ on a nonlifting parabolic-arc body of revolution having a thickness ratio of 1/12 in a transonic wind tunnel[107] and in a choked solid wall wind tunnel.[71,105]

parabolic-arc body of revolution with thickness ratio 1/12, tested in two different wind tunnels at NASA/Ames Research Center. The 14 ft wind tunnel is a conventional transonic wind tunnel with ventilated walls. Results are shown for freestream Mach number 1 as determined following usual practice for that wind tunnel, and also for a test Mach number of 1.018 to allow for effects of the walls according to one means of estimation. The 12 ft wind tunnel is a pressure-sealed solid-wall wind tunnel having sufficient power to operate in a fully choked manner when at somewhat less than atmospheric pressure. All of the results are very nearly the same over the forward 60% of the body length, and in good agreement with the theoretical pressure distribution provided by the local linearization method. Aft of that station, however, the three sets of experimental results diverge significantly. Examination of these differences and corresponding results for a number of other bodies of revolution having various thickness ratios and locations for the point of maximum thickness showed similar agreement over the forward parts of the bodies and differences over the rearward parts. The latter were found to be greater for thinner bodies and for those with the point of maximum thickness located forward rather than aft of the middle of the body.

The source of these differences over the rear of the bodies was explained in terms of the diagrams shown in Fig. 16.[71] The diagram on the left shows an abridged plot of the characteristic lines for an unbounded flow with freestream Mach number 1 about a parabolic-arc body of thickness ratio 1/6. The diagram was taken from Ref. 109 and was calculated by use of the parabolic method[98] to compute the conditions along the sonic line and a simplified method of characteristics based on the transonic small disturbance theory to compute the conditions in the supersonic region. The horizontal dashed line at some distance from the axis represents the position of the wall in the tests in the 12 ft pressure wind tunnel. The nearest point of the wall in the 14 ft transonic wind tunnel is 7/6 as far away from the axis of the body, and the nearest point of the wall in the tests illustrated in Fig. 15 is even farther from the body axis. The plot on the left shows that the wind-tunnel walls are sufficiently far away in the tests with the body of thickness ratio 1/6, both for those shown in Fig. 15 and for other tests in the Ames wind tunnels[71] not portrayed in this account, that none of the characteristic lines that striking the wall can reflect back and influence improperly the pressures on the rear of the body.

The diagram shown on the right of Fig. 16 presents the corresponding results for the body of thickness ratio 1/12. The results were obtained from those for the body of thickness ratio 1/6 by application of the transonic similarity rule for axisymmetric flow,[110] which states that the flowfield associated with sonic flow past an affinely related family of slender bodies of revolution can be

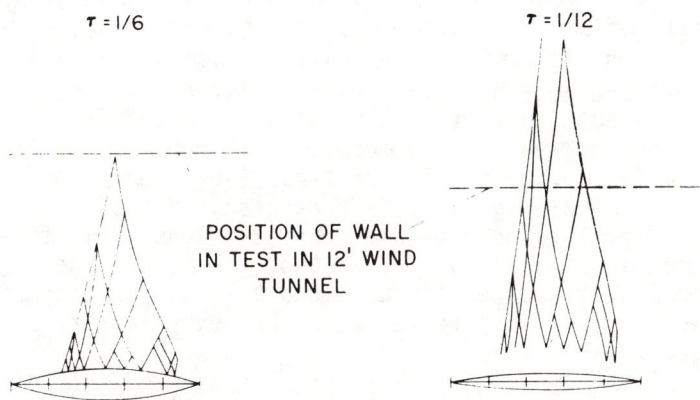

Fig. 16 Characteristics diagrams for the supersonic flow regions at $M_\infty = 1$ about parabolic-arc bodies of revolution illustrating the cause of localized wind-tunnel wall interference effects on the aft portions of a body.[17,105]

brought into coincidence by stretching the lateral coordinates in inverse proportion to thickness ratio. The position of the wall with respect to the model in the tests in the 12 ft pressure wind tunnel is again indicated by the dashed line. The diagram shows that the walls in both the 12 ft and 14 ft wind tunnels are not sufficiently far away to prevent characteristics reflecting from the walls to impinge on the rear of the body, and that the most forward reflected characteristic strikes the body at about 60% of the body length in the tests in the 12 ft pressure wind tunnel. This position coincides well with the point at which the various results presented in Fig. 15 begin to diverge from each other. The effect of reflected characteristics striking the body is to make the pressures more negative in the 12 ft wind tunnel, because the outgoing characteristics represent expansion waves that reflect from the solid wall of the tunnel as expansion waves. The sign of the corresponding effects in the 14 ft transonic wind tunnel is not so simple to ascertain, since the reflections from the solid part of the walls are expansion waves and those from the open parts are compression waves. For a variety of reasons, it was concluded that the influence of the reflections from the partly open walls of the 14 ft wind tunnel is very nearly equal in magnitude but opposite in sign to that of the reflections from the solid walls of the 12 ft pressure wind tunnel. This aspect of wind-tunnel wall interference imposes strong restrictions on the maximum length of bodies for which reliable results can be determined in either a transonic wind tunnel with ventilated walls or a solid wall wind tunnel operating in the choked condition. These restrictions become increasingly severe as the thickness ratio is diminished, and it is necessary to use models that are not only smaller in diameter but also smaller in length to prevent the reflected waves from impinging on the rear of the model. Although some of the other examples for which similar comparisons were made[71,105] showed substantially greater differences than those displayed in Fig. 15, and the same considerations apply to more complex configurations such as wing-body combinations, a lack of reliable and readily applicable means for assessing the effects of wall interference in transonic testing remains to this day.

Throughout the period to 1960, mathematical difficulties confined nearly all transonic analyses other than that of the similarity rules to either two-dimensional flows or to cases that could be reduced to one- or two-dimensional problems by use of axisymmetry or equivalence rule principles. A notable exception was the extension of the local linearization method to nonlifting wings at the very end of this time period.[105] Figure 17 illustrates plots of the pressure distributions calculated in this way for flow with $M_\infty = 1$ past three wings of rectangular planform of differing aspect ratio having parabolic-arc airfoil sections. No experimental data were acquired to evaluate the results, but they appear reasonable in all respects. As the aspect ratio approaches infinity, the pressure distribution in the midspan region approaches that indicated by the local linearization method for two-dimensional flow about the same airfoil, i.e., Eq. (39), whereas that at the wing tip is uniformly smaller by a factor $(1/2)^{2/3}$.

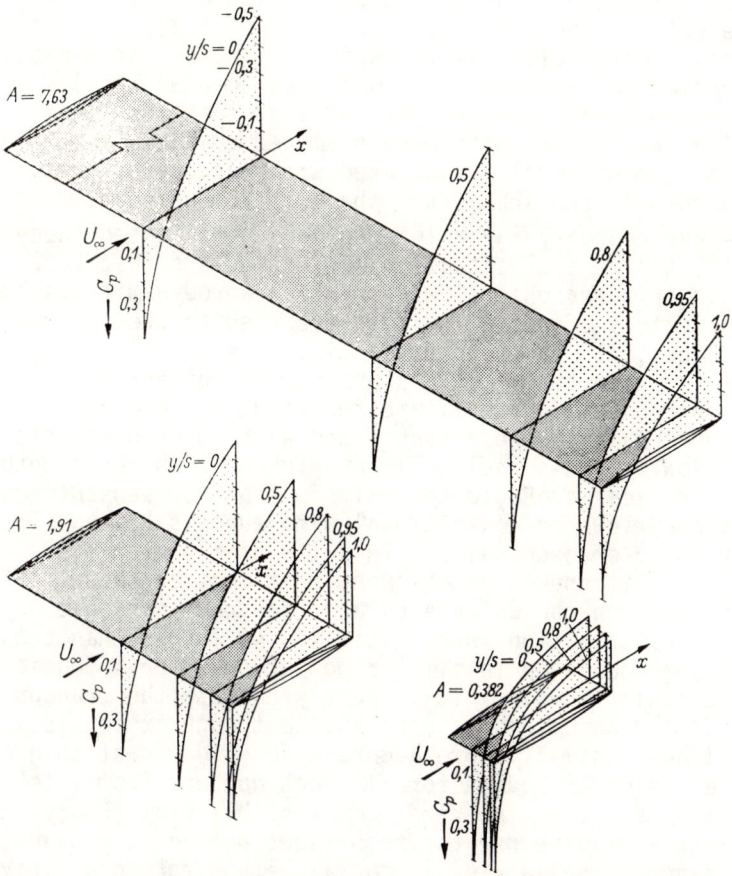

Fig. 17 Local linearization solutions for pressure distributions at $M_\infty = 1$ on three rectangular wings of aspect ratio A = 0.382, 1.91, and 7.63 having 6% thick parabolic-arc airfoils.

The magnitudes of the pressure are smaller in both regions for wings of finite aspect ratio. It was also shown that simple analytic expressions could be determined for the pressure distribution for $M_\infty = 1$ for these and other wings as the aspect ratio becomes small.

From the above account, it should be apparent that the period from 1945 to 1960 was indeed one of vigorous progress in transonic aerodynamics, both theoretically and experimentally. By 1960 there remained no question whatsoever that the transonic small disturbance theory provides an excellent basis for the analysis of many important transonic flows, even if it disregards completely all viscous phenomena; and that an even better description would be provided by the complete potential equation or the Euler equations for compressible flow used together with exact, as opposed to approximate, boundary conditions and pressure-velocity relationships. Many examples had been worked out and compared with experimental results, but numerous small differences and uncertainties were revealed when the results were examined critically. Some of these are obviously associated with viscous effects disregarded in the analysis. While phenomena such as shock-wave/boundary-layer interactions complicated these effects, the general situation was not unlike that for subsonic or supersonic flows for which aerodynamicists had a long history of using inviscid theories to predict real fluid behavior. Other uncertainties were equally obviously associated with the small disturbance and other approximations introduced to facilitate the solution, and still other uncertainites were associated with unknown effects of transonic wind-tunnel wall interference, which is both larger and less amenable to analysis than for subsonic or supersonic flows in conventional wind tunnels having either solid or open walls. In spite of the emphasis given here, it should be observed that many of these differences between the calculated and measured transonic pressure distributions were little or no larger than those normally encountered in similar comparisons for subonsic or supersonic flows. That is partly because the transonic small disturbance theory, approximate though it be, is overall substantially more accurate for flows past thin wings and slender bodies for all Mach numbers from 0 to about 1.5 than the linearized compressible flow theory then in almost exclusive use for such flows outside the transonic range. The lack of a general method for accurately solving the governing nonlinear partial differential equations of mixed elliptic-hyperbolic type prevented the rapid development of the theory, and particularly so for the three-dimensional and unsteady flows that were beginning to be considered by about 1960. Nearly a decade was to pass before such a method began to evolve, and it depended both on new concepts in numerical analysis and on substantially more powerful electronic computers than were available in 1960.

Transonic Aerodynamics in 1975

The period of the sixties were not one of rapid progress in transonic aerodynamics, but certain developmetns were under way that were to culminate in the almost explosive advance of the seventies. Practical interest in transonic aerodynamics was enhanced by the development of the shock-free supercritical airfoils by Pearcey[111] and Whitcomb[112] and the potential it provided for improved performance of aircraft in the lower part of the transonic speed range. New methods in numerical analysis including the concept of a dissipative finite-difference scheme,[113,114] were being developed, and vast improvements were being made in the speed and memory capabilities of computers.

Prior to the experiments of Pearcey[111] there was a general feeling among aerodynamicists that shock-free supercritical flows about airfoils probably could not be realized in practice. Experiments characteristically showed the presence of shock waves, and mathematical analysis had established the nonexistence of shock-free transonic potential flow solutions for a substantial variety of cases.[62,115,118] Rigorous mathematical proof was never established for the general case, however, and it remained for Pearcey to show experimentally the possibility of realizing almost shock-free transonic flow about a specially designed class of airfoils having a high suction peak over the forward part of the airfoil. The underlying principle is that supersonic expansion waves emanating from the high suction region would reflect from the sonic line back to the airfoil as compression waves and provide a useful amount of isentropic recompression, thereby reducing the strength of the shock wave that terminates the supersonic region. Pearcey proceeded by a laborious iterative series of experimental tests and rudimentary characteristics theory to design airfoils that had substantial regions of supersonic flow over the forward portions of the airfoil and essentaily shock-free deceleration to subsonic flow farther downstream. Within a few years, appropriate theory was developed to construct shock-free transonic flow solutions.[118-125] These solutions were obtained using the hodograph method applied to the exact equations for isentropic compressible flow and yielded the airfoil shape as part of the solution. Airfoils designed in this way were then tested in a wind tunnel and the validity of the shock-free solutions was established.[126] The supercritical performance of these airfoils was improved even more by introduction of the rear-loading concept.[112] The idea is to increase the pressure over the aft lower surface by locally thinning the airfoil, thus introducing aft camber. Since the effects are confined almost exclusively to the lower surface where the flow is normally subsonic, few additional considerations of transonic aerodynamics are involved.

Before proceeding to a more detailed discussion, some words are in order about the changes between 1960 and the present in the relationship between theory and experiment brought about by the dramatic increase in computing power during the period. In the precomputer era, aerodynamic theory and experiment lived in somewhat peaceful coexistence. In practical work, wind-tunnel experiment, and to a lesser extent flight testing, dominated. Theory provided useful guidance in many respects, but in only few special instances was it capable of providing reliable and complete solutions for configurations used in actual design. With a few exceptions, the available theories depended on the assumption of small perturbations and on the use of novel approximations, many of which were not readily amenable to rigorous evaluation or systematic improvement. In those few exceptions, such as Emmons' pioneering work[127] with the relaxation method and Vincenti and Wagoner's[47,58,60] series of transonic wedge flows solved in hodograph variables using relaxation and characteristics methods, the time required to perform the calculations for a single case was measured in weeks or even months, and the cases selected for solution were carefully chosen to facilitate the solution. These solutions were invaluable for establishing a theoretical foundation for calculating transonic flowfields, but too costly and far too limited in generality to be useful for predicting the aerodynamic properties of practical aircraft configurations. The computational methods now available will yield numerical solutions for increasingly realistic two- and three-dimensional configurations at a cost of a few minutes central-processor time of a large-size computer. This dramatic reduction in computation time masks an even more striking increase in total arithmetical operations required and also the many new mathmatical ideas involved. Many individuals and groups participated in this development, and we now turn to a summary of transonic aerodynamics as it appeared in 1975.[118,128-134] Highlights of the historical development from 1960 to 1975 are presented in less detail than in the preceding sections in deference to the more complete summaries of new numerical methods provided by the remainder of the papers in this book.

Whereas the great bulk of transonic aerodynamic theory prior to 1960 was based on the transonic small disturbance theory, and attention was often focused for simplicity on airfoils and bodies having pointed noses, the desire to design shock-free transonic airfoils having a round nose and a peaky pressure distribution led to the use of the exact partial differential equations for isentropic compressible potential flow which is as follows for two-dimensional flow:

$$(a^2 - \phi_x^2)\phi_{xx} - 2\phi_x\phi_z\phi_{xz} + (a^2 - \phi_z^2)\phi_{zz} = 0 \qquad (40)$$

where the velocity $q = \hat{i}U + \hat{j}W$ is related to the velocity potential ϕ by $q = \tilde{\nabla}\phi$, $a = [a_0 - (\gamma-1)q^2/2]^{1/2}$ is the local speed of sound, and a_0 is the stagnation speed of sound. Of the many methods for solving this equation the hodograph is unique in that it depends on a transformation that linearizes Eq. (40) without approximation by interchanging the dependent and independent variables. In this way, Eq. (40) transforms to

$$(a^2 - U^2)\Phi_{UU} - 2UW\Phi_{UW} + (a^2 - W^2)\Phi_{WW} = 0 \qquad (41)$$

in which Φ is the Lengendre potential related to the coordinates by $\Phi_u = x$ and $\Phi_w = z$. The Jacobian of the transformation cannot vanish where the flow is subsonic, but may vanish where it is supersonic, thereby signifying an inconvenient multiple mapping in the hodograph plane of a single point in the physical plane. As a result, use of this method is frequently confined to the subsonic region and a limited part of the supersonic region unless the latter is small, as at slightly supercritical Mach numbers. The remainder of the solution is then calculated by another method, such as characteristics or finite differences. The hodograph method was used extensively in the early development of transonic flow theory, but difficulties in imposing boundary conditions for an airfoil of specified shape restricted its usefulness. It came into a new era of

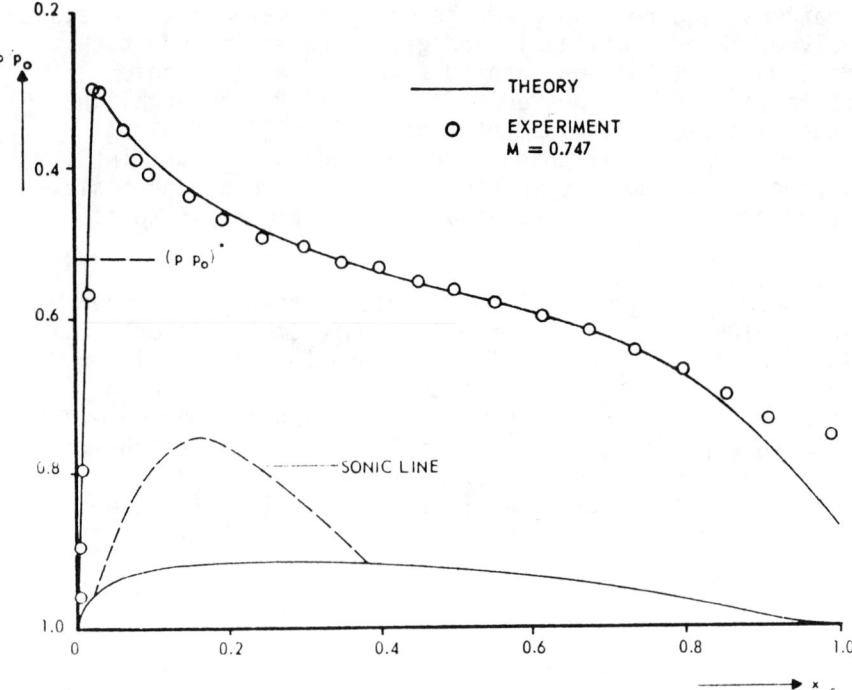

Fig. 18 Exact hodograph solution[119] and experimental confirmation[118,126] of a continuous supercritical pressure distribution on a nonlifting shock-free airfoil.

significance, however, in the design of shock-free airfoils for which it is particularly well suited.

Figure 18 shows the results of solving the hodograph equation using function theory and integral transforms[119] on both the subsonic and supersonic regions to design and calculate the pressure distribution on a nonlifting shock-free airfoil, together with experimental data from wind-tunnel tests. Figure 19 shows a corresponding comparision for a lifting airfoil.[118,122] The experimental results are in excellent agreement with the theory, except very near the trailing edge where the modest discrepancies were attributed to boundary-layer effects and model imperfections. The results are of fundamental importance because they indicate that the theoretical potential flow can in principle be approached arbitrarily closely, and that essentially shock-free transonic flows cannot universally be called "unstable" in any physically useful sense, as had been proposed so often previously.

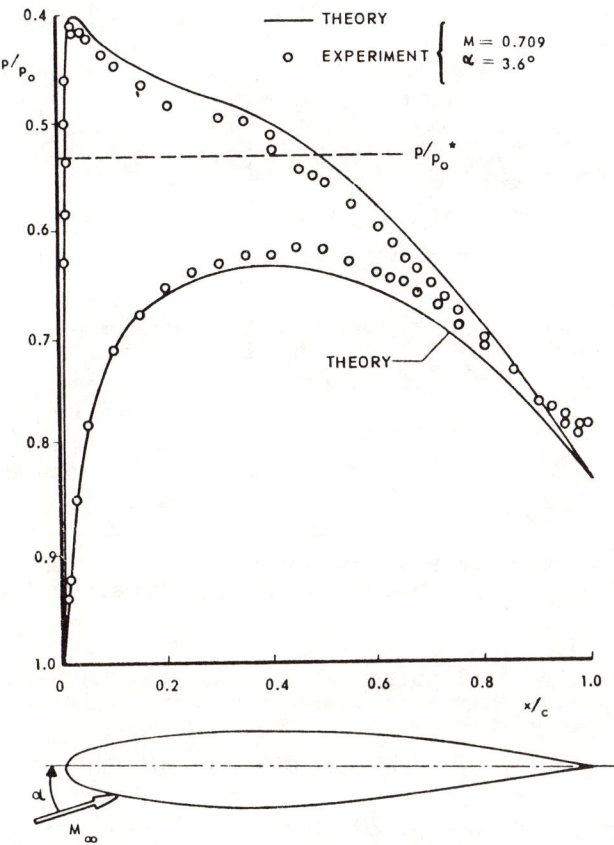

Fig. 19 Exact hodograph solution and experimental confirmation of a continuous supercritical pressure distribution on a lifting shock-free airfoil.[118,122]

Relaxation methods were used in the early transonic studies of Emmons and Vincenti and Wagoner cited above, but modern applications derive directly from Murman and Cole's[130] concept of using different difference expressions in the subsonic and supersonic regions in conformity with the respective domains of influence. The procedure was initially applied to Eq. (12) of the small disturbance theory in which ϕ_{zz} was approximated by centered differences, and ϕ_x and ϕ_{xx} were represented by centered differences in regions of subsonic flow and by one-sided upwind differences in regions of supersonic flow. The resulting large set of algebraic equations was solved numerically using a successive line overrelaxation (SLOR) procedure. It was not long before the same procedures were being

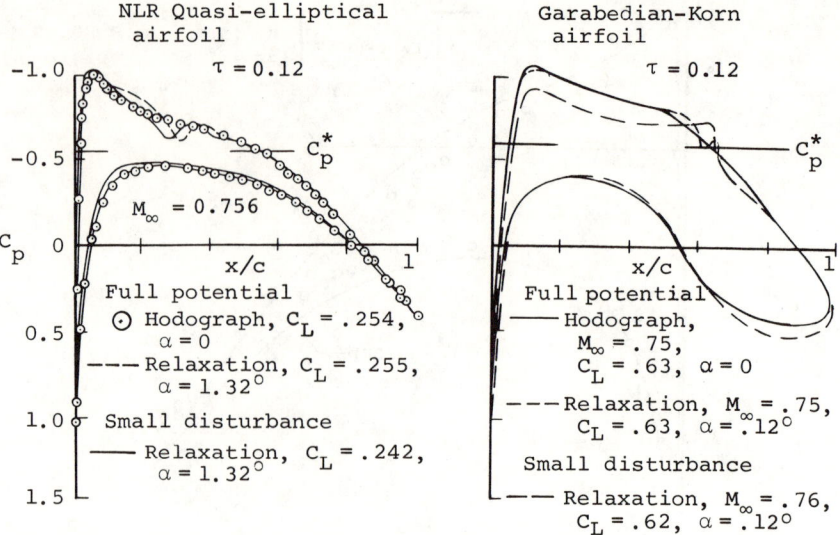

Fig. 20 Exact and approximate theoretical shock-free transonic distributions on two lifting supercritical airfoils.[132,134,137]

used to solve the full potential Eq. (40) for isentropic compressible flow.

The results of several calculations of this type[132,134,135] for two different shock-free supercritical airfoils are shown in Fig. 20 together with the results obtained by solving the full hodograph Eq. (41).[135,136] The pressure distributions provided by the relaxation and hodograph solutions of the full potential equations are almost indistinguishable and in nearly perfect agreement with the experimental data. That is as it should be since both methods should yield virtually exact numerical solutions of equivalent formulations of the same problem, and effects of the boundary layer disregarded in the potential theory analysis should be minimal for this case. Also included on the plots are the corresponding results provided by a relaxation solution of Eq. (12) of small disturbance transonic theory.[137] Although they differ slightly from the exact results for one of the airfoils, this comparison confirms once again that the small disturbance transonic theory can yield results of outstanding accuracy.

It was soon recognized that the use of only two types of operators for ϕ_x and ϕ_{xx} was insufficient since difficulties arose at the sonic line and shock wave where transitions between the two forms had to be made. The problem was resolved[138] by introducing a sonic finite-difference operator corresponding to the centered difference expression for ϕ_{zz} and a shock operator corresponding to the sum of the subsonic and supersonic operators, and applying these operators to Eq. (12) rewritten in conservation form

$$[(1-M_\infty^2)\phi_x - M_\infty^2(\gamma+1)\phi_x^2/(2U_\infty)]_x + (\phi_z)_z = 0 \qquad (42)$$

To distinguish between results obtained with the two types of differencing procedures, those employing the shock point operator, which assures conservation of mass at shock points, are designated FCR (fully conservative relaxation), and those which do not employ the shock point operator are termed NCR (not fully conservative relaxation). Further complications arise when the method is applied to complete potential Eq. (40) rewritten in conservation form. The centerline of the Mach forecone region of dependence in supersonic flow is no longer parallel to the x axis, but is in the local direction of the streamlines. The calculations fail to converge if the differencing is done in the same way as in the small disturbance theory since the region of dependence of the differencing scheme does not always include that of the differential equation. However, Jameson[139] showed that convergence could be achieved if the finite-difference elements were rotated

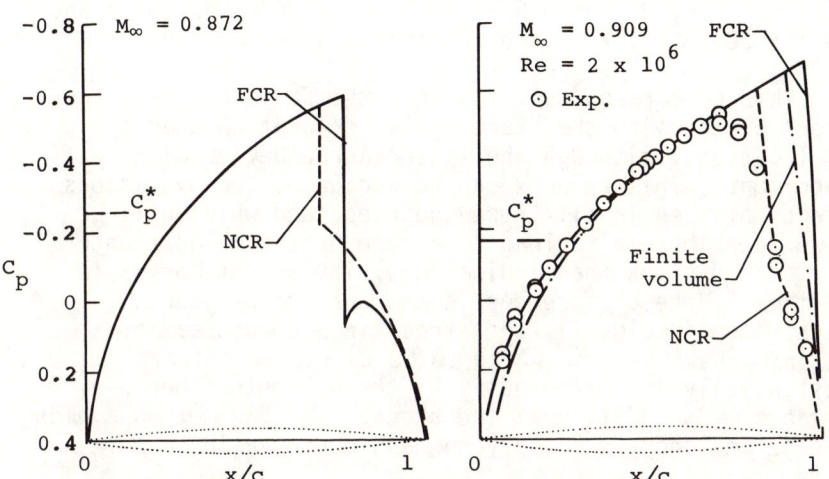

Fig. 21 Supercritical pressure distributions for a nonlifting 6% thick circular-arc airfoil indicated by the FCR and NCR solutions of the transonic small disturbance equations,[138] by the finite-volume solution of the Euler equations,[141] and by experiment.[140]

to allow for the change in direction. A more complete discussion of these points is given in this book in the article by Nixon and Kerlick, "Potential Flow Methods for Transonic Flow Prediction."

The difference between transonic pressure distributions calculated with the FCR and NCR procedures is illustrated in Fig. 21 for a nonlifting 6% thick circular-arc airfoil at two different freestream Mach numbers.[138] Away from the shock wave, the results are in agreement, but near the shock wave they are quite different. In particular, the NCR shock pressure jump does not approach the theoretical normal shock jump, whereas the FCR solution shows not only the correct condition is attained but the shock is followed by a well-defined re-expansion, as is proper when a shock intersects a convexly curved surface. The FCR solution also indicates a more downstream location for the shock wave than the NCR solution. Superposed on the results for $M_\infty = 0.909$ are experimental data[140] and a finite-volume solution[141] of the steady-state version of the Euler equations for inviscid flow

$$\frac{\partial}{\partial t}\begin{Bmatrix}\rho\\ \rho U\\ \rho W\\ E\end{Bmatrix} + \frac{\partial}{\partial x}\begin{Bmatrix}\rho U\\ \rho U^2+p\\ \rho UW\\ \rho U(E+p)\end{Bmatrix} + \frac{\partial}{\partial z}\begin{Bmatrix}\rho W\\ \rho UW\\ \rho W^2+p\\ \rho W(E+p)\end{Bmatrix} = 0 \quad (43)$$

in which ρ, U, W, and p have the same meaning as above and $E = e + (U^2 + W^2)/2$ is the total energy per unit volume.

This comparison confirms that the FCR shock location agrees better with the exact inviscid location than the NCR locations, although the latter agree better with experiment. This paradox can be accounted for by viscous effects omitted in all these theories, and which will be shown later herein to lead to a more upstream location of an embedded shock than indicated by inviscid theory. In addition to the differences in the surface pressures illustrated in Fig. 21, the streamlines downstream of the shock have been shown[142] to be displaced outward substantially and erroneously by the NCR method because of spurious mass addition at the shock. The FCR method avoids this deficiency by satisfying mass conservation everywhere.

Analogous results for the same airfoil in a slightly supersonic flow with $M_\infty = 1.15$ are shown in Fig. 22.[132,138] The calculations were performed using the two relaxation

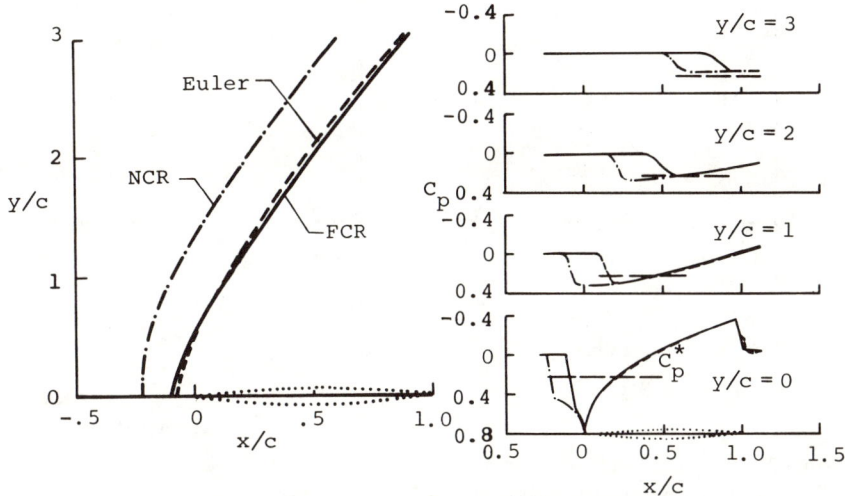

Fig. 22 Pressure distributions and bow wave locations for $M_\infty = 1.15$ for a nonlifting 6% thick circular-arc airfoil indicated by FCR and NCR solutions of the transonic small disturbance equations[132,138] and by a fully conservative time-dependent finite-difference solution of the Euler equations.[143]

methods and also a finite-difference solution of the Euler equations by a fully conservative time-dependent method.[143] All three methods indicate essentially the same pressure distribution on the airfoil, but the NCR method again indicates a location for the shock wave, this time detached, that is too far forward.

Although the integral equation method has been partially eclipsed by the hodograph and relaxation methods, recent re-evaluations[74,82,83] indicate that it has considerable merit. Figure 23 shows pressure distributions for a 6% thick circular-arc airfoil calculated using two versions of the integral equation method and the small disturbance FCR finite-difference method. The results of Kraft[82] were determined using a velocity profile to reduce the double integral of Eq. (24) to a single integral and iterating in the manner of Spreiter and Alksne.[51] Those of Nixon[74] were calculated using his extended integral equation method in which the double integral is evaluated by dividing the region of integration into a number of streamwise strips across which interpolation functions are used to express values for the integrand in terms of values along the strip edges. The two procedures yield virtually the same results as does the earlier procedure

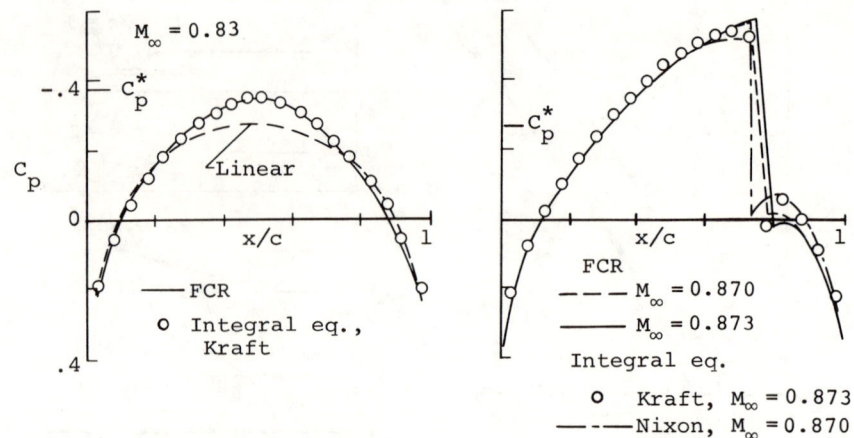

Fig. 23 Critical and supercritical pressure distributions for a nonlifting 6% thick circular-arc airfoil indicated by two different integral equation solutions and by FCR solutions of the transonic small disturbance theory.[74,82]

of Ref. 51. Similar results for the critical Mach number were also calculated by Ogana[83] who divided the region of integration into a large number of rectangles and evaluated the integral by quadrature at each step of an iteration process.

Corresponding results for a NACA 0012 airfoil are presented in Fig. 24 together with an essentially exact numerical solution of the Euler Eqs. (43)[144] for the subcritical case. The $x^{-1/2}$ singularity in u_L that would be present if the usual thin airfoil boundary conditions $(\phi_z)_{z=0} = U_\infty(dZ/dx)$ were employed was avoided in the integral equation results of Fig. 24 by using the relation $(\phi_z)_{z=Z} = (U_\infty + \phi_x)_{z=Z}(dZ/dx)$ for the boundary condition. Aside from modest differences near the shock, the integral equation results are virtually identical to those of the finite-difference relaxation method and obtained with considerably less computation.

Although the majority of analysis has been based on inviscid fluid theory, transonic flows are often highly influenced by viscous phenomena, particularly shock-induced boundary-layer separation. The most complete analysis of such a flow in the period under discussion is that of Diewert,[145,146] who used the finite-volume method to solve the time-dependent, Reynolds-averaged Navier-Stokes equations which, written in integral form,

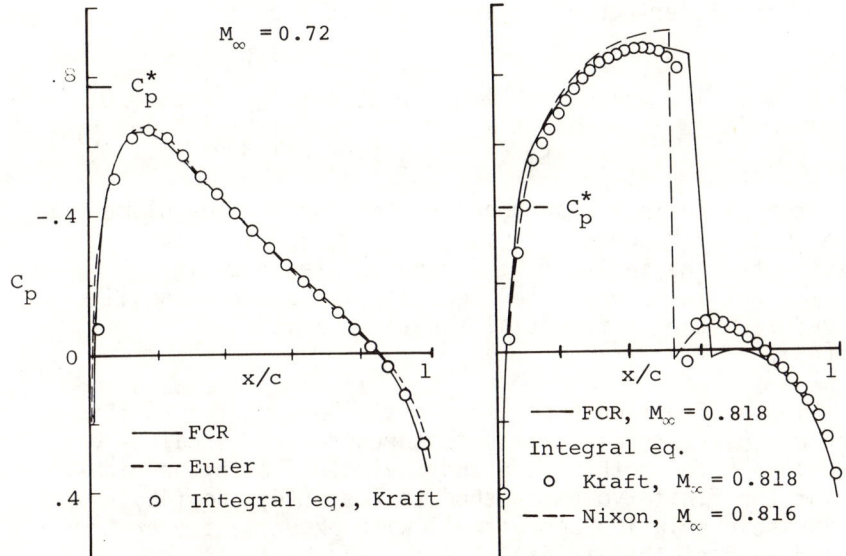

Fig. 24 Subcritical and supercritical pressure distribution for a nonlifting round-nosed NACA 0012 airfoil indicated by two different integral equation solutions by the FCR solutions of the transonic small disturbance theory,[74,82] and also by a numerical solution of the Euler equations for the subcritical case.[144]

are

$$\frac{\partial}{\partial t} \int_{vol} \begin{Bmatrix} \rho \\ \rho U \\ \rho W \\ E \end{Bmatrix} d\,vol + \int_S \begin{Bmatrix} \rho \underline{q} \\ \rho U \underline{q} + \bar{\underline{\tau}} \cdot \hat{i}_x \\ \rho W \underline{q} + \bar{\underline{\tau}} \cdot \hat{i}_y \\ E \underline{q} + \bar{\underline{\tau}} \cdot \underline{q} - k \bar{\nabla} T \end{Bmatrix} \cdot \hat{n}\, dS = 0 \qquad (44)$$

where

$$\underline{q} = \hat{i}_x U + \hat{i}_z W$$

$$\bar{\underline{\tau}} = \hat{i}_x \hat{i}_x \sigma_x + \hat{i}_x \hat{i}_z \tau_{xz} + \hat{i}_z \hat{i}_x \tau_{zx} + \hat{i}_z \hat{i}_z \sigma_z$$

in which κ is the thermal conductivity, ∇T is temperature gradient, σ and τ are normal and shear stress, $\hat{n}$ is a unit normal vector, S refers to the surface surrounding the volume considered in the first integral, and the other symbols have the same meaning as before. In addition to ordinary viscous stresses in τ, Diewert included turbulent Reynolds stresses estimated using four different algebraic eddy viscosity models. If the viscous and Reynolds stresses are disregarded, Eq. (44) reduces to the integral form of the Euler equations for inviscid flow

$$\frac{\partial}{\partial t} \int_{vol} \begin{Bmatrix} \rho \\ \rho U \\ \rho W \\ E \end{Bmatrix} d\,vol + \int_S \begin{Bmatrix} \rho \\ \rho U \\ \rho W \\ E \end{Bmatrix} \underline{q} \cdot \hat{n}\, dS + \int_S \begin{Bmatrix} 0 \\ p\hat{n} \\ p\hat{n} \\ p\underline{q}\cdot\hat{n} \end{Bmatrix} dS = 0 \qquad (45)$$

The integral forms are appropriate for the finite-volume method used by Diewert for viscous flow, and also by Rizzi[141] for the inviscid flow results presented in Fig. 21; but the other Euler equation methods are usually derived from the differential equations written in Eq. (43).

Figure 25 presents pressure distributions for an 18% thick circular-arc airfoil as measured[147] and as calculated[145,146] for both viscid and inviscid flow by application of the finite-volume method to Eqs. (44) and (45). The inviscid solution agrees well with experiment over the forward half of the airfoil, but is inaccurate in predicting shock strength and location, and the pressure level near the trailing edge. When the aft pressure recovery is strong, as for Reynolds number $Re = 2 \times 10^6$, the viscous solution agrees well with experiment. When the aft pressure recovery is weak, as for $Re = 10 \times 10^6$, it disagrees, probably because of inadequate turbulent modeling in the separated region. These differences should be remembered in evaluating comparisons between theory and experiment such as shown in Fig. 21.

We now turn attention to axisymmetric and related slender-body flows. Up to 1975, nearly all of the

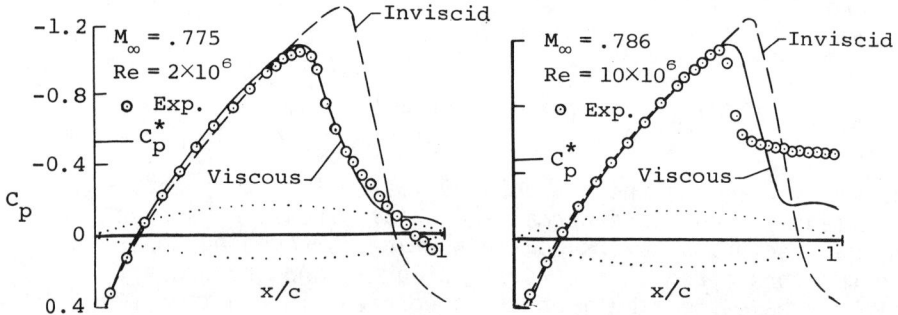

Fig. 25 Supercritical pressure distributions for a nonlifting 18% thick circular-arc airfoil indicated by finite-volume solutions of the Navier-Stokes and Euler equations for viscous and inviscid flow,[145,146] and by experiment.[147]

analyses were based on the three-dimensional transonic small disturbance theory described by Eq. (35). This theory has proven to be remarkably accurate for these applications, perhaps because the perturbations associated with slender bodies are typically substantially smaller than for the two-dimensional airfoil flows discussed above for which the small disturbance theory is already capable of producing remarkably useful results.

Figure 26 shows pressure distributions for a 10% thick parabolic-arc body of revolution for several M_∞ from 0.9 to 1.1 calculated using the local linearization[93] and finite-difference NCR[148] and FCR[133] methods together with experimental data.[107] All of these results are in good agreement over most of the body, but notable differences exist near the rear. Some of these may be attributed to viscous phenomena, but effects of the cylindrical sting model support and wind tunnel are also significant. To demonstrate the effects of the sting, NCR results[148] are presented for a shape that conforms to the test model and sting combination. The results are in improved agreement with the data, but significant differences remain.

To investigate the effects of the walls, NCR calculations were carried out[148] for the same body-sting com-

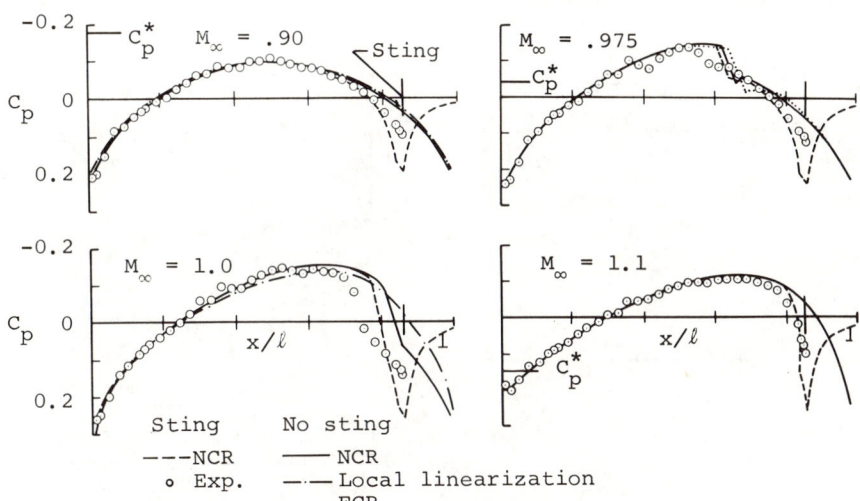

Fig. 26 Transonic pressure distributions for a nonlifting 10% thick parabolic-arc body of revolution indicated by NCR,[148] FCR,[133] and local linearization[93] solutions of the transonic small disturbance theory, and by experiment.[107]

bination in a circular wind tunnel having the same cross-section area relative to the body as the Ames 14 ft transonic wind tunnel where the tests were conducted. To simulate the ventilated walls of the test section, calculations were performed for a porous wall with various porosities. The results for a porosity of 0.5 are shown in Fig. 27 together with those for free air and for an open jet. Similar results have also been obtained for the porous wall using the alternating direction integration method.[149] The theory indicates that the walls produce an upstream shift of the shock wave, thereby bringing the calculated pressure distribution and pressure drag into virtually perfect agreement with measurements. Prudence should be exercised in accepting these results as definitive, however, since the boundary condition applied at the wall is a highly simplified representation of a complex local flow.

It has long been known, as discussed in connection with Fig. 8, that the transonic equivalence rule can be used to decompose approximately the solution for transonic flows past slender wings, bodies, and wing-body combinations into a number of solutions for two-dimensional incompressible flows plus the nonlinear solution for transonic flow past a nonlifting equivalent body of revolution having the same longitudinal distribution of cross-section area. Figure 28 presents comparisons with experiment[150] of such

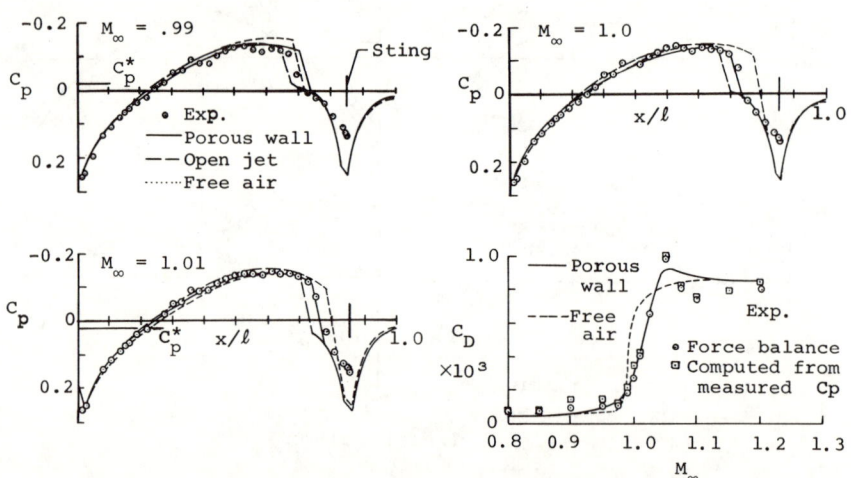

Fig. 27 Effects of wind-tunnel walls on the transonic pressure distribution and drag for a nonlifting 10% thick parabolic-arc body of revolution, as indicated by NCR solutions[148] of the transonic small disturbance theory and by experiment.[107]

an application for steady flow with $M_\infty = 1$ past two slender bodies with differing elliptic cross sections but the same longitudinal distribution of cross section as a parabolic-arc body of revolution of thickness ratio 1/12.[151] The solution for the equivalent body was determined using the local linearization method, the only method available at the time these results were originally published, and are for free airflow past a complete body, i.e., without a sting. After making allowances for the difference over the rear of the bodies that can be attributed to effects of wind-tunnel walls, model support, and differences between the local linearization and more nearly exact relaxation methods as illustrated in the two proceding figures, the results show that the equivalence rule provides a simple and accurate means for treating transonic flow past slender bodies of noncircular cross sections.

Figure 29 presents a similar comparison of theoretical and experimental results for $M_\infty = 1$ for one of the elliptic cross-section bodies of Fig. 28 at angles of attack of 2 deg, 4 deg, and 6 deg. Good agreement is again found along most of the body, with notable difference occurring on the rear. The latter could be rectified by replacing the solution for the equivalent body by a finite-difference relaxation solution like that of Fig. 27 to account for the effects of the model support and wind-tunnel walls.

The form of the transonic equivalence rule illustrated in Fig. 8 and used in the calculations illustrated in Figs. 28 and 29 is for the thickness-dominated case and is appropriate for bodies of nonvanishing thickness at small angles of attack. Extension of the rule to lift-dominated cases[152,153] has shown that it is necessary to include a two-dimensional dipole solution in the inner solution and

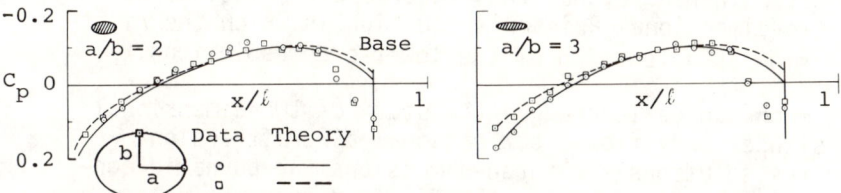

Fig. 28 Pressure distributions for $M_\infty = 1$ for two nonlifting parabolic-arc bodies of differing elliptic cross sections as indicated by the transonic equivalence rule and the local linearization solution for the equivalent body of revolution[151] and by experiment.[150]

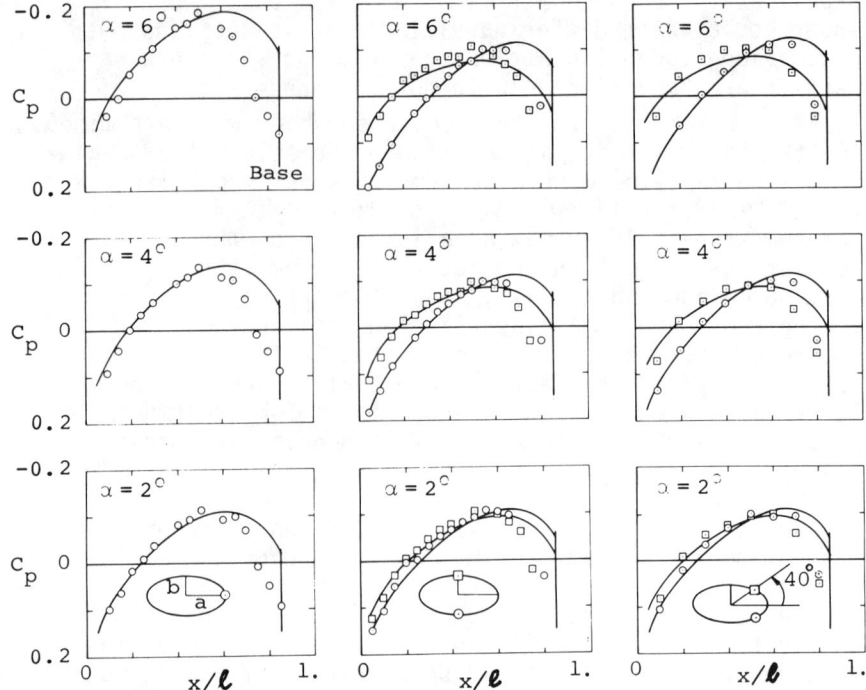

Fig. 29 Pressure distributions for $M_\infty = 1$ for a lifting parabolic-arc body of elliptic cross section as indicated by the transonic equivalence rule and the local linearization solution for the equivalent body of revolution[151] and by experiment.[150]

to replace the axisymmetric transonic flow solution with the transonic solution for an equivalent lifting body of revolution having the same longitudinal distribution of cross-section area and lift as the inner solution. The first-order thickness, lifting, source, and dipole inner solutions satisfy the two-dimensional Laplace equation as indicated in Fig. 8, but the higher-order solutions satisfy the two-dimensional Poisson's equation in which the right-hand side is a function of the lower-order solutions.

Much can be learned by the study of two-dimensional and slender-body flows, but rational aircraft design requires solutions for three-dimensional aerodynamic configurations. Fortunately, the finite-difference relaxation method can be generalized to these situations, and by 1975 computers and numerical methods had become sufficiently advanced to solve steady-state transonic small disturbance equations for certain important classes of

three-dimensional aerodynamic shapes. Figure 30 presents
a comparison of calculated and experimental results for
a swept ONERA M6 wing at an angle of attack of 3 deg
for $M_\infty = 0.84$.[154] Computations were performed using both
the NCR and FCR methods and required about 5 min of CPU
time on the CDC 7600 computer. As in the two-dimensional
and axisymmetric applications, the FCR method predicts a
more downstream location for the shock wave than the
NCR method, with the differences being more marked at higher Mach numbers. Again, the data agree better with the
NCR calculations, but this is believed to be due to a fortuitous cancellation of errors in the shock jumps by viscous
effects disregarded in the analysis.

In applications to swept wings such as shown in
Fig. 30, oblique shock waves occur that are not adequately
treated by the traditional Eq. (49) of the transonic small
disturbance theory. A substantial improvement can be made,

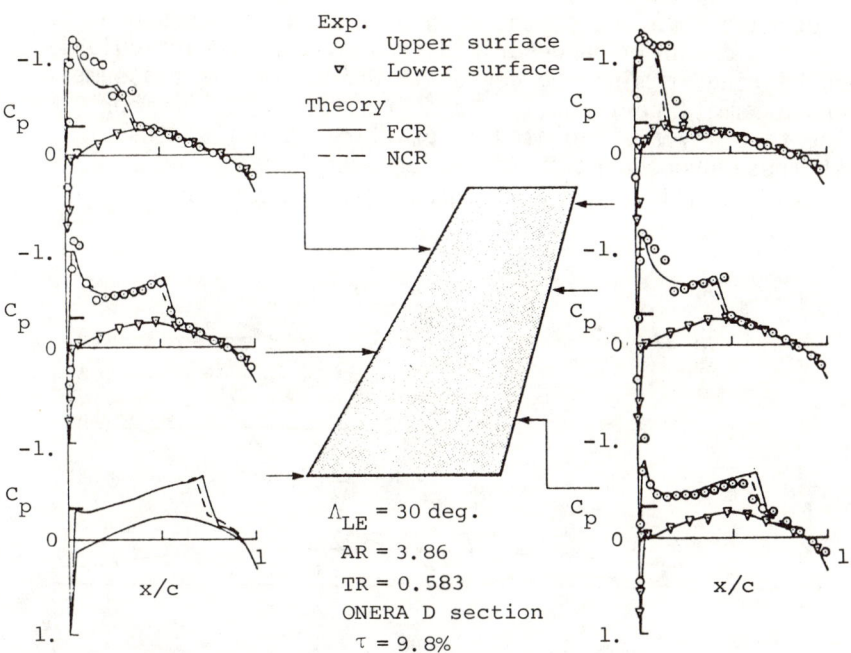

Fig. 30 Supercritical pressure distributions for $M_\infty = 0.84$ on a
lifting swept ONERA M6 wing of aspect ratio 3.86 having 9.8% thick
ONERA D airfoils at an 3 deg angle of attack as indicated by the
FCR and NCR solutions of the small disturbance theory and by
experiment.[154]

however, but adding two terms to obtain[132,144]

$$(1-M_\infty^2)\phi_{xx} + \phi_{yy} + \phi_{zz} = [M_\infty^2(\gamma+1)/U_\infty]\phi_x\phi_{xx} \qquad (51)$$
$$+ [M_\infty^2(\gamma-1)U]\phi_x\phi_{yy} + (2M_\infty^2/U_\infty)\phi_y\phi_{xy}$$

for steady transonic flow.

Moving closer to a complete airplane configuration, Fig. 31 presents experimental and NCR calculated pressures for a swept-wing-fuselage configuration at M_∞ = 0.93 and α = 0 deg.[154] The agreement with experiment is good along the fuselage centerline and the two inboard stations. In the computed results, the wing-root shock propagates laterally to y/s = 0.60, but the experimental shock dissipates before reaching this point. The source of the slight disagreement is not clear but may be due to viscous effects or to theoretical deficiencies associated with swept shock waves.

Overall, however, the accuracy generally attained by an accurate solution of the transonic small disturbance equations can only be described as remarkable, particularly when bearing in mind the degree to which the basic assumptions of small perturbations and freestream Mach number close to unity are violated in practical applications. With these procedures providing an even more accurate inviscid solution than aerodynamicists had been accustomed

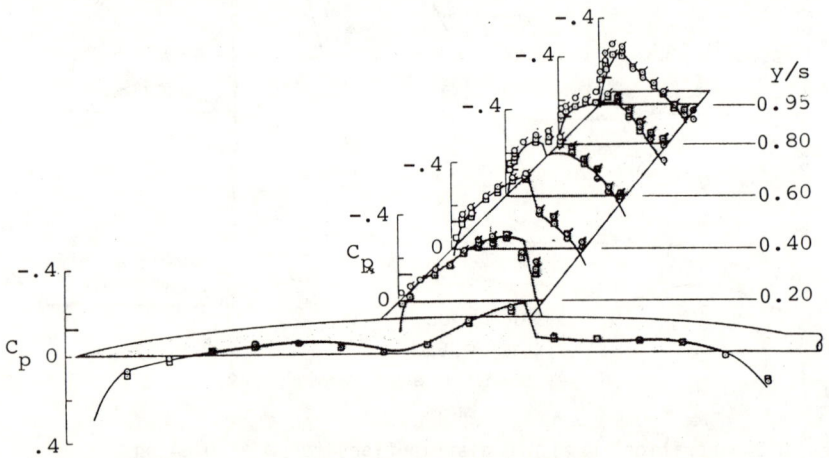

Fig. 31 Supercritical pressure distributions for M_∞ = 0.93 on a nonlifting swept wing-fuselage configuration as indicated by a NCR solution of the transonic small disturbance theory and by experiment.[154]

to having for subsonic or supersonic flows, the next obvious step in the realistic calculation of the pressure distribution and forces on an aerodynamic configuration is to provide an adequate account of the viscous effects of the boundary layer and wake. This is particularly true of the more modern wing designs that incorporate rear loading of the airfoils, for which, even at the same lift coefficient, the shape of the chordwise pressure or load distributions may be radically different in viscous and inviscid flows.

The general principles for calculating these effects have been well established for a long time. If it is assumed that methods are available for calculating 1) the development of the boundary layer and wake under a given external pressure distribution, and 2) the external inviscid flow with given internal boundary conditions, the matching between the two calculations can be arranged by considering an analytic continuation of the inviscid flow into the regions occupied by the viscous layers (in which regions this fictitious inviscid flow will, of course, differ from the real flow). Many procedures of varying complexity and accuracy have evolved through the years, particularly for subsonic flows,[156] and attention was turning by 1975 to the development of appropriate theories for dealing with transonic flows. Already discussed herein in connection with Fig. 24 are the two-dimensional finite-difference relaxation results of Diewert.[145,146] An excellent account of other approaches under development at the U. K. Royal Aircraft Establishment has been given by Lock.[155] By 1975 notable progress was being made in accounting quantitatively for viscous effects, even in some cases where the effects were substantial (more than 20 deg on the lift). In general, good agreement between the experimental data and the viscous theory was attained for flows that were largely subcritical. For flows in which a shock wave is present, the agreement, although still adequate for most purposes was not quite so good; but it is still remarkable considering both the approximations in the theory and the uncertainties with regard to wind-tunnel interference when large regions of supercritical flow are present.

As an alternative to the finite-difference relaxation approach to three-dimensional transonic flow analysis, mention should be made of the alternating direction integration method of Sedin and Karlsson,[149] which has been shown to be capable of providing plausible pressure distributions on wings and wing-body combinations. This method involves elements that are closely related to those of approximate factorization procedures that are finding an increasing role in the more efficient finite-difference relaxation solutions developed over the last few years.

It should be quite clear from the foregoing that substantial progress in the accurate prediction of both steady and unsteady transonic, aerodynamic properties of wing and body configurations resembling modern aircraft had been made by 1975, and that a firm foundation had been laid for further progress as better numerical algorithms and electronic computers became available. By that date, analysis had also commenced on unsteady transonic flow,[175] and on two important related applications, namely helicopter rotors[131,158] and rotating turbomachinery.[159] With respect to the former, it may be noted that the onset of transonic flow over the outer portion of the rotor blades is one of the primary conditions that sets the performance limits of modern helicopters. A transonic regime near a blade tip is inherently nonlinear, three-dimensional and, with forward flight, unsteady. The results available in 1975 were more suggestive than complete, but they illustrated the significance of transonic aerodynamics for helicopter rotors and the need for a more comprehensive study.

With repect to transonic rotating turbomachinery analyses, only a few results were available in 1975, and they were rather provisional because of the relatively small number of grid points used in the finite-difference relaxation calculations. They were sufficient to demonstrate a recognizable resemblance to experimental data, particularly with respect to rather complex shock configurations between the blades, and to encourage further development of analysis of transonic flow through high-speed turbomachinery. The flows are very complex, however, and it is evident that continued effort will be required for some time before accurate and economical methods will be available for the routine solutions of these problems.

Concluding Remarks

The preceding discussion has provided an account of the development of transonic aerodynamics from its beginnings at the turn of the century to 1975. It is intended to provide a background against which the modern developments described in the remainder of this book can be viewed. A large number of references are cited, but many significant contributions are not, or are hidden in references to summary papers. In many instances, the selection has been made on the basis of interlocking relations that help to evaluate the various results and to provide a continuing base from which further discussion can proceed smoothly.

Overall, it was clear by 1975 that tremendous advances were being made in the analysis of transonic flows of practical interest, and that these problems no longer appeared so formidable as they had only a few years before. Indeed, some of the simpler two-dimensional and axisymmetric steady flows could be considered solved, with even alternative methods available to choose among. By 1975, the research frontier was moving to more complex steady and unsteady three-dimensional flows past wings, wing-body combinations, helicopter rotors, and through rotating turbomachinery fans and compressors. The modern computer has brought immense calculating power to bear on these problems, and the goal of replacing the wind tunnel with a computer is beginning to look more achievable than ever before. However, all will not fall into place by itself. Much work remains to be done, but the directions are clearly indicated and the rewards of improved aerodynamic design and analysis are sufficient to demand the effort be made.

Acknowledgments

Acknowledgment for partial support of the preparation of this paper is made to NASA/Ames Research Center where the author was assigned during part of his sabbatical leave from Stanford University under the provisions of the Assignment Agreement of the Intergovernmental Personnel Act of 1970, and to Bernard Halliwell and Carol Sevilla of Nielsen Engineering & Research, Inc., of Mountain View, Calif., for typing of the final manuscript and preparation of the figures. Specific acknowledgment should go to my colleague Dr. Stephen S. Stahara, with whom I have discussed many aspects of transonic flow during the preparation of this paper, and to Dr. David Nixon for his encouragement to undertake this account of a long-standing and centrally important problem of aeronautics.

References

[1] Molenbroeck, P., "Über eininge Bewegungen eines Gases bei Annahme eines Geschwindigdeits-potentials," Archiv. Math. und Phys., Series 2, Vol. 9, 1890, pp. 157-195.

[2] Chaplygin, S. A., "On Gas Jets," NACA TM 1063, 1944.

[3] Douglas, G. P. and Wood, R. M., "The Effects of the Tip Speed Airscrew Performance," British Aeronautical Research Council, R&M 884, 1924.

[4] Briggs, L. J., Hull, G. F., and Dryden, H. L., "Aerodynamic Characteristics of Airfoils at High Speeds," NACA 207, 1925.

[5] Briggs, L. J., and Dryden, H. L., "Pressure Distribution over Airfoils at High Speeds," NACA Rep. 255, 1927.

[6] Stanton, T. E., "A High Speed Wind Tunnel for Tests on Aerofoils," British Aeronautical Research Council, Rep. 1130, 1928.

[7] Busemann, A., Profilmessungen bei Geschwindigkeiten nahe der Schallgeschwindigkeit (im Hinblick auf Luftschrauben), Jahrbuch der Wissenschaftlichen Gesellschaft fur Luftfahrt, 1928, pp. 95-99.

[8] Jacobs, E. N., and Shoemaker, J. M., "Test on Thrust Augmentors for Jet Propulsion," NACA TN 431, 1932.

[9] Stack, J., "The NACA High Speed Wind Tunnel and Tests of Six Propeller Sections," NACA Rep. 463, 1933.

[10] Stack, J., and von Doenhoff, S. E., "Tests of 16 Related Airfoils at High Speeds," NACA Rep. 492, 1934.

[11] Stack, J., Lindsay, W. F., and Littell, R. E., "The Compressibility Burble and the Effects of Compressibility on Pressures and Forces Acting on an Airfoil," NACA Rep. 646, 1938.

[12] Stack, J., "Compressible Flows in Aeronautics. Eighth Wright Brothers Lecture," Journal of Aeronautical Sciences, Vol. 12, April, 1945, pp. 127-148.

[13] Glauert, H., "The Effect of Compressibility on the Lift of Airfoils," British Aeronautical Research Council, R&M 1135, 1927.

[14] Prandtl, L., "Über Stromungen, deren Geschwindigkeiten mit der Schallgeschwindigkeit vergleichbar sind," Journal of the Aeronautical Research Institute of the University of Tokyo, No. 65, 1920.

[15] Tsien, H.-S., "Two-Dimensional Subsonic Flow of Compressible Fluids," Journal of Aeronautical Sciences, Vol. 6, Oct. 1939,

[16] Ackeret, J., "Luftkräfte an Flugeln, die mit grösserer als Schallgeschwindigeit bewegt werden," Z. Flugtechn. Motorluftsch., Vol. 16, No. 3, 1925, pp. 72-74.

[17] Busemann, A., "Aerodynamischer Auftrieb bie Uberschall geschwindigkeit," *Luftfahrtforschung*, Vol. 12, Oct. 1935, pp. 210-220.

[18] Spreiter, J. R. and Steffen, P. J., "Mach-Reynolds Relation Probed," *SAE Journal*, Vol. 55, Jan. 1947, pp. 33-39. See also NACA TN1044, 1946.

[19] Hartman, E. P., "Adventures in Research," *A History of Ames Research Center 1940-1965*, NASA SP-4302, 1970, 555 pp.

[20] Gray, G. W., *Frontiers of Flight*, Alfred Knopf, New York, 1948, 362 pp.

[21] Spreiter, J. R. and Galster, G. M., "Observations of Aileron Flutter on the Bell P.63A-6 Airplane in Flight at High Mach Numbers," NACA ACR GB08, 1946.

[22] Göthert, B., "Ebene and raumliche Strömung bei hohen Unterschallgeschwindigkeiten," *Jahrb. der deut. Luftfahrtforschung*, 1941, pp. I156-I157. Also *Bericht 127 der Lilienthal-Gesellschaft fur Luftfahrtforschung*, 1940, pp. 97-101.

[23] Jones, R. T., "Wing Plan Forms for High-Speed Flight," NACA TN 1033, 1946.

[24] Jones, R. T., "Properties of Low-Aspect-Ratio Pointed Wings at Speeds below and above the Speed of Sound," NACA TN 1032, 1946.

[25] Wright, R. H. and Ward, V. G., "NACA Transonic Wind Tunnel Test Sections," NACA RM L8J06, 1948.

[26] Whitcomb, R. T., "A Study of the Zero-Lift Drag-Rise Characteristics of Wing-Body Combinations Near the Speed of Sound," NACA Rep. 1273, 1956. (Supersedes NACA RM L52H08.)

[27] Leipmann, H. W. and Puckett, A. E., *Introduction to Aerodynamics of a Compressible Fluid*, John Wiley & Sons, New York, 1947, 262 pp.

[28] Sauer, R., *Introduction to Theoretical Gas Dynamics*, Edwards, 1947, 222 pp.

[29] Sears, W. R., "Small Perturbation Theory," *General Theory of High Speed Aerodynamics*, edited by W. R. Sears, Princeton University Press, Princeton, N. J., 1954, pp. 61-121.

[30] Heaslet, M. S. and Lomax, H., "Supersonic and Transonic Small Perturbation Theory," *General Theory of High Speed Aerodynamics*, edited by W. R. Sears, Princeton University Press, Princeton, N.J., 1954, pp. 122-344.

[31] Robinson, A. and Laurmann, J. A., Wing Theory, Cambridge University Press, England 1956, 569 pp.

[32] Lighthill, J., "Higher Approximations," General Theory of High Speed Aerodynamics, edited by W. R. Sears, Princeton, University Press, Princeton, N. J., 1954, pp. 345-489.

[33] Kuo, Y. H. and Sears, W. R., "Plane Subsonic and Transonic Potential Flows," General Theory of High Speed Aerodynamics, edited by W. R. Sears, Princeton University Press, Princeton, N.J., 1954, pp. 490-582.

[34] Howarth, L. Ed., Modern Developments in Fluid Dynamics, High Speed Flow, Vols.1 and 2, Oxford University Press, Oxford, England, 1953, 875 pp.

[35] Ferri, A., Elements of Aerodynamics of Supersonic Flows, MacMillan Co., New York, 1949, 434 pp.

[36] Spreiter, J. R., "The Aerodynamic Forces on Slender Plane- and Cruciform-Wing and Body Combinations," NACA Rep. 962, 1950. (Supersedes NACA TN 1662 and 1987.)

[37] Heaslet, M. A., Lomax, H., and Spreiter, J. R., "Linearized Compressible-Flow Theory for Sonic Flight Speeds," NACA Rep. 956, 1950. (Supersedes NACA TN 1824.)

[38] Oswatitsch, K and Wiegardt, K., "Theoretische Untersuchungen uber stationäre Potentialströmungen und Grenzschlichten bei hohen Geschwindigkeitin," Lilienthal-Gesellschaft fur Luftfahrtforschung, Ber 13/1, 1942, pp. 7-24. (Also available as NACA TM 1189.)

[39] Busemann, A. and Guderley, K. G., "The Problem of Drag at High Subsonic Speeds," British M.A.P., Rep. and Trans. 184, 1947.

[40] Guderley, K. G., "Considerations of the Structure of Mixed Subsonic - Supersonic Flow Patterns," AAFAMC Tech. Rep. F-TR-2168-ND, 1947.

[41] Guderley, K. G., "On the Transition from a Transonic Potential Flow to a Flow with Shocks," AAFAMC Tech. Rep. F-TR-2160-ND, 1947.

[42] von Kármán, T., "The Similarity Law of Transonic Flow," Journal of Mathematics and Physics, Vol. 26, Oct. 1947, pp. 182-190.

[43] von Kármán, T. "Supersonic Aerodynamics - Principles and Applications," Journal of Aeronautical Science, Vol. 14, July 1947, pp. 373-402 (discussion 403-309).

[44] Spreiter, J. R., "Theoretical and Experimental Analysis of Transonic Flow Fields," NACA-University Conference on Aerodynamics, Construction, and Propulsion, Vol. II, Aerodynamics, a compilation of papers presented at Lewis Flight Propulsion Laboratory, Cleveland, Ohio, Oct. 20-22, 1954.

[45] Cole, J. D., "Drag of Finite Wedge at High Subsonic Speeds," Journal of Mathematical Physics, Vol. 30, July 1951, pp. 79-93.

[46] Guderley, G. and Yoshihara, J., "The Flow Over a Wedge Profile at Mach Number 1," Journal of Aeronautical Sciences, Vol. 17, Nov. 1950, pp. 723-735.

[47] Vincenti, W. G. and Wagoner, C. G., "Transonic Flow Past a Wedge Profile with Detached Bow Wave," NACA Rep. 1095, 1952. (Supersedes NACA TN 2339 and 2588.)

[48] Liepmann, H. W. and Bryson, A. E., Jr., "Transonic Flow Past Wedge Sections," Journal of Aeronautical Sciences, Vol. 17, Dec. 1950, pp. 745-755.

[49] Bryson, A. E., Jr., "An Experimental Investigation of Transonic Flow Past Two-Dimensional Wedge and Circular-Arc Sections Using a Mach-Zehnder Interferometer," NACA Rep. 1094, 1952. (Supersedes NACA TN1560.)

[50] Spreiter, J. R., "On the Application of Transonic Similarity Rules to Wings of Finite Span," NACA Rep. 1153, 1953. (Supersedes NACA TN 2726.)

[51] Spreiter, J. R. and Alksne, A. Y., "Theoretical Prediction of Pressure Distributions on Nonlifting Airfoils at High Subsonic Speeds," NACA Rep. 1217, 1955. (Supersedes NACA TN 3096.)

[52] Liepmann, H. W., "The Interaction Between Boundary Layer and Shock Waves in Transonic Flow," Journal of Aeronautical Sciences, Vol. 13, Dec. 1946, pp. 623-637.

[53] Ackeret, J., Feldman, F., and Rott, N., "Untersuchungen au Verdichtungsstossen und Grenzschichten in schnell bewegten Gasen," Mitteilungen aus dem Institute für Aerodynamik, Nr. 10, Zurich, 1946. (Also availabe as NACA TM 1113.)

[54] Wood, G. P. and Gooderum, P. B., "Investigation with an Interferometer of the Flow Around a Circular-Arc Airfoil at Mach Numbers Between 0.6 and 0.9," NACA TN 2801, 1952.

[55] Vincenti, W. G., Dugan, D. W., and Phelps, E. R., "An Experimental Study of the Lift and Pressure Distribution on a Double-Wedge Profile at Mach Numbers Near Shock Attachment," NACA TN 3225, 1954.

[56] Humphreys, M. D., "An Investigation of a Lifting 10 Percent-Thick Symmetrical Double Wedge Airfoil at Mach numbers up to 1," NACA TN 3306, 1954.

[57] Guderley, G and Yoshihara, H. "Two-Dimensional Unsymmetrical Flow Patterns at Mach Number 1," Journal of Aeronautical Sciences, Vol. 20, Nov. 1953, pp. 757-768.

[58] Vincenti, W. G. and Wagoner, C. G, "Theoretical Study of the Transonic Lift of a Dobule-Wedge Profile with Detached Bow Wave," NACA TN 2832, 1952.

[59] Willmarth, W. W., "The Lift of Thin Airfoils at High Subsonic Speeds," Ph.D. Dissertation, Case Institute of Technology, Cleveland, Ohio, 1953.

[60] Vincenti, W. G., Wagoner, C. G., and Fisher, N. H., Jr., "Calculations of the Flow Over an Inclined Flat Plate at Free Stream Mach Number One," NACA TN 3723, 1956.

[61] Guderley, K. G., Theorie schallnaher strömungen, Springer-Verlag, Berlin, 1957. Available in translation (by J. R. Moszynski) as The Theory of Transonic Flow, Addison-Wesley, Reading, Mass., 1962, 376 pp.

[62] Ferrari, C. and Tricomi, F. G., Aerodinamica Transonica. Edizioni Cremonese, 1962. Available in translation (by R. H. Cramer) as Transonic Aerodynamics, Academic Press, New York, 1968, 653 pp.

[63] Ward, G. N, "Approximate Methods," Modern Developments in Fluid Dynamics - High Speed Flow, Vol. I, edited by L. Howarth, Oxford University Press, Oxford, England, 1953, pp. 267-324.

[64] Hantzsche, W. and Wendt, H., "Der Kompressibilitätseinfluss für dunne wenig gekrumpte Profile bei Unterschallgeschwindigkeit," Z.a.M.M., Vol. 22, April 1942, pp. 72-86.

[65] Imai, I, "Approximation Methods in Compressible Fluid Dynamics," Institute for Fluid Dynamics and Applied Mathematics, University of Maryland, College Park, Technical Note BN-95, 1957.

[66] Lighthill, M. J., "Higher Approximations," General Theory of High Speed Aerodynamics, edited by W. R. Sears, Princeton University Press, Princeton, 1954, pp. 345-489.

[67] Van Dyke, M. D., "Second-Order Subsonic Airfoil Theory Including Edge Effects," NACA Rep. 1274, 1956.

[68] Asaka, S., "Application of the Thin-Wing Expansion Method to the Flow of a Compressible Fluid Past a Symmetrical Circular-Arc Aerofoil," Journal of Physical Society of Japan, Vol. 10, June 1955, pp. 482-492; Errata, Ibid., Vol. 10, July 1955, p. 593, and Vol. 13, Jan. 1958, p. 115.

[69]Kaplan, C., "The Flow of a Compressible Fluid Past a Curved Surface," NACA Rep. 768, 1943.

[70]Kaplan, C. "On a Solution of the Nonlinear Differential Equation for Transonic Flow Past a Wave-Shaped Wall," NACA TN 2383, 1951

[71]Spreiter, J. R., "Aerodynamics of Wings and Bodies at Transonic Speeds," Journal of Aerospace Sciences, Vol. 26, Aug. 1959, pp. 465-487.

[72]Oswatitsch, K., "Die Geschwindigkeitsverteilung bei lokalen Überschallgebieten an flachen Profilen," ZAMM, Vol. 30, Nr. 1/2, 1950, pp. 17-24.

[73]Oswatitsch, K., "Die Geschwindigkeitsverteilung au symmetrischen Profilen beim Auftreten lokaler Überschallgebiete," Acta Physica Austriaca, Vol. 4, Nr. 2/3, 1950, pp. 228-271.

[74]Nixon, D. and Hancock, G. J., "Integral Equation Methods - a Reappraisal," Symposium Transsonicaum II, edited by K. Oswatitsch and D. Rues, Spreinger, New York, 1976, pp. 174-182.

[75]Ogana, W. and Spreiter, J. R., Derivation of an Integral Equation for Transonic Flows," AIAA Journal, Vol. 15, Feb. 1977, pp. 281-283.

[76]Spreiter, J. R., Alksne, A. Y., and Hyett, B. J., "Theoretical Pressure Distributions for Several Related Nonlifting Airfoils at High Subsonic Speeds," NACA TN 4148, 1958.

[77]Gullstrand, R. R., "The Flow Over Symmetrical Aerofoils without Incidence in the Lower Transonic Ranges," Royal Institute of Technology, Stockholm, Sweden, KTH Aero TN 20, 1951.

[78]Gullstrand, T. R., "The Flow Over Symmetrical Aerofoils without Incidence at Sonic Speed," Royal Institute of Technology, Stockholm, Sweden, KTH Aero TN 24, 1952.

[79]Gullstrand, T. R., "A Theoretical Discussion of Some Properties of Transonic Flow over Two-Dimensional Symmetrical Aerofoils at Zero Lift with a Simple Method to Estimate the Flow Properties," Royal Institute of Technology, Stockholm, Sweden, KTH Aero TN 25, 1952.

[80]Gullstrand, T. R., "The Flow Over Two-Dimensional Aerofoils at Incidence in the Transonic Speed Range," Royal Institute of Technology, Stockholm, Sweden, KTH Aero TN 27, 1952.

[81]Zierep, J., "Die Integralgleichungsmethode zur Berechnung schallnaher Strömungen," Symposium Transsonicum, edited by K. Oswatitsch, Springer Verlag, Berlin/Göttingen,Heidelberg, 1964, pp. 92-109.

[82] Kraft, W. M., "An Integral Equation Method for Boundary Interference in Performated-Wall Wind Tunnels at Transonic Speeds." Ph.D. Dissertation, University of Tennessee, Knoxville, 1975.

[83] Ogana, W., "Computation of Steady Two-Dimensional Transonic Flows by an Integral Equation Method," Ph.D. Dissertation, Stanford University, Stanford, Calif., 1975.

[84] McDevitt, J. B., "A Correlation by Means of Transonic Similarity Rules of Experimentally Determined Characteristics of a Series of Symmetrical and Cambered Wings of Rectangular Plan Form," NACA Rep. 1253, 1955. (Supersedes NACA RM A51L17b and A53G31.)

[85] Page, W. A., "Experimental Determination of the Range of Applicability of the Transonic Area Rule for Wings of Triangular Plan Form," NACA TN 3872, 1956.

[86] Oswatitsch, K., "Die Theoretischen Arbeiten uber Schallnahe Strömungen am Flugtechnischen Institut der Kungl." Tekniska Hokskolan, Stockholm, Proceedings of the Eighth International Congress on Theoretical and Applied Mechanics, 1953.

[87] Oswatitsch, K.,"The Area Rule," Applied Mechanics Reviews, Vol. 10, Dec., 1957, pp. 543-545.

[88] Whitcomb, R. T., "A Study of the Zero-Lift Drag-Rise Characteristics of Wing-Body Combinations Near the Speed of Sound," NACA Rep. 1273, 1956. (Supersedes NACA RM L 52H08).

[89] Heaslet, M. A. and Spreiter, J. R., "Three-Dimensional Transonic Flow Theory Applied to Slender Wings and Bodies. NACA Rep. 1318, 1957. (Supercedes NACA TN 3717.)

[90] Yoshihara, H., "On the Flow over a Cone-Cylinder Body at Mach Number One," WADC Tech. Rep. 52-295, 1952.

[91] Heaslet, M. A. and Spreiter, J. R., "Reciprocity Relations in Aerodynamics," NACA Rep. 1119, 1953. (Supersedes NACA TN 2700, 1952.)

[92] Page, W. A., "Experimental Study of the Equivalence of Transonic Flow About Slender Cone-Cylinders of Circular and Elliptic Cross Section," NACA TN 4233, 1958.

[93] Spreiter, J. R., and Alksne, A. Y., "Slender Body Theory Based on Approximate Solution of the Transonic Flow Equation," NASA TR-R-2, 1958.

[94] Berndt, S. B., "On the Drag of Slender Bodies at Mach Number one," Zeitschrift Angewandte Mathematik und Mechanik, Vol. 35, Heft 9/10, 1955, p. 362.

[95]Berndt, S. B., "On the Drag of Slender Bodies at Sonic Speed," FFA Rep. 70, Aeronautical Research Institute of Sweden, 1956.

[96]Spreiter, J. R., "On the Range of Applicability of the Transonic Area Rule," NACA TN 3673, 1956.

[97]Drougge, G., "Some Measurements on Bodies of Revolution at Transonic Speeds," Actes, IXe Congrès International de Mechanique Appliquée, Tome II, 1957, pp. 70-77.

[98]Oswatitsch, K. and Keune, F., "The Flow Around Bodies of Revolution at Mach Number 1," Proceedings of the Conference on High Speed Aeronautics, Polytechnic Institute of Brooklyn, N. Y., 1955.

[99]Maeder, P. R., "The Linear Approximation to the Transonic Small Disturbance Equation," Symposium Transsonicum, edited by K. Oswatitsch, Springer-Verlag, Berlin/Göttingen/Heidelberg, 1964, pp. 112-125.

[100]Hosokawa, I., "A Simplified Analysis for Transonic Flows Around Thin Bodies," Symposium Transsonicum, edited by K. Oswatitsch, Springer-Verlag, Berlin/Göttingen/Heidelberg, 1964, pp. 184-199.

[101]Spreiter, J. R. and Alksne, A. Y., "Thin Airfoil Theory Based on Approximate Solution of the Transonic Flow Equation," NACA Rep. 1359, 1958. (Supersedes NACA TN 3970).

[102]Alksne, A. Y. and Spreiter, J. R., "Theoretical Pressure Distributions on Wings of Finite Span at Zero Incidence for Mach Numbers Near 1," NACA TR R-88, 1961.

[103]Spreiter, J. R., "The Local Linearization Method in Transonic Flow Theory," Symposium Transsonicum, edited by K. Oswatitsch, Springer-Verlag, Berlin/Göttingen/Heidelberg, 1964.

[104]Michel, R., Marchaud, R., and Le Gallo, J., "Etude des Ecoulements Transsoniques Autour des Profils Lenticulaires, á Incidence Nulle," ONERA Pub. 65, 1953.

[105]Spreiter, J. R., Smith, D. W., and Hyett, B. J., "A Study of the Simulations of Flow with Free-Stream Mach Number 1 in a Choked Wind Tunnel," NASA TR R-73, 1960.

[106]Berndt, S. B., "Theory of Wall Interference in Transonic Wind Tunnels," Symposium Transsonicum, edited by K. Oswatitsch, Springer-Verlag, Berlin/Göttingen/Heidelberg, pp. 288-309.

[107]Taylor, R. A. and McDevitt, J. B., "Pressure Distributions at Transonic Speeds for Parabolic-Arc Bodies of Revolution Having Fineness ratios of 10, 12, and 14," NACA TN 4243, 1958.

[108] McDevitt, J. B. and Taylor, R. A., "Pressure Distributions at Transonic Speeds for Slender Bodies Having Various Axial Locations of Maximum Diameter," NACA TN 2480, 1958.

[109] Oswatitsch, K., "Die Berechung wirbelfrier achsensymmetrischer Uberschallfelder," Osterreichisches Ingenieur Archiv, Vol. 10, Heft 4, 1956, pp. 359-382.

[110] Oswatitsch, K. and Berndt, S. B., "Aerodynamic Similarity at Axisymmetric Transonic Flow Around Slender Bodies," KTH Aero TN 15, Royal Institute of Technology, Stockholm, Sweden, 1950.

[111] Pearcey, H. H, "The Aerodynamic Design of Section Shapes for Swept Wings," Advances in Aeronautical Sciences, Vol. 3, Pergamon Press, New York, 1962, pp. 277-322.

[112] Whitcomb, R. T., "Review of NASA Supercritical Airfoils," Proceedings of the Ninth International Congress of the International Council of the Aeronautical Sciences (ICAS), 1974.

[113] Lax, P. and Wendroff, B., "Systems of Conservation Laws," Communications on Pure and Applied Mathematics, Vol. 13, May 1960, pp. 217-237.

[114] Richtmyer, R. D. and Morton, K. W., Difference Methods for Initial-Value Problems, 2nd ed., Interscience, New York, 1967, 445 pp.

[115] Morawetz, C. S., "On the Non-Existence of Continuous Transonic Flows Past Profiles," Communications on Pure and Applied Mathematics, Vol. 9, No. 1, Feb. 1956, pp. 45-68; Vol. 10, No. 1, Feb. 1957, pp. 107-131; Vol. 11, No. 1, Feb. 1958, pp. 129-144; Vol. 17, No. 3, Aug. 1964, pp. 357-367.

[116] Bers, L., Mathematical Aspects of Subsonic and Transonic Gas Dynamics, John Wiley & Sons, New York, 1958, 164 pp.

[117] Manwell, A. R., The Hodograph Equations, Oliver and Boyd, Edinburgh, 1971, 476 pp.

[118] Nieuwland, G. Y and Spee, R. B., "Transonic Airfoils: Recent Developments in Theory, Experiment, and Design," Annual Reviews of Fluid Mechanics, Vol. 5, edited by M. D. Van Dyke and W. G. Vencenti, Annual Reviews, Inc., Palo Alto, Calif, 1973, pp. 119-150.

[119] Nieuwland, G. Y., "Transonic Potential Flow Around a Family of Quasi-Elliptical Aerofoil Sections," NLR TR 172, 1967.

[120] Takahashi, S., "A Method of Obtaining Transonic Shock-Free Flow Around Lifting Aerofoils," Transactions. Japan Society for Aeronautical and Space Science, Vol. 16, No. 34, 1973, pp. 246-263.

[121] Boerstoel, J. W., "A Survey of Symmetrical Transonic Potential Flows Around Quasi-Elliptic Aerofoil Sections," NLR-TR T. 136, 1967.

[122] Boerstoel, J. W. and Uijlenhoet, R., "Lifting Aerofoils with Supercritical Shockless Flow," Proceedings of the Seventh Congress of the International Council of the Aeronautical Sciences (ICAS), 1970.

[123] Boersteol, J. W., "Review of the Application of Hodograph Theory to Transonic Aerofoil Design and Theoretical and Experimental Analysis of Shock-Free Aerofoils," Symposium Transsonicum II, edited by K. Oswatitsch and D. Rues, Springer, Berlin/Heidelberg/New York, 1976, pp. 109-133.

[124] Korn, D. G., "Computation of Shock-Free Transonic Flows for Airfoil Design," NYU Rep. NYO-1480-125, Courant Institute of Mathematical Sciences, 1969.

[125] Bauer, F., Garabedian, P., and Korn, D., "Supercritical Wing Sections, I and II," Lecture Notes in Economics and Mathematical Systems, Vols. 66 and 108, Springer, Berlin/Heidelberg/New York, 1972, 211 pp and 296pp.

[126] Spee, B. M., "Investigations on the Transonic Flow Around Aerofoils," NLR TR 69122 U, 1969.

[127] Emmons, H. W., "Flow of a Compressible Fluid Past a Symmetrical Airfoil in a Wind Tunnel and in Free Air," NACA TN 1746, 1948.

[128] Oswatitsch, K. and Rues, D. Symposium Transsonicum II, Springer-Verlag, Berlin/Heidelberg/New York, 1976, 574 pp.

[129] Jameson, A., Transonic Flow Calculations, von Kármán Institute Lecture Series on Computaitonal Fluid Dynamics, 1976.

[130] Murman, E. M, "Transonic Aerodynamics," AIAA Professional Study Series, AIAA, New York, 1975, 211 pp.

[131] Ballhaus, W. F., Some Recent Progress in Transonic Flow Computation, von Kármán Institute Lecture Series on Computational Fluid Dynamics, 1976.

[132] Bailey, F. R., On the Computation of Two- and Three-Dimensional Steady Transonic Flows by Relaxation Methods, von Kármán Institute Lecture Series on Progress in Numerical Fluid Dynamics, 1974.

[133] Spreiter, J. R. and Stahara, S. S., "Developments in Transonic Steady and Unsteady Flow Theory," Proceedings of the Tenth Congress of the International Council of the Aeronautical Sciences (ICAS), 1976.

[134] Lomax, H. and Steger, J. L., "Relaxation Methods in Fluid Mechanics," Annual Reviews of Fluid Mechanics, Vol. 7, edited by M. D. Van Dyke, W. G. Vincenti, and J. V. Wehausen, Annual Reviews Inc., Palo Alto, Calif., 1975, pp. 63-88.

[135] Garabedian, P. R., and Korn, D. G., "Analysis of Transonic Airfoils, Communication in Pure and Applied Mathematics, Vol. 24, No. 6, Nov. 1971., pp. 841-851.

[136] Lock, R. C., "Test Cases for Numerical Methods in Two-Dimensional Transonic Flows," AGARD Rep. N01 575, 1970.

[137] Krupp, J. A. and Murman, E. M., "Computation of Transonic Flows Past Thin Lifting Airfoils and Slender Bodies," AIAA Journal, Vol. 10, July 1972, pp. 880-886.

[138] Murman, E. M., "Analysis of Embedded Shock Waves Calculated by Relaxation Methods," AIAA Journal, Vol. 12, May 1974, pp. 626-633.

[139] Jameson, A., "Iterative Solution of Transonic Flows Over Airfoils and Wings, Including Flows at Mach 1," Communications in Pure Applied Mathematics, Vol. 27, 1974, pp. 283-309.

[140] Knechtel, E. D., "Experimental Investigation at Transonic Speeds of Pressure Distributions over Wedge and Circular-Arc Airfoil Sections and Evaluations of Perforated-Wall Interference," NASA TN D-15, 1959.

[141] Rizzi, A, "Transonic Solutions of the Euler Equations by the Finite Volume Method," Symposium Transsonicum II, edited by K. Oswatitsch and D. Rues, Springer, Berlin/Heidelberg/New York, 1976, pp. 567-574.

[142] Newman, P. A. and South, J. C., Jr., "Conservative Versus Nonconservative Differencing: Transonic Streamline Shape Effects," NASA TMX-72827, 1976.

[143] Magnus, R. M., "The Direct Comparison of the Relaxation Method and the Pseudo-Unsteady Finite Difference Method for Calculating Steady Planar Transonic Flow," TN-73-SP03, Convair Aerospace Divsion of General Dynamics, 1973.

[144] Sells, C. C. L., "Plane Subcritical Flow Past A Lifting Airfoil," Proceedings of the Royal Society, London, Vol. 308 (Series A), 1968, pp. 377-401.

[145] Diewert, G. S., "Numerical Simulation of High Reynolds Number Transonic Flow," AIAA Journal, Vol. 13, Oct. 1975, pp. 1354-1359.

[146] Diewert, G. S., "Computation of Separated Transonic Turbulent Flow, AIAA Journal, Vol. 14, June 1976, pp. 735-740.

[147] McDivitt, J. B., Levy, L. L., Jr., and Diewert, G. S., "Transonic Flow About a Thick Circular-Arc Airfoil," AIAA Journal, Vol. 14, May 1976, pp. 606-613.

[148] Bailey, F. R., "Numerical Calculation of Transonic Flow About Slender Bodies of Revolution," NASA TN D-6582, 1971.

[149] Sedin, Y. C-J. and Karlsson, K. R., "Some Numerical Results of a New Three-Dimensional Transonic Flow Method," Symposium Transsonicum II, edited by K. Oswatitsch and D. Rues, Springer-Verlag, Berlin/Heidelberg/New York, 1975, pp. 487-494.

[150] McDevitt, J. B. and Taylor, R. A., "Force and Pressure Measurements at Transonic Speeds for Several Bodies Having Elliptical Cross Sections," NACA TN 4362, 1958.

[151] Spreiter, J. R. and Stahara, S. S., "Aerodynamics of Slender Bodies and Wing-Body Combinations $M_\infty = 1$," AIAA Journal, Vol. 9, Sept. 1971, pp. 1784-1791.

[152] Cheng, H. K. and Hafez, M. M., "Transonic Equivalence Rule: A Nonlinear Problem Involving Lift," Journal of Fluid Mechanics, Vol. 72, Part 1, 1975, pp. 161-187.

[153] Barnwell, R. W., "Transonic Flow About Lifting Configurations," AIAA Journal, Vol. 11, May 1973, pp. 764-766.

[154] Bailey, F. R. and Ballhaus, W. F., "Comparisons of Computed and Experimental Pressures for Transonic Flows About Isolated Wings and Wing-Fuselage Configurations," NASA SP-347, 1975.

[155] Lock, R. C., "Research in the UK on Finite Difference Methods for Computing Steady Transonic Flows," Symposium Transsonicum II, edited by K. Oswatitsch and D. Rues, Springer-Verlag, Berlin/Heidelberg/New York, 1975, pp. 457-486.

[156] Schlichting, H., Boundary-Layer Theory, McGraw-Hill Book Co., New York, 1968, 747 pp.

[157] Spreiter, J. R. and Stahara, S. S., "Unsteady Transonic Aerodynamics - An Aeronautics Challenge," Unsteady Aerodynamics, Vol. II, edited by R. B. Kinney, University of Arizona Press, Tucson, 1975, pp. 553-581.

[158] Caradonna, F. X. and Isom, M. P, "Numerical Calculation of Unsteady Transonic Potential Flow over Helicopter Rotor Blades," AIAA Journal, Vol. 14, April 1975, pp. 482-488.

[159] Erdos, J., Alzner, E., Kalben, P., McNally, E., and Slutsky, S., "Time-Dependent Transonic Flow Solutions for Axial Turbomachinery," NASA SP-347, 1975.

Chapter II.

Commercial Transports—Aerodynamic Design for Cruise Performance Efficiency

F. T. Lynch[*]
*Douglas Aircraft Company, McDonnell Douglas Corporation,
Long Beach, Calif.*

Introduction

Means of increasing the cruise (fuel) efficiency of commercial transports are usually centered around ideas for reducing engine-specific fuel consumption, airframe drag, and airframe weight. All of these ideas are receiving much attention today, in view of the large impact that rapidly rising fuel prices are having on airline direct operating costs. For example, weight reductions through the use of composite and other advanced materials, load control or alleviation, and active controls to reduce empennage size are all being pursued. Means to improve basic engine-specific fuel consumption and to reduce deterioration are also being developed. Likewise, methods to reduce airframe drag, including the use of supercritical airfoils, are being investigated. Pertaining to this last category, the purpose of this paper is to examine the subject of the aerodynamic design of commercial transports for cruise performance efficiency, focusing on some of the real problems facing the aircraft designer today and the related difficulties encountered in predicting or measuring the corresponding aerodynamic parameters.

First, to set the stage, the differences in the transonic flow problems that are faced by the respective designers of commercial airline transports and military aircraft, both transports and high-performance fighters, are outlined.

Presented at the Transonic Perspective Symposium, NASA/Ames Research Center, Moffett Field, Calif., Feb. 18-20, 1981.
Copyright © American Institute of Aeronautics and Astronautics, Inc., 1981. All rights reserved.
   [*]Principal Program Engineer, R&D Programs, Aerodynamics Subdivision.

Then, the cruise drag efficiency characteristics of several in-service commercial transports are examined in order to determine areas where drag improvements are possible, and correspondingly, where emphasis should be placed in future designs. Following this, some of the factors that must be taken into account in the design of an efficient transport wing are discussed, including the use of supercritical airfoil technology, along with the subsonic and transonic flow problems that are encountered in the wing design process. Both computational and experimental aspects are addressed. Interference drag design problems and their corresponding computational and experimental aspects are then discussed.

Finally, recommendations are provided relative to developments in both computational and experimental technologies that are vitally needed by the commercial transport designer in order to improve the cruise performance efficiency of future designs and, at the same time, reduce the development risk.

Commercial Transports vs Military Aircraft

The transonic flow design problems that need to be solved on military aircraft and commercial airline transports differ greatly in many aspects due to fundamental differences in requirements - even for the transports. Furthermore, the financial climate in which a commercial transport is developed differs very much from that for military aircraft. This difference can significantly influence the level or type of advanced technology incorporated in the respective aircraft.

Transports - Commercial vs Military

Commercial and military transports are typically designed for two fundamentally different functions. Whereas the typical military transport must provide an airdrop capability in addition to providing a large-volume airlift of many kinds of supplies and equipment, some outsize motorized vehicles and some high density (such as tanks), into and out of all kinds of airfields in response to emergencies, the commercial transport's function is to return a profit to the airline operator by safely carrying passengers over routes into and out of existing airports at a high frequency almost every day. Differing requirements resulting from this very diverse usage lead to quite dissimilar looking configurations with somewhat different transonic flow design problems to solve. For example, the military transport typically has both a high wing and an undesirably shaped (from an aerodynamics viewpoint) upswept aft fuselage resulting from the requirement to load and unload its cargo at airfields not necessarily equipped to handle cargo. The airdrop requirements can further compromise the aft fuselage shape. An unfavorable consequence of the high wing location is the need for relatively large external gear pods to house the main landing gear, which, by comparison, would normally be well hidden (buried) in the wing root of a low-wing airplane. Another complication sometimes present in military transports is the need to operate from rather short airfields which can require a powered lift system with the attendant reduced cruise performance due to a larger tail and heavy flap, plus the concern over a more interference-prone engine installation. The YC-15, shown in Fig. 1, characterizes a military transport configuration resulting from these requirements.

An outgrowth of all of these aerodynamically nonoptimum configuration features of a typical military transport is a practical limitation on the cruise Mach number that can be realistically achieved without incurring large performance (fuel burned and payload) penalties. Since there is no large productivity influence on direct operating costs (DOC) with military transports due to the low utilization (relative to commercial transports), the result is a correspondingly lower wing sweepback and a somewhat easier transonic wing design problem. This easier wing design problem on military transports is, however, offset to a degree by a more difficult configuration integration task considering, just as a minimum, the influence of the large gear pods and the adverse aft fuselage shape.

Fig. 1 YC-15 military transport.

For a commercial airline transport, the emphasis in design, particularly at transonic cruising conditions, is on direct operating costs (DOC), fuel efficiency, aircraft productivity, return on investment (ROI), and safety. Today's customer for the commercial transport is no longer represented by the self-made airline executive who as a young man loved flying and "had a dream." Instead, today's airline president is typically a highly educated businessman who probably knows more about finance and hotels than he does about airplanes, and who, assuming all else equal, is primarily interested in the bottom line, i.e., what is the return on the investment? His job has also been complicated in recent years by inflation, deregulation, rapidly escalating fuel prices, and the imposition of stringent noise requirements. Fuel efficiency is of utmost importance today in the design of a commercial transport. It is projected that fuel costs will soon account for nearly two-thirds of the direct operating costs of an airline using current state-of-the-art aircraft.[1] By comparison, fuel costs over the life of a military transport are a small fraction of those for a commercial transport due to the much lower utilization, i.e., approximately 3 h a day compared to 10-12 h a day airline utilization. Also, the primary mission for a military transport is carrying a defined payload a given distance or training to fight a war, not minimizing fuel burned and returning a profit as it is for the commercial transport.

Turning now to general configuration differences, in contrast to the typical high-wing military transport, the commercial airline transport, which is designed primarily for the transportation of passengers with only a secondary consideration given to the carrying of cargo to enhance the utility of the airplane,[2] typically has no requirement that would lead to the need for either a high wing (with external gear pods) or an undesirably shaped upswept aft fuselage. As a result, the typical high-performance commercial transport, represented by the DC-10 shown in Fig. 2, is a low-wing configuration with no serious fuselage shaping concerns. The low wing is better aerodynamically and also from a crashworthiness standpoint. Furthermore, commercial transports are not normally encumbered with any short field requirements that would necessitate more interference-prone engine installations.

Consequently, the commercial transport speed capability is normally not limited by the geometry characteristics

Fig. 2 DC-10 commercial transport.

that often place a practical limit on the cruise Mach number of military transports. This is fortunate because the DOC economics of a commercial transport point to the need for higher cruise speeds than normal for the military application due to the much higher utilization of the airline airplane. Insurance costs and depreciation contributions to DOC are reduced directly as the utilization is increased. Crew costs and maintenance (airframe and engine) costs are also reduced as the flight time is reduced. The net result is that the DOC typically decreases with increasing Mach number (increased productivity) up to the point where the airframe weight and cost increases associated with further increases in wing sweep (or reductions in thickness) plus the increased fuel burned with increasing Mach number eventually outweigh the reductions in time-related costs. This effect is illustrated in Fig. 3. For a new transcontinental range airplane, the minimum DOC point is near 0.8 Mach number, significantly faster than any "modern" military transports.

Commercial Transports vs High-Performance Fighters

In contrast to carrying either passengers or cargo, high-performance fighters are designed for engaging in air combat and for weapons delivery. These two very different functions result in the need for the designer of combat aircraft to solve some quite complex transonic flow design problems not faced by the designers of commercial transports. Some of the fundamental differences between the two types of configurations that directly influence the type of transonic flow problems to be solved are listed in Table 1.

Probably the most important difference between the two types of aircraft is the difference in maneuvering requirements at transonic conditions. Whereas the commercial transport is designed for a maximum limit design load factor of 2.5 g, but even 0.1 g produces passenger discomfort, the designer of the high-performance fighter is typically looking for 7.5 to 9 g maneuvering capability for climbing and turning. The emphasis on transonic maneuverability for high-performance fighter aircraft without compromising mission radius requirements or supersonic acceleration capability has stimulated the development of various concepts for producing additional lift under separated flow conditions at high angles of attack where vortex lift is very important. Such concepts as the use of leading edge slats transonically, close coupled wing-canards, wing plus strakes, three surface configurations, variable camber surfaces, spanwise blowing, forward sweep, aeroelastically tailored wings, etc., have been pursued by the fighter designer as more efficient means of increasing the maximum lift capability and hence maneuvering performance at transonic conditions than brute force increases in thrust-to-weight ratio.[3-6] When maneuvering at high angles of attack (up to over 50 deg), the high-performance fighters experience nonlinear aerodynamic characteristics because of the flow separation.[7] Consequently, since they require high-performance and maneuverability near operating limits, fighter aircraft have typically led the way in developing advanced control concepts such as fly-by-wire systems, etc. in order to handle these aerodynamic characteristics.[8] Also, operating at these very high angles of attack leads to the need to address departure and spin characteristics since it is currently impossible to achieve a departure proof or spin proof high-performance fighter aircraft without unduly detracting from agility in combat.[9] Now,

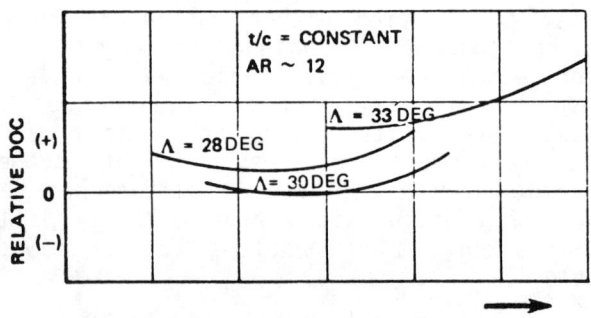

Fig. 3 DOC optimization for new transcontinental range commercial transport.

Table 1 High-performance fighters versus commercial transports

Item	High performance fighters	Commercial transports
Mission	Air combat Weapons delivery	Carrying passengers DOC, fuel burned, ROI
Maneuverability and design loads	Approximately -3 to +7.5 to 9 g High angles of attack (40 deg to 50 deg desired) Stall and spin Greater need for advanced control systems	0.1 g produces passenger discomfort 2.5 g is limit design load factor 1.3 g margin to buffet Resulting low α
Speed capability	Supersonic combat or intercept	Strictly subsonic
Size	Smaller Lower Reynolds number (some laminar flow)	Bigger Higher Reynolds number (typically all turbulent)
Configuration layout	Wide fuselage relative to wing span Area ruling Irregular shapes Airframe - propulsion integration (high thrust-to-weight ratio) Low aspect ratio thinner wing ($\leq 6\%$) Weapons carriage Closely coupled - interacting	Narrower fuselage (to wing) No overall area ruling High aspect ratio thicker wings Aerodynamically desirable shapes Not nearly as closely coupled (typically)
Other	Low radar cross section desirable	Safety

in complete contrast to this drive to achieve increased maximum lift and improved high-lift drag polars at transonic conditions, the commercial transport designer has no use for this vortex lift and its consequences. In fact, as will be discussed later in this paper, the commercial transport designer wants to minimize any vortex lift transonically, as it adversely effects the wing weight without providing any performance benefit.

The difference in maximum speed requirements and vehicle size between fighters and transports also significantly influences the type of transonic design flow problems that need to be solved for both. For example, the supersonic combat requirement for the fighters leads to the need for area ruling and has prompted studies of forward swept and oblique wing configurations as a result of their potential lower wave drag at transonic and supersonic speeds. This area ruling requirement also influences the resulting propulsion-airframe integration design solution. Fighter wings are also significantly thinner and have sharper leading edges because of the supersonic requirements. Engine inlets are much different. Control surface designs are also affected, i.e., fighters typically require all-moving fully powered tails for stability and control due to the ineffectiveness of hinged elevators at supersonic conditions. In contrast to this supersonic dash requirement, the efficient commercial transport would very rarely if ever exceed 0.9 Mach number in its airline life, and probably never even 0.85 Mach number for efficient cruising conditions. A flight Mach number of typically no more than 0.95, the design condition for an upset and resulting overspeed condition, would be the maximum encountered and it would only be demonstrated by the manufacturer in the airplane flight development program, but probably never experienced by an airline. Consequently, little or no consideration of area ruling per se <u>other than for local interference</u> problems is necessary in the design of a commercial transport. With regard to size differences, due to its larger size and consequently significantly higher Reynolds number, the flow on the wing of a typical commercial transport at high-speed conditions is normally completely turbulent starting right from the leading edge attachment line. On the other hand, with the fighter there will typically be at least a small region of laminar flow near the wing leading edge that must be reckoned with at transonic conditions.

The requirements for a high-performance fighter, including weapons delivery, are typified by the F-15

Fig. 4 F-15 high-performance fighter.

configuration as shown in Fig. 4. It is a very closely coupled, interacting configuration with some irregular shapes. The fuselage (engine)-to-wing-span ratio is quite large compared to a transport due to the higher thrust-to-weight ratio and lower wing aspect ratio. It seems very obvious that the analysis and design of this type of configuration at transonic conditions must consider the vehicle as a whole due to its highly interacting features. Again comparing this situation to a typical commercial transport (depicted by a DC-10 in Fig. 2), the transport design is not nearly as highly interacting as the fighter. Although there are transonic flow situations at design conditions outside the airline cruising regime where wing-tail interactions are very important, most of the transonic flow design problems faced by the commercial transport designer can be reasonably segregated and attacked individually. These include the wing design, the wing-pylon-nacelle interference problem, wing-fuselage interaction, fuselage afterbody-tail surfaces inter-relationship, fuselage nose and cockpit design, nacelle inlet-cowling design, any third engine installation, etc. No predominant interactions exist.

One other significant difference between fighters and transports that influences the incorporation of advanced technologies is the issue of safety. The commercial transport is required to meet stringent safety standards dictated by government agencies, and is the result of an astounding number of design requirements, criteria, and considerations. Safety is also foremost in the minds of the airlines. The airline criteria is that in case of a tradeoff in design criteria, first safety and then good operating economics shall override.[2] Hence, whereas such advanced technologies as full-time fly-by-wire flight control systems are being used now on high-performance fighters and would have considerable potential for application to civil aircraft, their use on commercial transports would be delayed by safety requirements.

Further discussion of the unique aspects of fighter airplane design is contained in the paper by R.G. Bradley in this book.

Commercial Transports - Cruise Drag Efficiency

With fuel costs becoming such a large part of the airlines' direct operating costs, one of the aerodynamic designer's major challenges for any current or new airplane program is to minimize the drag at cruising conditions. The large economic impact of making a drag improvement is illustrated in Fig. 5 for representative short- and long-range airplanes. The impact is greatest on the larger, longer range aircraft due to the greater use of fuel. As fuel prices exceed $1/gal, the savings in fuel on a DC-10 for a 5% drag reduction are approaching one-half million dollars/airplane/year. The totals become very large when the number of airplanes and typical number of years in service are considered.

The classic means of evaluating the aerodynamic efficiency of transport aircraft has been in terms of M x L/D (Mach number times lift-to-drag ratio) since it is proportional to range. Aerodynamic efficiency

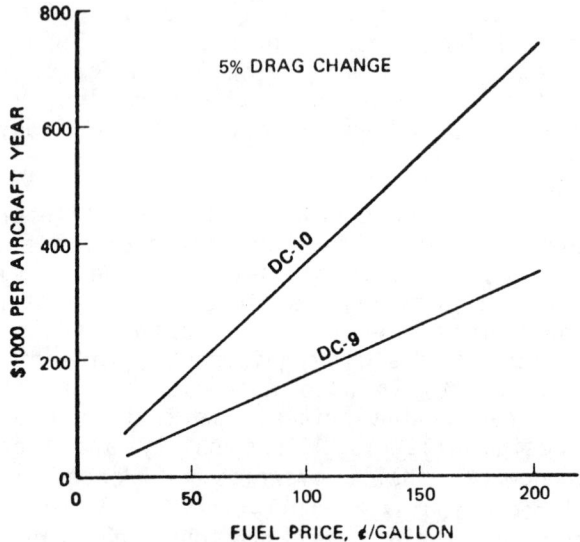

Fig. 5 Economic impact of cruise drag.

characteristics of several in-service commercial transports expressed in this manner are presented in Fig. 6.[10] The M x L/D for each airplane increases with Mach number until the drag buildup due to the occurrence of supersonic flow and resulting shock waves offsets the Mach number increase. Airlines typically cruise the airplanes at a Mach number just beyond the peak M x L/D - at about 99% of the peak value, the result of a tradeoff between reduced fuel efficiency and increased productivity.

One very obvious characteristic illustrated in Fig. 6 is the big difference (∿25%) in the described aerodynamic efficiency between long- and short-range airplanes resulting from their very different design requirements. However, it is important to note that this difference is not a measure of how well the aerodynamic designers of each type of airplane has done his job. Neither is the difference in M x L/D between any two airplanes designed to "similar" requirements since differences in fuselage configuration and wing geometry (area, aspect ratio, sweep, and thickness ratio) can strongly influence the resulting M x L/D characteristics. A more meaningful measure of the transonic <u>aerodynamic design efficiency</u> is an evaluation of either how close did the drag level at prescribed cruising conditions come to

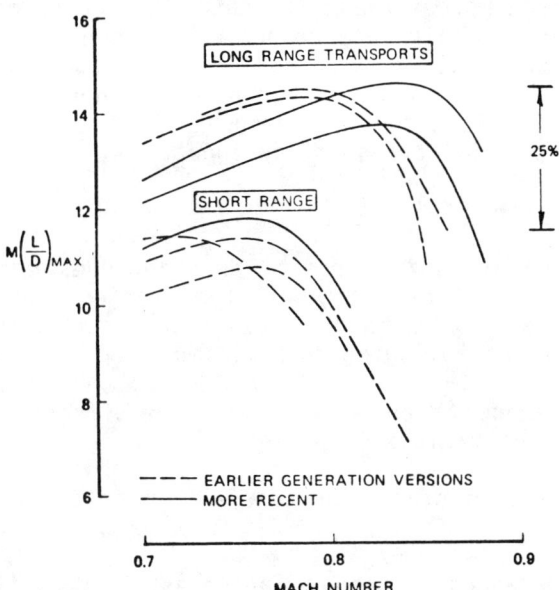

Fig. 6 $M(L_D)_{MAX}$ efficiency of representative commercial transports.

the predicted characteristics, or how close is the drag level at these conditions to the minimum possible for the particular airplane layout selected, as a result of many trade studies, as the most appropriate configuration to satisfy the specific requirements. To assess how the transonic cruise drag for a particular airplane compares to the minimum possible or practical, it is convenient to separately examine the following three parts of the total drag:

Basic parasite drag at zero lift ⎫
Lift-dependent induced drag ⎬ before Mach effects (typically $M \leq 0.6$ or 0.7)
⎭

Compressibility drag - drag increase with Mach number

Significant excessive drag in any of these components can lead to an inefficient and uncompetitive airplane. Before looking at these three parts separately, it is instructive to look at the total drag breakdown at cruise conditions. The typical relative magnitude of the three parts for a long-range transport at cruising conditions is illustrated in Fig. 7 using the DC-10 as an example. Basic parasite drag accounts for over half of the total, and the induced (vortex) drag accounts for the rest except for a little over 5% for compressibility drag. A corresponding breakdown by primary airplane components, shown in Fig. 8, demonstrates that two-thirds of the total drag is associated with the wing. Hence, the normal large emphasis that is placed on the aerodynamic design of the wing by the commercial transport designer.

Basic Parasite Drag Efficiency

The category of basic parasite drag includes not only the friction and form drag of the various airplane components (treated as though they were each isolated), but also all of the following as a minimum:

1) Interference drag - addition (or offsetting) of supervelocities of various components.

2) Intersection drag - boundary-layer interaction in junctures.

3) Protuberances - canopy, antennas, vents, lights, etc.

TRANSPORTS—DESIGN FOR CRUISE PERFORMANCE

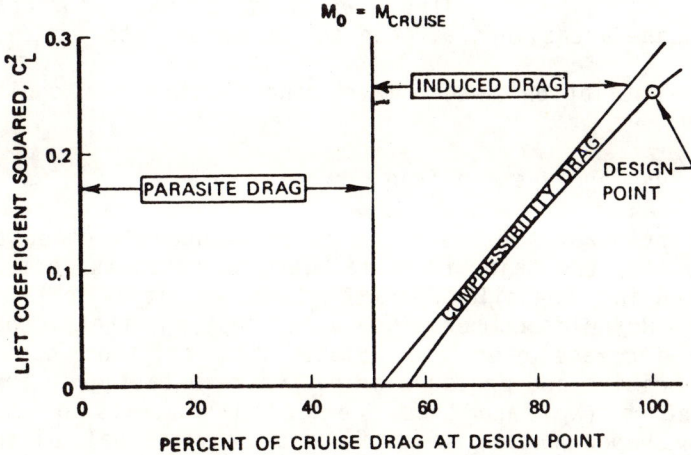

Fig. 7 Relative magnitude of drag components.

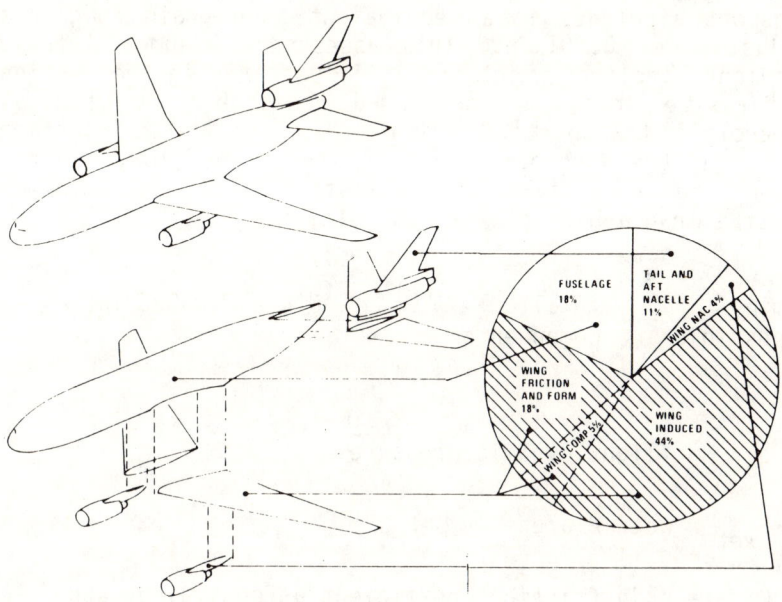

Fig. 8 Typical drag breakdown of transport aircraft at cruise conditions.

4) Excressences (dirt) - gaps, steps, protruding rivets, base areas, nonfair control surfaces, etc.

5) Ventilation drag - air conditioning and cooling airflows.

6) Fuselage upsweep drag.

Prediction of the basic (no interference) friction and form drag for the major identifiable components is normally done by using so-called form factors, normally obtained from high Reynolds number wind-tunnel testing, that account for the increase over flat plate skin friction due to supervelocities and pressure drag. Representative of these form factors (K_F) are those derived from tests of many fuselage-shaped bodies (Fig. 9) and conventional airfoil shapes (Fig. 10). Airplane surface finish (roughness) conditions, excluding all of the above listed items, are deduced from airplane drag measurements in conjunction with a <u>very detailed</u> analysis of all of the identifiable contributors to the drag. A typical surface condition resulting from this type of analysis on a few different transport airplanes is an equivalent sand grain roughness of just under 0.001 in. This assessment assumes a fully turbulent airplane, which, as is illustrated in the section on Parasite Drag, is considered reasonable for typical commercial transports. With this realistic surface condition, the assumption of fully turbulent flow, and the appropriate form factors, an estimate of the minimum parasite drag achievable is possible, i.e.

$$D_{P_{min}} = q_0 \sum_{\substack{\text{all possible} \\ \text{components}}} S_{wet} \times C_{f_{@\ K\ =\ 0.00095\ in.}} \times K_F$$

where

q_0 = flight dynamic pressure

S_{wet} = wetted area

C_f = skin friction coefficient at equivalent sand grain roughness of 0.00095 in.

By comparing the parasite drag determined from flight measurements to this minimum achievable value at the same flight conditions, a measure of the parasite drag design efficiency (or lack thereof) is obtained. This "excess" parasite drag, attributable to some or all of the items listed at the beginning of this section, and expressed in terms of a fraction of the minimum achievable value, is presented in Fig. 11 for several commercial transports. These airplanes are grouped by their design range requirements. The "excess" parasite drag varies from as little as 8% (for the latest version of DC-10-30) to as much as 30% (for a current twin-engine medium-range transport). For consistently designed airplanes, the excess should be inversely proportional to the range since

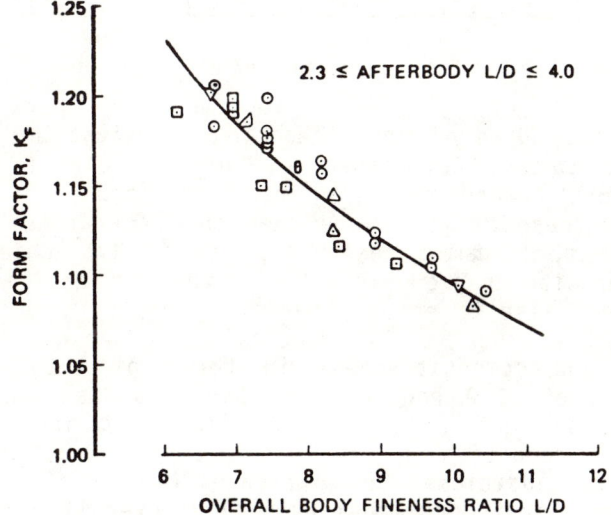

Fig. 9 Fuselage form factors.

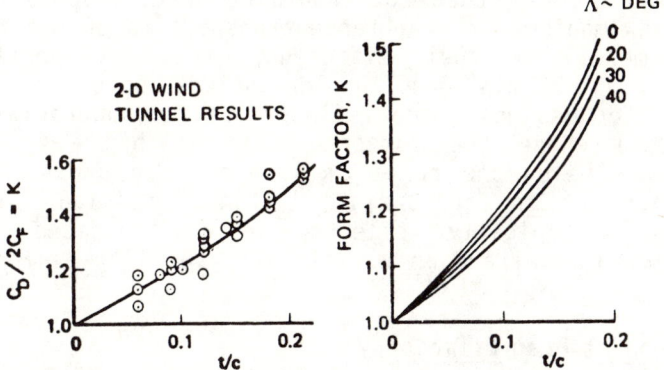

Fig. 10 Wing form factors for conventional-type aircraft shapes.

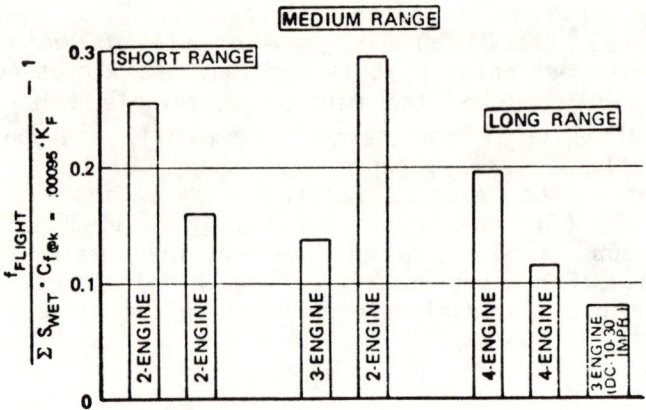

Fig. 11 Basic parasite drag efficiency.

design trade studies of drag reduction vs weight and cost for protuberances, excressences, etc. typically are weighted more toward the drag reduction aspect as the range is increased. It can be seen that this trend does exist. The aerodynamic designer's job is to achieve a level of parasite drag consistent with the airplane mission and cost, but also to ensure that the parasite drag level is minimized as much as is practical in order to assure efficiency and competitiveness in terms of total fuel burned, DOC, etc. A reduction in parasite drag has the same effect on performance as a similar, but often more "glamorous," reduction in the transonic compressibility drag. Every additional 2% increase in parasite drag increases the total drag (fuel burned) by over 1%. Hence, considerable attention and effort must be focused on minimizing the parasite drag. An excess parasite drag fraction of no more than 10% should be an achievable goal for any new medium- or long-range transport to be built in these times when fuel efficiency is so important. Exclusions from this goal would include specific drag penalties for designs (nacelle interference, supercritical wing, etc.) where the parasite drag is high at lift coefficients below the range of interest for cruise, but is minimized at the appropriate lifting condition. This situation is typically reflected in an indicated high parasite drag level and an artificially low lift-dependent drag.

Lift-Dependent Drag Efficiency

Flight test results for several airplanes show that the lift-dependent drag prior to the onset of compressibility effects typically has a parabolic variation from a lift coefficient of near 0.25 (lowest point typically obtained) up to at least 0.5, and usually 0.6 or more. For convenience in definition of terms, the parabolic drag polar is normally extrapolated to zero lift. Over this total range of lifting conditions the lift-dependent drag (generally called induced drag) is defined as follows:

$$C_{D_i} = C_L^2 / \pi e AR$$

where the airplane efficiency factor, e, is the correction term that is derived to account for the drag increase over and above the theoretical minimum induced drag of a wing with an ideal elliptic span load distribution. The primary sources of lift-dependent drag that cause the actual airplane efficiency factors to be less than the theoretical maximum are additional vortex drag due to a nonelliptic span load distribution, and a viscous drag increase with increasing lift on the wing and other components.

Deviations from the idealized elliptic load distribution always occur as a result of the presence of the fuselage. Additional deviations are often the result of trade studies involving wing drag and stability characteristics vs wing weight and cost considerations. Fortunately, the additional vortex-drag penalties due to these nonelliptic span load distributions can be reliably predicted using current lifting surface programs. However, the viscous drag contribution is much more difficult to accurately predict. One means of assessing the magnitude and variability of the viscous drag contribution to the lift-dependent drag is to deduce the value from flight data on current commercial transports. To do this, the various airplane lift-dependent drag increases over and above the theoretical minimum vortex drag for an elliptic span load for each has been determined at a lift coefficient of 0.5, and then adjusted by the known vortex-drag penalties. The resulting viscous contribution obtained in this manner (Fig. 12) varies from a minimum of about 10 drag counts to a maximum of about 18 counts. The difference between the best and the worst amounts to approximately 3% of the total drag of a long-range airplane at cruising conditions.

Hence, the viscous contribution to the lift-dependent drag must be considered and taken into account in the

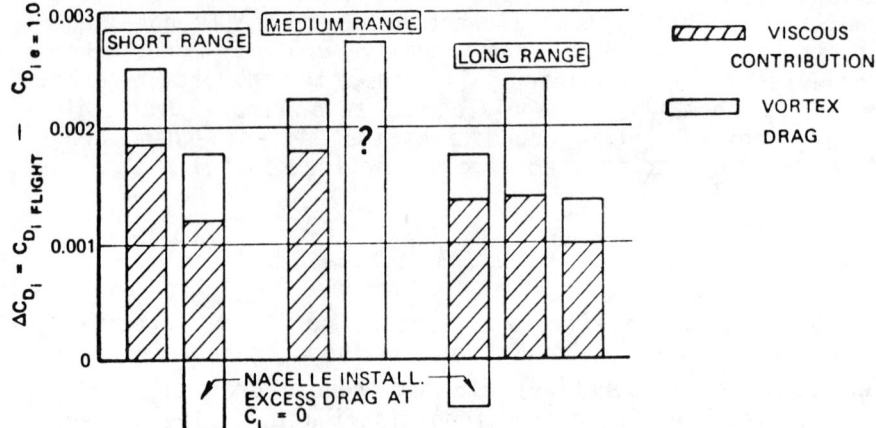

Fig. 12 Nonelliptic lift-dependent drag breakdown.

selection of the wing airfoil section in particular, but also in the overall wing design (taper ratio, etc.) and total configuration design. For example, it is known that the use of fences to control wing stalling characteristics results in a significant increase in the viscous contribution to the lift-dependent drag. Consequently, their use should be avoided if at all possible.

Compressibility Drag Efficiency

Increases in airplane drag coefficient as Mach number is increased toward the cruising point are usually referred to as compressibility drag. Compressibility drag is conventionally associated with the formation of local regions of supersonic flow on the wing and other parts of the airplane (see Fig. 13), but, in reality, includes any variation of the parasite and vortex drag with Mach number in addition to the shock wave drag and any drag due to shock-induced separations.

The development of the compressibility drag with increasing Mach number can be conveniently separated into three segments. First, there is the very gradual drag increase that can occur before any supersonic velocities arise. Next, there is a somewhat more rapid drag rise that occurs as regions of supersonic velocity (and hence some shock waves) form and start to spread out. Finally, there is the very steep drag rise that usually starts when the Mach number just ahead of shock wave on the upper surface

TRANSPORTS—DESIGN FOR CRUISE PERFORMANCE

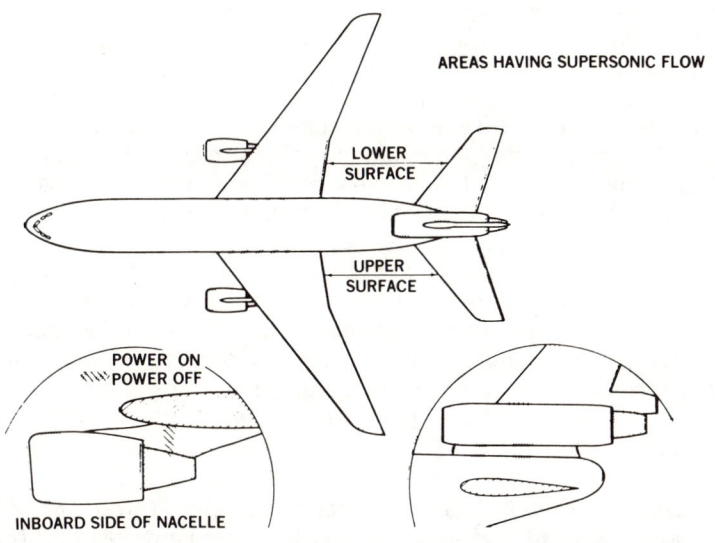

Fig. 13 Extent of supersonic flow at high-speed cruise conditions.

of the wing exceeds approximately 1.2. The point at which this steep drag rise starts can also be influenced by the existence of interference drag situations such as a poor nacelle-pylon-wing installation, etc.

At Mach numbers below those at which shock waves exist, the airplane parasite drag is composed mostly of the sum of skin friction drag and the form drag, both of which vary with Mach number. The skin friction drag coefficient decreases with increasing Mach number, because of the thickening of the boundary-layer due to increased boundary-layer heating. On the other hand, the form drag increases with increasing Mach number because pressure coefficients (absolute values) tend to be increased by increases in Mach numbers. Thus supervelocities, adverse pressure gradients, and boundary layer displacement thicknesses become more pronounced. Consequently, whether or not there is any drag rise in the absence of any supersonic velocities, and, if so, how great, depends on the balance between these two components for a given configuration. There are airplanes for which there is no drag increase until supersonic velocities arise. On the other hand, a majority of airplanes do have a gradual rise before supersonic velocities occur.

Supersonic velocities usually occur first on the upper forward part of the wing, on the forward part of the nacelle, and on the afterbody of a short-duct nacelle. The consequent drag rise is at first small because the extent of any shock waves is very limited at first. As the supersonic velocities start to spread aft on the wing to areas where the surface curvature is more gradual, the drag rise becomes more pronounced. This rapid increase is related to the fact that the velocity gradients normal to the surface are reduced as the surface curvature becomes more gradual, and, as a result, the induced velocities (hence shock waves) extend further from the surface for a given velocity at the surface. Finally, a definite increase in the airplane drag coefficient usually occurs when the shock Mach number of the wing exceeds about 1.2. Further increases in Mach number produce very large drag increases that are caused by stronger shock waves and also shock-induced separation.

Because of the mixed character of the flow and the strong influence of viscous effects, current analytical methods cannot reliably predict the compressibility drag, but can only indicate trends. Hence, empirical correlations must be used, based on both wind-tunnel and flight data. Ideally, the compressibility drag of an airplane should be estimated separately for all of the components and then adjusted for interference effects. However, since the goal for a well-designed commercial transport is to design and arrange the components other than the wing to have a drag divergence Mach number higher than that of the wing, the dominant factor in predicting and achieving the appropriate airplane compressibility drag characteristics at cruising conditions is the wing. Consequently, commercial transport total airplane compressibility drag prediction methods are characteristically tied to the wing characteristics. This assumes, of course, that interference drag effects will be largely eliminated in the development program (not always a good assumption). The type of method often used is shown in Fig. 14 for airplanes with wings designed with "conventional" airfoil sections, i.e., not aft-loaded. The wing compressibility drag based on flight data of existing McDonnell Douglas commercial jet transports is referenced to a drag divergence Mach number that is a function of wing sweep, average thickness ratio, and design lift coefficient. This characteristic is then adjusted for the compressibility drag of the fuselage, nacelles, pylons, tail surfaces, and any interference and trim drag effects to obtain the predicted compressibility drag for the total airplane. It is then left to the aerodynamic designer to

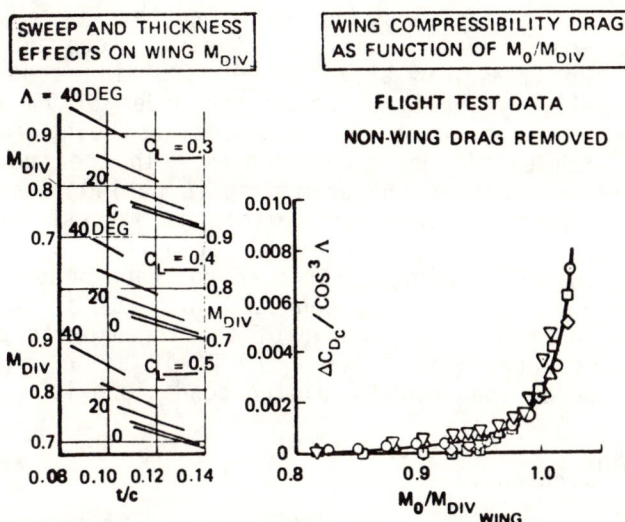

Fig. 14 Wing compressibility drag prediction - conventional airfoils.

develop and tailor the total configuration to achieve the predicted characteristics. Relative degrees of success for two MDC long-range transports and two other currently popular commercial transports are illustrated in Fig. 15. The best that has been achieved are the two MDC airplanes, with airplane drag divergence Mach within 1% of the predicted wing-alone characteristics. On the other hand, compared to the same prediction criterion, cruise speed penalties of from 4-5% are not uncommon. In these latter cases, the airplane performance consistent with the wing weight incorporated (sweep and thickness influence) has not been achieved.

Trim Drag

Trim drag is the difference between the drag of the airplane in pitch equilibrium and the drag of the airplane in the nonequilibrium condition for which there is no moment provided by the pitch control surface (normally the horizontal tail). The lift is required to be the same at both conditions. It can be shown that the trim drag consists of the sum of the change in the airplane tail-off induced drag due to the lift (usually downward) on the horizontal tail, the horizontal tail induced drag, and the component of the horizontal tail lift in the direction of the downwash at the tail surface. Trim drag is most strongly influenced by the airplane (wing-body) pitching moment characteristics and the tail length. For current airplanes with wings designed using "conventional" airfoils, the trim drag at <u>efficient</u> cruising conditions is quite small, typically being of the order of 1% of the total drag or less, with the lower values being for the airplanes with the longer tail lengths. In the customary drag breakdown, part of the trim drag is buried in each of the three major drag components with the largest part in the compressibility drag. However, for airplanes designed using aft-loaded airfoils, trim drag can become more important and is often accounted for as a separate category. The typical breakdown of the three parts of the trim drag is noteworthy. Using the DC-10 as an example, the breakdown at long-range cruising conditions is

$$\left. \begin{array}{l} C_{D_{wing\text{-}body}} = +5.5\% \\ C_{D_{tail}} = +0.3\% \\ C_{L_T} \sin\epsilon = -4.7\% \end{array} \right\} \text{net } C_{D_{trim}} = +1.1\%$$

Several controversies have existed regarding the magnitude of the trim drag. Specific DC-10 flight testing has been

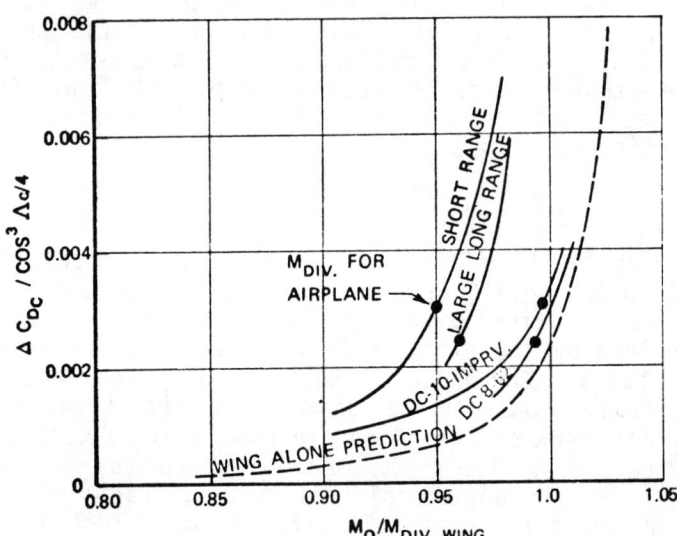

Fig. 15 Total airplane compressibility drag versus wing alone prediction.

dedicated to resolving these issues, and has confirmed the above evaluation.

Commercial Transport Transonic Wing Design

The design and resulting performance of the wing is the essence and heart of a commercial transport. The cruise (clean) wing configuration sets the tone for airplane cleanliness, it is overwhelmingly predominant in determining the lift-dependent drag, and only if the wing design is good can the airplane compressibility drag be as desired. Some of the factors that must be taken into account in the design of the wing are discussed in this section along with some of the subsonic and transonic flow problems that are encountered in the wing design process. The latter includes both computational and experimental aspects.

General Design Considerations

Some of the criteria that must be met in the design of any efficient transonic transport wing include:

1) Good drag characteristics (parasite, induced, compressibility) over range of lift coefficients, i.e. $CL_{DESIGN} \pm 0.1$ at M_{CRUISE}.

2) No excessive penalties for installation of nacelle-pylons, fairings, etc.

3) Buffet boundary high enough (1.3 g margin required) to permit cruising at design lift coefficients.

4) No pitch-up tendencies near stall, buffet boundary.

5) Control surface effectiveness must be maintained.

6) No unsatisfactory off-design performance.

7) Must have sufficient fuel volume for design range.

8) Must be structurally efficient (to minimize weight).

9) Must provide sufficient space to house main landing gear.

10) Must be compatible with selected high-lift system.

11) Must be consistent with airplane design for relaxed static stability.

12) Must be manufacturable at a reasonable cost.

Parameters that are weighed in trade studies leading up to a basic design to satisfy the above criteria include, as a minimum:

- planform
- span
- aspect ratio
- area
- weight
- airfoil types, thickness
- design lift coefficient
- desired speed
- sweepback angle
- taper ratio
- choice of high-lift system
- low-speed performance
- landing gear location

In addition to these, many other design variables must be considered which, almost without exception, have both favorable and unfavorable effects on the total airframe system. These include the spanwise distribution of lift and maximum thickness (see Fig. 16), spanwise airfoil variations, etc.

In the past, including the design of the DC-10, 747, and L1011, the wing design procedure was almost totally based on the use of linearized inviscid computational methods in conjunction with a somewhat cut-and-try experimental program to refine the wing configuration once the planform, type of "conventional" airfoils, average thickness, and design lift coefficient had been determined from a design optimization study. Unfortunately (for the particular manufacturer and airline customer) less than optimum designs have often been the result of this design approach. Now, however, recent advances in nonlinear transonic computational techniques have made it possible to efficiently address many more design objectives than was previously possible, and concurrently reduce the design cycle flow time. These advances, in conjunction with the development of the aft-loaded (so-called "supercritical") airfoils, have revolutionized the transonic wing design process, and, in conjunction with the influence of rapidly rising fuel prices, the resulting design solutions, i.e., transport wings being designed today, really look different from previous designs.

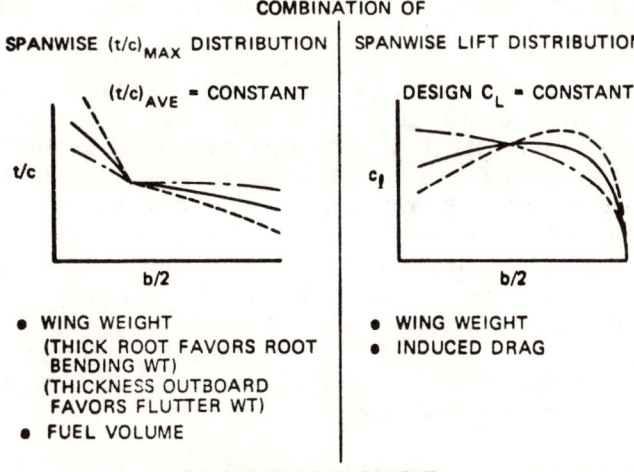

Fig. 16 Spanwise lift and thickness considerations.

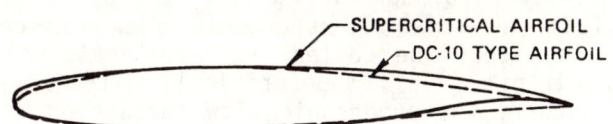

Fig. 17 Supercritical vs conventional airfoil shapes.

<u>Supercritical Airfoils-Wings</u>. Research accomplished over the past few years on supercritical airfoils, and on wings designed using these airfoils, has shown conclusively that there is a definite performance advantage to be obtained from the use of this technology.[11] However, the degree to which this advantage is used, and the manner in which it is used, are functions of many variables, not the least of which is the always present risk element. Before examining the various ways in which the supercritical airfoil's advanced technology can be utilized, it is instructive to look at how this type of airfoil derives its improvement over the "conventional" airfoils. Fundamentally, the supercritical airfoil is capable of generating greater amounts of lift for a given thickness and drag at high subsonic Mach numbers. Relative to a "conventional" airfoil, the distinguishing geometric characteristics of a <u>full</u> supercritical airfoil are a

slightly blunter nose, a flatter upper surface, and a highly cambered thin trailing edge (see Fig. 17).

The camber distribution of a full supercritical airfoil is compared in Fig. 18 to a "conventional" (DC-10-type) airfoil. Also illustrated are the camber distributions for two often considered variations of the supercritical airfoil, one where the aft camber is less than half of that of the full supercritical airfoil, and the other (referred to as quasisupercritical) where the camber distribution over the forward part of the airfoil is very similar to the conventional airfoil. It will be subsequently shown that neither of these variations is nearly as effective as the "full" supercritical approach in terms of improved performance potential.

The increased lift capability of the supercritical airfoil comes from the aft camber and the increased chordwise extent of "supercritical" (locally supersonic) flow over the airfoil's upper surface (see Fig. 19). It is from the latter feature that its name is derived. The name is technically somewhat misleading in that all subsonic transports operating at their most efficient conditions in terms of speed, and lift-to-drag ratio, have sizeable local regions of supersonic flow; hence, all transport wings could be called "supercritical." The new technology airfoil (i.e., supercritical) differs only in that it can carry the supersonic flow further back on the chord, hence increasing the lift. The maximum local Mach number at cruise is no greater for a supercritical airfoil

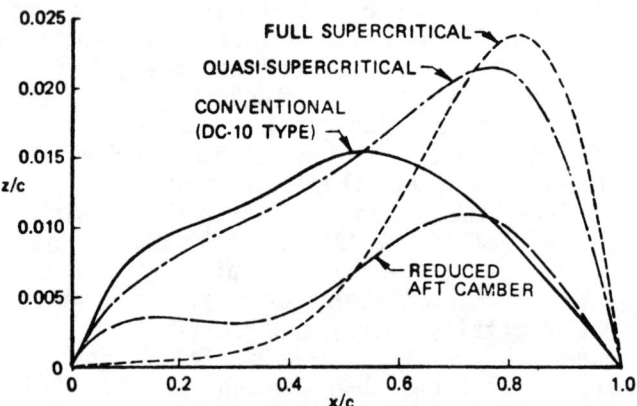

Fig. 18 Airfoil camber distributions.

than for today's airfoils. It can be seen from Fig. 19 that at least one-half of the supercritical airfoil's gain comes from an increase in lift due to the cambered trailing edge. Most airfoils, since the beginning of aviation, could have benefited from this type of aft loading. However, until recently, a design constraint which called for relatively slab-sided surfaces in the region of the controls has been prevalent throughout industry. This constraint was established because of manufacturing cost and complexity as well as a concern, aerodynamically, for the asymmetry in the region of the control surfaces.

Two-dimensional airfoil tests at high Reynolds numbers, many wind-tunnel tests at moderately high Reynolds numbers of complete three-dimensional wing-body configurations, and flight testing of the Douglas YC-15 have all verified the potential performance advantage of the supercritical airfoil-wing concept. The magnitude of the improvement demonstrated in two-dimensional airfoil tests at quite high Reynolds numbers is illustrated in Fig. 20. The improvement in lifting capability (reduced upper surface velocities at a given lift) is manifested by either an increased drag divergence Mach number for a given airfoil maximum thickness, or an increased airfoil thickness for a fixed drag divergence Mach number relative to a

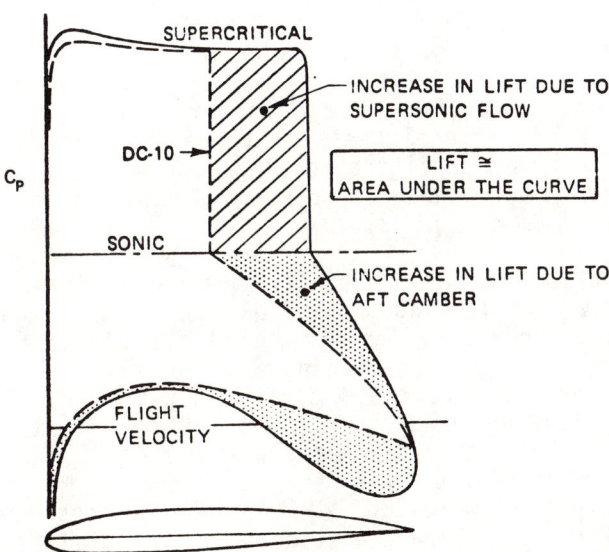

Fig. 19 Sources of increased lifting capability.

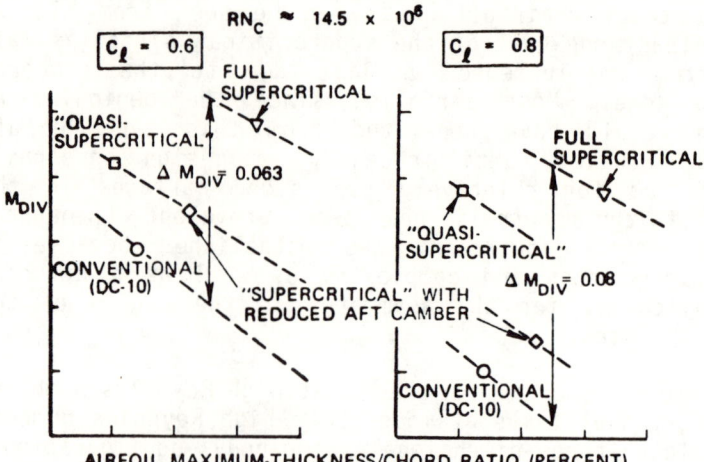

Fig. 20 Two-dimensional airfoil test results.

conventional airfoil. It can be seen that the improvement increases with increasing lift coefficient, a characteristic that makes the supercritical airfoil concept attractive for application to high aspect ratio wings. The much smaller improvements achieved with the two variations of the supercritical airfoil introduced in Fig. 18 are also apparent. Considering the reduced aft camber concept, the test results clearly point out that the potential improvement is not proportional to the amount of aft camber (see Fig. 21) at the higher lift coefficients needed if higher aspect ratio wings are to be efficient. Going beyond two-dimensional airfoil characteristics, results from wind-tunnel tests of three-dimensional (full) supercritical wing configurations have shown that the gain is similar to the improvement demonstrated two-dimensionally (see Fig. 22).

A selection of parameters for a supercritical airfoil is made by considering both low-speed and high-speed requirements for the design mission of the aircraft. Some of the pros and cons concerning leading edge radius and the choice of aft camber are spelled out in Fig. 23. The effect of leading edge radius variations on the drag divergence Mach number and drag creep characteristics of supercritical airfoils is further illustrated in Fig. 24. For any airplane employing leading edge devices for low-speed performance characteristics, the decision in favor of the reduced leading edge radius is clear in view of the difference in drag creep prior to the drag divergence Mach

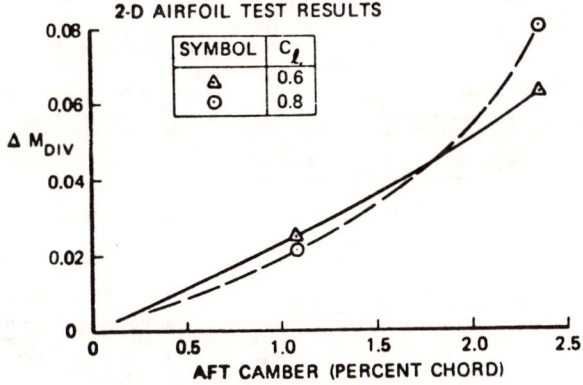

Fig. 21 Supercritical airfoil improvement versus aft camber.

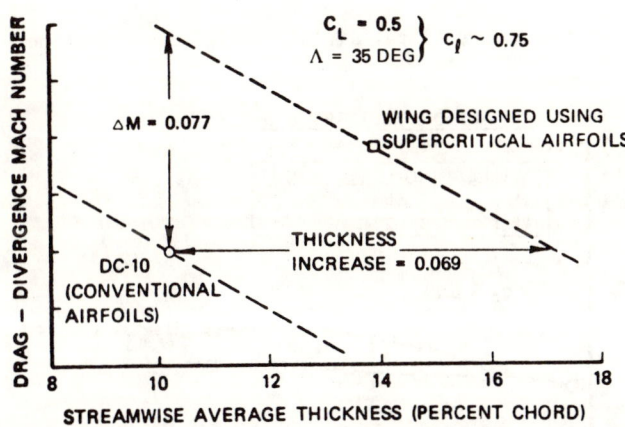

Fig. 22 Wind-tunnel results for three-dimensional supercritical wing plus fuselage.

number. This creep is caused by the presence of shocks near the leading edge of the blunter configurations at relatively low Mach numbers. Judging from the number of different approaches being pursued currently by the different manufacturers, the choice of the proper amount of aft camber is not nearly as obvious. In favor of an aft camber in excess of 2% of the wing chord is the disproportionately higher improvement available (Fig. 21), particularly at the higher lift coefficients needed for higher aspect ratio wings. Weighed against the supercritical airfoil-wing design with this high aft camber are the aerodynamic concerns associated with trim drag and the steep adverse pressure gradients on both the

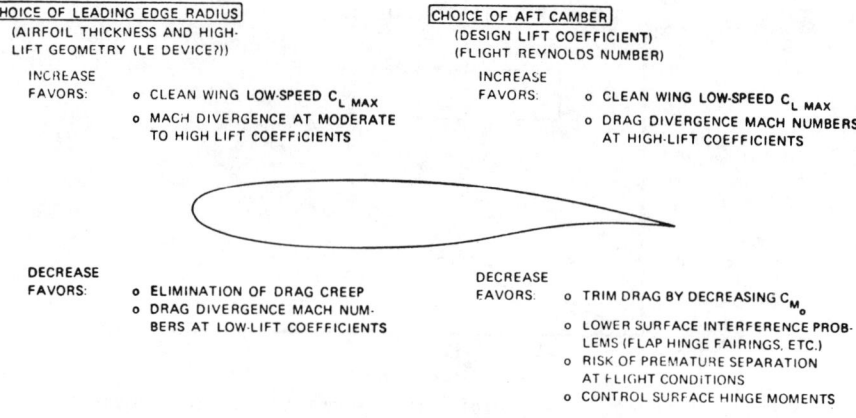

Fig. 23 Some supercritical airfoil design considerations.

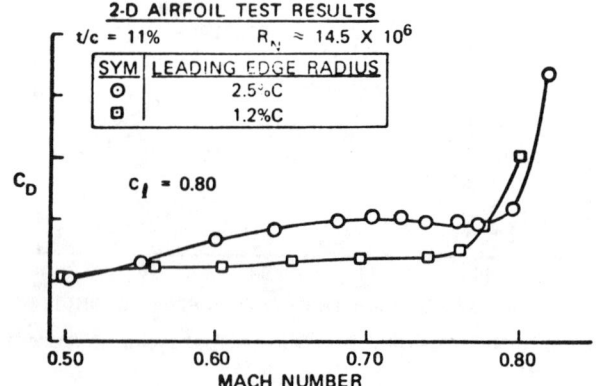

Fig. 24 Leading edge radius effects on drag characteristics of supercritical airfoils.

upper and lower surfaces, plus the structural and packaging concerns associated with the relative thinness of the rearward part of the airfoil. The aerodynamic designers' job is to select and recommend the appropriate compromise.

The increased lifting capability of the supercritical airfoil-wing can be used to increase lift or speed, to increase thickness or aspect ratio, or to decrease sweep. Because of the emphasis on fuel efficiency, the current trends for commercial transports are to use the

TRANSPORTS—DESIGN FOR CRUISE PERFORMANCE

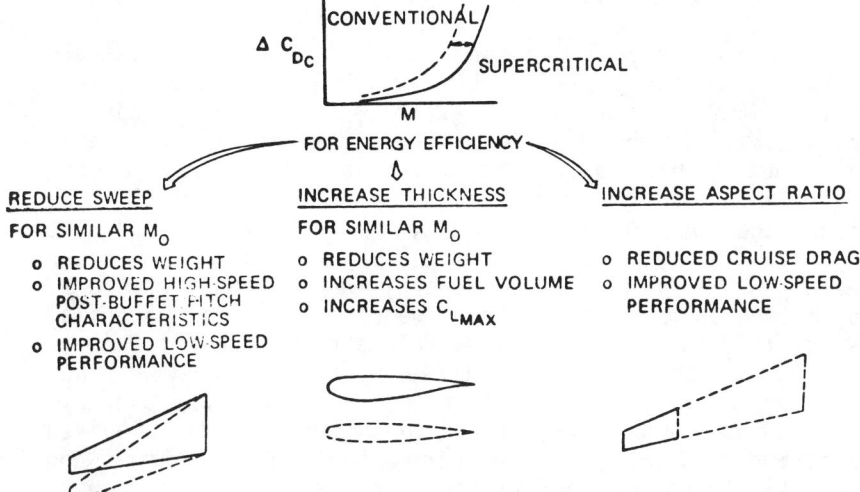

Fig. 25 Application of supercritical wing technology to new wing design.

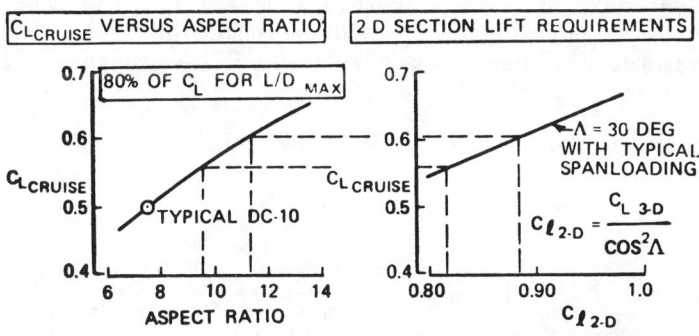

Fig. 26 Effect of aspect ratio on required cruise lift coefficient.

supercritical wing's benefit to reduce sweep and increase thickness, both of which reduce weight, and then to use this weight reduction and improved structural efficiency to increase aspect ratio which, in turn, requires the higher cruise lift capability of the supercritical airfoil-wing. Some of the advantages realized from this combination are illustrated in Fig. 25, the most influential of which is the drag reduction resulting from the increased aspect ratio. The higher aspect ratio wing makes use of (requires) the increased lifting capability of the supercritical airfoil-wing since the desired (most efficient) cruise lift coefficient increases directly with aspect ratio (see Fig. 26).

Once the general planform, airfoil types, approximate average thickness, design lift coefficient, and desired cruise Mach number are established from a design optimization study, the aerodynamic designer's job starts in earnest. His goal is to design a minimum weight wing consistent with achieving as much of the airfoil two-dimensional potential as possible, minimizing any parasite or induced drag penalties and satisfying all of the other design criteria listed in General Design Considerations. To achieve as much of the two-dimensional airfoil potential as possible typically requires maintaining straight isobars parallel to the wing sweep (keeping the shock swept) and avoiding any premature separations or excessive boundary-layer thickening. Many iterations are usually required involving tailoring of the wing planform, chordwise and spanwise airfoil characteristics such as thickness, camber, twist, etc. This is the process which, in the past, involved much cut-and-try wind-tunnel testing with often less-than-superb final designs (see Fig. 15). This procedure has now been revolutionized by the current transonic computational methods, including the inverse (design) methods now commonly used.[12] A representative

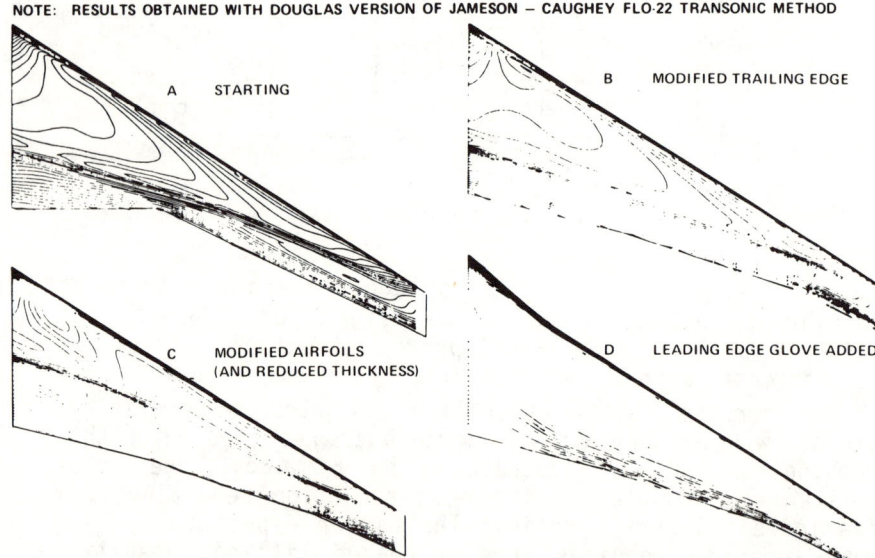

Fig. 27 Predicted upper-surface isobars for several configuration modifications.

example of how these new methods were used to modify a high aspect ratio supercritical wing design that wind-tunnel test results showed was approximately 0.02 deficient in drag divergence Mach number to a design that achieved its full potential performance is illustrated in Fig. 27.[13] The initial configuration was deficient due to a premature separation at the trailing edge break station caused by an aft shock location and the attendant steep adverse pressure gradient from the shock to the trailing edge. In a relatively short period of time, 40 different wing configurations involving 100 flow solutions with the Douglas version of the Jameson-Caughey FLO-22 program were studied. The effects of planform, airfoil type, thickness, and twist variations were systematically determined. It was found that in order to suppress the development of the strong aft shock over the inboard part of the wing, careful tailoring of thickness, planform trailing edge sweep, and airfoils was required. For example, an increase in chord at the trailing edge break station was fundamental to effecting an improvement in the three-dimensional flow characteristics over the inboard panel. A somewhat lower thickness ratio in the inboard region was also determined to be essential. The use of an inboard leading edge extension gave a significant improvement through elimination of the inboard shock by a reduction in section lift coefficient and an increase in isobar sweep. Other effects investigated included variations in airfoil leading edge and aft-camber geometries to determine the effects on predicted drag creep and drag divergence Mach number. The final result was a wing design that, when wind-tunnel tested, achieved the full potential drag divergence Mach number even though the resulting shock patterns and detailed pressure distributions were a little different than predicted. As a result of successes such as this, these advanced computational techniques used in conjunction with wind-tunnel development and verification testing form the basis for current design efforts, and make it possible to refine wing shapes to an extent that would be impracticable by even the most skillful application of a cut-and-try experimental approach.

Some Related Computational and Experimental Problems

Although current computational capabilities are very much improved over what they were when the last of the currently operational commercial transports were designed, and likewise, experimental techniques have been improved, many problems are still encountered by the aerodynamic designer that constantly point out the need for improved experimental capabilities. Some of these problems that are faced by the wing aerodynamic designer are discussed in the following sections, grouped by the performance parameter most affected. The problems discussed are by no means intended to be all inclusive, just representative.

   Parasite Drag. Aerodynamic designers have long been plagued with the uncertainty as to whether test results obtained at typical wind-tunnel Reynolds numbers will accurately represent or predict full-scale aircraft characteristics. While there are admittedly countless examples of excellent agreement between wind-tunnel and flight results, particularly as the wind-tunnel Reynolds number is increased, history has also provided numerous examples of transport-type aircraft configurations tested in conventional transonic wind-tunnel facilities where conditions at flight Reynolds number have either been more adverse or less adverse than those observed in the wind tunnel. The major concern of the aerodynamic designer (and his management) is the case where flight results are more adverse than those observed in the wind tunnel, since this situation, if the flight results are not acceptable or competitive, can lead to the need to make very costly hardware modifications. Although it did not involve production hardware, a situation encountered that has potentially very broad implications involved a drooped outboard aileron configuration on the DC-10 that was being investigated as a means of providing some reduction in drag through the use of a moderate amount of aft loading and by a reduction in the lift-dependent drag through an improvement in the spanload distribution. Wind-tunnel test results at a mean aerodynamic chord (MAC) Reynolds number of 6 million had indicated that nearly a 2% drag reduction was achievable with this concept along with a 1/3 deg reduction in aircraft pitch attitude. When this configuration was flight tested, the result, as seen from Fig. 28, compared to the wind-tunnel results, was a drag increase and no change in airplane pitch attitude. Tuft observations revealed that the differences were due to a flow separation that existed on the aileron upper surface in flight that had not occurred in the wind tunnel. Although not confirmed by surface pressure measurements, it is rationalized that the pressure peak on the upper surface at the rotation point of the drooped aileron would be increased and the resultant imposed aft gradient steepened at inflight conditions due to the

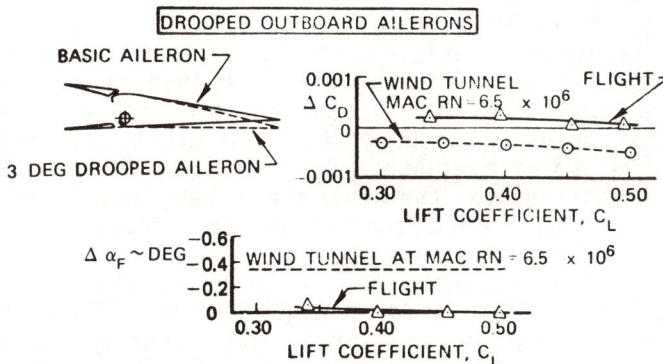

Fig. 28 DC-10 flight versus wind tunnel discrepancy.

nondimensionally thinner boundary-layer displacement effect.

Two ways to avoid this type of occurrence would be to either test the configuration up to full-scale Reynolds numbers or have a computational method that could properly analyze this type of situation as a function of Reynolds number. Until such time as the National Transonic Facility (NTF) is completed, debugged, and operational (a year or two from now), no experimental solution seems possible. Current transonic computational methods are incapable of accurately assessing this situation due to inadequacies of the transonic potential flow methods, particularly in the trailing edge region, and due to the lack of a good three-dimensional inviscid-viscous interaction model for this region. The three-dimensional aspect, due to the spanwise flow component near the trailing edge and hence greater susceptibility to boundary-layer separation, will most likely be predominant for this type of problem and others involving separation near the trailing edge of a swept wing.

The thought of wind-tunnel testing at higher Reynolds numbers in order to avoid discrepancies such as those illustrated by the DC-10 drooped aileron example is very appealing to the designer, but accomplishing it (once the NTF is operational) is not as straightforward and simple as it might appear. One problem facing the aerodynamic designer is the question of boundary-layer transition location at Reynolds numbers higher than currently attainable, but lower than full scale. This is a very practical concern due to the anticipated limited availability of the NTF to any one manufacturer during a commercial transport development program. Consequently, most of the development testing will continue to be done in existing transonic wind tunnels, and it will be necessary to correlate results obtained in the NTF with the results obtained in these other tunnels. To do this will require testing over the Reynolds number range between what is obtained in current tunnels and full-scale values. The difficulties and complexities that will be involved in accomplishing this appear to be much greater than presently encountered in today's wind tunnels. The added difficulties are due to the very small roughness sizes involved, cross flow stability effects, and heat-transfer effects on transition Reynolds number.

With regard to the small roughness sizes, two areas of concern are the implications of the very small critical roughness heights that can influence the transition location and the small size of roughness elements required to fix transition at a predetermined location. Critical roughness sizes are very small, i.e., 10^{-4} to 2×10^{-4} in., at unit Reynolds numbers of the order of 60×10^6/ft, and could lead to several difficulties such as:

1) Small model imperfections or roughness that might be caused by an abrasive material in the air could cause the location of transition to be very irregular with time on any model.

2) The existence of rows of static pressure orifices could lead to irregular transition patterns across the span of a wing.

3) Typical sublimation techniques used for detecting transition in conventional wind tunnels would not be acceptable for use in the cryogenic tunnel (even if compatible with the low temperatures) since the size of the pigment in the fluid would influence the location of transition. Consequently, new methods must be developed for detecting transition location on models in the cryogenic tunnel.

Regarding the second concern, a straightforward application of Braslow's criteria[14,15] for sizing transition tripping devices on wind-tunnel models (see Fig. 29) shows that the minimum roughness height to fix transition at the tripping device at the somewhat less

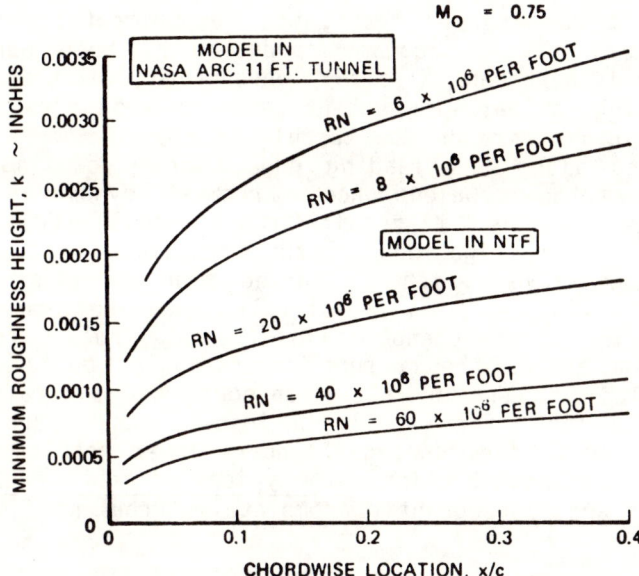

Fig. 29 Minimum roughness heights to fix transition.

than full-scale Reynolds numbers of interest in the NTF is much smaller than the corresponding value appropriate for use on models in today's conventional transonic wind tunnels. Whereas the typical roughness height used regularly today is about 0.003 in., the minimum roughness height to be used on models in the NTF at a Reynolds number 60×10^6/ft is approximately 0.0005 in. The requirement to install and maintain a 0.0005 in. transition fixing device on a wing in a repeatable manner is expected to be difficult from a handling, application, and inspection standpoint. It is anticipated that a considerable effort will be required to develop an acceptable technique.

Cross flow stability effects are very important when testing at Reynolds numbers between those obtained in current tunnels and the full-scale values. During flight tests conducted nearly 30 yrs ago at the Royal Aircraft Establishment[16,17] it was discovered that at certain conditions the boundary-layer transition location on a swept wing, initially located well aft of the leading edge, moves forward very rapidly to the leading edge once a critical Reynolds number is exceeded. This behavior was initially attributed to the instability of sweep-induced cross flows along the leading edge attachment line.[18]

However, it was learned later, when experimental laminar flow control flight programs conducted by both Handley Page[19] and Northrop Norair[20] encountered difficulties in achieving laminar flow except near the wing tips, that the limitations were due to turbulence from the wing root propagating along the leading edge, thereby causing the whole flow to be turbulent once a critical Reynolds number was exceeded. As a result of numerous additional experimental investigations of the transition process of the boundary layer which is formed along a swept wing attachment line,[21-23] a relatively good semiempirical understanding of the phenomena is in hand. Poll's recent paper[23] summarizes the current understanding quite well, and concludes that for all transport-type aircraft in service, except for small business jets, turbulent contamination by the fuselage boundary layer results in a turbulent attachment line <u>at typical flight cruising conditions</u> and, consequently, completely turbulent flow on the wing.

The parameter most frequently used to define the critical leading edge Reynolds number is the Reynolds number of the boundary layer along the attachment line based on the momentum thickness (R_θ). It has been shown that there are two different boundaries, the lower ($100 < R_\theta < 120$) being applicable when there is contamination by the fuselage boundary layer, and the upper ($240 < R_\theta < 280$) corresponding to conditions with natural transition on a <u>smooth</u> leading edge. At values of R_θ above these limits the attachment line is fully turbulent. Poll showed that relaminarization in the favorable pressure gradient between the attachment line and the point of peak suction pressure is unlikely at flight Reynolds numbers. On the other hand, there is some evidence that relaminarization can occur on wind-tunnel models when R_θ is somewhat above the lower limit. Below the lower limit, the flow will stay laminar or revert to laminar flow even if turbulent contamination occurs at the wing-fuselage juncture. Transition to turbulent flow then occurs further aft on the wing. Relative to the upper limit, it is possible to have a laminar attachment line and a subsequent chordwise region of laminar flow up to this limit if the fuselage boundary-layer disturbances are remote or eliminated, and the leading edge is maintained in a smooth and clean condition. However, the actual limit for a swept three-dimensional tapered wing in a transonic wind tunnel will be lower due to the influence of tunnel noise and freestream turbulence, and will also vary spanwise according to the proximity to the fuselage. Heat-transfer effects would also influence this critical Reynolds number at which the transition location will suddenly move forward to the leading edge of the wing.

To illustrate the differences in attachment line conditions encountered under different testing conditions, starting with those of a current transonic tunnel, then at the higher Reynolds numbers to be encountered in the NTF, and finally at full-scale flight conditions, calculations were made for the DC-10 wing configuration at representative cruising conditions. The variation in R_θ with test Reynolds number is shown in Fig. 30 for four spanwise stations on the wing, starting at the root. It can be seen that, at the highest Reynolds number attainable in the NASA ARC 11 ft tunnel, only the area of the wing close to the root would be expected to have a turbulent attachment line. This prediction is in agreement with the observed location of natural transition at these conditions which was right at the leading edge at the wing root, but well aft of the leading edge over the rest of the span. Looking next at the attachment line conditions at

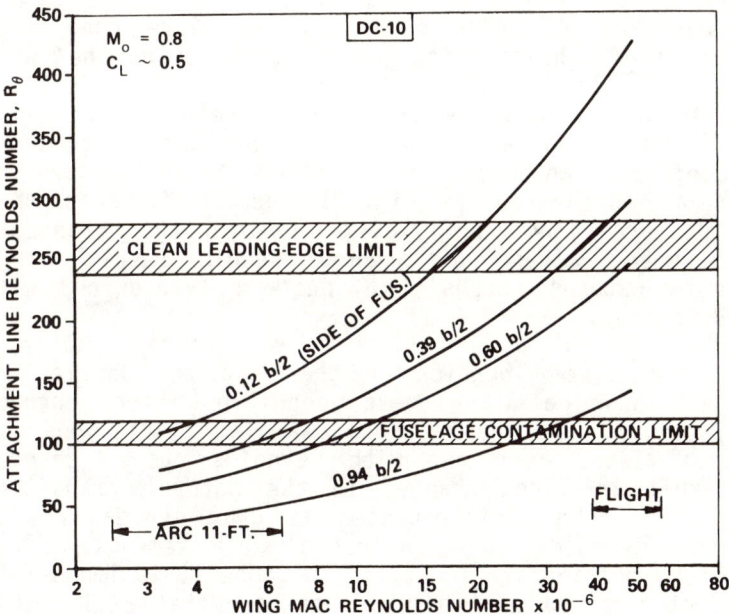

Fig. 30 Variation of DC-10 attachment line parameter with Reynolds number.

flight Reynolds numbers, most of the wing, except the area near the tip, exceeds the clean leading edge limit and hence a turbulent attachment line is a foregone conclusion in these areas. For the area near the tip, although below the clean leading edge limit, it is expected that the combination of surface roughness (including bugs) and the influence of bursts of turbulence from adjacent parts of the wings would also result in a turbulent attachment line condition with no chance for relaminarization.

Unfortunately, interpretation of the consequences of attachment line conditions at Reynolds numbers less than flight Reynolds numbers is not nearly as obvious as either the flight or the lower Reynolds number wind-tunnel conditions. One thing that does appear certain, however, is that the natural transition pattern resulting from the attachment line conditions illustrated in Fig. 30 will vary significantly over the test Reynolds number range with the spanwise extent of the turbulent attachment line increasing as the Reynolds number is increased. The exact variation in the natural transition pattern will be influenced by many factors including the model surface condition, tunnel noise, freestream turbulence characteristics, and heat transfer. Many of these may vary as tunnel conditions (Mach number, Reynolds number, etc.) are changed. Also the attachment line conditions will vary from configuration to configuration as the sweep and leading edge geometry are changed. To illustrate typical differences that would result, the attachment line Reynolds number variation was estimated for an advanced wing configuration with quite different airfoils. The results are presented in Fig. 31 and it is seen that conditions are significantly different over the outer part of the wing. These differences should result in a different natural transition pattern variation with Reynolds number.

The transition location on the wing, in particular, must be known at all test conditions for correct interpretation of data. For example, variations in attachment line conditions with Reynolds number changes that result in large changes in the natural transition location would be misinterpreted as drag increases with increasing Reynolds number unless these movements in the transition location are identified. Since it is impossible to predict the natural transition pattern that will result on configurations tested in the NTF cryogenic tunnel, since sublimation techniques will not be acceptable for

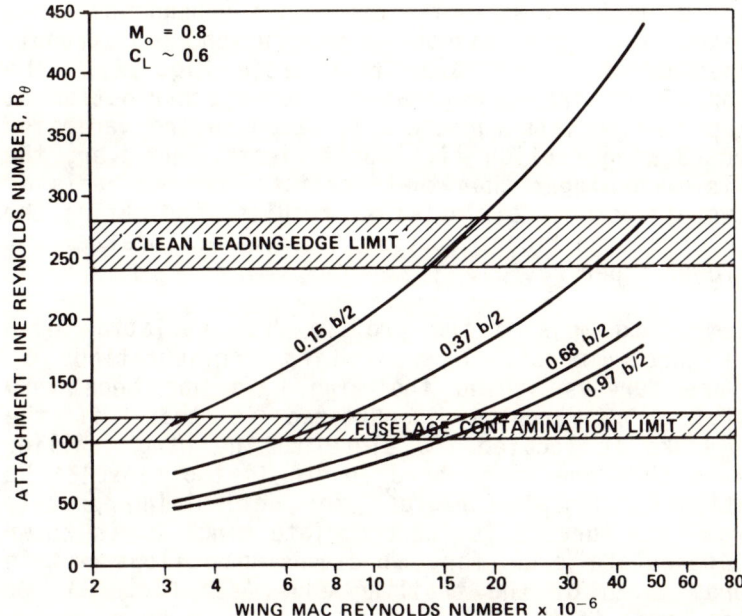

Fig. 31 Attachment line parameter for advanced wing configuration.

detecting it, and since artificial fixing of the transition pattern is neither practical nor even feasible at all test conditions, instrumentation must be developed and incorporated to permit the determination of the transition pattern across the span of the wing without interfering with the flow itself. These measurements must be obtained at the same time as the usual force and pressure measurements in order to avoid discrepancies due to changes in model surface condition, heat transfer, and tunnel operating conditions. One very promising approach to satisfy this requirement is the unique configuration of vacuum deposited hot film gages being developed by Fancher.[24]

Lift-Dependent Drag. While vortex-drag penalties due to nonelliptic span-load distributions can be reliably predicted using current lifting surface programs, the viscous drag increase with increasing lift, which comes primarily from the wing, remains difficult to compute or to measure in the wind tunnel. This component needs to be assessed in order to properly evaluate airfoil-type differences and effects of wing taper ratio changes. For example, a correlation of the full-scale airplane-deduced viscous contributions to the lift-dependent drag illustrates that the viscous contribution is strongly influenced by the wing taper ratio (see Fig. 32). The direction of the trend, a greater viscous contribution at lower taper ratios, is fundamental, based on the variation in outboard wing section lift coefficients. However, the effect is much bigger than that predicted on the basis of using two-dimensional airfoil test results, indicating the presence of a significant adverse three-dimensional effect at the lower taper ratios.

Attempts to measure the profile drag variation with lift on three-dimensional swept wing configurations by using wake surveys behind the wing have not been very successful. One problem encountered is that, if the traverse rake is located too close to the wing trailing edge (less than one wing chord length is too close), the application of Jones' formula[25] for determining profile drag from wake surveys is inappropriate since it is known to be inaccurate even for two-dimensional flows within one chord length of the trailing edge (see Fig. 33). On the other hand, if the traverse rake is located far enough aft of the trailing edge to avoid this problem, then it is very difficult to segregate the drag sources, particularly for the outer part of the wing near the wing-tip vortex. Another question to be answered regarding the accuracy of wake surveys behind swept wings has to do with identifying the inaccuracy due to the sizeable (and varying) spanwise flow components that exist at the traverse stations. Measurements made by Michel et al.[26] behind a swept wing show that spanwise flow angles up to 10 deg are not uncommon. These spanwise flow angles in the near wake vary across the span due to planform and airfoil effects

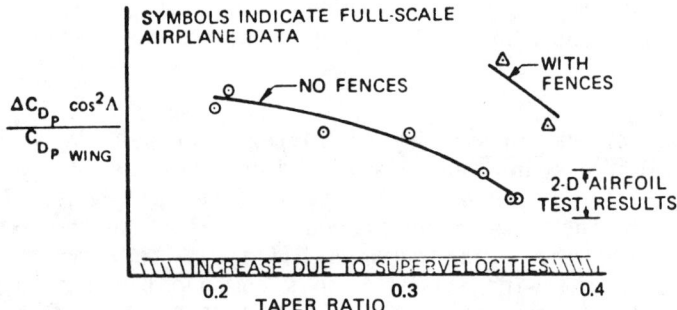

Fig. 32 Effect of taper ratio on viscous contribution to lift-dependent drag.

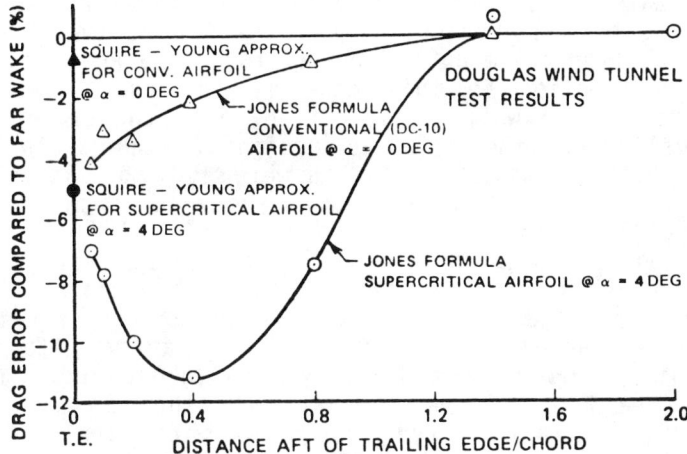

Fig. 33 Errors associated with Jones formula and Squire-Young approximation.

as well as due to any differences in the position of the traverse rake relative to the wing trailing edge. They also vary with changes in lifting conditions. If wake surveys behind swept wings are to be used, then these three-dimensional effects must be accounted for if the results are to be used for anything other than qualitative assessments of areas of high drag.

Presently available computational methods offer little help in predicting this elusive viscous contribution to the lift-dependent drag for practical three-dimensional swept wing configurations. It is probably even optimistic to expect that the two-dimensional airfoil characteristics would be correctly predicted by available computational methods since a reliable viscous-inviscid interaction model for the trailing edge to wake region is essential. As illustrated in Fig. 33, use of the Squire-Young approximation, even when starting with the boundary-layer flow properties measured at the trailing edge station, can lead to errors of 5% in the calculated drag relative to the far wake derived value. However, even if the two-dimensional airfoil characteristics could be accurately calculated, that would not be sufficient since the problem of interest clearly has significant three-dimensional flow effects. What is needed is a combined viscous-inviscid method for swept three-dimensional wings including an appropriate three-dimensional interaction model for the trailing edge to wake region.

Uncertainties are also encountered in the wind-tunnel measurement of lift-dependent drag using complete three-dimensional models mounted on a force balance. The unresolved question in this case is in regard to the appropriate tunnel wall configuration, both the extent of the porosity and the detailed design of the slots or holes. The results of a systematic study of porosity effects was undertaken by Douglas using a DC-10 model in the Rockwell International 7 ft transonic wind tunnel. The wall porosity (normal holes) was varied by taping the walls to reduce the porosity below the nominal 19% configuration. The model was tested upright and inverted in order to account for any flow angularity effects. Appropriate sting support tares were also applied. The results (see Fig. 34) showed that even the 19% maximum wall porosity was not open enough to match the flight results in terms of induced-drag efficiency factor. In fact, considering that the wind-tunnel measured efficiency factor should probably be lower than the flight value due to Reynolds number effects, the discrepancy is even greater. These results help to point out that a much better understanding of transonic tunnel wall effects is needed before wind-tunnel measured induced-drag characteristics can be used with any confidence.

Compressibility Drag. The efficient cruise condition for most current commercial transports occurs when the wing compressibility drag is between 20 and 25 drag counts on a two-dimensional basis, i.e., $\Delta C_{D_c}/\cos^3\Lambda$ (see Fig. 14). corresponds to approximately 13 to 16 drag counts on a 30 deg swept wing or even less for higher sweep angles. In order to describe the flow conditions that would typically exist at this condition, two-dimensional airfoil compressibility drag characteristics measured at relatively high Reynolds numbers for several different airfoil types were correlated as a function of the local Mach number just ahead of the shock on the airfoil upper surface. The correlation for a full supercritical airfoil is shown in Fig. 35. It can be seen that at the typical cruise condition, the local Mach number just ahead of the shock is about 1.22 to 1.23, not high enough for the shock wave of itself to cause any boundary-layer separation at meaningful Reynolds numbers. Also shown in Fig. 35 is the wave drag of the shock as derived from the measured wake profiles for this configuration by Elfstrom's method.[27] The wave drag is predominant in determining the drag rise shape near the divergence Mach number, whereas the viscous drag (total minus wave) controls the amount of drag rise

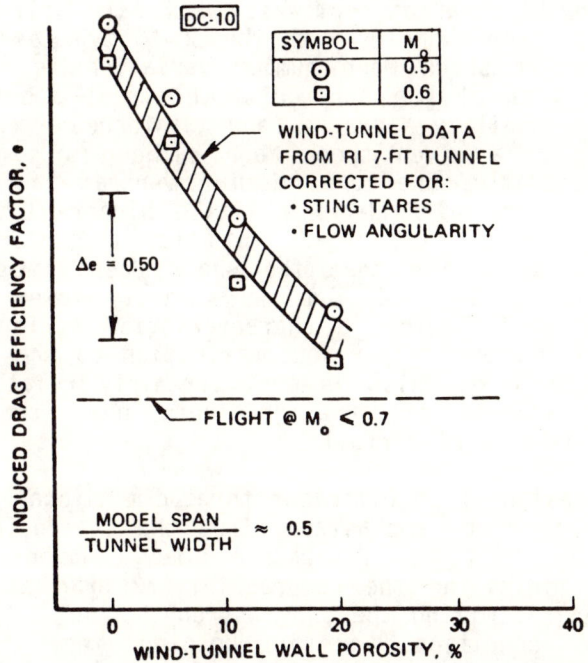

Fig. 34 Wind-tunnel wall porosity effects on induced drag.

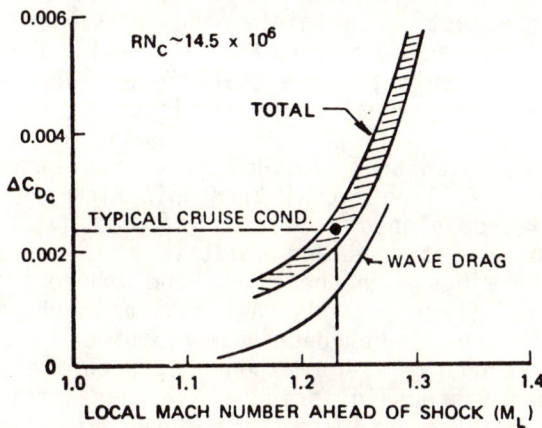

Fig. 35 Supercritical airfoil compressibility drag.

up to this point. This situation should hold true for any well-designed configuration that is void of premature boundary-layer separation. At the design point for this supercritical airfoil, the compressibility drag due to viscous effects and that due to wave drag are nearly equal.

This relationship varies somewhat with the different airfoil types. For example, with the DC-10 type airfoil, the wave drag vs shock Mach number variation is nearly identical to that of the supercritical airfoil, but the total compressibility drag (and inferred viscous contribution) at a given shock Mach number is somewhat lower. Consequently, the local Mach number at the shock for the DC-10 type airfoil is a little higher (0.01 to 0.02) at the cruise condition. This is in conformance with the popular notion that the shock Mach number at representative cruise conditions is slightly lower on a supercritical airfoil-wing, a characteristic attributable to the higher viscous drag rise contribution to the total compressibility drag, which is most certainly related to the steeper adverse pressure gradients over the aft portion of this type of airfoil.

For the design of an efficient three-dimensional swept wing, the areas to be emphasized, from the standpoint of attaining the desired divergence Mach number and concurrently minimizing the compressibility drag at this point, are in minimizing the shock strength over the wing at a given condition and in keeping the viscous contribution due to continuously increasing adverse pressure gradients under control (no excessive boundary-layer thickening or separation). This latter area is where the greatest uncertainty and risk is involved, considering current computational and experimental capabilities available for the analysis and development of advanced three-dimensional configurations.

Current computational design procedures based on the Jameson-Caughey full potential transonic flow method, such as the Douglas-developed version,[12,28] are quite good in predicting shock patterns and relative shock strengths on transport-type wings. On the other hand, there are several shortcomings of these methods that must be overcome before these methods can be productively combined with three-dimensional boundary-layer methods to accurately predict the combined viscous-inviscid flow characteristics of three-dimensional swept wings at transonic conditions. While some very useful results related to separation prediction on three-dimensional wings have been obtained by combining these potential flow and three-dimensional boundary-layer programs in their current state, many obstacles must be overcome before meaningful <u>drag calculations</u> or less restrictive applications of separation prediction can be performed. Some of the shortcomings of the current full potential transonic methods that must be addressed lie in the fuselage representation, the shock-wave representation, and the mesh spacing near the trailing edge.

The problem with the fuselage representation, apart from the need to more accurately account for the effect of a finite-length fuselage on the overall wing flowfield to replace some of the current approximate corrections being used in the various design methods, is that the prediction of unrealistically severe adverse pressure gradients over the inner rearward part of the wing (see Fig. 36) must be corrected in order to prevent prediction of excessively high drag and premature separation.

With regard to the shock-wave representation, the problem is that the Jameson-Caughey inviscid representation of the shock-wave pressure gradient is smeared over a much greater chordwise distance than actually occurs. The shock jump, from the point of maximum Mach number to a point where the flow is slightly subcritical, is spread over a distance (several computational mesh widths) that is well

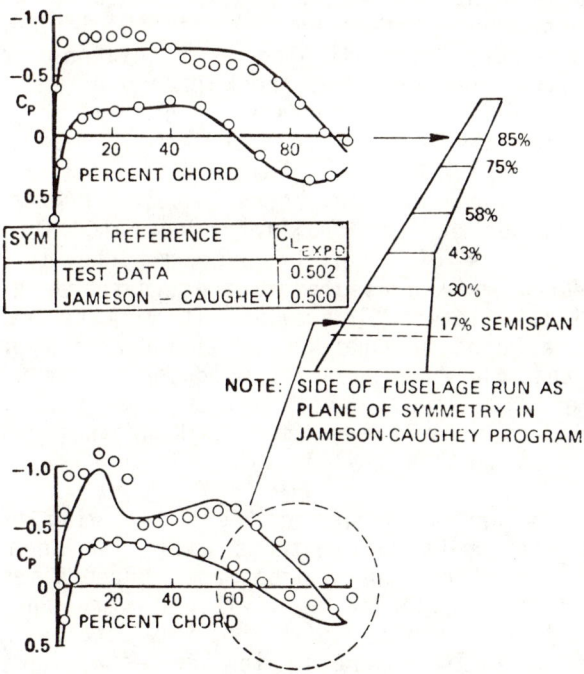

Fig. 36 Calculated versus experimental wing pressure distributions.

in excess of 10% of the local wing chord in many cases. Depending on the spanwise location, Reynolds number, and shock location, this distance over which the shock jump is spread can easily be 20 or more local boundary-layer thicknesses. This is in contrast to a distance of the order of 5 or 6 boundary-layer thicknesses that has been shown as being required to match experimental results at moderate shock strengths. To illustrate the effect that this nonrepresentative smearing has on the calculated boundary-layer characteristics, the skin friction and displacement thickness characteristics predicted with the two-dimensional version of the Douglas CKR method[29] are shown in Fig. 37 for the upper surface of an aft-loaded airfoil configuration at a condition that was approaching trailing edge separation experimentally. The shock Mach number of 1.27 is close to that for which a separation bubble should occur at the foot of the shock. The pressure distributions used are experimental measurements except right in the region of the shock where they are modified to permit a variation in the smearing effect. With the shock smeared over 20 boundary-layer thicknesses, the predicted skin-friction characteristics at the end of the shock pressure gradient show no tendency toward separation. At 6 boundary-layer thicknesses, the predicted boundary layer is just about separated, as it should be. The calculated boundary-layer displacement thickness at the airfoil trailing edge when the shock pressure gradient is spread over 6 boundary thicknesses is nearly 50% greater that it is when the pressure gradient is spread over 20 boundary-layer thicknesses. This difference in the calculated displacement thickness would probably be magnified even more if the trailing edge calculation were handled correctly. In order to overcome the difficulty with the smeared shock pressure gradient in the development of an appropriate interaction model, it will be necessary to investigate means to sharpen up the inviscid pressure gradient so that the true smearing produced by the boundary layer can be calculated. A capability analogous to that incorporated in the two-dimensional TAIR program developed by Holst at NASA Ames is needed.

The details and mesh spacing near the wing trailing edge in the Jameson-Caughey methods will also need to be improved upon before an appropriate three-dimensional inviscid-viscous interaction model for this region can be developed. It has been shown[13] that the details in this region are extremely important in the establishment of

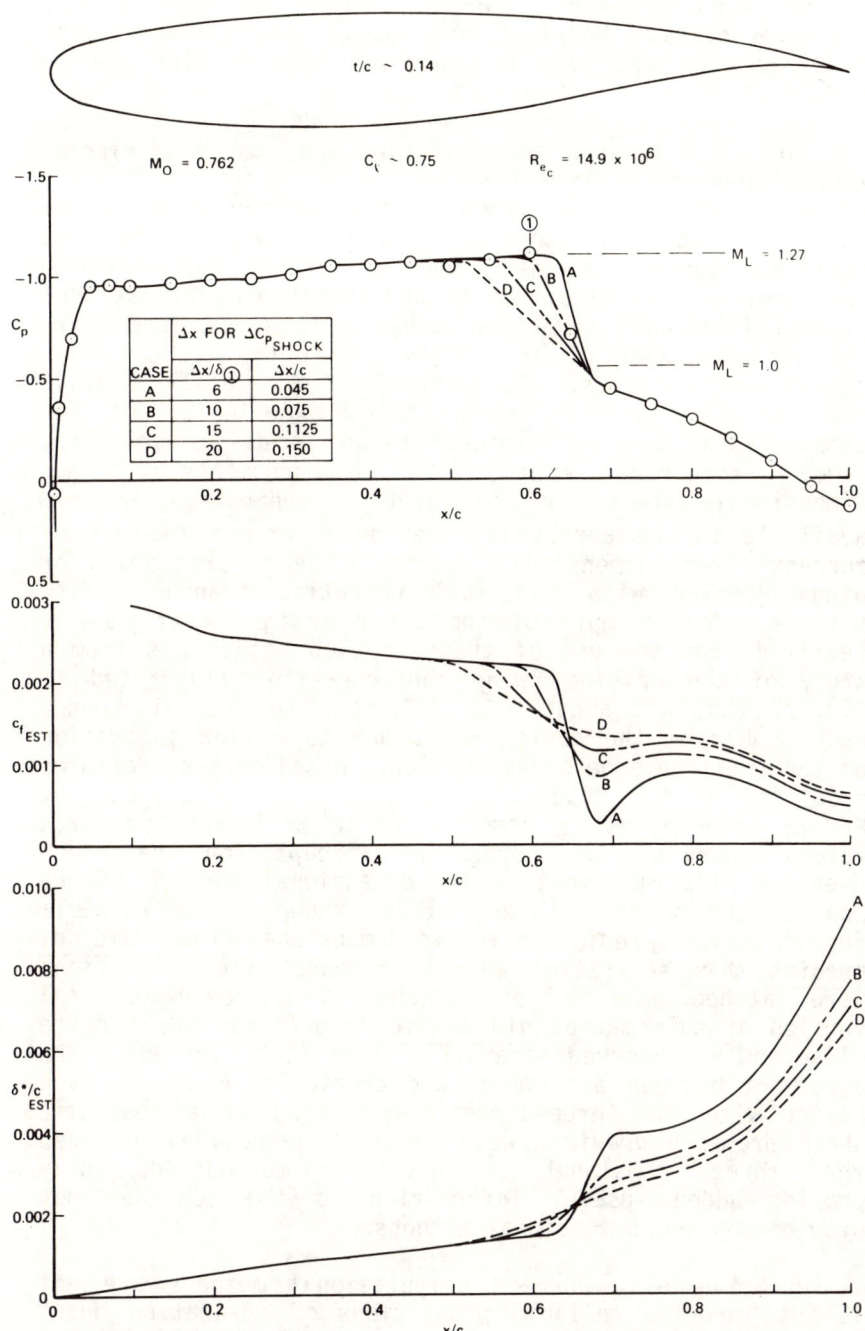

Fig. 37 Effect of shock pressure-gradient smearing on calculated boundary-layer characteristics.

boundary-layer growth (drag) and separation onset characteristics. Melnik[30] has shown that these details are very important even in the prediction of airfoil-wing lift.

Turning now to the viscous and inviscid-viscous interaction portions of the computational capability needed to predict the viscous drag and separation onset characteristics of three-dimensional swept wings at transonic conditions, many improvements in the boundary-layer methods, particularly in terms of the development of inviscid-viscous interaction models, are still needed for the ultimate goal of accurately predicting viscous drag rise characteristics. Realistically, this goal is still several years away considering the complexity of the computational and experimental tasks to be accomplished. However, some very useful qualitative assessments of three-dimensional effects on the onset of separation are now available to the aerodynamic designer through the use of current three-dimensional boundary-layer programs for wings coupled with the full potential transonic flow methods. One example of the added insight that can be realized from the use of these coupled methods is from a study of the original wing configuration illustrated in Fig. 27 that was about 0.02 deficient in drag-divergence Mach number at the design point due to a flow separation at the trailing edge break station. Based on the pressures predicted by the Douglas version of the Jameson-Caughey FLO-22 program, which compared quite well with measured values ahead of the separation (except for the shock pressure rise smearing), a two-dimensional version of the Douglas CKR boundary-layer method, known to be reliable for separation prediction in two-dimensional flows, <u>did not predict</u> this separation when wind-tunnel flow conditions were matched. On the other hand, the three-dimensional version of this method <u>did predict</u> separation onset quite close to the observed separation point (see Fig. 38). The agreement between experiment and prediction would be even better with the three-dimensional method if a realistic shock pressure gradient were imposed. Hence, it is seen that three-dimensional boundary-layer methods do today provide added, useful information to the designer not provided by two-dimensional methods.

Unfortunately, current computational methods are not able to provide reliable predictions of separation onset for marginal trailing edge conditions or quantitative predictions of drag due to viscous effects on three-

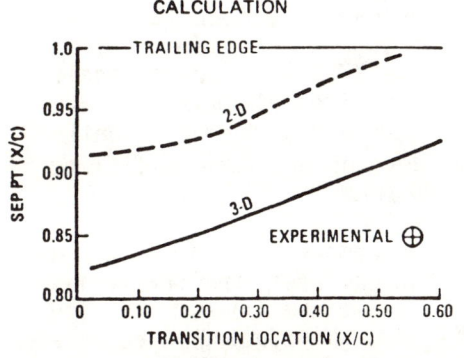

Fig. 38 Calculated versus experimental separation location.

dimensional swept wings at transonic conditions, which is really what the designer needs in the development of advanced wing configurations where trailing edge separation due to three-dimensional effects at flight Reynolds numbers could be a concern. Furthermore, examples such as the DC-10 drooped aileron wind tunnel vs flight comparison (Fig. 28) cause the designer to be concerned with the reliability of wind-tunnel results obtained in today's conventional transonic wind tunnels for configurations prone to trailing edge separation. If a configuration is not prone to separation, the Reynolds number effects are not important. Likewise, if a configuration is very poor at today's wind-tunnel Reynolds numbers, then chances are it will not be dramatically improved at flight Reynolds numbers. However, if a configuration is marginal in terms of trailing edge separation, then test results at higher Reynolds numbers are very important since experience has shown that Reynolds number effects can be either favorable or unfavorable in going to flight conditions, depending on the particular flow situation. The NTF, when operational, will provide the vitally needed experimental capability to investigate these Reynolds number effects and hence greatly reduce the risk involved in developing more efficient commercial transport wing configurations.

Before leaving the subject of compressibility drag, a quick look at the subject of "shock-free" airfoils-wings is appropriate. The <u>concept</u> is attractive from the standpoint of eliminating or greatly reducing the wave drag due to the wing shock (Fig. 35) by eliminating the shock or making it acceptably weak. In theory, the improvement in aerodynamic efficiency could be used to reduce weight by decreasing the wing sweep or by increasing the thickness of the wing. There are (at least) two very practical problems that loom as major obstacles to ever incorporating this concept on a commercial transport. First, the commercial transport is not a point design configuration. Due to the assignment of cruise altitudes and the limited number of step altitude changes possible during a given flight, the cruise lift coefficient throughout a flight typically varies by as much as ±0.1. Secondly, many wing designs which analysis codes have predicted to be shock free at certain transonic conditions have, in fact, not been shock free when wind-tunnel tested. Obviously, viscous and inviscid-viscous interaction effects have a large influence on the success of this concept.

Until such time as these three-dimensional viscous effects can be properly accounted for, an enormous risk in the form of a built-in performance penalty if shock-free conditions do not materialize will be associated with trying to take advantage of this concept to reduce wing sweep or increase wing thickness.

<u>Buffet Boundary</u>. Buffeting is defined as the structural response to the aerodynamic excitation produced by separated flows. For a commercial transport, the normal encounter with buffeting at transonic cruise conditions occurs when the airplane encounters a strong gust which increases the effective angle of attack to the point where an upper surface flow separation occurs, typically somewhere along the wing trailing edge. The buffet boundary is usually defined as the first appearance of a "significant" area of separated flow and is expressed in terms of a lift coefficient vs Mach number boundary.

This boundary can at times be nearly as important a performance parameter as drag is since the maximum cruising lift coefficient is limited by the requirement to maintain a 1.3 g margin-to-buffet onset. As illustrated in Fig. 26, the desired cruise lift coefficient for maximum efficiency increases with increasing wing aspect ratio. If the buffet boundary for a particular design is not high enough to maintain the 1.3 g margin over the desired cruise lift coefficient, then the full potential of the wing aspect ratio cannot be utilized. Furthermore, the performance loss due to an inadequate buffet boundary can be disproportionately higher than just the buffet boundary deficit due to the relatively large increments in assigned cruise altitudes for a given flight direction and the difficulties sometimes encountered in receiving clearance to change altitudes.

The onset of buffeting in flight is often difficult to quantify when pilot impressions are involved, because these can be misleading if the pilot sits at or close to a node of the predominant modes being excited.[31] To better quantify airplane buffet onset, normal acceleration traces are often obtained at both the captain's seat and the airplane center of gravity. Typically, at low normal accelerations (less than ±0.1 g) there are some differences between the two locations, but for normal accelerations of about ±0.1 g, which has become an accepted buffet boundary definition, the buffet intensity at the pilot's seat usually coincides with that measured at the center of gravity. A representative variation in measured buffet intensities is illustrated in Fig. 39. It can be seen that there is a 10% difference in buffet onset lift coefficient between a normal acceleration of ±0.03 g (which is perceivable if you're looking for it) and the somewhat standard buffet boundary definition of ±0.1 g. Clearly there is flow separation on the wing when the normal acceleration is ±0.03 g. The problem then is how do you determine the separation intensity, etc. necessary for a ±0.1 g normal acceleration to be felt at either the captain's seat or at the airplane center of gravity.

Although the onset of separation on three-dimensional wings can sometimes be predicted by the combination of transonic potential flow methods and three-dimensional boundary-layer methods, these predictions will not be reliable enough until proper inviscid-viscous interaction models are developed and verified. However, even if

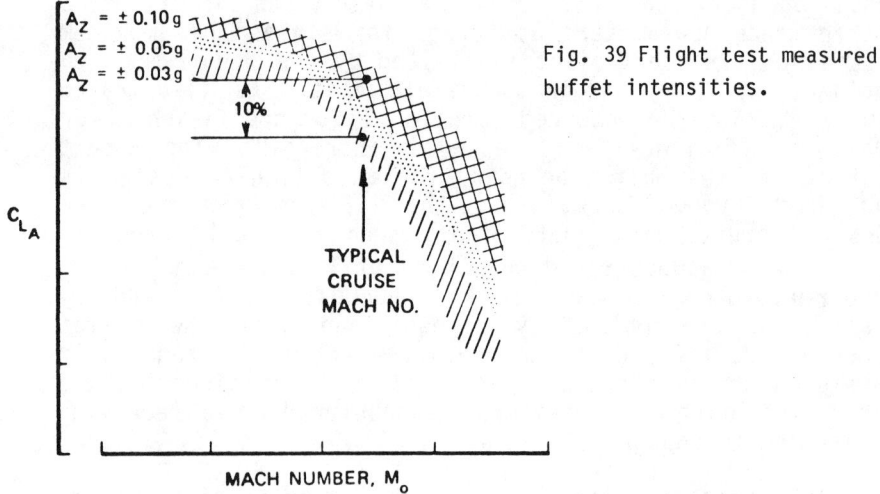

Fig. 39 Flight test measured buffet intensities.

separation onset could be accurately predicted today, there are no means available to predict the aerodynamic excitation after separation.[31] Consequently, the prediction of the onset and intensity of airplane buffeting will continue to be based on model tests in wind tunnels with all of the attendant limitations. Probably the biggest limitation today is the concern over inadequate Reynolds number simulation, since it is well known that separation onset is the aerodynamic parameter most susceptible to sometimes unpredictable Reynolds number effects. Fortunately, this concern will be largely eliminated once the NTF is available in a few years. The other limitations that need to be addressed are the uncertainties involved in wind-tunnel data interpretation. The methods that have customarily been used in the past to define buffet onset are based on either defining the break in the lift curves or trailing edge pressure divergence. While these methods have often yielded good predictions of flight buffet onset, there are some exceptions. It has been shown, for example, that on a three-dimensional swept wing the initial onset of separation and loss of lift on one area of the wing may be associated with a compensating increase in lift on another area of the wing, so that there may be no break in the lift curves at buffet onset.[31] If buffet onset is due to other than a trailing edge separation, then trailing edge pressure divergence will occur after buffet onset. Also, while either one of these methods can be successful in predicting separation onset, they provide no obvious means of predicting separation (buffet) intensity. Recent improvements in wind-tunnel test techniques directed at the problem of predicting the severity of buffet include the use of unsteady wing-root bending measurements and unsteady pressure measurements. However, significant development and correlation with flight measurements on transport wings will be necessary in order to verify the accuracy of these techniques.

In order to appreciate the difficulty involved in developing accurate computational methods to predict buffet boundaries for three-dimensional swept wings at transonic conditions, it is instructive to consider the flow situation that is typical for buffet (trailing edge separation) onset. A correlation of the local Mach number just ahead of the shock and the shock position when trailing edge pressure divergence occurs is presented in Fig. 40 based on two-dimensional test results obtained for several types of airfoils at relatively high Reynolds numbers. It can be seen that for representative shock positions at buffet onset (near 50% chord) the local Mach number at the shock is near 1.4, high enough to cause a significant separation zone at the foot of the shock. The correlation with shock position was selected to illustrate the importance of the distance aft of the shock separation that the boundary layer has to recover in before encountering the steeper adverse pressure gradient near the trailing edge. Hence, the required computational method must account for the separation at the shock-wave

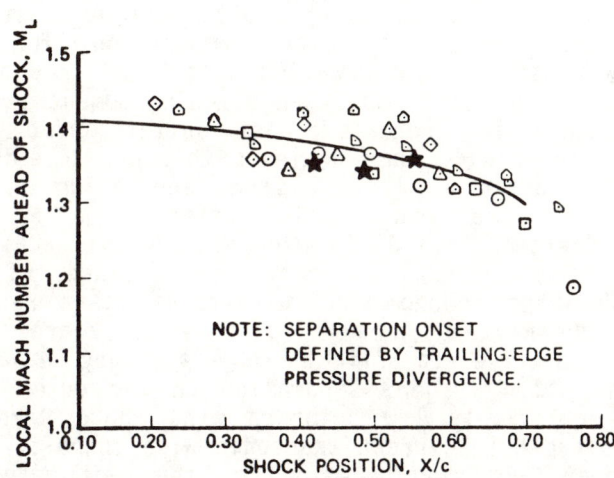

Fig. 40 Shock wave parameter correlation at buffet onset.

interaction and its influence downstream in order to properly predict separation onset near the trailing edge. Testing to determine turbulence and interaction details in this region will be imperative in developing the necessary interaction models. These models will have to be developed for two-dimensional flows at first and then extended to account for three-dimensional effects. After that, it will be time to worry about predicting the details of the unsteady flows that will be necessary to determine the aerodynamic excitation characteristics necessary for estimation of buffet intensity.

Post-Buffet Characteristics. The aerodynamic designer must also be concerned with the transonic flow characteristics of the wing at angles of attack beyond buffet onset up through maximum lift conditions where large areas of flow separation are encountered. First, it is now a requirement that the airplane not exhibit any tendencies to pitch up at these conditions. Hence the wing-airplane combination must either be designed to preclude pitch up at these conditions or the avionics and control systems developed and incorporated to limit the airplane from reaching these conditions. Another parameter to be considered, which has become important with the advent of high aspect ratio supercritical wings, is the dynamic influence of gust loadings on the maximum lift coefficient experienced by the wing. This is important from the standpoint of structural loads and the resultant wing weight.

With regard to the post-buffet pitch characteristics, the wing and horizontal tail combination for an aerodynamically stable configuration must together provide a continuously increasing nose down pitching moment as the angle of attack is increased past buffet onset (see Fig. 41). A swept wing that achieves this characteristic would normally avoid an initial large separation on the outer part of the wing. The dilemma facing the aerodynamic designer lies in knowing when he has achieved this characteristic for flight conditions. Since this is a viscous-dominated phenomenon, wind-tunnel test results obtained at the Reynolds number available today are often pessimistic relative to flight in that they have an earlier, and often larger, separation on the outer wing due to the relatively low outboard wing chord Reynolds numbers attainable in current transonic wind tunnels. The designer does not want to make the wing overly conservative (higher drag and weight) by increasing taper ratio and/or twist to avoid a "low" Reynolds number

TRANSPORTS—DESIGN FOR CRUISE PERFORMANCE

phenomenon not necessarily realistic at flight conditions. Likewise, he does not necessarily want to resort to a T-tail arrangement to aid this situation. On the other hand, he does not want to risk assuming that Reynolds number effects will clear up any wind-tunnel indicated problem, because, if they do not, the only solutions are to incur a sizeable drag penalty due to installing a stall-fixing device (see Fig. 32) or to suffer an unacceptable schedule slippage to develop the necessary black boxes to limit the airplane from reaching these conditions. This uncertainty that often exists is one of the commercial transport designers' big incentives to have the NTF full-scale Reynolds numbers capability available as soon as possible. In light of the pending availability of the NTF, and considering the extremely complex nature of this three-dimensional problem involving large regions of separated flow, the wing-tail interaction, etc., it does not seem prudent to even think about developing the methods to analyze this flow situation at this time. Higher priority should be given to developing the methods needed for the prediction of three-dimensional drag characteristics and buffet onset conditions where the need and likelihood of success is greater.

Commercial transport wing loads for structural design have typically been set by maneuvering load criteria. However, due to the aerodynamic characteristics of supercritical airfoils and wings designed using these airfoils, imposed gust loadings on the wing can be critical in determining the wing structural design and weight. A standard dynamic loads analysis for gust

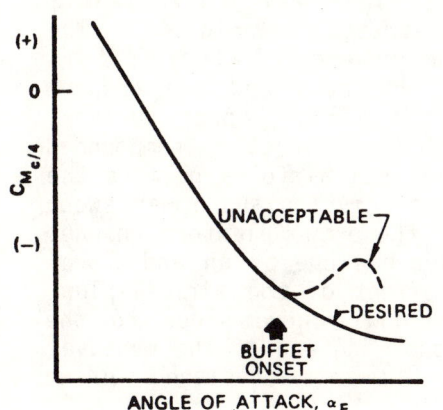

Fig. 41 Desired post-buffet pitch characteristics.

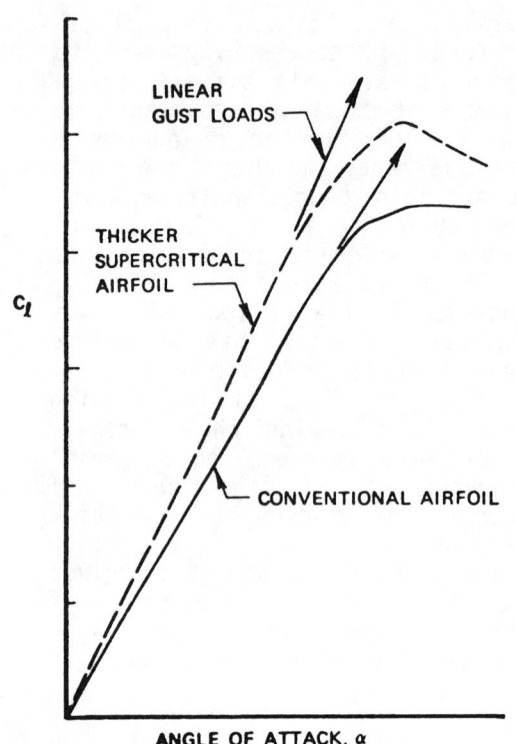

Fig. 42 Airfoil lift curves at transonic conditions.

effects starts with the static loads analysis and then adds a dynamic effect represented by a <u>linear</u> set of aerodynamic coefficients. Consequently, the lift curve slope at the point that the dynamic influence is imposed becomes crucial. A comparison of the two-dimensional lift curve slopes at transonic conditions for a conventional type (DC-10) airfoil and a somewhat thicker full supercritical airfoil is illustrated in Fig. 42. The greater lift curve slope of the supercritical airfoil is attributable to the increased thickness and the greater extent of supersonic flow on the airfoil upper surface. The increase in airfoil lift curve slope is compounded further when the use of supercritical airfoils permits the design of wings with higher aspect ratios and lower sweep angles. The combined effect of these two planform changes can easily increase the lift curve slope by an additional 0.01. As a result, the total effect of the airfoil-plus-planform changes increases the lift curve slope to the point where the imposed gust loads instead of the maneuver loads are often critical on wings designed using supercritical airfoils.

TRANSPORTS—DESIGN FOR CRUISE PERFORMANCE

Faced with this situation, the aerodynamic designer would like to determine if this linear gust loads analysis is overly conservative, resulting in unnecessary wing weight increases. A review of the status of research on airfoil dynamic stall characteristics[32-34] indicates that whereas the linear approximation is probably reasonably representative for low Mach numbers, it does appear to be overly conservative for application at typical transonic cruise Mach numbers. Unfortunately, it is difficult for the aerodynamic designer to prove this using the wind tunnel due to the scaling parameters that need to be matched, i.e., the product of pitch rate and wing chord. In the absence of this proof, the linear analysis must be used.

Interference Drag Problems

While the goal for the design of an efficient commercial transport is to define and arrange the components other than the wing to minimize any interference drag and to have a drag divergence Mach number higher than that of the wing, this goal is often not attained due to interference flow situations not presently amenable to calculation and often not accurately represented at typical wind-tunnel Reynolds numbers. The sources of the greatest uncertainty are typically nacelle-pylon-wing-type interference problems, which characteristically are worse at flight Reynolds number than they are in the wind tunnel, and flows along intersections, which also are often worse at flight conditions. These two flow situations are discussed in this section.

Excess Drag at Intersections

Flight flow visualization studies on representative transport aircraft have clearly shown that separation due to the viscous effect in flows along intersections (corners) represents a significant source of excess drag at cruise conditions on most current configurations. For example, each of the four wide-bodied jet transports has had problems of flow separation at the trailing edge of the nacelle-pylon juncture region (see Fig. 43). Flow separation problems have also been encountered on several aircraft at the wing-fuselage juncture (see Fig. 44). Wind-tunnel and flight investigations of winglet installations have likewise indicated instances of flow separation at the wing-winglet juncture detracting from the full theoretical benefit of winglets. The standard means that the designer has of addressing these potential problems is through the use of flow visualization techniques in the wind tunnel. Unfortunately, several instances have been experienced where flow separations that existed in flight did not exist at the wind-tunnel Reynolds numbers attainable today. The examples of the separations at the nacelle-pylon juncture and on the fuselage tail cone are representative cases. In order to avoid these types of problems in the future, the aerodynamic designer needs both a computational method that will analyze compressible viscous flows in juncture regions and a test facility (NTF) that will allow simulation of flight Reynolds numbers. Appropriate flow visualization techniques for these cryogenic test conditions will also be needed.

Nacelle-Pylon-Wing Interference

The shape of the streamlines on a swept wing (Fig. 45) is such that an effective convergent-divergent channel is

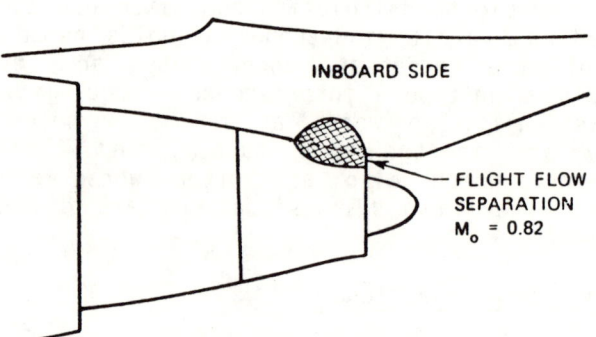

Fig. 43 Flow separation at nacelle-pylon juncture.

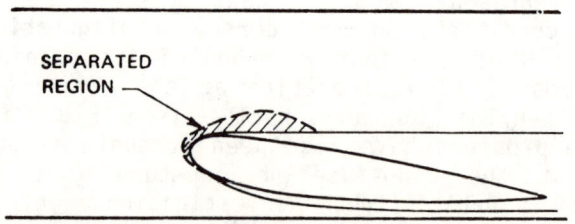

Fig. 44 Flow separations at wing-fuselage juncture.

TRANSPORTS—DESIGN FOR CRUISE PERFORMANCE

Fig. 45 Swept wing streamlines.

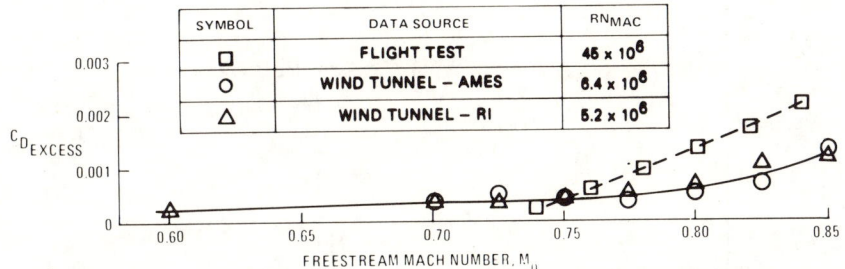

Fig. 46 Interference drag of DC-8 prototype long duct nacelle installation.

formed under the wing just aft of the wing leading edge on the inboard side of the nacelle-pylon installation. This channel provides the potential for large interference penalties due to supersonic flow and the resulting shock waves and separation unless the pylon and nacelle are placed properly. Apart from three-dimensional subsonic potential flow programs and some relatively simple local area ruling methods, the aerodynamic designer who is trying to design this installation to minimize interference drag must rely primarily on wind-tunnel measurements. However, to make the problem even more difficult, this type of installation has characteristically been plagued with having conditions at flight Reynolds numbers being worse than those measured in the wind tunnel. One of the first examples of this occurred during a prototype flight investigation at a long duct nacelle installation for the DC-8. Flight results obtained for a configuration with the nacelle afterbody well back under the wing clearly showed a much greater interference drag penalty than had been indicated in the wind tunnel. The comparison of the flight and wind-tunnel measured interference drags is shown in Fig. 46 and it can be seen that the interference drag penalty in flight is more than

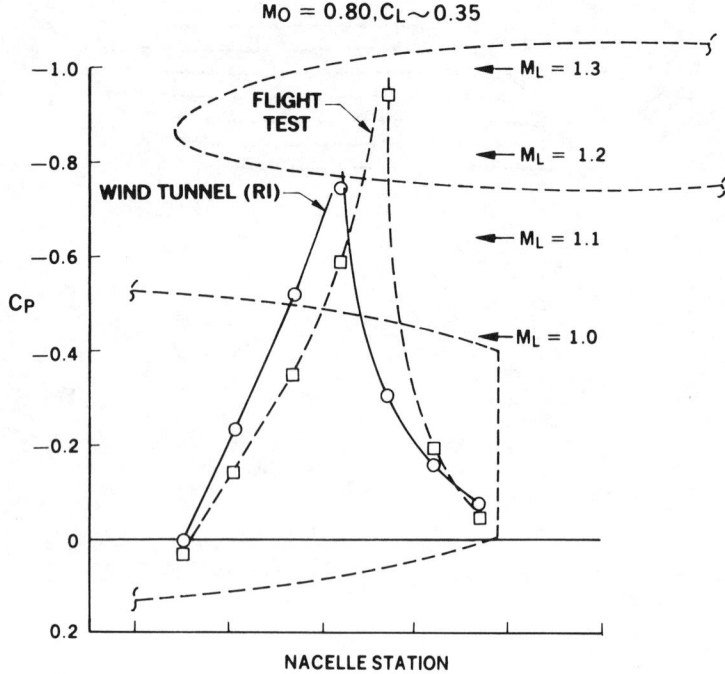

Fig. 47 Comparison of wind tunnel and flight pressure distributions for DC-8 prototype long duct nacelle installation.

double the wind-tunnel value at representative cruise conditions. Examination of the pressure distribution on the nacelle in the channel between the wing and nacelle as depicted in Fig. 47 explained the cause for the difference in that the shock in the channel was stronger and further aft in flight than in the wind tunnel. This behavior is thought to be due to the nondimensionally thicker boundary-layer displacement buildup at wind-tunnel Reynolds numbers on the components in the channel which causes a reduction in the effective surface (diffuser) curvature aft of the sonic throat. Application of Whitcomb's local area role concept[35] to this critical flow condition shows that only a small change in curvature is required to account for the difference between the flight and wind-tunnel results.

A very similar situation, but to a lesser degree, was encountered on early versions of the DC-10. The excess drag due to the flow separation illustrated in Fig. 43 did not exist in the wind tunnel because the adverse pressure gradients in the wind tunnel were not as severe as they were in flight and hence no separation was present at wind-tunnel Reynolds numbers.

The installation of a relatively large nacelle (high bypass ratio engine) and pylon on a high aspect ratio supercritical wing requires even greater care than it has on lower-aspect-ratio conventional wings. This type of installation (see Fig. 48) is more difficult for the designer because the supercritical-wing lower surface has higher velocity peaks and steeper adverse pressure gradients than a conventional wing, and the relatively short wing chord tends to result in a short-chord/low-fineness-ratio pylon with high-velocity peaks and strong adverse pressure gradients that superimpose with those of the wing in the wing-pylon juncture region. Somewhat offsetting these concerns are a typically lower design cruise speed than prevails for existing (DC-10, L1011, 747, etc.) aircraft, and the fact that the designer is much more knowledgeable about this potential problem area today than he was when the last airplanes were designed.

Advances in both computational and experimental techniques are needed to assist the designer with this installation interference problem in order to help minimize the risk involved in future aircraft design programs. A computational capability that can account for the transonic nonplanar aspects together with viscous approximations and an appropriate simulation of the fanjet flow would help the designer in the positioning and tailoring of the components. Ultimately, more accurate modelling of the viscous and inviscid-viscous interaction effects including the corner flow aspect should be added to permit realistic predictions of separation onset over a range of Reynolds numbers. The need for full-scale Reynolds number experimental simulation should be <u>partially</u> satisfied by the NTF. However, the NTF will not be of much use for this interference problem that now typically involves short-fan-duct installations of fuel efficient high bypass ratio engines unless appropriate

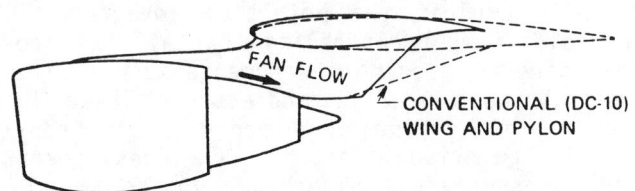

Fig. 48 Nacelle-pylon installation on supercritical wing.

engine simulators are developed to permit representation of the high-velocity fan flow which is crucial to this problem.

Recommendations

Some of the subsonic and transonic flow problems encountered in the aerodynamic design of commercial transports for cruise performance efficiency have been reviewed. While the advances made in recent years in the development of computational methods, particularly the nonlinear transonic methods, have made it possible to efficiently address many more design objectives than was previously possible, much more work in the development of both computational and experimental techniques is necessary if the risk involved in developing more fuel efficient commercial transports is to be significantly reduced. In particular, the following research and development efforts are urged.

1) In order to permit the development of the computational design procedures needed to accurately calculate the drag and buffet onset characteristics of representative three-dimensional swept wings, a close collaboration is necessary between the developers of the transonic potential flow techniques and those developing the required viscous and inviscid-viscous interaction techniques. Modifications must be incorporated in the transonic potential flow methods to make them compatible with the inputs required for the inviscid-viscous interaction models. Of particular interest are the shock-wave and trailing edge area representations. Further, to guide the continuing development of the necessary inviscid-viscous interaction models, first for two-dimensional flows and then, eventually, for three-dimensional flows, extensive experimental data will be needed to describe both the mean flow and turbulence characteristics in these interaction regions for a wide range of flow conditions.

2) To fully exploit the potential advantage of the very high Reynolds number capability that will be provided by NTF, many issues regarding instrumentation requirements and testing techniques need be addressed. These include the development of engine simulators plus transition detection and flow visualization techniques compatible with the high-pressure/low-temperature environment of the NTF cryogenic wind tunnel. Also, requirements for model

thermal conditioning (minimizing heat-transfer effects) must be realistically established.

3) To permit a more accurate analysis of interference flow situations, particularly for nacelle-pylon installations on the wing, a computational capability is needed that can account for transonic nonplanar aspects together with viscous approximations and an appropriate simulation of the engine fan-jet flow. Ultimately, more accurate modelling of the viscous and inviscid-viscous interaction effects including the corner flow aspect should be developed to permit realistic predictions of separation onset.

References

[1] Swihart, J.M., "The Next Generation of Commercial Aircraft - The Technology Imperative," presented to 12th Congress of ICAS, Oct. 1980.

[2] Overend, W.J., "Design Criteria for Airline Operations," AIAA Paper No. 79-1849, presented at AIAA Aircraft Systems and Technology Meeting, New York, Aug. 1979.

[3] Agnew, J.W. and Hess, J.R., "Benefits of Aerodynamic Interaction to the Three Surface Configuration," AIAA Paper No. 79-1830, presented at AIAA Aircraft Systems and Technology Meeting, New York, Aug. 1979.

[4] Robinson, M.R. and Silverman, S.M., "From HiMAT to Future Fighters," AIAA Paper No. 79-1816, presented at AIAA Aircraft Systems and Technology Meeting, New York, Aug. 1979.

[5] Anglin, E.L. and Satran, D., "Effects of Spanwise Blowing on Two Fighter Airplane Configurations," AIAA Paper No. 79-1663, presented at AIAA Atmospheric Flight Mechanics Conference for Future Space Systems, Boulder, Colo., Aug. 1979.

[6] Frink, N.T. and Lamar, J.E., "An Analysis of Strake Vortex Breakdown Characteristics in Relation to Design Features," AIAA Paper No. 80-0326, presented at AIAA 18th Aerospace Sciences Meeting, Pasadena, Calif., Jan. 1980.

[7] Ericson, L.E., "Technical Evaluation Report of the Fluid Dynamics Panel Symposium on Dynamic Stability Parameters," AGARD-AR-137, April 1979.

[8] Rediess, H.E., "Impact of Advanced Control Concepts on Aircraft Design," presented to 12th Congress of ICAS, Oct. 1980.

[9] Mello, J. and Agnew, J., "McAIR Design Philosophy for Fighter Aircraft Departure and Spin Resistance," SAE Paper No. 791081, Dec. 1979.

[10] Bowes, G.M., "Aircraft Lift and Drag Prediction and Measurement," AGARD-LS-67, May 1974.

[11] Steckel, D.K., Dahlin, J.A., and Henne, P.A., "Results of Design Studies and Wind Tunnel Tests of High-Aspect Ratio Supercritical Wings for an Energy Efficient Transport," NASA CR 159332, Oct. 1980.

[12] Henne, P.A., "An Inverse Transonic Wing Design Method," AIAA Paper No. 80-0330, presented at AIAA 18th Aerospace Sciences Meeting, Pasadena, Calif., Jan. 1980.

[13] Lynch, F.T., "Recent Applications of Advanced Computational Methods in the Aerodynamic Design of Transport Aircraft Configurations," presented to 11th Congress of ICAS, Sep. 1978.

[14] Braslow, A.J. and Knox, E.C., "Simplified Method for Determination of Critical Height of Distributed Roughness Particles for Boundary-Layer Transition at Mach Numbers from 0 to 5," NASA TN 4363, Sep. 1958.

[15] Braslow, A.L., "Use of Grit-Type Boundary-Layer Transition Trips on Wind Tunnel Models," NASA TN D-3579, Sep. 1966.

[16] Gray, W.E., "The Effect of Wing Sweep on Laminar Flow," RAE TM 225 (ARC 14,929), 1952.

[17] Gray, W.E., "The Nature of the Boundary-Layer Flow at the Nose of a Swept Wing," RAE TM 256 (ARC 15,021), 1952.

[18] Owen, P.R. and Randall, D.G., "Boundary-Layer Transition on a Sweptback Wing," RAE TM 277 (ARC 15,022), 1952.

[19] Gaster, M., "On the Flow Along Swept Leading Edges," Aeronautical Quarterly, Vol. XVIII, May 1967, pp. 165-184.

[20] Pfenninger, W., "About Some Flow Problems in the Leading Edge Region of Swept Laminar Flow Wings," Northrop Norair Report BLC-160, 1964.

[21] Gregory, N. and Love, E.M., "Laminar Flow on a Swept Leading Edge - Final Progress Report," NPL Aero. Memo. 26, 1965.

[22] Pfenninger, W., "Laminar Flow Control - Laminarization," Special Course on Concepts for Drag Reduction, AGARD Report 654, March 1977.

[23] Poll, D.I.A., "Transition in the Infinite Swept Attachment Line Boundary Layer," Aeronautical Quarterly, Vol. 30, Nov. 1979, pp. 607-629.

[24] Fancher, M.F., "A Hot-Film System for Boundary-Layer Transition Detection in Cryogenic Wind Tunnels," presented to Euromech Colloquium 132 on Hot-Wire, Hot-Film Anemometry and Conditional Measurement, July 1980.

[25] Jones, B.M., "The Measurement of Profile Drag by the Pitot Traverse Method," ARC RM 1688, 1936.

[26] Michel, R., Mignosi, A., and Quemard, C., "The Induction Driven Tunnel T2 at ONERA-CERT: Flying Qualities, Testing Techniques and Examples of Results," AIAA Paper No. 78-767, presented at AIAA 10th Aerodynamic Testing Conference, San Diego, Calif., Apr. 1978.

[27] Elfstrom, G.M., "Extraction of Wave Drag from Airfoil Wake Measurements," AIAA Paper No. 81-0291, presented at AIAA 19th Aerospace Sciences Meeting, St. Louis, Mo., Jan. 1981.

[28] Henne, P.A. and Hicks, R.M., "Transonic Wing Analysis Using Advanced Computational Methods," AIAA Paper No. 78-105, presented at AIAA 16th Aerospace Sciences Meeting, Huntsville, Ala., 1978.

[29] Cebeci, T., Kaups, K., and Ramsey, J.A., "A General Method for Calculating Three-Dimensional Compressible Laminar and Turbulent Boundary Layers on Arbitrary Wings," NASA CR-2777, 1977.

[30] Melnik, R.E., Chow, R., and Mead, H.R., "Theory of Viscous Transonic Flow Over Airfoils at High Reynolds Numbers," AIAA Paper No. 77-680, presented at AIAA 10th Fluid and Plasmadynamics Conference, Albuquerque, New Mex., June 1977.

[31] Mabey, D.G., "Prediction of the Severity of Buffeting - Structural Response to the Aerodynamic Excitation Produced by Separated Flow," AGARD-LS-94, Feb. 1978.

[32] McCroskey, W.J., "Some Current Research in Unsteady Fluid Dynamics," presented to ASME Winter Annual Meeting, Dec. 1976.

[33] Ericson, L.E. and Reding, J.P., "Scaling Problems in Dynamic Tests of Aircraft-like Configurations," AGARD-CP-227, Sep. 1977.

[34] McCroskey, W.J., McAlister, K.W., Carr, L.W., Pucci, S.L. and Lambert, O., "Dynamic Stall on Advanced Airfoil Sections," presented at the 36th Annual Forum of the American Helicopter Society, May 1980.

[35] Kutney, J.T. and Piszkin, S.P., "Reduction of Drag Rise on the Convair 990 Airplane," AIAA Paper No. 63.276, presented at AIAA Summer Meeting, Los Angeles, Calif., June 1963.

Chapter III.

Practical Aerodynamic Problems—Military Aircraft

Richard G. Bradley*
General Dynamics Corporation, Fort Worth, Texas

Introduction

The design of tactical military aircraft presents a unique challenge to the aerodynamicist because of the vast spectrum of operational requirements encompassed by today's military scenario. The designer is faced with a multitude of design points throughout the subsonic-supersonic flow regimes plus many off-design constraints that call for imaginative approaches and compromises. Transonic design objectives are often made more difficult by restraints imposed by subsonic and supersonic requirements. For example, wings designed for efficient transonic cruise and maneuver must also have the capability to accelerate rapidly to supersonic speeds and exhibit efficient performance in that regime.

The design problem is further complicated by the fact that weapon systems of today are required to fill multiple roles. For example, an aircraft designed to fill the basic air superiority role is often used for air-to-ground support, strike penetration, or intercept missions. Thus, carriage and delivery of ordinance and carriage of external fuel present additional key considerations for the aerodynamicist. The resulting aircraft flowfield environment encompasses a complex mixture of interacting flows.

Presented at the Transonic Perspective Symposium, NASA/Ames Research Center, Moffett Field, Calif., Feb. 18-20, 1981. Copyright © 1981 by General Dynamics. Published by the American Institute of Aeronautics and Astronautics with permission."
*Manager, Aerodynamics.

This paper presents some of the aerodynamic problems associated with tactical military aircraft in the transonic flow regime. The limitations of current design and analysis methods are noted, and some of the considerations resulting from realistic configuration constraints are highlighted. Emphasis is placed on the fighter class of aircraft since the current operational scenario leads to a most challenging array of aerodynamic considerations.

Performance Environment

Today's tactical aircraft is configured for multiple design points. Figure 1 shows two typical mission profiles for a fighter. As can be seen, the performance requirements are specific and require optimization in design for a multiplicity of flight conditions. The aircraft may be required to take off in a short distance and perform an efficient subsonic cruise to a designated point of loiter or reconnaissance, and it may engage in transonic air combat at any point along the way. Further, the aircraft may be required to accelerate rapidly — to either escape or overtake the adversary — and may have to penetrate supersonically to reach an interior target and, then, may be required to maneuver, execute an efficient turn, and return after weapons delivery.

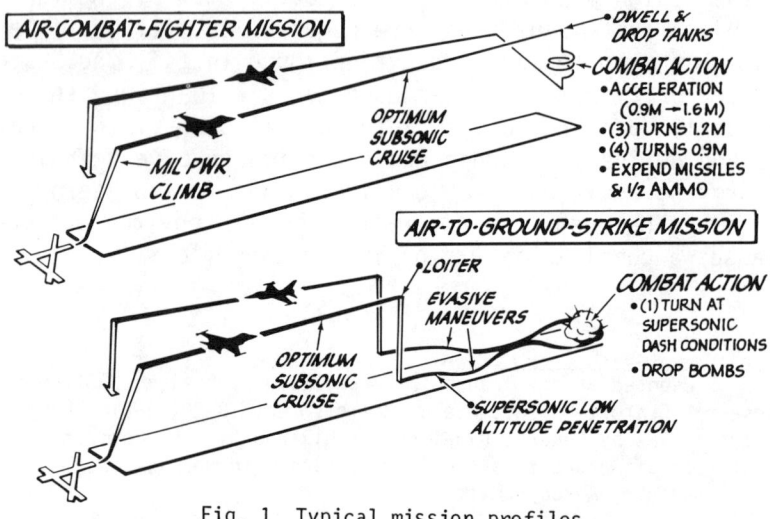

Fig. 1 Typical mission profiles.

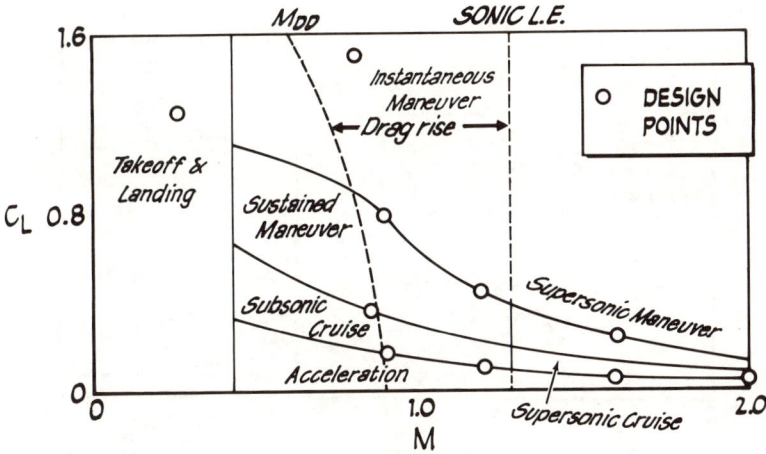

Fig. 2 Typical performance map for fighters.

The design goals, then, for a tactical weapons system may include efficient cruise at both subsonic and supersonic Mach number, superior maneuverability at both subsonic and supersonic Mach number, and rapid acceleration. And, of course, the aircraft must be controllable throughout the flight spectrum. Also, since weapons delivery is a key feature, low-drag weapons carriage and accurate release and delivery of those weapons is a prime consideration. Further, in today's environment, the aerodynamicist's problem is complicated by the requirement for invisibility — from electronic, optical, and thermal standpoints — which further complicate the design process.

The multiple design point requirement turns out to be the major driver for the designer of fighter aircraft. The aerodynamic requirements for each of the design points often present conflicting requirements. For example, the need for rapid acceleration to supersonic flight and efficient supersonic cruise calls for thin wing sections with relatively high sweep and with camber that is designed to trim out the moments resulting from aft ac movement at supersonic flight. However, these requirements are contrary to those requirements for efficient transonic maneuver, where the designer would prefer to have thicker wing sections designed with camber for high C_L operation and a high-aspect-ratio planform to provide a good transonic drag polar. De-

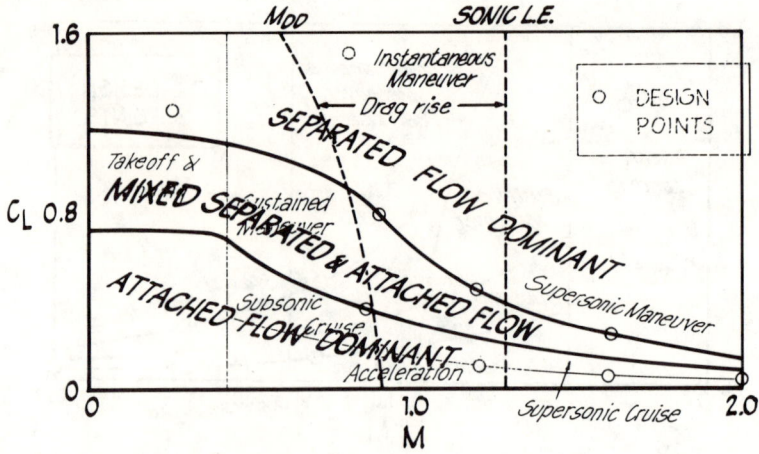

Fig. 3 General flow regions encountered.

signers are thus faced with a situation of compromise. These conflicting requirements suggest the obvious solution of variable geometry, that is, variable sweep wing and/or variable camber. Although this is a satisfactory aerodynamic solution, in many cases the resultant weight increases to a configuration can be prohibitive.

A typical performance spectrum corresponding to mission requirements is illustrated in Fig. 2, which presents a map of lift coefficient versus Mach number. The low Mach end of the spectrum throughout the C_L range is typical of takeoff and landing for the configuration. The subsonic cruise and supersonic cruise portions are noted in the moderate lift range. Acceleration to high supersonic speeds occurs at the low lift coefficients. Sustained maneuver takes place in the C_L range of less than one for most fighter configurations. Above this lift coefficient, the aircraft is in the instantaneous maneuver regime. Drag rise occurs, depending on the wing geometry, in the range of 0.8 to 1.2 Mach.

The particular flow conditions that correspond to the flow map are shown in Fig. 3. At the cruise and acceleration points, the aircraft designer is dealing primarily with attached flow, and his design objective is to maintain attached flow for maximum efficiency. At the higher C_L's, corresponding to instantaneous maneuver, separated flow becomes the dominant feature. Current designs take advantage of the separated flow by forming vortex flows in this range. Intermediate C_L's corresponding to sustained maneuver are usually a mixture of separated and attached flow. Consequently, if the aircraft is designed with camber to minimize separation in the maneuver regime, the configuration will have camber drag and may have lower surface separation, which increases drag at the low C_L's needed for acceleration.

Thus, the aircraft design is a compromise to achieve an optimal flow efficiency considering the numerous design points associated with the mission objectives. It is easily seen that the transonic flow problems that must be addressed in fighter design are driven to a very large extent by the constraints imposed at the other design points — supersonic and subsonic. This mix of flow characteristics poses a challenge in analyzing, evaluating, and designing an aircraft in the transonic regime. The following paragraphs deal with some of the specific problems that are incurred and some of the tools and approaches that are necessary for addressing these problems.

Transonic Wing Design Problem

The design of wings for military aircraft involves complex trades between propulsion system, weight, wing loading, and aircraft sizing. A detailed discussion of these parametric trades is beyond the scope of the present paper. Instead, the discussion concentrates on some of the aerodynamic considerations that become a necessary part of such trades.

Supersonic design enjoys relatively precise computation and optimization thanks to wide applicability of linearized theory. As a result, the optimization problem is well posed and experimental verification is relatively straightforward. On the other hand, in the transonic regime, computational methods are not fully reliable and wind-tunnel testing is most difficult. Aircraft optimization is, at best, heuristic in the absence of exact opti-

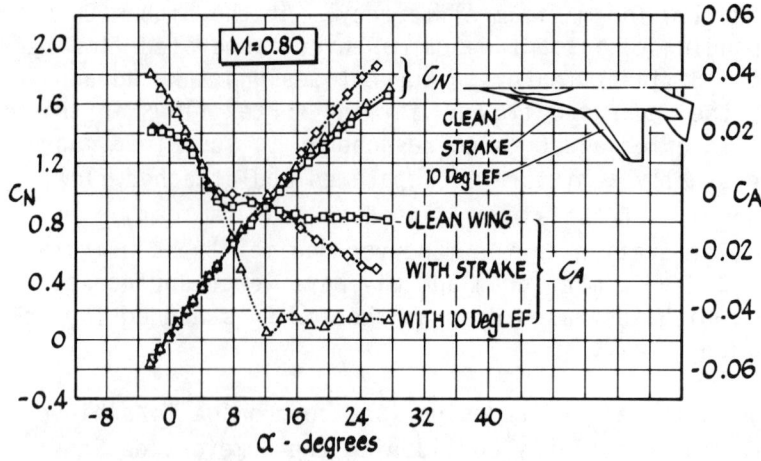

Fig. 4 Axial and normal force characteristics.

mization criteria and depends to a great extent on the experience and intuitive skill of the aerodynamicist.

The aircraft design objectives for both transonic and supersonic conditions may be summed up in terms of axial force and normal force relationships. Typical curves of axial and normal force variations with angle of attack are shown in Fig. 4 for a fighter configuration at 0.8 Mach. The break in the downward trend of axial force with angle of attack, as shown best for the uncambered wing data, signals leading edge separation which, in turn, is accompanied by a decrease in normal force and, sometimes, by the onset of buffet. The obvious goal is to maximize normal force slope while minimizing axial force.

The design of efficient transonic configurations may proceed from two conceptual schools of logic. One acknowledges that the optimum low-drag flow must accelerate rapidly over the airfoil to supercritical flow and decelerate in a nearly isentropic manner, avoiding strong shocks and/or steep gradients that can lead to significant regions of separation. This approach sets attached flow or near fully attached flow as an intuitive design goal and is typically used for aircraft that permit strong emphasis to be placed on transonic cruise or sustained maneuver design points. The leading edge flap data shown in Fig. 4 illustrate the favorable effect of leading edge camber in reducing the axial force.

The second school of thought recognizes the inevitability of significant flow separations at design conditions and adopts a philosoply of controlling certain regions of separation through vortex flows to complement other regions of attached supercritical flow. The strake-on data in Fig. 4 illustrate the favorable effects of vortex flow on normal and axial force at high angle of attack. This approach is appropriate for configurations constrained by multiple design points that emphasize added supersonic requirements. Current tactical fighters that rely on high wing loadings for transonic performance are good examples. The F-16 and F-18 employ a combination of controlled vortex flow and variable camber to achieve maneuverability.

Attached Flow Design

Transonic flow is, of course, inherently nonlinear and highly three-dimensional for wing-body combinations that are typical of military tactical aircraft. The analytical design of such wing-body combinations requires the solution of nonlinear partial differential equations and the careful control of surface shapes in such a manner so as to minimize the drag associated with the transonic flows. The nonlinear inviscid flow and its resultant shock waves interact with the boundary-layer displacement thickness. Strong shock-wave/boundary-layer interactions cause rapid increases in drag and lead to buffet, which reduce maneuverability and combat effectiveness. The shock-induced boundary-layer separation is perhaps the dominant problem associated with the shock waves on the wing upper surface.

The general goals for attached flow transonic wing design, then, may be stated very simply. First, one would like to maximize the supersonic flow occurring on the upper surface of the wing to produce the desired lift coefficient, and, second, one must minimize the shock strength and/or extent of the viscous separation that occurs. Two general approaches have evolved to accomplish these simplified design goals. Target pressure distributions for airfoil sections that illustrate these approaches are shown in Fig. 5.

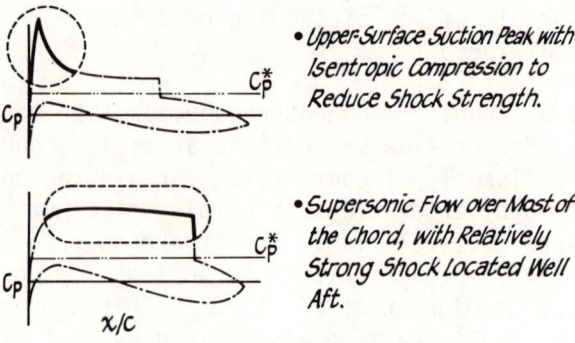

Fig. 5 Airfoil pressure distributions.

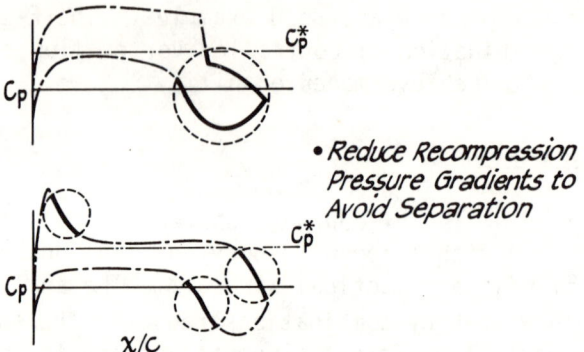

Fig. 6 Boundary-layer separation considerations.

The first approach uses the "peaky" pressure distribution. The section is designed to maintain an upper surface suction peak near the leading edge followed by an isentropic compression to minimize the shock strength, thus reducing the interaction of the shock wave with the boundary layer and the subsequent separations. The second approach uses the "flat top" distribution. The objective of this approach is to maintain a large area of supersonic flow over most of the airfoil chord and accept a relatively strong shock wave located well aft on the configuration. Both of these approaches have met with success in improving the drag divergence characteristics for airfoil sections.

Any design approach must be subject to restraints on the severity of the pressure gradients in order to avoid separation, as illustrated in Fig. 6. For the aft-loaded airfoil with a flat top pressure distribution, the pressure gradients aft of the shock wave and on the cambered lower surface near the trailing edge are strong. The aft camber is increased to increase total section lift without decreasing the drag divergence Mach number due to flow separations. In the case of a "peaky" distribution, recompression pressure gradients are critical near the leading edge on the upper surface and on both the lower and upper surface near the trailing edge.

Selection of pressure distribution goals for finite wing design is dependent on wing planform parameters, specifically sweep, taper, and aspect ratio. Generally speaking, the peaky distribution has some advantages for wings with higher sweep where recovery of leading edge suction at transonic maneuver C_L is possible. The flat top distribution is more appropriate for lower sweep where the component of velocity normal to the leading edge is higher.

The development of airfoil sections to meet the aforementioned design goals has progressed rapidly over recent years. This work has resulted from experimental development as well as analytical treatment. (See, for example, Refs. 1 and 2.) Recent developments in computing turbulent interactions on airfoils which consider displacement effects on the airfoil, wake, curvature effects, shock-wave/boundary-layer interactions, and trailing edge/boundary-layer interactions are summarized in a paper by Melnik.[3]

A discussion of the details of the various two-dimensional methods is not appropriate for the current paper. The concern here is with practical applications to the design of fighter wings. However, it is noteworthy that the development of mathematical models for calculation of two-dimensional airfoil flows is somewhat clouded by the absence of reliable two-dimensional experimental data for comparison. Effects such as three-dimensional wall effects, wind-tunnel blockage, etc., cast a degree of uncertainty on the validation of methods that are developed. Recent work sponsored by Stanford University to define experimental data baselines that are reliable and well understood should improve this situation. (For example, see a paper by Kline[4].)

The application of optimized two-dimensional airfoil sections to the design of finite wings is appropriately based on concepts derived from simple sweep theory, strictly valid for infinite yawed wings. The strongly three-dimensional effects of the wing body juncture, taper, and wing tip must be treated by local modifications to the geometry. A classic analytical approach to the problem has been given by Lock and Bridgewater[5]. Whitcomb (e.g., Ref. 1) has pioneered the application of experimental skills to adapt two-dimensional section designs to finite wing configurations.

Wings for fighter aircraft pose a formidable design task since the planforms are often of low aspect ratio, are highly tapered, and are small relative to the fuselage size. Figure 7 shows a typical fighter wing planform. The effects of the body, taper, and tip tend to dominate the transonic flow over the wing such that application of optimized two-dimensional sections may provide only a starting point for wing design. A research configuration, sketched in Fig. 7, is chosen to illustrate the point. An advanced transonic wing has been designed and tested for this configuration. Results are given in Ref. 6. Two-dimensional theory (Bauer, Garabedian, Korn, and Jameson (BGKJ) with boundary layer[2]) was used as a starting point for the design. A three-dimensional code (Jameson[7]) was used to refine the design for the finite wing.

Test results revealed that the resulting wing performed exceptionally well. An improvement of 7% in sus-

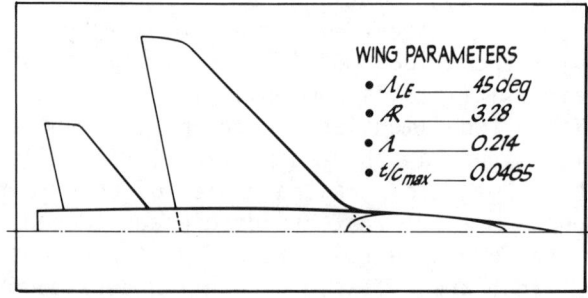

Fig. 7 NASA fighter configuration.

PRACTICAL AERODYNAMIC PROBLEMS—MILITARY AIRCRAFT

tained maneuver lift coefficient was observed when compared with another wing of the same planform optimized experimentally with smooth leading edge and trailing edge flaps.

Post-test analysis has been made to test the validity of simple sweep considerations for the planform. At the 63% span station the equivalent two-dimensional airfoil geometry was defined and pressure distributions for that section were computed with the BGKJ procedure. The equivalent two-dimensional pressure distribution was determined from the measured pressure data with simple sweep equations, as illustrated in Fig. 8. The reduced three-dimensional pressures are compared with two-dimensional theory for two section lift values in Fig. 9.

The comparisons of Fig. 9 reveal that the theory does not adequately represent the data. This lack of agreement does not imply that the two-dimensional theory is inadequate. The BGKJ method has been well substantiated for two-dimensional transonic, unseparated flows. (See, for example, Ref. 1.) The comparison does, however, illustrate the shortcomings of the simple-sweep assumption for the planform, even at a midspan station where tip and body effect should be minimized. What is needed, then, for fighter

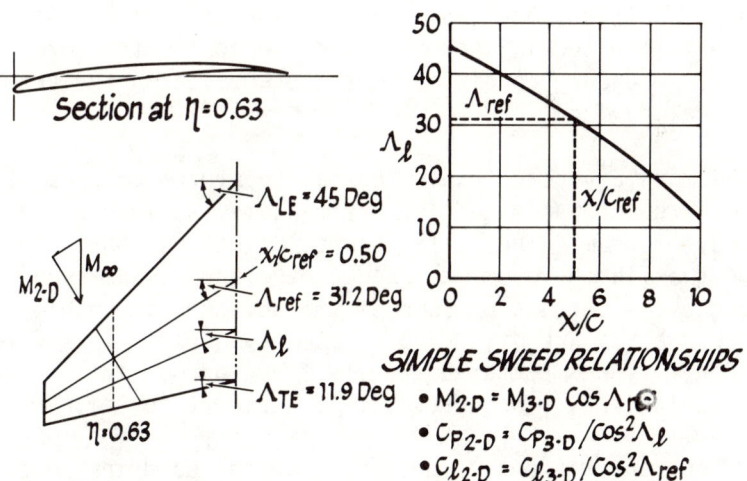

Fig. 8 Conversion from three-dimensional to equivalent two-dimensional data.

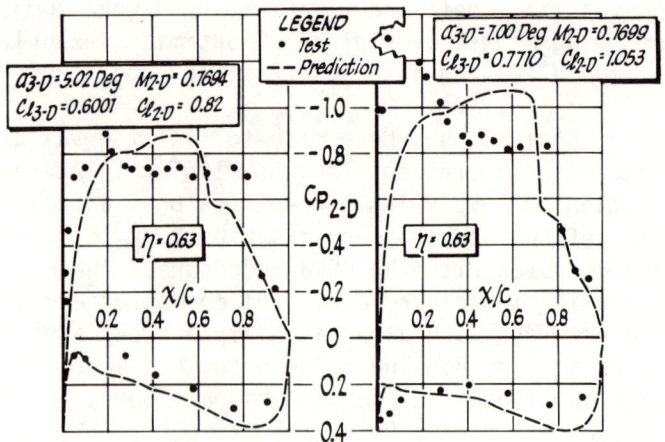

Fig. 9 Equivalent two-dimensional pressure comparisons.

wing design is not further refinement to two-dimensional methods with detailed viscous considerations but, rather, refinement to three-dimensional computational methods.

During the past several years, three-dimensional inviscid transonic flow computational techniques have been developing at a rapid pace. The developments have followed two complementary approaches. One uses the modified trantransonic small disturbance equation with the geometrical boundary conditions applied on a mean surface. The other uses the full potential equation and, hence, is the more accurate. Boundary conditions must be applied on the actual geometrical surface.

The more advanced transonic small disturbance computer programs are extensions of the basic Bailey-Ballhaus[8] wing-body program. The full potential equation approach follows the finite volume technique developed by Jameson and Caughey.[9] Details of some of the advances in both of these approaches are discussed in other papers in this volume. Bhateley has compared results from some of the available computational procedures with experiment in Ref. 10.

Of course, viscous effects must be included in the computational models for a complete description of the flow about arbitrary configurations. A review of some recent work in viscous interactions may be found in the AGARD Conference Proceedings.[11] However, since attached flow wing design is concerned with developing geometry that avoids severe interactions and separations, complex modeling of strong viscous phenomena may not be essential. Design visibility may be obtained intuitively without costly viscous computer programs.

An approach to the design of transonic wings making use of three-dimensional transonic theory has been treated Mann.[12] The procedure developed by Mann provides the designer with a set of guidelines for the systematic alteration of wing profile shapes to achieve a desired pressure distribution. Application of the method has been made to the research wing-body configuration shown in Fig. 7.

The application of numerical optimization techniques to wing design has been pioneered by Hicks and Vanderplaats.[13] An effort to assess the feasibility of performing computerized wing design by numerical optimization is reported in Ref. 14. A design program is described which combines a full potential inviscid aerodynamics code of Jameson with a conjugant gradient optimization algorithm. Three design problems were selected to demonstrate the technique. Results from this approach are promising, although problems remain concerning the consistency and accuracy of numerical aerodynamic calculations and with the establishment of precise criteria and constraints for the optimization code.

An experimental approach to optimization has been reported by Levinsky and co-workers[15,16]. Although many mechanical difficulties remain, their self-adaptive wind-tunnel model approach is attractive because measured aerodynamic drag can be used as a firm figure of merit.

The fundamental problem facing the aerodynamic designer in transonic flow is the absence of precise optimization criteria on which to base geometry modifications. Of course the true figure of merit is minimum drag for a selected design lift coefficient. A significant shortcoming of three-dimensional transonic codes to date is their inability to accurately predict drag for the configuration. This leaves the designer faced with the problem of specifying other intuitive design requirements, such as specified pressure distributions that are expected to result in near minimum drag. Agreement on exactly what distributions result in truly minimum drag is open to interpretation and as many proposed solutions may be found as there are designers who approach the problem. The ultimate goal, then, for improvement of computational methods has to be the accurate calculation of aerodynamic drag for the configuration. The computational task is a formidable one because the details of leading edge pressure distribution (a function of Reynolds number) and details of shock/boundary-layer interactions and separations, as well as the shock drag itself, become important considerations.

In the interim some carefully controlled wind-tunnel and analytical experiments are needed to develop a data base for relating pressure distribution criteria to drag optimization for thin wings. Design criteria could then be developed to use with existing computational methods to provide the designer with a consistent set of optimization goals.

The tactical military aircraft design problem is made more difficult by the supersonic acceleration requirement. The wings must be as thin as structurally feasible to reduce drag, but fixed camber suitable for optimum transonic maneuver is not practical because of supersonic camber drag. An obvious solution is a smoothly varying wing camber design. However, structural and actuation system weights prove to be prohibitive for thin wings. Simple leading edge and trailing edge flaps often prove to be the most practical compromise for high-performance, multiple design point configurations.

The extreme possibilities are apparent. One may design a wing with optimized transonic maneuver camber and twist and attempt to decamber the wing with simple flaps for supersonic flight. On the other hand, one may design the wing with no camber or with a mild supersonic camber and attempt to obtain transonic maneuver with simple flaps.

The supersonic, camber drag penalty associated with transonic maneuver camber designs is illustrated in Fig. 10.

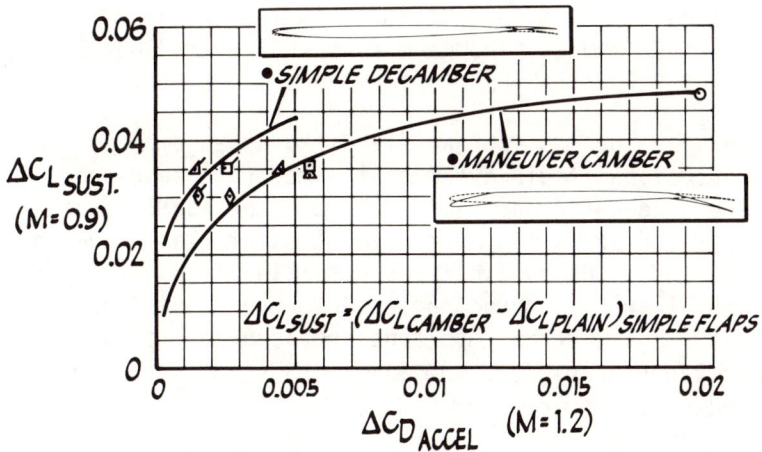

Fig. 10 Supersonic drag penalty trends.

The curves relate experimental trends of the M = 1.2 drag penalty associated with sustained maneuver increments above those possible with scheduled leading edge and trailing edge flaps. Experience shows that simple flap decambering of the maneuver wing can only reduce the drag partially because of camber and twist that remains in the wing structural box.

For a typical fighter the potential gain in sustained load factor of 4% may lead to a penalty in acceleration from 0.9 to 1.6 Mach of 11% even with a simple flap decambering for supersonic flight. The trends of Fig. 10 indicate that wings designed for greater gains in maneuver C_L result in disproportionately large supersonic drag penalties because the more extreme wing box camber and twist cannot be removed with simple flap decambering.

These trends indicate that in many cases a more favorable compromise for fighter design must start with an essentially uncambered wing or a wing with mild supersonic camber and proceed with simple flaps whose planform and deflection are optimized for maneuver. Because of new emphasis on STOL, the flap area will be larger than present trends. Weights, because of the flap design, will be a critical item. Optimization considering low-speed, high-lift, transonic speed maneuver, and supersonic acceleration clearly is the design of the future. The larger flap area may require hingeline skew or double hingeline designs for good transonic performance.

The above design approach leads to a new emphasis for the computational fluid dynamicist. Accurate transonic codes are needed for describing three-dimensional viscous flows with surface slope discontinuities at flap hingelines and with surface discontinuities resulting from part span flaps and ailerons. Aerodynamic performance, of course, is important. But just as important are accurate loads at the optimum flap schedule and the limit loads for structural and actuator system design.

One further comment on computational methods must be added. Aerodynamic designers are, in general, not intimately familiar with the details of convergence and stability of complex transonic codes. These codes give the designer visibility into the details of the interacting flowfield. In order to be useful, transonic codes <u>must be usable</u>. And they can only be usable if care is taken during development to ensure that geometry input and solution output is sufficiently simple and reliable.

<u>Separated Flow Design</u>

Even though the designer has worked carefully to design a wing for attached flow and to optimize its performance, he is inevitably faced with the problem of predominately separated flows at some point in the sustained and instantaneous maneuver regime. The manner in which the wing separates and how the separation develops over the configuration will strongly affect the vehicle's drag and its controllability at the higher C_L's. The current generation of fighter aircraft — the F-16 and F-18 — employ strakes or leading edge extension devices to provide a controlled separated flow. Controlled vortex flow can then be integrated with variable camber devices on the wing surface to provide satisfactory high-lift, stability and control, and buffet characteristics. The resulting flowfield is a complex one combining attached flows over portions of the wing with the vortex flow from the strake.

PRACTICAL AERODYNAMIC PROBLEMS—MILITARY AIRCRAFT 165

Vortex control devices can improve the transonic drag polar, but highly integrated designs are required to avoid other aerodynamic problems and also to minimize the potential weight penalties. The F-16 forebody strake vortex system is clearly visible as it interacts with the wing flow field in Fig. 11. Forebody strakes and canards have similar aerodynamic effects — both good and bad. For example, the F-16 forebody strake design required extensive integration with the wing variable leading edge flap system and empennage to achieve significant aerodynamic improvements. The aerodynamic improvements achieved with the forebody strake/variable leading edge flap combination are illustrated in Fig. 12 for the YF-16 aircraft. The dashed curve depicts the lift curve and drag polar for the wing without the strake and variable leading edge flaps. The strake with no leading edge flap significantly improves the static aerodynamic characteristics. Further improvement is seen with the variable leading edge flap in combination with the strake configuration. Figure 13 shows the effectiveness of the forebody strake and the variable camber leading edge in reducing buffet intensity, σ_{cg}, for the configuration.

These aerodynamic benefits do not come without very extensive design integration to avoid the potentially serious

Fig. 11 Strake vortex system on an F-16.

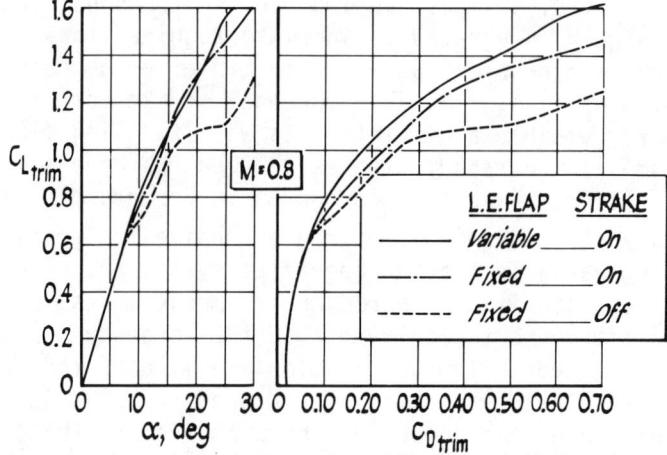

Fig. 12 Aerodynamic benefits of combined forebody strake variable camber system.

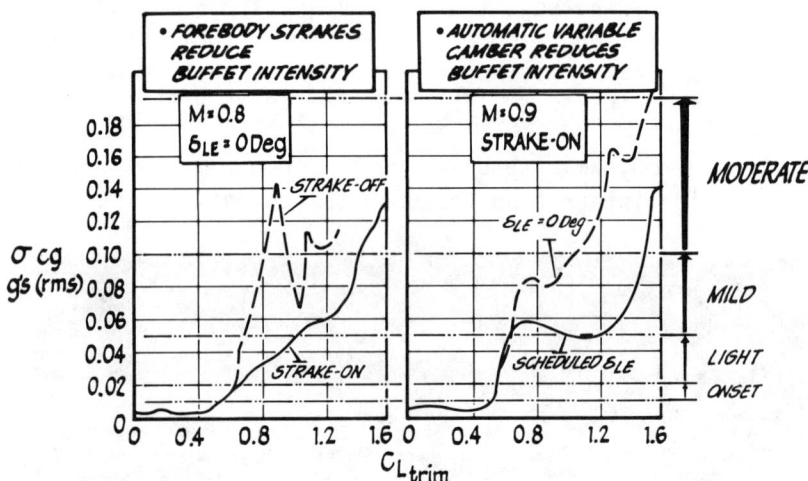

Fig. 13 Buffet characteristics.

problems of pitchup, deep stall, and unsatisfactory lateral directional stability that can accompany strake/canard devices. An example of the sensitive nature of forebody strake design is shown in Fig. 14, which illustrates how a very small change in strake planform can create serious deep stall problems. Figure 15 illustrates the sensitivity of lateral and directional stability to changes in forebody strake geometry, and leading edge flap deflection.

PRACTICAL AERODYNAMIC PROBLEMS—MILITARY AIRCRAFT

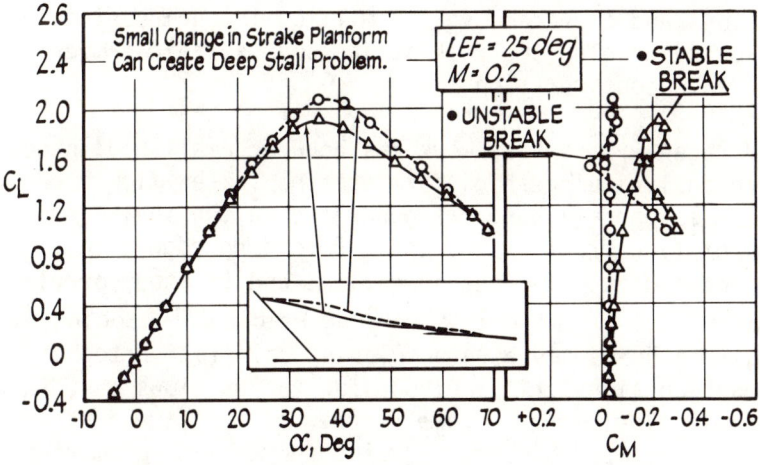

Fig. 14 Influence of strake geometry on stall characteristics.

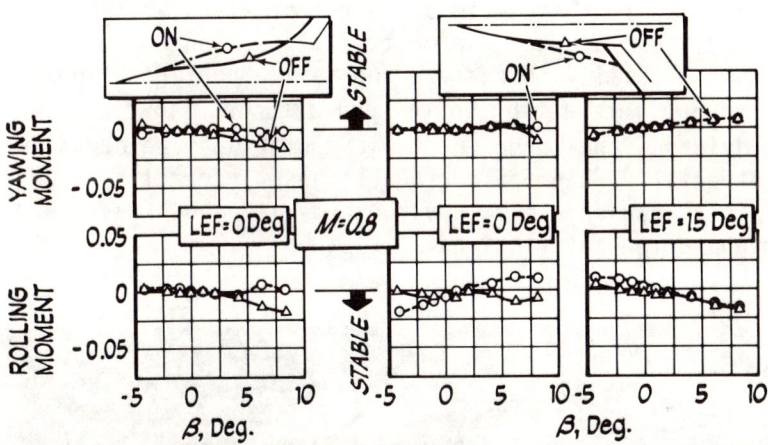

Fig. 15 Influence of strake geometry and leading edge flap on lateral/directional stability, M = 0.8.

The sensitivity of the strake design and its strong interaction with other configuration variables are representative of the magnitude of the problem inherent in theoretical prediction techniques in the separated flow regime. Adequate theoretical methods have not been available for analyzing these complex interacting flowfields and, as a result, it has been necessary in the past to make extensive use of expensive wind-tunnel tests to develop fighter aircraft. For example, General Dynamics tested 109 strake configurations and expended over $4 million in engineering and test time to develop the F-16 strake and variable camber leading edge configuration.

In recent years some work has been accomplished in developing analytical methods for treating vortex flow problems. A large data base was generated in the YF-16 and YF-17 prototype programs and the F-5 leading edge extension work. These data have been summarized and limited correlations presented by Smith et al.[17] and Headley.[18] Boeing, working with NASA, has been developing calculation techniques based on singularity methods for vortex flows over rather simple planforms.[19] Some strake design guidelines have been developed by Lamar[20] using vortex lattice methods coupled with the suction analogy and a quasivortex lattice method with separated flow is being developed by Mehrotra and Lan.[21] While these methods represent steps in the right direction, a great gap must still be spanned in developing the ability to analyze and design configurations with a strong interacting vortex and attached flowfields. This provides a challenge for the computational aerodynamicist, especially when one considers these interacting flowfields in a transonic flow environment with shock waves and local separation interactions.

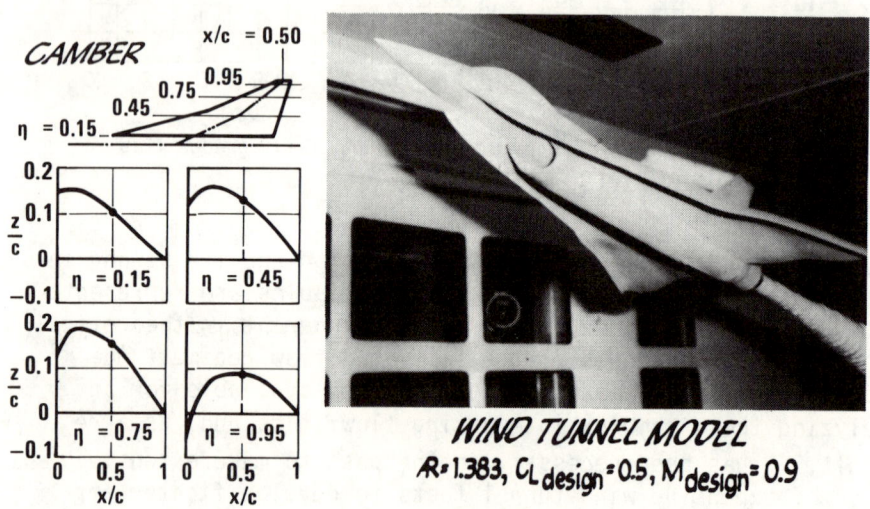

Fig. 16 Vortex camber design.

We should not pass from the separated flow wing design discussion without mentioning a new concept of controlled vortex flow for designing military aircraft having emphasis on supersonic configurations. Wing planforms for supersonic cruise aircraft have higher leading edge sweep and generally lower aspect ratios; these planforms develop vortex flows at relatively low angle of attack. As a result, the transonic drag characteristics are lacking in the maneuver regime since drag polars generally reflect very little leading edge suction recovery. Recently, wings of this type have been designed to take advantage of the separated vortex flows rather than to try to maintain attached flow to higher C_L values. Some early work in this area is reported in Ref. 22, where Schemensky, Lamar, and Subba attempted to design a vortex camber for a highly swept wing planform. The objective was to design a camber that would form a vortex flow at a transonic maneuver point and capture this vortex flow on the leading edge of the cambered surface to provide a leading edge thrust component, thus improving the drag polar. Figure 16 shows the planform and resulting camber design that was incorporated into a research model and tested in the NASA/Langley 7x10-Foot High-Speed Tunnel. The transonic polar characteristics for the wing with the vortex camber design are compared with a planar wing of the same planform with and without leading edge flaps in Fig. 17. It is seen that the camber design was successful at the design point; however, this extreme camber has very poor

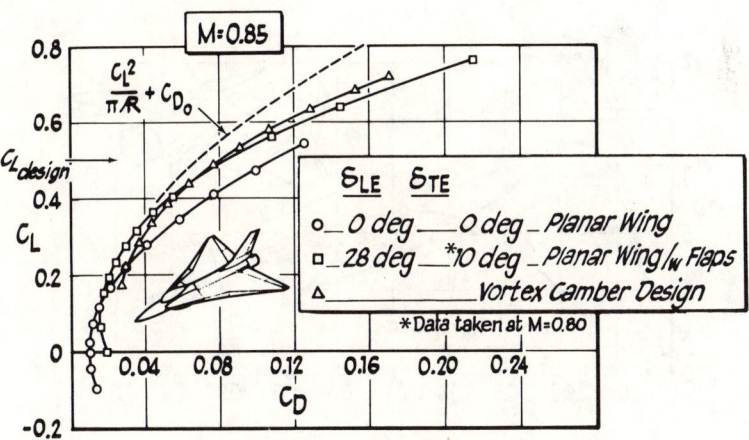

Fig. 17 Vortex camber benefits.

characteristics at off-design (low C_L) conditions. The practicality of building such a cambered wing that could be decambered for supersonic flight restricts the usefulness of this design approach. However, it does provide some insight as to the efficiency that can be obtained by design for separated flow. The good drag characteristics of the planar wing with simple leading edge and trailing edge flaps are especially attractive in view of the simplicity of the design. The leading edge flap planform is noted to be rather unconventional.

Following this initial effort, work has been done by a number of investigators, including Yoshihara, Lamar, and Rao, in conceiving leading edge flap systems which are designed to hold a vortex on the wing leading edge (vortex flaps), thus improving transonic polars. This work, which is continuing by NASA, General Dynamics, Boeing, and others, holds a great deal of promise. NASA and General Dynamics have jointly developed vortex flaps for a cranked wing configuration, as reported in Ref. 23. Typical vortex flap effectiveness is shown in Fig. 18. This work is continuing because of the potential benefits to transonic maneuver without penalty to supersonic cruise performance. In fact, the whole concept of designing aircraft for separated flows rather than trying to maintain attached flows for high sweeps in the transonic regime is of growing importance and receiving a great deal of attention.

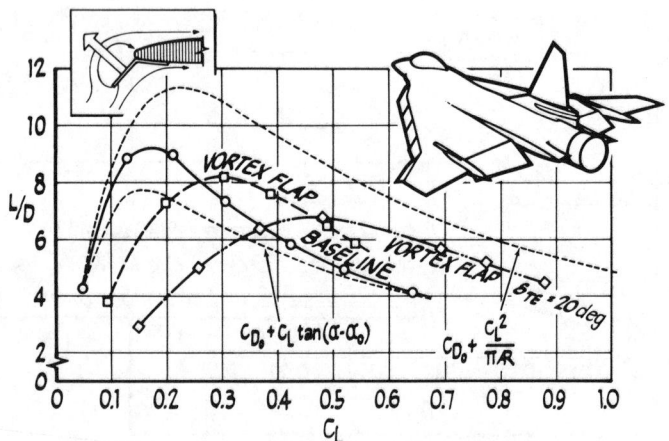

Fig. 18 Vortex flaps on research model.

PRACTICAL AERODYNAMIC PROBLEMS—MILITARY AIRCRAFT

Aeroelastic Tailoring

A growing trend today is an increased use of composites in military aircraft because of the significant weight savings they provide. In addition to the weight adsign opportunities, one of which is the control of aeroelastic characteristics through a process referred to as aeroelastic tailoring. The unique feature of aeroelastic tailoring is that beneficial aeroelastic characteristics such as increased twist and camber and increased flutter speed are actively sought and controlled during the design process rather than being the analytical consequence of a design based on strength and flutter considerations alone. As shown in Fig. 19, the use of aeroelastic tailoring results in considerably less compromise from the designer by providing the capability to obtain camber and twist under high-load maneuver conditions while not paying the camber/twist drag penalty at 1 g cruise and acceleration conditions.

Aeroelastically tailored wings can be designed by use of an automated design method developed by General Dynamics under contract to the Air Force Flight Dynamics Laboratory.[24] The procedure is an interdisciplinary design program combining aerodynamic, static aeroelastic, flutter, and structural calculations to provide optimum composite and metal wing skin thickness distributions and laminate orientations that satisfy specified design constraints, as illustrated in Fig. 20. The designs are determined by use of nonlinear programming techniques that incorporate an interior penalty function for the optimization.

Past analytical studies reveal that, while aeroelastic tailoring can yield aeroelastic benefits, less than close scrutiny of the design objectives and design constraints can result in unacceptable aerodynamic characteristics. This point is illustrated in Figs. 21 and 22, which show the aeroelastic drag benefit or penalty as a function of C_L resulting for specific design objectives. All of the wings were designed at Mach 0.9, but there is a large variation in the results, depending on the design objective sought. Allowing the ply orientation angles to be a variable provides the best design at Mach 0.9, but it is inferior to

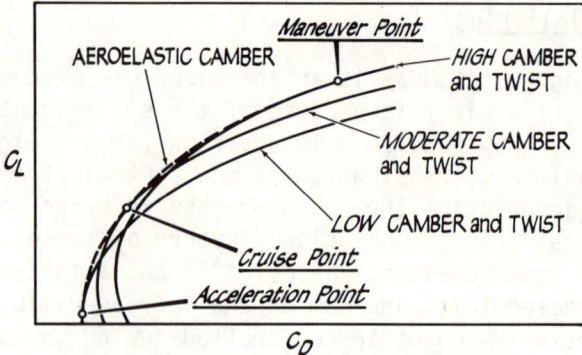

Fig. 19 Effect of favorable aeroelastic deformation.

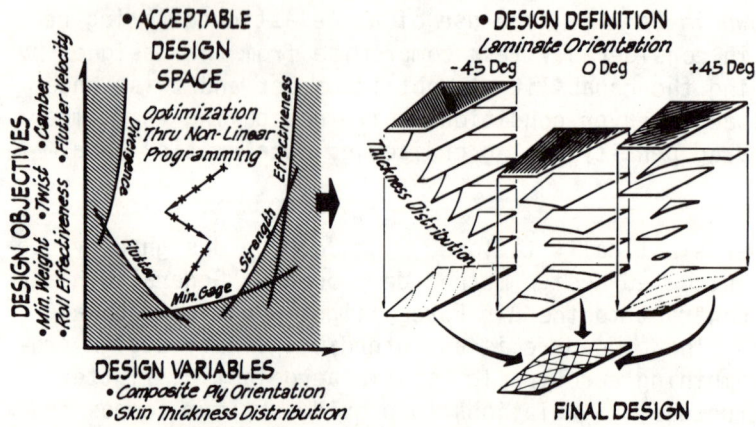

Fig. 20 Aeroelastic design approach.

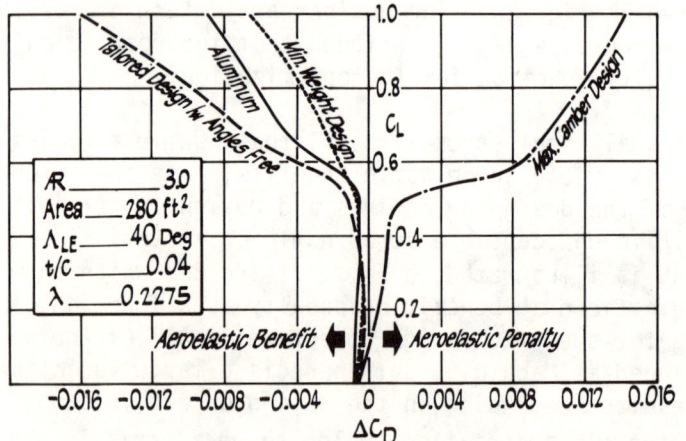

Fig. 21 Aeroelastic tailoring effect on drag, M = 0.9.

PRACTICAL AERODYNAMIC PROBLEMS—MILITARY AIRCRAFT 173

the aluminum design at Mach 1.2. The poor results shown by the maximum camber design are due to the wing tip twisting up (washin), which more than negates the benefit sought with aeroelastic camber. The minimum weight design is typical of state-of-the-art composite lifting surfaces. These results show that the aerodynamicist must take an active part in the design process to avoid finding out after it is too late that the performance gains due to weight savings have been erased by poor aeroelastic characteristics.

Aeroelastic tailoring is only one example of how the disciplines must work closely together at an early stage in the design process in order to extract the benefits that each can offer the other. In today's climate, optimization is becoming a necessity rather than a remote goal. If optimization is to be achieved, the designer must have accurate design tools. As an example, consider the aeroelastic tailoring design loop, which is by nature an iterative process. The design procedure often requires as many as 3000 iterations between the structural skin design and resulting aeroelastic load distribution. Linear aerodynamic theory is often used in design, partly because the transonic finite-difference procedures are too costly and time-consuming at present to use when so many design iterations are required. While aeroelastically tailored wind-tunnel models have been designed and tested that clearly validate the pre-

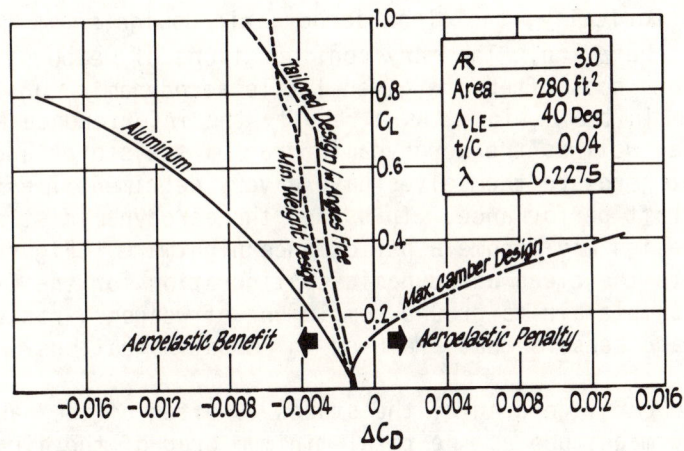

Fig. 22 Aeroelastic tailoring effect on drag, M = 1.2.

dicted benefits[25] the use of linear theory may well prevent definition of the optimum design. This is illustrated in Fig. 23, which schematically shows, as a function of C_L, the amount of aeroelastic twist required to prevent flow separation and the accompanying drag increase. Another boundary represents the maximum achievable twist available through the use of composites when all constraints (strength, flutter, etc.) have been met. Only a small region of design space may be available for achieving the goal of no flow separation at the design point. Errors in the methodology, such as use of linear theory at transonic Mach numbers, may prevent the best design from emerging. The need for computationally efficient transonic codes is apparent.

Store Carriage and Separation

Perhaps the greatest irony for the tactical aircraft designer results from the fact that an aircraft designed to be the ultimate in aerodynamic efficiency throughout a performance spectrum is often used as a "truck" to deliver armaments. Aircraft that are designed in a clean configuration are often used operationally to carry an assortment of pylons, racks, missiles, fuel tanks, bombs, designator seeker pods, launchers, dispensers, and antennaes that are attached to the configuration at any conceivable location.

Carriage

Historically, the clean aerodynamic design seems to stop at the pylon. The many configurations of weapons, racks, and pods often are given little aerodynamic consideration in the design. As a result, the interference effeffect as well as the aerodynamic drag of the stores and attachment hardware themselves have a very detrimental effect on aircraft performance. In usage, the aerodynamicist's dream design may become a performance nightmare. Figure 24 contrasts the clean aerodynamic configuration for the F-16 with a sample air-to-ground complement of weapons, racks, tanks, and sensors that are used in combat situations.

The carriage drag of the stores is often of the same order of magnitude as the total minimum drag of the aircraft itself. For example, the minimum drag of the F-16

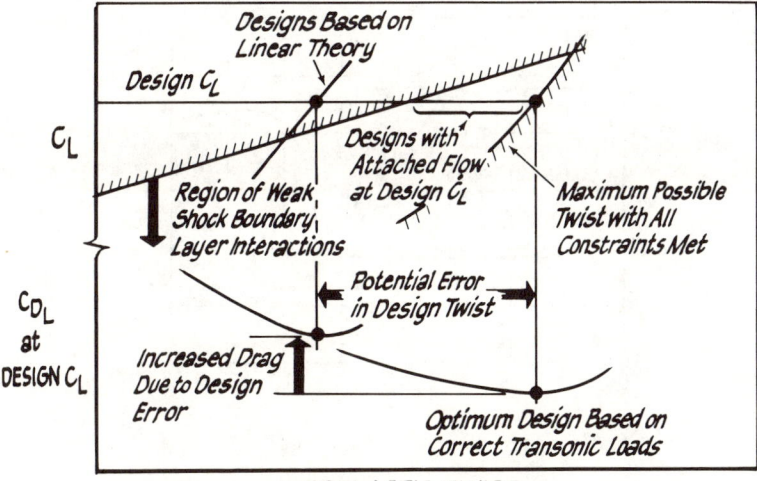

Fig. 23 Aeroelastic design space.

Fig. 24 Typical store loading.

aircraft is compared in Fig. 25 with and without the air-to-ground weapons load illustrated in Fig. 24. It is readily seen that the store drags themselves present as large a problem to the aircraft designer as the drag of the clean configuration. Store carriage on modern tactical aircraft is extremely important, particularly as one approaches the transonic regime, where the interference effects of the stores and pylons are highest and most detrimental to performance.

Not only is the cruise of a fully loaded aircraft important, but it is highly desirable for the vehicle to be

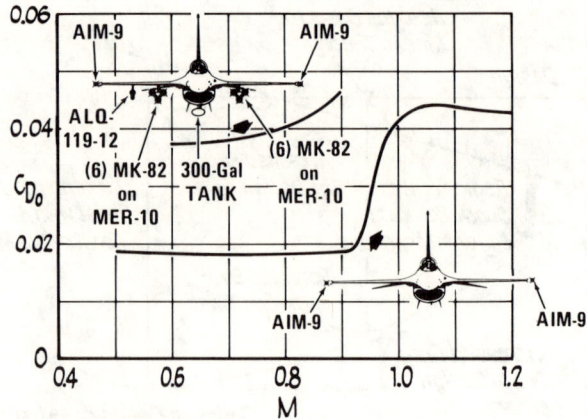

Fig. 25 Stores drag effect.

able to maneuver under loaded conditions. For example, if the aircraft is engaged by an aggressor during the initial phase of the mission, it is important for the fighter to be able to maneuver for self-defense without having to jettison the stores, thus scrubbing its air-to-ground mission. The adverse interference effects associated with store carriage not only produce large increases in drag, but can severely limit usable lift, adversely affect the stability of the aircraft, lower buffet boundaries, and be detrimental to store separation trajectories. These considerations lead the designer to search for ways to efficiently carry his weapon load while seeking, where possible, to minimize the carriage drag by minimizing interference effects. The problem is complicated by the fact that today's tactical aircraft must be versatile enough to carry an almost endless array of existing stores in every conceivable combination. For example, the F-16 is capable of carrying 60 different external stores in various combinations. Figure 26 illustrates some of these stores.

Haines recently[26] illustrated some of the sources of store interference and some potential ways to minimize that effect. An example of favorable interference is shown in Fig. 27, where carriage of missiles on wing tips is seen to improve the drag polar in the maneuver range. The missile and launcher serves to effectively extend the aspect ratio or wing span of the configuration, thus resulting in improved drag characteristics at lift.

PRACTICAL AERODYNAMIC PROBLEMS—MILITARY AIRCRAFT   177

Fig. 26 Store carriage possibilities.

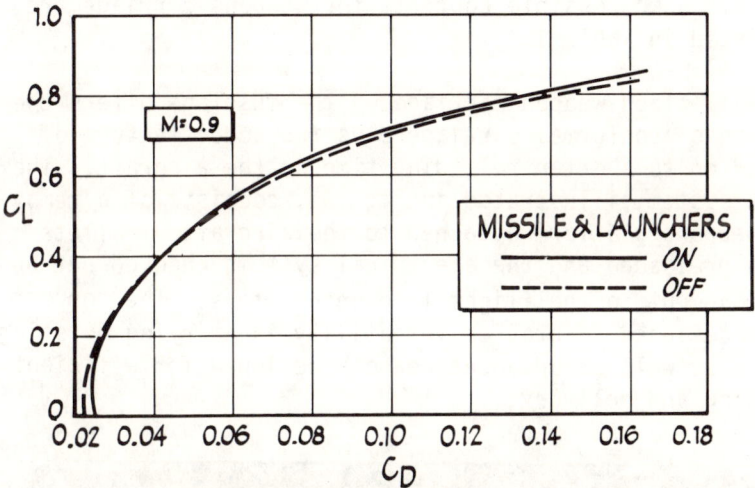

Fig. 27 Influence of wing-tip missiles and launchers on drag.

One approach for minimizing the penalties that normally exist with pylon/multiple bomb rack configurations is shown in Fig. 28. In this approach, low-drag weapons carriage concepts are adapted to a configuration with a long-chord, highly swept wing planform. Weapons are mounted conformally on short individual ejector pylons and positioned to take full advantage of tandem shielding and stagger. Figure 29 illustrates a conventional multiple bomb rack loading on the same model.

Wind-tunnel testing was conducted to obtain a direct comparison between conformal and conventional installations of MK-82 bombs, as shown in Figs. 28 and 29. Conformal carriage in this case permits one to carry 14 MK-82 bombs at substantially less drag than a 12 MK-82 pylon/multiple bomb rack mounting, as shown in Fig. 30. Significant benefits for the conformal carriage approach, in addition to increased range, are realized: increased number of weapons and carriage flexibility; increased penetration speed; higher maneuver limits; and improved supersonic persistence. Lateral directional stability is actually improved with weapons on.

Military aircraft of the future will have added emphasis on store carriage and release early in the design process. Some possible concepts for weapons carriage are contrasted in Table 1.

Palletized weapon carriage on the fuselage offers the low drag of conformal carriage plus the added feature of preloading to shorten reloading time on the aircraft. The pallet system, illustrated in Fig. 31, consists of a removable weapons platform attached to the aircraft. Pallets can be preloaded and the electrical systems checked out before movement to the flight line on bolsters. The concept can be designed to provide versatility in carrying existing weapons as well as advanced weapons designed for efficient packaging and delivery.

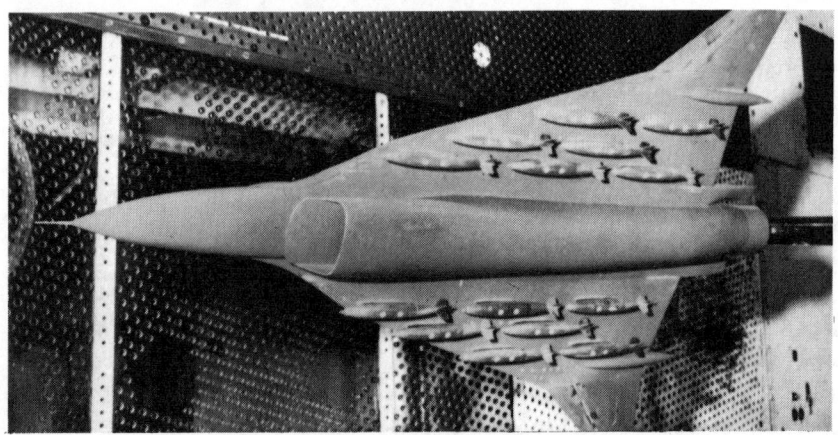

Fig. 28 Conformal carriage example.

Fig. 29 Conventional multiple bomb rack carriage.

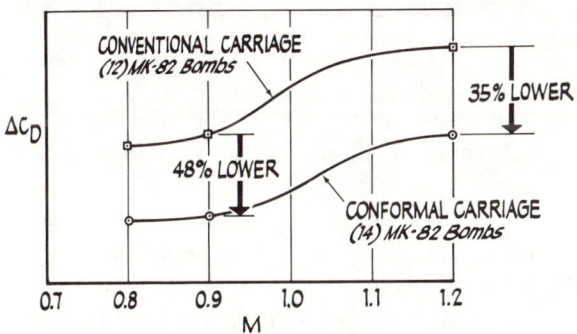

Fig. 30 Drag comparison — conformal vs conventional carriage.

Separation

Not only must external stores be carried efficiently by tactical aircraft, but they must release cleanly and follow a predictable trajectory through the vehicle's flowfield. The store trajectory is governed by the highly unsteady forces and moments acting on the store produced by the nonuniform flowfield about the configuration and the aerodynamic characteristics and motions of the store itself. The problem is complicated by realistic combat requirements for jettison or launch at maneuver conditions and multiple release conditions where the weapons must not

Table 1 Weapon carriage concepts

Store carriage concepts	Advantages	Disadvantages
1. Wing pylon carriage	Most flexible carriage mode — large payloads, inefficient store shapes	High drag / High radar cross section
2. Internal carriage	Low drag / Low radar cross section	Limited weapon flexibility / Increased fuselage volume
3. Semisubmerged carriage	Low drag / Low radar cross section	"Holes" must be covered up after weapon drop / Severely restricted payload flexibility
4. Conformal carriage	Most flexible of low drag carriage concepts	Size restrained

"fly" into one another. A typical multiple weapon release sequence in 1 g flight is illustrated in Fig. 32.

At transonic conditions the interactions between the stores and the aircraft flowfield are magnified by local regions of supersonic flow where shock waves impinge on the surfaces and resultant boundary-layer separations can adversely affect the trajectory. Historically the weapons separation problem has been approached experimentally through wind-tunnel testing with drop models and captive trajectory techniques. Exact analytical evaluation of the unsteady, transonic weapons separation phenomena has not been feasible because of the absence of sufficiently accurate models for the complex flowfields and, also, because of the lack of the necessary computer capability. However, some significant progress in this direction is being made.

In Ref. 27 Stahara has reported recent progress in an extensive theoretical and experimental program for establishing a predictive method for three-dimensional transonic flowfields about the aircraft and loading distributions on external stores located in the flowfield. The theoretical method, designed for rapid calculations, relies on the classical transonic equivalence rule, which is then extended to account for three-dimensional cross-flow effects. One modification replaces the linear two-dimensional cross-flow solutions with a linear three-dimensional solution obtained by panel methods. Another correction employs a three-dimensional nonlinear finite-difference solution of the small disturbance equations.

A method for analyzing store separation characteristics has been given by Deslandes in Ref. 28. The problem is simplified by quasilinearization of the time dependence in the separation dynamics, with a flow angularity technique employed to evaluate first-order interference effects. The method can use theoretical or experimentally determined flowfield data for the interference calculations. Good correlation with drop model tests is demonstrated for subsonic Mach number with an angle-of-attack variation. High-transonic Mach number applicability has not been demonstrated in the literature.

The precise analytical treatment of the store carriage and separation problem is a complex one that, most likely,

Fig. 31 Palletized carriage concept.

Fig. 32 Multiple stores release.

can only be approached with future computer systems. Three-dimensional finite-difference codes will require the ability to model complex, multiple bodies and complement of course and fine grid systems to define the interacting flow phenomena. Current work at General Dynamics is directed toward solving the transonic small disturbance equations for stores with multiple fins in the presence of wing-body configurations.

Concluding Remarks

The preceding discussion has attempted to highlight some of the transonic flow problems that are associated with the design of high-performance military aircraft. The problems discussed are by no means all inclusive. One problem area for design of military aircraft that has not been treated is the multiple surface flowfield effects. An aircraft with close coupled canard-wing or wing-tail arrangement presents added complexity to analytical representation of the flowfield. The interaction of the wake of the leading surface and, in particular, the vorticity shed from the tip influences the flowfield over the trailing surface. In transonic flow these free vortex systems can interact with shock waves that are forming on the aft surface to give vortex burst.

Another problem that has not been discussed is the one concerning interactions of the propulsive system with the flowfield. The calculation of flows in the vicinity of an inlet with its supersonic region and shock waves is another complex one. This is particularly true in modern aircraft where the inlets may be nonsymmetric and of complex geometric shape designed to reduce radar cross section. Interacting effects of nozzle flow with the afterbody of a nacelle or fuselage and the subsequent plume effects on separation presents another concern. In some cases the propulsion system is arranged to give favorable induced effects on lifting surfaces. Such effects as jet flap, vectored thrust near the wing trailing edge, and spanwise blowing on the transonic flowfield about the wing present another situation that requires renewed thinking in developing and applying computational methods.

The effects that have been discussed lead to the following conclusions:

1) The tactical military aircraft design problem is one of compromise because of the multiple design point requirements associated with missions that require good low-speed, transonic maneuver and cruise, and supersonic maneuver and cruise characteristics for the vehicle.

2) The wing design problem for tactical aircraft is heavily influenced by supersonic design requirements. The gurations requires that three-dimensional methods be emphasized since simple sweep theory is inapplicable in many instances.

3) For good supersonic performance, wings of minimum thickness and low camber are desirable. New emphasis needs to be placed on accurate methods for analytically designing wings that rely on simple leading edge and trailing edge flaps for their transonic maneuver capability.

4) Mixed flows which combine attached flow on the outer wing panel with controlled vortex flows over other portions of the wing are particularly attractive for fighters. Analytical methods that can accurately predict vortex flow phenomena are required.

5) Aeroelastic tailoring provides a new approach to providing favorable camber and twist for thin winged fighter configurations. In fact, composite wings may require tailoring to avoid significant losses in performance due to adverse camber and twist effects.

6) Transonic store carriage and separation phenomena are important features of military aircraft. Analytical methods are generally not satisfactory for precise calculation of operational store loads and separation trajectories.

The present paper has highlighted some of the transonic aerodynamic challenges associated with design and operation of tactical military aircraft. It is hoped that these design considerations will serve as a stimulus to the experimental and computational aerodynamicist. Paramount in the development of computational methods for use by the designer are simplicity of input and understandability of output. The designer must use computational methods to give him visibility. This visibility into the details of the flow provide the tool that the designer can use to solve problems and ultimately evolve superior tactical aircraft for the future.

Acknowledgments

The author is indebted to the members of the Aeroanalysis Group at the Fort Worth Division of General Dynamics for providing technical assistance and consultation. Especially, the contributions of B. D. Miller, W. W. Braymen, C. W. Smith, H. J. Sherrer, and A. E. Sheridan are acknowledged. Text prepared by Bev Yohner, CPS.

PRACTICAL AERODYNAMIC PROBLEMS—MILITARY AIRCRAFT

References

[1] Whitcomb, R.T., "Review of NASA Supercritical Airfoils," ICAS Paper 74-10, Haifa, Israel, Aug. 1974.

[2] Bauer, F., Garabedian, P., Korn, D., and Jameson, A., *Lecture Notes in Economics and Mathematical Systems - Supercritical Wing Sections II*, Vol. 108, Springer-Verlag, New York, 1975.

[3] Melnik, R.E., "Turbulent Interactions on Airfoils at Transonic Speeds — Recent Developments," *AGARD Conference on Computation of Viscous-Inviscid Interactions*, CP-291, Paper 10, U.S. Air Force Academy, Colorado Springs, Colo., Oct. 1980.

[4] Kline, S.J., "The 1980-81 AFOSR-HTTM-Stanford Conference on Complex Turbulent Flows: Comparison of Computation and Experiment — A Progress Report," *AGARD Conference on Computation of Viscous-Inviscid Interactions*, CP-291, Paper 22, U.S. Air Force Academy, Colorado Springs, Colo., Oct. 1980.

[5] Lock, R.C. and Bridgewater, J., "Theory of Aerodynamic Design for Swept Winged Aircraft at Transonic and Supersonic Speeds," *Progress in Aeronautical Sciences*, Vol. 8, Pergamon Press, Oxford, 1967, pp. 139-228.

[6] Hadley, S.K., Mann, M.J., and Ferris, J.C., "Design and Validation of a High-Lift Transonic Airfoil," NASA ATAR Conference (Classified Session), NASA/Langley Research Center, Va., 1978.

[7] Newman, P.A. and Davis, R.M., "Input Description for Jameson's Three-Dimensional Transonic Airfoil Analysis Program," NASA TM X-71919, Feb. 1974.

[8] Bailey, F.R. and Ballhaus, W.F., "Comparisons of Computed and Experimental Pressures for Transonic Flows about Isolated Wings and Wing-Fuselage Configurations," *Aerodynamic Analyses Requiring Advanced Computers*, NASA SP-347, Part II, Mar. 1975, pp. 1213-1232.

[9] Jameson, A. and Caughey, D.A., "A Finite Volume Method for Transonic Potential Flow Calculations," AIAA Paper No. 77-635, Albuquerque, New Mex., June 1977.

[10] Bhateley, I.C., Mann, M.J. and Ballhaus, W.F., "Evaluation of Three-Dimensional Transonic Methods for the Analysis of Fighter Configurations," AIAA Paper 79-1528, Williamsburg, Va., July 1979.

[11] AGARD Conference on Computation of Viscous-Inviscid Interactions, CP-291, U.S. Air Force Academy, Colorado Springs, Colo., Oct. 1980.

[12] Mann, M.J., "The Design of Supercritical Wings by the Use of Three-Dimensional Transonic Theory," NASA TP-1400, Feb. 1979.

[13] Hicks, R.M. and Vanderplaats, G.N., "Design of Low-Speed Airfoils by Numerical Optimization," SAE Paper 750524, 1975.

[14] Hicks, R.M. and Henne, P.A., "Wing Design by Numerical Optimization," AIAA Paper 77-1247, Seattle, Wash., Aug. 1977.

[15] Levinsky, E.S., Schappelle, R.H., and Pountney, S., "Airfoil Optimization through the Adaptive Control of Camber and Thickness," General Dynamics CASD-NSC-75-004, Sept. 1975.

[16] Levinsky, E.S., Palko, R.L., McClain, A.A., Schappelle, R.H., Clay, T.H., and Lohr, A.D., "Semispan Wind Tunnel Test Evaluation of a Computer-Controlled Variable-Geometry Wing," Arnold Engineering Development Center, AEDC-TR-78-51, Jan. 1979.

[17] Smith, C.W., Ralston, J.W., and Mann, H.W., "Aerodynamic Characteristics of Forebody and Nose Snakes Based on F-16 Wind Tunnel Test Experience," NASA CR-3053, July 1979.

[18] Headley, J.W., "Analysis of Wind-Tunnel Data Pertaining to High-Angle-of-Attack Aerodynamics," Air Force Flight Dynamics Lab, AFFDL-TR-78-94, July 1978.

[19] Johnson, F.T., Lu, P., Tinsco, E.N. and Epton, M.A., "An Improved Panel Method for the Solution of Three-Dimensional Leading-Edge Vortex Flows — Vol 1. Theory Document, NASA CR-159173, 1979.

[20] Lamar, J. E., "Strake-Wing Analysis and Design," AIAA Paper 78-1201, Seattle, Wash., July 1978.

[21] Mehrota, S.C. and Lan, C.E., "A Theoretical Investigation of the Aerodynamics of Low-Aspect-Ratio Wings with Partial Leading-Edge Separation," NASA CR-145304, Jan. 1978.

[22] Schemensky, R.T., Lamar, J.E., and Subba, C.S., "Development of a Vortex-Lift-Design Method and Application to a Slender Maneuver-Wing Configuration," AIAA Paper 80-0327, Pasadena, Calif., Jan. 1980.

[23] Smith, C.W., Campbell, J.F., and Huffman, J.K., "Experimental Results of a Leading-Edge Vortex Flap on a Highly-Swept Cranked

Wing," NASA Tactical Aircraft Research and Technology Conference, NASA Langley Research Center, Va., Oct. 1980.

[24] Lynch, R.W., Rogers, W.A., and Braymen, W.W., "Aeroelastic Tailoring of Advanced Composite Structures for Military Aircraft — Vol I. General Study, Air Force Flight Dynamics Lab, AFFDL-TR-76-100, April 1977.

[25] Rogers, W.A., Braymen, W.W., Murphy, A.C., Graham, D.H., and Love, M.H., <u>Validation of Aeroelastic Tailoring by Static Aeroelastic and Flutter Tests</u>, Air Force Wright Aeronautical Laboratories, AFWAL-TR-81-XXXX, Wright-Patterson Air Force Base, Ohio, to be published.

[26] Haines, A.B., "Prospects for Exploiting Favourable and Minimizing Adverse Aerodynamic Interference in External Store Installations," <u>AGARD Conference on Subsonic/Transonic Configuration Aerodynamics</u>, CP-285, Paper 5, Munich Neubiberg, Germany, May 1980.

[27] Stahara, S.S., "Study of Transonic Flow Fields about Aircraft: Application to External Stores," <u>AGARD Conference on Subsonic/Transonic Configuration Aerodynamics</u>, CP-285, Paper 8, Munich Neubiberg, Germany, May 1980.

[28] Deslandes, R., "Evaluation of Aircraft Interference Effects on External Stores at Subsonic and Transonic Speeds," <u>AGARD Conference on Subsonic/Transonic Configuration Aerodynamics</u>, CP-285, Paper 6, Munich Neubiberg, Germany, May 1980.

Chapter IV.

Experimental Testing at Transonic Speeds

James A. Blackwell Jr.[*]
Lockheed-Georgia Company, Marietta, Ga.

Nomenclature

b	=	wing span
c	=	wing chord
c_d	=	section drag coefficient
c_l	=	section lift coefficient
c_n	=	section normal force coefficient
c_{n_a}	=	slope of section normal force vs angle-of-attack curve
c_m	=	section pitching moment coefficient
C_D	=	aircraft drag coefficient
C_L	=	aircraft lift coefficient
C_P	=	pressure coefficient
$C\mu$	=	blowing coefficient
M	=	Mach number
P_T	=	total pressure
q_o, q_{oo}	=	freestream dynamic pressure

Presented at the Transonic Perspective Symposium, NASA/Ames Research Center, Moffett Field, Calif., Feb. 18-20, 1981. Copyright © 1982 by the American Institute of Aeronautics and Astronautics. All rights reserved.
[*]Senior Staff Specialist, Flight Sciences Division.

R_N	=	Reynolds number based on reference chord
x, y, z	=	Cartesian coordinates
α	=	angle of attack
β	=	angle of yaw
δ^*	=	boundary-layer displacement thickness
Δ	=	difference in two quantities
τ	=	wind-tunnel wall porosity (open area/closed area), %
η	=	nondimensional semispan (2y/b)

Subscripts

B	=	blockage
c	=	corrected
M	=	Mach number
SH	=	shock location
T.E.	=	trailing edge
T	=	transition location

Introduction

Over the last 50 years, the demands for increased performance of new aircraft have resulted in the exponential growth of wind-tunnel test time required to develop these aircraft. This growth is illustrated in Fig. 1 (adapted from a figure appearing in Ref. 1). Concerns have been expressed that should the exponential trend shown in Fig. 1 continue, the cost due to experimental testing of new aircraft in the 1980s and beyond would be prohibitive, and the test time required impractical.

It is the author's opinion that the exponential growth in wind-tunnel test time will not persist for the near future. Instead, the wind-tunnel test time requirements will experience a leveling trend. This leveling off of test time requirements will come for several reasons:

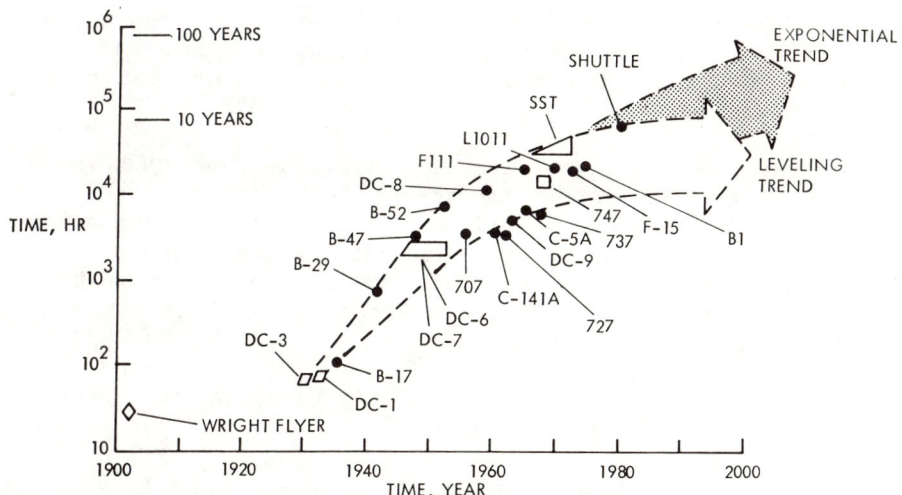

Fig. 1 Total wind-tunnel test hours for development of various aircraft.

limitations on available manpower and wind-tunnel resources; schedule and cost constraints on new aircraft programs; and increasing costs of wind-tunnel test time and models.

Although the wind-tunnel test time used for new aircraft development may level off, the future requirements for aerodynamic design information will not level off but will probably continue to increase rapidly. More and more design information will be needed to cope with increased configurational complexity, demands for improved aerodynamic performance, and the need to reduce the risks involved in new aircraft design.

If the demand for aerodynamic information continues to increase, obviously, there will be a requirement for experimental data beyond that which current levels of wind-tunnel testing are able to supply. Today, this deficit in information is being partially filled by the increased usage of computational methods. The increased use of the computer, however, places even more burdens on the available wind-tunnel test time. For example, the formulation of advanced theoretical models requires physical information in more detail and of higher accuracy than ever before. A further example is the increasing requirement for experimental test data for verification and calibration of computational design and analysis methods to determine their range of usefulness.

Clearly, to meet the expanded needs for aerodynamic information in the future, the quality and quantity of experimental data produced for each wind-tunnel test hour must be increased.

The roles of experimental testing in the 1980s and beyond will include the traditional role of new aircraft development testing, as well as the new emphasis fostered by the increased use of computational methods. A summary of the specific requirements for future experimental testing related to aircraft research and development are listed below:

1) Basic experimental data will be needed for formulating advanced theoretical models.

2) Experimental data will be required for theoretical code verification and calibration.

3) Insight and understanding into the physics of the flow that leads to the development of new aerodynamic concepts will be needed.

4) Experimental data will be required for aircraft configuration development and performance verification.

In order to meet the future requirements for experimental data with the resources available to the aerodynamic community, it is important that each new experiment be designed to produce the maximum amount of technical information. This requires that the engineer planning the experiment have a thorough understanding of aerodynamic testing technology. In particular, he must be knowledgeable about the capability of various wind-tunnel facilities, instrumentation, and test procedures. Also, he needs to be aware of potential problem areas in wind-tunnel testing and possible solutions.

A general understanding of aerodynamic testing technology is also needed by the engineers who use the experimental data to satisfy one of the above data requirements. It is important that the limitations and origins of each experiment be recognized and the application of the experimental data be made accordingly.

In summary, aerodynamic engineers must have a good understanding of aerodynamic testing technology in order to maximize the amount of useful information produced during an experiment and to knowledgeably use the data. The purpose of this article is to contribute to the general understanding of aerodynamic testing technology by presenting a state-of-the-art review of experimental testing at transonic speeds with particular emphasis on specific problems arising in experimental testing at transonic speeds and ways that are currently used to minimize or solve the problems.

To provide a framework for the discussion of problems/solutions related to various facets of experimental testing at transonic speeds, the events associated with designing an experimental test will be described first. This will be followed by sections dealing with problems/solutions in specific areas of experimental testing.

Experimental Test Process

The sequence of events that occur in designing an experimental test to satisfy a specific data requirement are presented in Fig. 2. Once a need for an experiment is established, the first step is to visualize the experiment. Next, a wind-tunnel facility is tentatively selected based on test section size, wall configuration, and availability. The experimental apparatus (model plus support system) and the range of flow conditions (Mach number, Reynolds number, etc.) for the test are then selected and compared to the

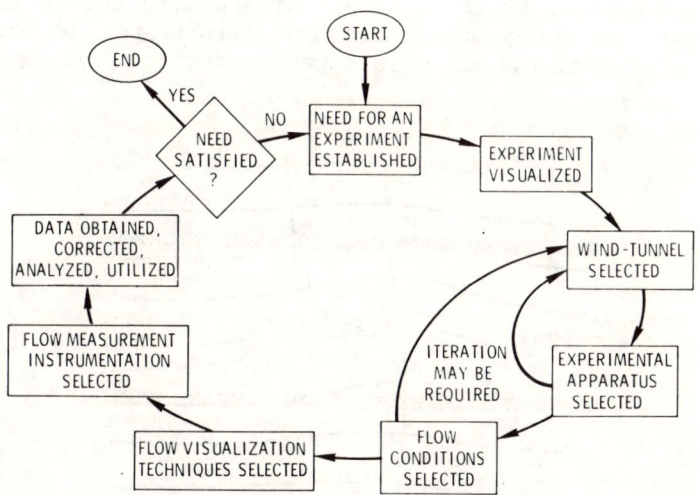

Fig. 2 Experimental test cycle.

capability of the chosen wind-tunnel to determine if they are consistent. If they are not satisfactory, an alternate wind-tunnel is selected, or the experiment is redesigned to minimize the unsatisfactory aspects.

Having established the test configuration, the flow visualization techniques and the flow measurement instrumentation are selected to yield the desired aerodynamic data. After the data have been obtained, the question, Has the need for data been satisfied? can be answered. If the experiment has not satisfied the need completely, additional experiments may be required.

In the sections that follow, the problems/solutions associated with various facets of the experimental test process shown in Fig. 2 will be discussed. Specifically, the areas of experimental testing that will be addressed include: wind-tunnel wall interference (a factor in wind-tunnel selection); wind-tunnel models and support system; Reynolds number simulation (a factor in flow condition selection); aerodynamic flow visualization; and aerodynamic flow measurements.

Wind-Tunnel Wall Interference

Wall Interference -- A Significant Problem

In selecting a wind-tunnel facility for an experiment, careful consideration should be given to minimizing the problems created by the presence of the wind-tunnel walls. Problems caused by the walls are illustrated in Fig. 3 where a wind-tunnel wall is superimposed on a "free-air"

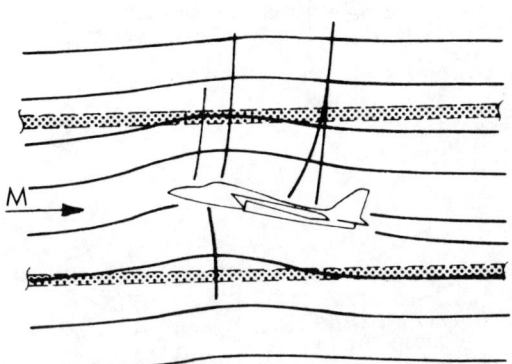

Fig. 3 Wind-tunnel walls imposed on free-air flowfield.

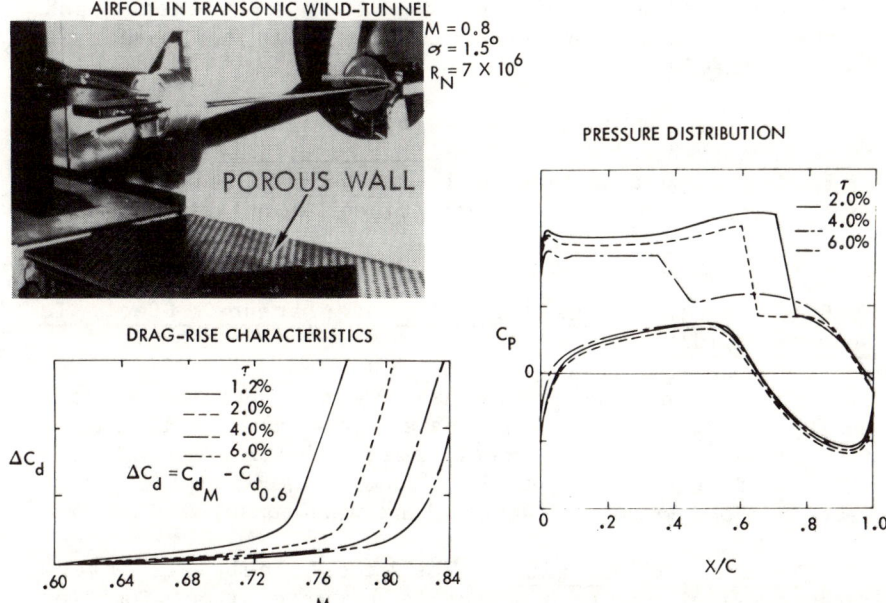

Fig. 4 Transonic wall interference effects on a 10% supercritical airfoil.

type flowfield. Obviously, the wall will influence the streamline pattern and interfere with the extent of the shock waves.

The major aerodynamic problems associated with wind-tunnel wall interference at transonic speeds are: blockage effects; lift interference effects; sonic line interference for M < 1; and shock-wave reflection.

At transonic speeds, the blockage and lift interference effects can be quite large. This is illustrated in Fig. 4 (from Ref. 2) where the wall interference effects on the aerodynamics of the NASA 10% thick supercritical airfoil are shown. As can be seen, the effect of varying the wall configuration of the Lockheed Compressible Flow Wind Tunnel on the aerodynamic forces and pressures are substantial.

For some experiments (e.g., airfoils), the sonic line in the flowfield generated by the model may reach the wind-tunnel wall. This condition can occur for Mach numbers significantly less than one. If the sonic line does intersect the wind-tunnel wall, obviously a normal "free-air" type flowfield would not be present, and substantial wall interference effects could be present in the aerodynamic data.

Problems due to shock-wave reflection occur when the shock waves generated by a model are of sufficient strength to reach the test section walls and reflect back onto the model. This phenomenon is most evident when testing near a Mach number of 1.³

Technology for Minimizing Wall Interference Effects Is Growing Rapidly

The current ways being used by the aerodynamic community to minimize or eliminate wind-tunnel wall interference are: reducing model size, utilizing ventilated-wall wind tunnels; applying theoretical/empirical corrections; and using self-correcting wind-tunnel walls.

Reducing model size. The wind-tunnel wall interference effect due to blockage is a strong direct function of the ratio of model cross-sectional area.⁴ Obviously, to minimize these effects the model should be made as small as possible. A reduction of model size would also reduce the possibility of a sonic line intersecting the wind-tunnel wall or of a shock wave reflecting back onto the model. However, reducing the model size is often not consistent with achieving high Reynolds number data (which is usually desired) or fabricating highly accurate wind-tunnel model contours.

Utilizing ventilated wall wind tunnels. Blockage and lift interference effects at transonic speeds for reasonable size models can be minimized through the use of ventilated walls. Ventilated walls generally fall into two categories: porous walls and slotted walls.

The object of the ventilated walls is to allow the streamlines near the test section wall to flow through the walls in order to generate streamline patterns representative of an infinite flowfield.

Most of the transonic facilities in use today with ventilated walls were designed with the aid of linear theory. That is, the amount of porosity for a porous tunnel or the slot size for a slotted tunnel was selected on the basis of linear theory considerations aided by experimental tests. However, even for subsonic flows for which linear theory is applicable, no one porosity or slot configuration will minimize both blockage interference and lift interference effects at the same time.

For two-dimensional tests the wall ventilation is generally configured to minimize blockage interference which predominately affects the Mach number of the freestream. The lift interference effects on angle of attack are generally so large that the experimental angle of attack is disregarded and section normal force coefficient is emphasized.

For three-dimensional tests, the wall ventilation is generally configured to minimize lift interference effects since angle of attack is of great importance in aircraft configuration experiments. The model blockage in three-dimensional tests is substantially lower than for airfoil experiments, thus allowing this approach. In fact, for reasonable size models at moderate lift coefficients the wall interference corrections predicted by linear theory in a ventilated wind tunnel are nearly negligible. Hence, most aircraft configuration experiments in ventilated wind tunnels do not include any corrections for wall interference effects.

The above approach worked long as linear wall interference theory was applicable to the experiment. With the advent of advanced airfoils and wings in the late 1960s

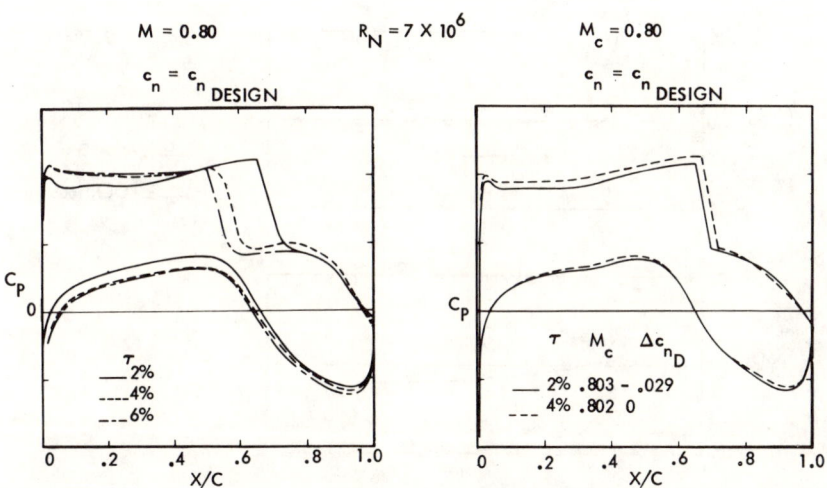

Fig. 5 Corrections applied for transonic wall interference effects on a 10% supercritical airfoil.

which had substantial regions of supercritical flow, the linear wall interference theory used for selecting the amount of wall ventilation was no longer directly applicable. Also, the aerodynamics on these configurations were much more sensitive to small changes in Mach number resulting from blockage interference.

Ventilated walls also substantially reduce the effects of shock-wave reflection. This is particularly so for the porous walls[3].

Applying theoretical/empirical corrections. Since the walls of most transonic wind tunnels in use throughout the world are fixed in geometry and since linear theory is inadequate to predict wall interference corrections when substantial supercritical flow is present on the model, new procedures for minimizing or correcting the aerodynamic data for wall interference effects are needed.

Recent research on wall interference effects at transonic speeds have indicated that a first-order correction to the freestream Mach number and angle of attack is sufficient for a large number of flows. This is illustrated in Fig. 5 (from Ref. 5) using experimental data for the same 10% supercritical airfoil shown in Fig. 4. First, the lift interference effects on angle of attack are minimized by comparing pressure distributions at a constant normal force coefficient (compare to Fig. 4). Second, a

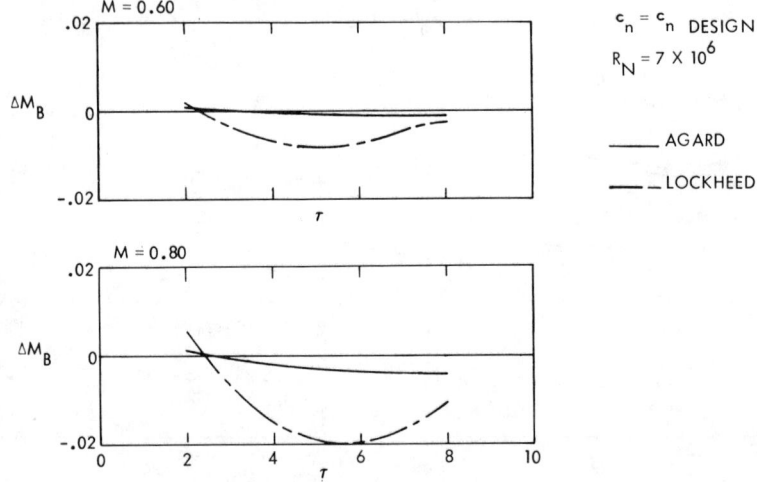

Fig. 6   Wind-tunnel blockage corrections for 10% supercritical airfoil.

simple Mach number correction to the freestream Mach number is applied (using the empirical method of Ref. 5) to minimize blockage interference effects. In general, it can be said that the agreement shown for the corrected data in this figure using simple first-order corrections is good considering the small difference in normal-force coefficient.

The inadequacy of linear theory to provide the correct blockage corrections is demonstrated in Fig. 6 for the airfoil of Fig. 4. The corrections, obtained using the Lockheed method of Ref. 5 which utilizes experimental pressure measurements at the tunnel walls, indicate the

Fig. 7 Wind-tunnel interference correction for 10% supercritical airfoil.

variation of the Mach number correction with wall porosity to be substantially larger in magnitude than that predicted by linear theory.[4]

Linear theory is also inadequate to provide realistic corrections for lift interference effects on angle of attack for transonic flows. This is illustrated in Fig. 7 for the 10% supercritical airfoil of Fig. 4. Obviously, the linear AGARD corrections of Ref. 4 do not collapse the normal force vs angle-of-attack slope data obtained at various wall porosities to that predicted by the Bauer theoretical program of Ref. 6 at transonic speeds. However, as expected, the AGARD corrections are reasonable at subsonic speeds.

A recent publication by Mokry and Ohman[7] which also utilizes experimental wall pressures shows promise for providing theoretical first-order corrections for both Mach number and angle of attack on airfoils at transonic speeds. Additional work related to determining wall interference corrections using measured wall pressures and linear theory is found in Ref. 8.

A more sophisticated approach to determining first-order wall interference corrections at transonic speeds has recently been proposed by Stahara and Spreiter[9] for axisymmetric nonlifting bodies and by Hinson and Burdges[10] for wings. This approach utilizes measured wind-tunnel wall pressures combined with three-dimensional transonic analysis methods. The correction is determined by computing the solution around the model in unconfined flow and comparing the result to a solution computed using the measured wind-tunnel wall pressures as boundary conditions in the transonic solution process. Having quantified the influence of the wind-tunnel walls on the theoretical solution, the theoretical free-air solution is recomputed while varying freestream conditions to obtain the best possible match with the computed solution using the experimentally measured wall boundary conditions. When a match in pressure distributions is obtained, the difference in angle of attack and freestream Mach number for the two solutions are the first-order wall interference corrections. These implied wall corrections are then applied to the experimental wind-tunnel model data.

To demonstrate the effectiveness of utilizing measured wall pressures combined with transonic analysis codes to produce first-order wall interference corrections, an example is presented for a transport supercritical wing model tested by Hinson and Burdges.[10] The model, shown in Fig. 8, is mounted on the floor of the Lockheed Compressible Flow Wind Tunnel. The solid blockage for this model is high for transonic testing and is equal to approximately 1.8% of the tunnel cross-sectional area. The open area of the porous wall is 4%. Rails are shown mounted to the tunnel walls for obtaining measured wall pressures.

The experimentally measured wall pressures using the rails in Fig. 8 are presented in Fig. 9 and are compared to computed free-air results (using Ref. 11) that would occur

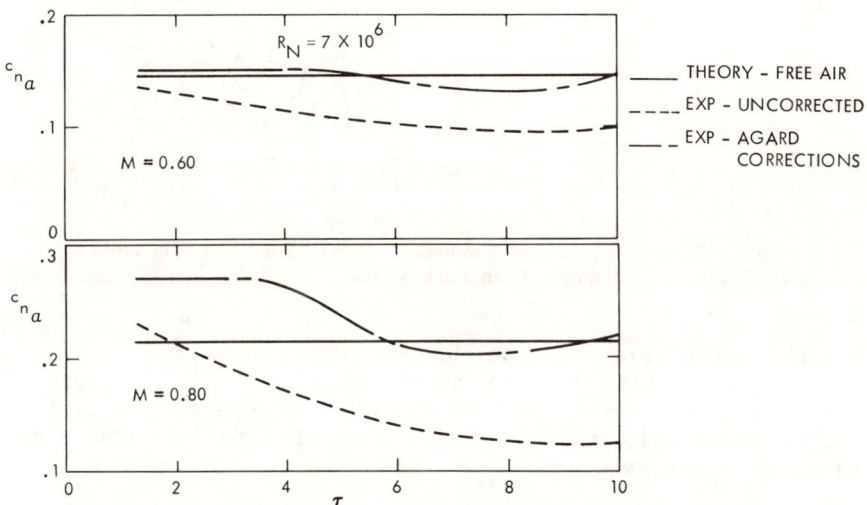

Fig. 8 Supercritical wing model in Lockheed CFWT.

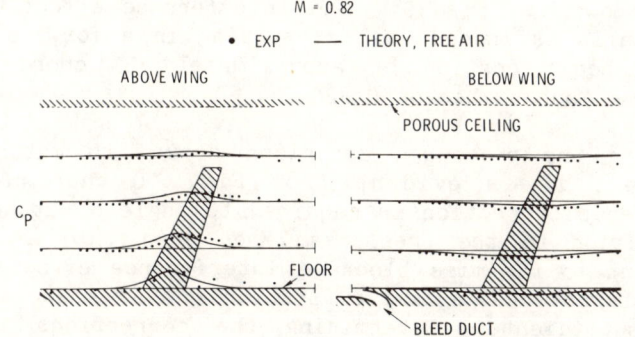

Fig. 9 Flowfield pressures for advanced transport wing.

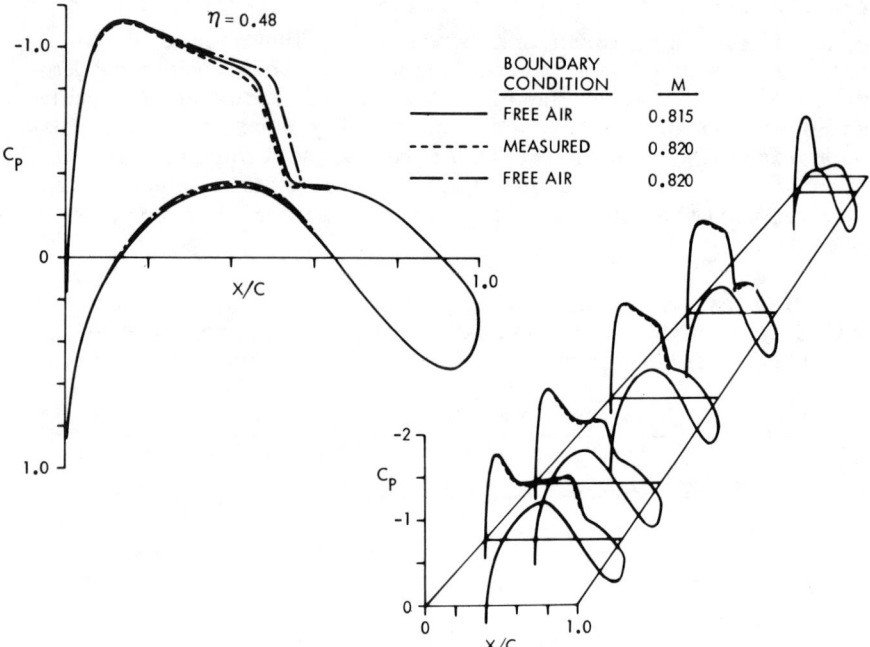

Fig. 10 Effect of outer boundary conditions on theoretical calculations for advanced transport wing.

in the flowfield at the tunnel wall locations. The difference in measured and theoretical free-air pressures near the tunnel wall for the transport wing are relatively small, indicating that the wall-induced interferences are small for this case.

When the experimental far field pressures (Fig. 9) are incorporated as boundary conditions in the transonic code,[11] the effect on the computed wing surface pressures are as shown in Fig. 10. The interference effect of the tunnel walls is indeed small resulting in a forward shift in the shock location of approximately 5% chord for a freestream Mach number of 0.820.

Using the procedure described above, to match wing pressures, it is evident from Fig. 10 that no lift interference correction is required to angle of attack and a correction to the freestream Mach number of -0.005 is sufficient to minimize blockage interference effects. It is of interest that a simple first-order correction appears to be sufficient. Determining the corrections in the manner of Fig. 10 for various test conditions, the results can then be applied to the experimental data.

Since most current approaches to developing theoretical/empirical corrections for wall effects at transonic speeds utilize experimental wall pressures, it is recommended that instrumentation to record wall pressures be strongly considered where it is suspected that wall interference effects are large. However, extreme care should be given to the way the wall pressures are obtained. For instance, it has been found that pressure rails such as those shown in Fig. 8 can cause a change in the test section Mach number on the same order of magnitude as the wall interference corrections. The "empty tunnel" calibration must include the effect of the rails on the Mach number level and distribution.

Using self-correcting or self-streamlining wind-tunnel walls. For situations where lift interference and blockage effects are expected to be large (e.g., big model in small test section), the simple wall interference corrections discussed above are not applicable. One solution to minimize wall interference for this situation is to use active rather than passive wind-tunnel walls. The use of

Fig. 11 Test section of Calspan self-correcting wind tunnel.

active walls allows the simulation of unconfined flow streamlines at the wind-tunnel walls. Generally, active walls fall into two categories: self-correcting walls and self-streamlining walls.

Self-correcting walls produce interference-free flows by controlling the flowfield in the vicinity of the walls through suction or blowing. This is accomplished by first measuring the components of the disturbance velocity near the wall. These measured velocities are then analyzed to see if they are consistent with the boundary conditions in the flowfield for unconfined flow. If they are not consistent, new values of wall suction and blowing are estimated. The iteration is continued until the measured wall velocities are consistent with the boundary condition for unconfined flow.

An example of a self-correcting wind-tunnel is the Calspan 1-Foot Wind Tunnel described in Ref. 12. This continuous flow facility has a two-dimensional test section with porous top and bottom walls of 22.5% open area. The plenum chambers behind the porous walls are divided into 18 segments, 10 on the top and 8 on the bottom. Each segment is connected to a pressure and a suction source through individual control valves.

A NACA 0012 airfoil model mounted in the Calspan Self-Correcting Wind Tunnel and test section instrumentation for measuring wall velocities are shown in Fig. 11. Typical results for this installation are presented in Fig. 12 (from Ref. 12).

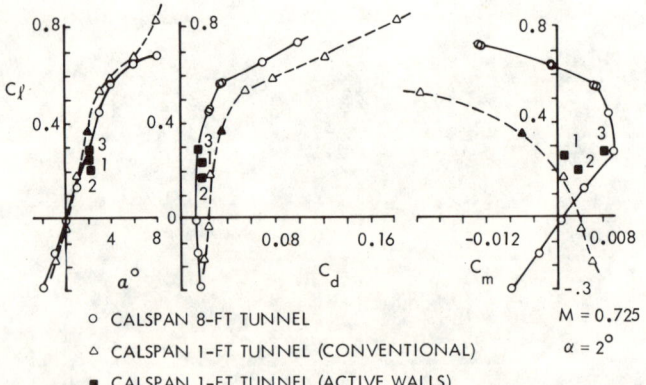

Fig. 12 Effects of wall control in self-correcting wind tunnel on NACA 0012 airfoil.

In Fig. 12 "interference-free" data obtained in the Calspan 8-Foot Wind Tunnel for the NACA 0012 airfoil are presented for reference (0.75% blockage). A second series of data is shown for the airfoil in the Calspan 1-Foot Tunnel with the ventilated walls passive (6.0% blockage). Obviously, the wall interference effects are considerable. For an angle of attack of 2 deg, the self-correcting mode is illustrated for three iterations. As can be seen, the third iteration is very close to the "interference-free" data.

A second approach to produce interference-free flow is to use self-streamlining walls. The concept is based on the wind-tunnel walls being flexible, allowing them to be moved to follow streamlines above and below the model. The criteria for determining the wall shape are based on static pressure measurements on the wall and theoretical potential flow calculations for the flowfield properties along the streamline path. If the measured and theoretical pressures at the wall show a pressure differential, the wind-tunnel walls are recontoured until the pressure imbalance is zero all along the wall. This fulfills the required boundary condition that the static pressure on either side of a streamline be equal.

Fig. 13 Test section of University of Southampton self-streamlining wind tunnel.

An example of a self-streamlining facility is the University of Southampton Transonic Wind Tunnel shown in Fig. 13. The facility test section is 6 in. wide with a variable height of 3 to 9 in. The airfoil model shown in the facility has a chord of 6 in. Results from this facility are presented in Ref. 13 and show considerable promise for minimizing wall interference effects.

Wind-Tunnel Models and Support Systems

Wind-Tunnel Models Must Be Tailored To Meet Objective of Experiment

The selection of the type of model to be used is dependent on the objective of the experiment and the chosen wind tunnel. Generally, models can be classified as: basic research models; concept development models; code verification models; configuration development models; or production aircraft models.

Basic research models. Basic research models are designed to produce experimental data of a fundamental

Fig. 14 Lockheed shock/boundary-layer interaction model installation in Lockheed CFWT.

nature -- usually for modeling of physical phenomena. For example, a basic research model specifically designed to investigate transonic shock-wave/boundary-layer interactions on a supercritical airfoil (from Ref. 14) is shown in Fig. 14. This half-airfoil is mounted on a special boundary-layer removal system in the floor of the Lockheed Compressible Flow Wind Tunnel. A motorized probe translates vertically to measure the airfoil flowfield.

The important point to be made about research models is that to obtain a unique set of data, the model may have to be unique in design and be integrated with the wind-tunnel facility.

Concept development models. Concept development models are configured generally to highlight a particular aerodynamic concept. For instance, the new winglet concept for reducing induced drag developed by Dr. Richard T. Whitcomb of NASA is shown in Fig. 15 (from Ref. 15) on a subsonic transport mounted in the NASA 8-Foot Transonic Wind Tunnel. This particular model installation (semispan) was selected to achieve a Reynolds number on the winglet sufficient to adequately evaluate its aerodynamic performance. To repeat, the model installation was selected to highlight an aerodynamic concept and evaluate its performance.

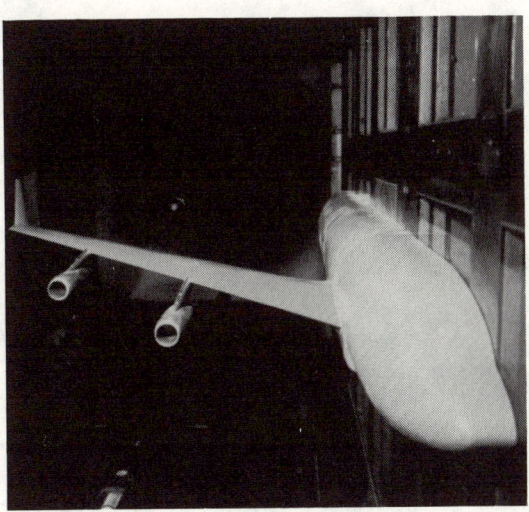

Fig. 15 NASA winglet model in NASA 8-foot wind tunnel.

Code verification models. Code verification models are designed to verify or calibrate theoretical aerodynamic analysis codes to allow their general use in aerodynamic design work. The wing model previously shown in Fig. 8 was developed solely for evaluating new transonic wing analysis codes.

Models for code verification tests should be as simple as possible. This greatly facilitates code correlations. However, to investigate simple configurations often leads to complicated test setups. Although not shown, the wing model in Fig. 8 is mounted on a special test wall that

Fig. 16 Grumman ground-attack/air-to-air advanced fighter configuration.

removes the tunnel wall boundary layer to promote realistic flows at the wing root.

Configuration development models. Configuration development models are models that are designed to demonstrate and refine the latest developments in advanced technology on aircraft that have definable future missions and markets. As an example of a configuration development model, an advanced dual role (ground-attack/air-to-air) fighter recently developed by Grumman[16] is presented in Fig. 16. This model was designed for testing in the 16T and 16S wind tunnels at AEDC. The configuration is designed to have efficient supersonic capability, excellent transonic maneuverability, and near STOL field performance.

The model test installation for configuration development models is selected to provide good performance data at the design conditions.

Production aircraft models. Production aircraft models are designed with an emphasis on complete aircraft performance throughout its operating range. A production model of the Lockheed C-5A mounted in the Calspan Wind Tunnel is shown in Fig. 17. Great care is taken to minimize any type of wall interference or support interference.

Model Support System Effects Must be Minimized

The selection of the model support system to be used in an experiment is dependent on the objective of the

Fig. 17 Lockheed C-5A aircraft configuration in Calspan wind tunnel.

experiment and the geometry of the model. Generally, support systems fall into two categories: wall mounts and sting mounts.

Wall mount. For wall mounted configurations, the principal problem is how to eliminate the effect of the tunnel wall boundary layer on the model.

For two-dimensional airfoil models attached to the tunnel walls, the wall effects can be eliminated by having a large span model (e.g. Fig. 4) or by a wall boundary-layer removal system.

For semispan aircraft models mounted to the tunnel wall, a variety of methods have been used to minimize the wall boundary-layer effects. One method is to use a splitter plate to divert the wall boundary layer. Such an installation is shown in Fig. 18 (from Ref. 17) for a

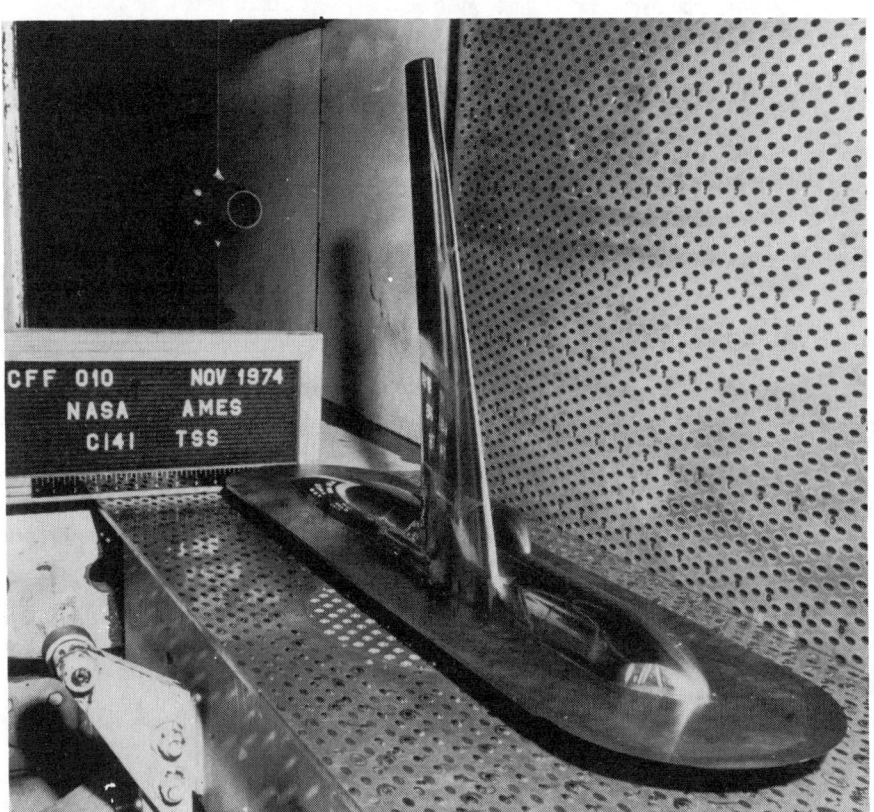

Fig. 18 Lockheed C-141 aircraft semispan model in Lockheed CFWT.

Lockheed C-141 model nonmetrically mounted in the Lockheed Compressible Flow Wind Tunnel. In order for this installation to produce satisfactory data, great care must be exercised to avoid spillage around the edges of the plate.

In Fig. 15, a second type of semispan installation is shown. The fuselage of the winglet model is nonmetric and the wing is metric. Also, the fuselage is displaced from the wall a distance approximately equal to the thickness of the tunnel wall boundary layer. The major problem encountered in this setup is the interference between the metric and nonmetric portions of the model.

A third wall mount installation is shown in Fig. 19. The wall boundary layer is diverted out of the tunnel just prior to the model location. The entire transport configuration is metric for this installation in the Lockheed Compressible Flow Wind Tunnel. Care must be taken

Fig. 19 Lockheed advanced transport aircraft model in Lockheed CFWT.

in this setup to prevent leakage between the model and wall.

Sting mount. Several problems arise when a sting is used as a model support. First, a rear entry sting for some configurations can cause substantial modifications to the aircraft fuselage geometry. This is evident in Fig. 20 where an advanced Lockheed $M = 0.95$ transport is shown mounted in the NASA 11-Foot Transonic Wind Tunnel. The aft fuselage geometry has been modified substantially from that which would exist on a real aircraft. The same can be said of the Grumman fighter configuration in Fig. 16. For both these configurations, the aerodynamics of the aft fuselage are not realistically simulated.

A second problem with a sting mount is that its presence interferes with the flowfield around the aircraft. In Fig. 17, the presence of the rear entry sting obviously alters the flowfield around the upswept fuselage afterbody.

Fig. 20 Lockheed ATT-95 transport aircraft in NASA 11-foot wind tunnel.

EXPERIMENTAL TESTING AT TRANSONIC SPEEDS

Fig. 21 Lockheed advanced transport aircraft configuration in Calspan wind tunnel.

One solution to minimize the sting interference is to use a blade sting. In Fig. 21, a blade sting is used on an advanced Lockheed transport model in the Calspan 8-Foot Wind Tunnel. This installation allows the model to be attached to the sting in an area where its effect on the model aerodynamics is minimized. However, care must be taken with a blade sting design to avoid a large increase in the tunnel blockage.

For experiments near a Mach number of 1, the blade sting has the disadvantage of altering the configuration area distribution. This could result in spurious drag data for conditions where area considerations affect wave drag.

For production aircraft experiments, more than one sting installation should be used to allow the effects of sting interference to be isolated. A typical study of this type is found in Ref. 18.

Power Plant Simulation Must Be Done With Care

The selection of the method of power plant simulation on the wind-tunnel model must be done with great care in order to obtain realistic results in the wind tunnel. For configurations where the power exhaust is not expected to

Fig. 22 NASA advanced turboprop transport model.

interfere significantly with the rest of the configuration, flow through jet simulators can be used. Typical of these engine installations are the fighter and transport configurations shown in Figs. 16 and 17, respectively.

For aircraft configurations where the effects of the power plant on the aircraft flowfield are expected to be significant, the power plant must be correctly simulated on the model. An example, where a powered model is required, is presented in Fig. 22. This figure shows an advanced NASA turboprop mounted on a supercritical wing[19] in the NASA Ames 14-Foot Transonic Wind Tunnel. Obviously, for this configuration, there will be significant effects of the turboprop flowfield on the wing aerodynamics.

The primary difficulty encountered with powered models is the accurate accounting of the various forces and tares involved with using a powered simulator. Careful calibration of the powered simulator is required in isolation and in the presence of the model.

Reynolds Number Simulation

Reynolds Number Scale Effects Can Cause Significant Problems

The Reynolds number capability of today's operational large transonic wind tunnels is substantially less than that required for simulating the aerodynamics of current aircraft. This is illustrated in Fig. 23 (after Kuhn). As larger and larger aircraft are developed, the Reynolds number gap continues to grow.

An example of the difficulty that occurs when Reynolds number is not correctly simulated is shown in Fig. 24 from Ref. 20. For the C-141 aircraft, the full-scale flight pressure data in Fig. 24 is significantly different from that measured in the wind tunnel at low Reynolds number. Clearly, this gives the aircraft designer a substantial problem when he must use low Reynolds number data to establish the performance of his configuration.

The major problems resulting from Reynolds number scale effects are: pressure gradient-induced boundary-layer separation; changes in boundary-layer thickness; and shock-induced boundary-layer separation.

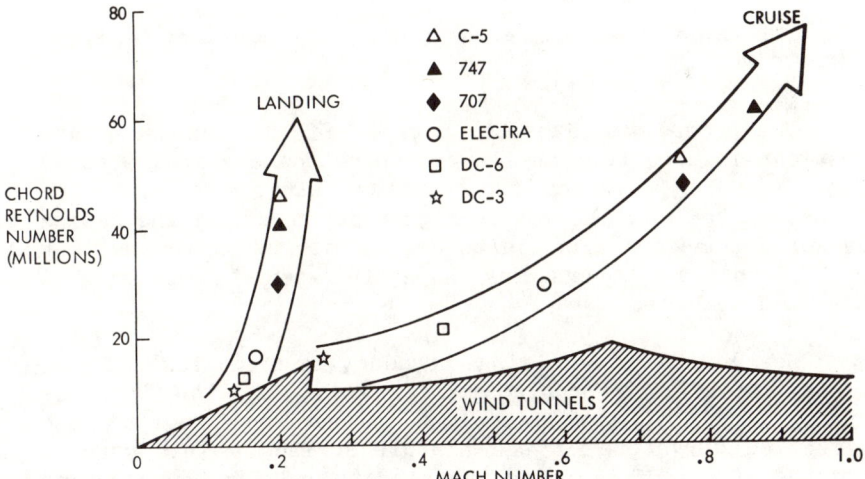

Fig. 23 The growing Reynolds number gap.

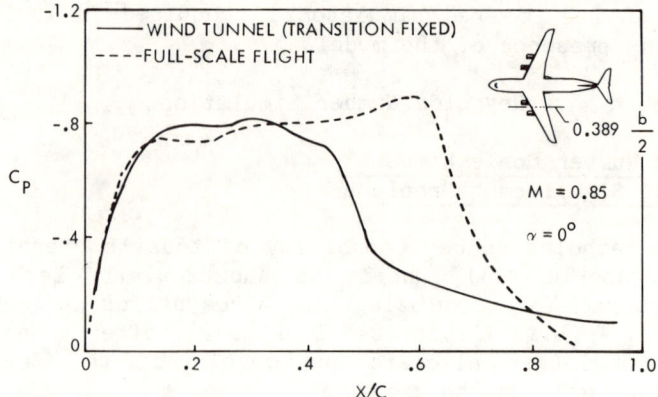

Fig. 24 Lockheed C-141 pressure distributions.

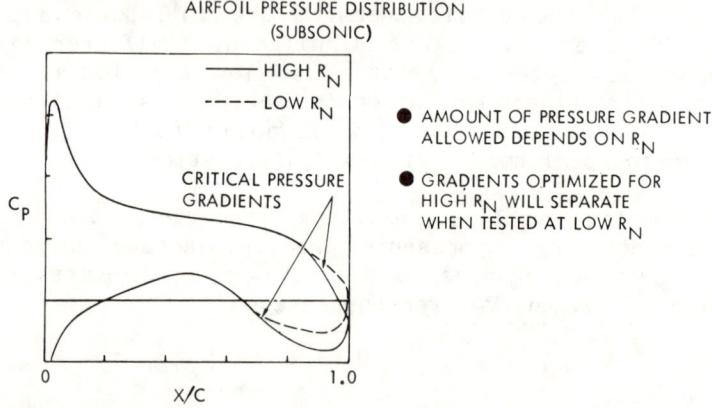

Fig. 25 Scale effect on pressure gradient induced boundary-layer separation.

The problems due to scale effects on pressure gradient-induced boundary-layer separation are indicated in Fig. 25. For advanced supercritical wings the amount of pressure gradient allowed is highly dependent on the design Reynolds number. For instance, gradients optimized for high Reynolds numbers may separate when tested at low Reynolds numbers.

Scale effects on wing boundary-layer thickness can cause significant problems (Fig. 26). At low Reynolds numbers the boundary layer is thicker and reduces the effective curvature of the airfoil surface. This usually has the effect of reducing the airfoil or wing camber. For advanced airfoils, the low Reynolds number data will

EXPERIMENTAL TESTING AT TRANSONIC SPEEDS 217

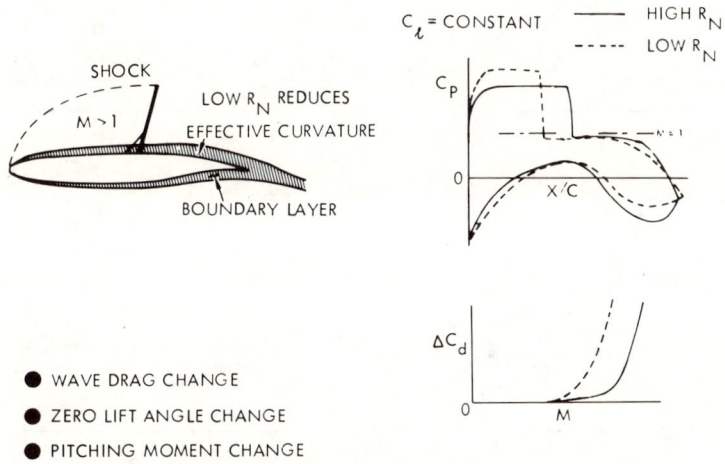

Fig. 26 Scale effect on boundary-layer thickness.

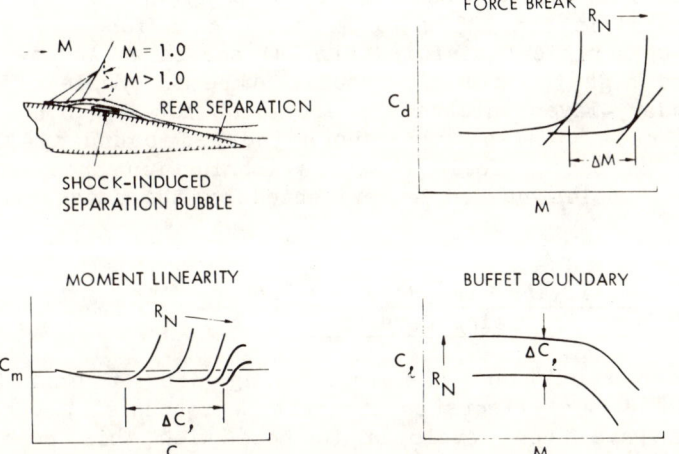

Fig. 27 Scale effect on shock-induced boundary-layer separation.

produce a higher shock drag (at constant lift), a more positive zero lift angle of attack, and more positive pitching moment coefficient relative to that experienced at high Reynolds numbers.

Scale effects on the wing pressure distribution due to shock-induced boundary-layer separation can be substantial as has already been shown in Fig. 24. The changes in force data due to scale effects on shock-induced boundary-layer separation are summarized in Fig. 27.

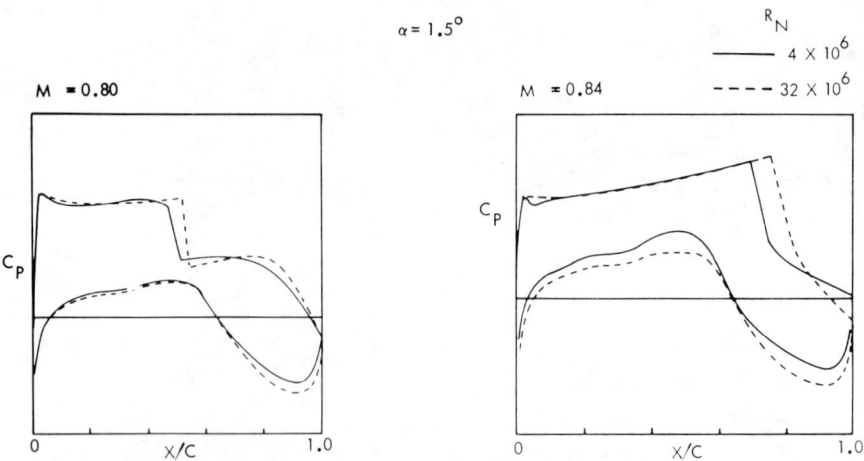

Fig. 28 Scale effect on NASA 10% thick supercritical airfoil.

Reynolds number scale effect problems are further illustrated in Fig. 28 where pressure data for a NASA 10% thick supercritical airfoil (Fig. 4) are shown at low (4 x 10^6) and high (32 x 10^6) Reynolds numbers. Scale effects on boundary-layer thickness are evident for M = 0.80, the design case. Also, the shock-induced boundary-layer effects on the airfoil pressure distribution are changed due to Reynolds number as evidenced by the data for M = 0.84.

Ways of Quantifying High Reynolds Number Scale Effects Are Being Developed

As a result of the increasing need to quantify Reynolds number effects on aircraft performance, a variety of ways have been developed to cope with this problem: fixed transition at model leading edges; natural transition; vortex generators; fixed transition aft on the model; panel model; component testing; and high Reynolds number facilities.

Fixed transition at model leading edges. The earliest method of scale effect simulation was to produce an all turbulent boundary layer over the model at low Reynolds numbers. This eliminated any laminar boundary-layer shock-induced separations which are different in character from those in turbulent boundary layers. Also, a turbulent boundary layer greatly facilitated extrapolation of low Reynolds number skin friction drag data to full-scale

Reynolds number conditions. Unfortunately, the problems noted in Figs. 25 to 27 were still present.

Natural transition. The objective of this approach is to approximate the high Reynolds number boundary-layer characteristics at low Reynolds numbers by allowing the boundary layer to transition naturally aft on the model. This has the overall effect of thinning the low Reynolds number boundary layer, thereby more closely simulating the flight boundary-layer characteristics in regions sensitive to pressure gradient-induced separation (Fig. 25) and in the shock/boundary-layer interaction region (Fig. 27). Furthermore, the low Reynolds number viscous uncambering effects are reduced (Fig. 26).

An example of the above approach is shown in Fig. 29 for a Lockheed advanced cargo transport at low Reynolds numbers. For transition fixed at the wing leading edge, the boundary layer on both the upper and lower surfaces are separated. For natural transition, the transition point moves aft on the wing for low Reynolds numbers resulting in a thinner turbulent boundary layer over the aft part of the wing. As can be seen, for natural transition, the separation bubble and trailing edge separation disappear, and the aft loading is substantially increased. Significant improvements are also seen in the drag data.

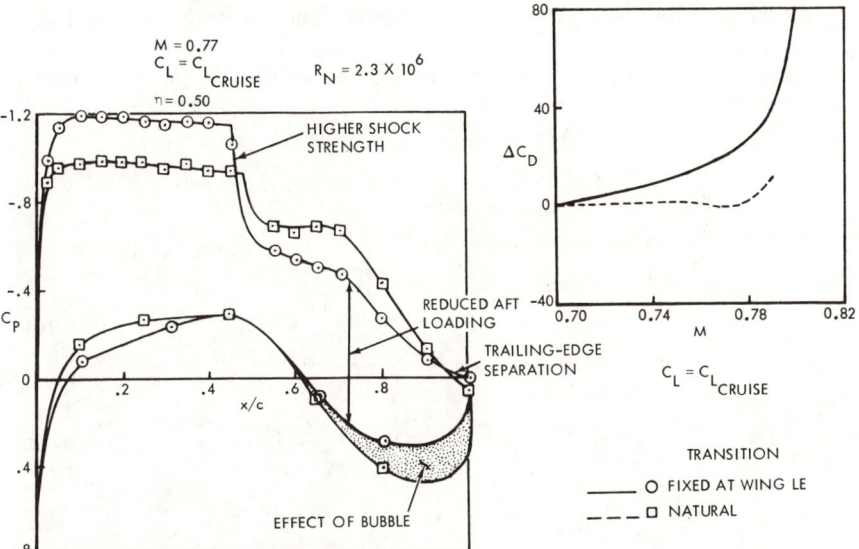

Fig. 29 High R_N simulation on advanced transport using natural transition at low R_N.

Unfortunately, there are several drawbacks to this approach. First, considerable chordwise variation of transition location is possible when the shape of the pressure distribution changes with Mach number or angle of attack. This introduces considerable problems in data analysis. Furthermore, since the transition location on the wing is not generally known for the test conditions, difficulties arise in correcting the low Reynolds number drag level (skin friction effects) to full-scale values. Another problem is the possibility of a laminar shock/boundary-layer interaction which is considerably different in character and sensitivity to the turbulent shock/boundary-layer interaction experienced in flight.

Vortex generators. The objective of this approach is to re-energize the low Reynolds number turbulent boundary layer following a transonic boundary-layer interaction using vortex generators such that the pressure gradient-induced separation at flight conditions is simulated. Wind-tunnel data was obtained using this approach on the Lockheed C-141 aircraft with the vortex generators located at 55% wing chord. These results are presented in Fig. 30. Reasonable correlation of the wing shock location can be seen with flight tests using this approach.

There are several problems with the use of this method. First is the actual design of the vortex generators, their spacing, and their location on the wing. The second is their effect on drag level which is an unknown quantity. A third problem is that they become ineffective if they are covered by a shock-induced separation bubble (Fig. 27).

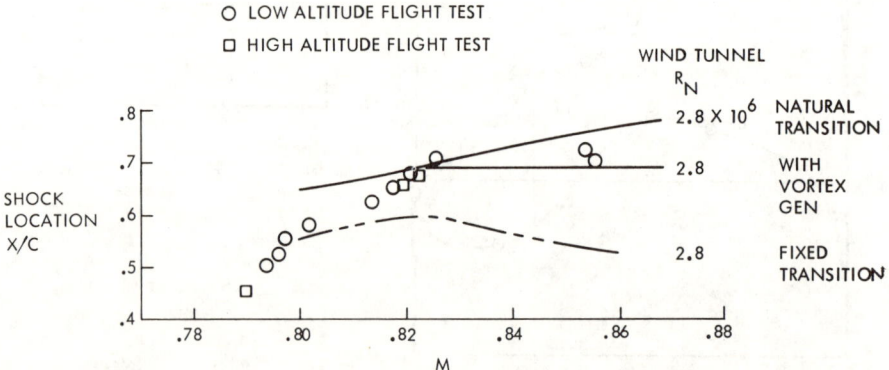

Fig. 30 High R_N simulation on Lockheed C-141 using vortex generators.

Fixed transition aft on the model. This scale effect simulation concept was proposed by Blackwell[22] to alleviate some of the problems associated with the natural transition approach previously discussed. The objective of this approach is to determine a fixed transition location on the wing at low Reynolds numbers that results in boundary-layer conditions at the wing trailing edge similar to that experienced in flight. This is known as the trailing edge criteria. The rearward movement of the transition location is constrained such that the boundary layer is tripped prior to the shock waves and hence a turbulent shock/boundary-layer interaction occurs. An example of this low Reynolds number and high Reynolds number trailing edge boundary-layer matching procedure is illustrated in Fig. 31. The method is heavily weighted toward airfoil designs that experience shock-induced boundary-layer separation at the wing trailing edge.

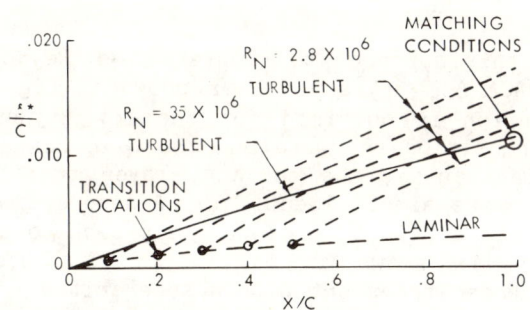

Fig. 31 High R_N simulation using variable location artificial transition.

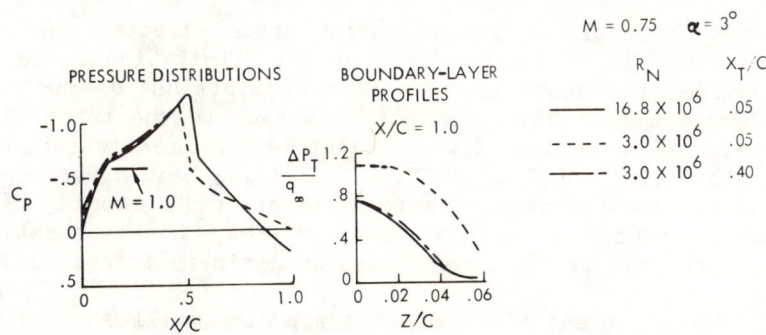

Fig. 32 Transition simulation of scale effects on NACA 65_1-213 airfoil.

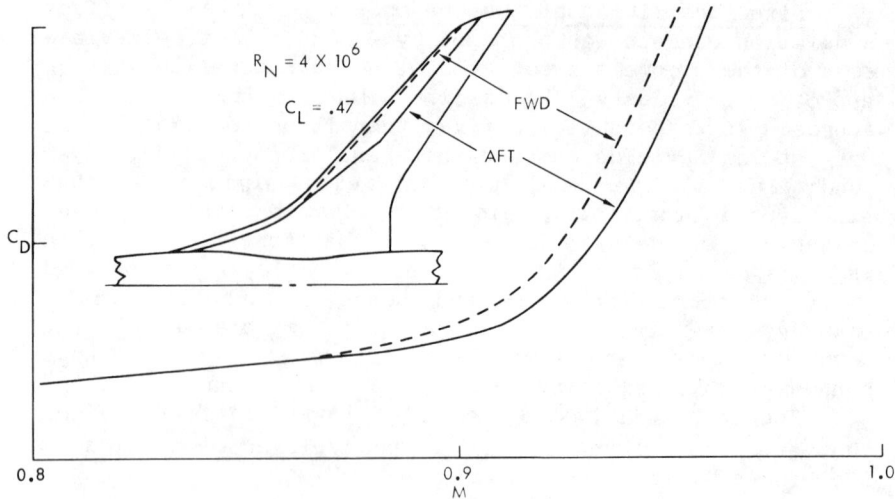

Fig. 33 High R_N simulation on ATT-95 transport using fixed aft transition at low R_N.

Use of this approach to simulate high Reynolds number data on a NACA 65_1-213 airfoil is shown in Fig. 32. Very good correlation is obtained for the low Reynolds number data with transition rearward and the high Reynolds number data. In particular, the boundary-layer profiles at the trailing edge are almost identical. A second example which illustrates the application of this approach to an aircraft configuration is shown in Fig. 33. The model is an advanced Lockheed transport with a supercritical wing. The aft transition location can be seen to produce an effect on airfoil drag rise similar to that which would be expected from scale effects due to boundary-layer thickening (Fig. 26).

For general use in simulating scale effects, the aft transition fixing approach has several limitations. To be effective, the shape of the pressure distribution must be conducive to promoting laminar flow back to the transition strip. This is not always possible, in particular for "supercritical" or "peaky" airfoils at low Mach numbers and all airfoils with considerable angle-of-attack loading. If natural transition occurs ahead of the fixed transition location, uncertainty in the data analysis is introduced.

The aft transition location approach still appears to be the best available procedure (as opposed to natural transition and vortex generators) to determine the

EXPERIMENTAL TESTING AT TRANSONIC SPEEDS 223

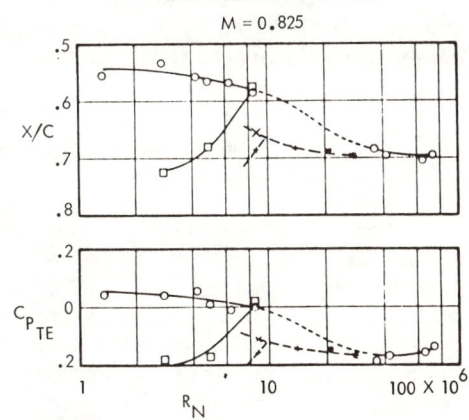

Fig. 34 High R_N simulation on Lockheed C-141 using panel model.

sensitivity of a wing to scale effects when the aerodynamicist is forced to evaluate his design in a low Reynolds number wind tunnel. However, only selective tests should be conducted for conditions where the wing boundary layer is approaching separation or a shock/boundary-layer interaction occurs and only then when there is assurance that the wing pressure distribution will promote laminar flow back to the transition strip.

Panel model. The panel model approach to simulating wing scale effects was developed by Cahill et al. of Lockheed.[23] This method consists of taking a wing section representative of an aircraft wing and building a large chord panel model of constant section to test in a large conventional wind tunnel. The objective of this approach is to determine the scale effects on a panel model in the hope that they will be indicative of the scale effects on the wing.

An example of this approach is indicated in Fig. 34. The panel model shown in this figure was developed using a wing section (0.389 b/2) representative of the Lockheed C-141 aircraft wing. Contoured end plates (Fig. 34) to approximate infinite span conditions were necessary before a reasonable scale effect simulation could be obtained.

A comparison of the panel model results to data for a complete wing is presented in Fig. 34. As can be seen, the panel model approach does not correctly simulate the magnitude of the scale effects present on the wing. It is felt that this discrepancy is primarily due to the lack of proper simulation by the panel model of the complete three-dimensional wing environment.

Clearly, before this procedure can be used, the ability of the experiment to reproduce the physics of the actual flow to be simulated must be assured.

Component testing. Component testing is a method whose use is rapidly increasing for determining Reynolds number scale effects. This method consists of testing small models in relatively small wind tunnels capable of generating high Reynolds numbers either by operating at high total pressures and/or cryogenic temperatures. Testing small models representing aircraft components such as airfoils, wings, fuselages, empennages, and pylon/nacelles allows the sensitivity of an aircraft design to Reynolds number to be assessed by analyzing the sensitivity of its components.

The merit of the component test approach using small high Reynolds number facilities has already been demonstrated for airfoils in Fig. 28. A more stringent test of the approach is shown in Fig. 35 (from Ref. 17)

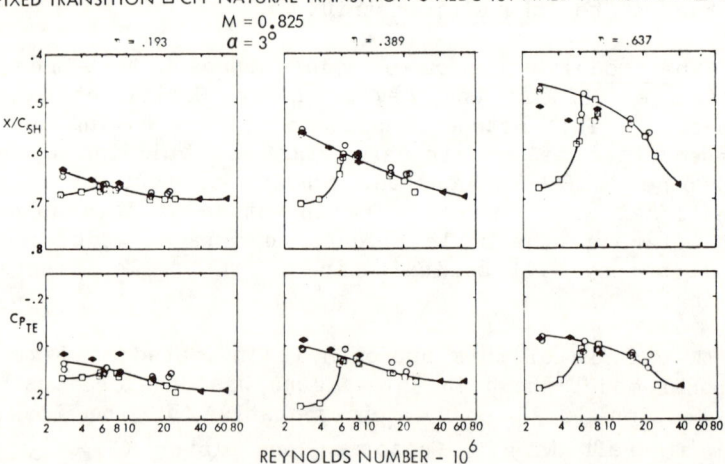

Fig. 35 High R_N simulation on Lockheed C-141 using semispan model.

where data for the C-141 semispan wing-body model (Fig. 18) is compared to low Reynolds number wind-tunnel data (AEDC 16T) and high Reynolds number flight data for the complete aircraft. Both shock location and wing trailing edge pressure data are presented. The correlation of the semispan model data from the other sources can be seen to be quite good over the entire Reynolds number spectrum.

High Reynolds number facilities. Currently most high Reynolds number transonic wind tunnels in operation or being planned utilize either high pressures and/or cryogenic temperatures.

Typical of new facilities that can test a complete aircraft model at fullscale Reynolds numbers is the National Transonic Facility.[24] This facility is under construction at NASA/Langley and is scheduled for completion in 1982. An artist's perspective of the transonic wind tunnel is shown in Fig. 36. The test section is 2.5 m^2. Using both pressurization and cryogenic temperatures, a chord Reynolds number of approximately 120 million can be achieved for a typical aircraft model (compare to Fig. 23).

If scale-model tests can be conducted at flight Reynolds numbers, it would appear that the Reynolds number

Fig. 36 Perspective of NASA National Transonic Facility.

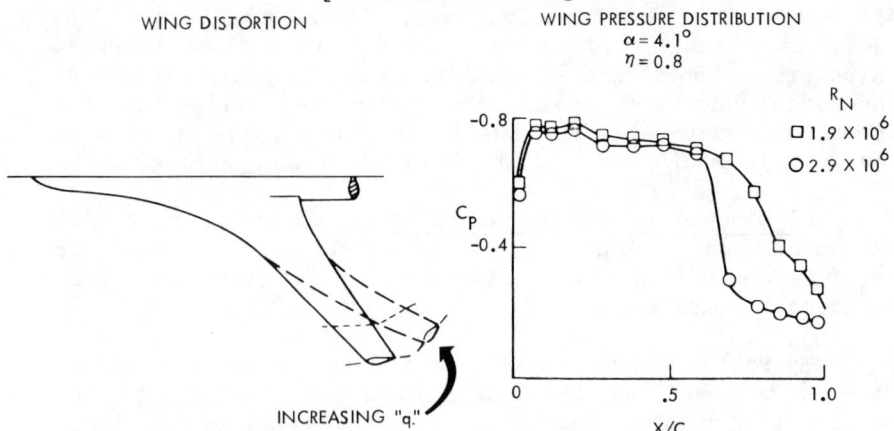

Fig. 37 Aeroelastic problems in high pressure wind tunnels.

simulation problems would be solved. Unfortunately, this is not so, since several problems accompany high Reynolds number testing. Typical of these problems are:

1) Model structural design for cryogenic temperatures and high model loads.

2) Model instrumentation in a cryogenic environment.

3) Model smoothness criteria for a highly thinned boundary layer.

4) Sting size relative to model size for high model loads.

5) Aeroelastic effects.

Each of these problems are discussed in detail in the High Reynolds Number Research Workshop proceedings.[24]

As an example of the problems encountered in high Reynolds number testing, the problem due to aeroelastic effects is illustrated in Fig. 37 for an advanced NASA supercritical wing.[24] As Reynolds number is increased on this wing, the increased dynamic pressure causes the wing tip to unload driving the shock wave forward. This is in direct conflict to the way the shock wave moves for a rigid model with increasing Reynolds number (see Fig. 28).

Aerodynamic Flow Visualization

Flow Visualization Is Important To Understanding the Physics of the Experiment

In designing an experiment to yield a maximum amount of technical information, the importance of flow visualization cannot be minimized. Too often, the aerodynamicist does not have sufficient information to render the proper interpretation of the data. For instance, two-dimensional understanding of flow separation is often applied to three-dimensional phenomena. Also, based on model surface flowfield data (i.e., oil flows), erroneous conclusions to the nature of the global flowfield have been made.

Flow Visualization Can Be Used To Highlight Aerodynamic Flow Phenomena

Flow visualization techniques generally fall into three categories:

1) Boundary-layer transition visualization.

2) Surface streamline visualization.

3) Flowfield visualization.

Boundary-layer transition. The location of the boundary-layer transition on the model must be accurately known in order to properly interpret the model data. In particular, if artificial transition strips are used, it is important that the transition strips be sized to ensure that transition does indeed occur at the strip location.

The technique principally used to detect the transition location is sublimation. A mixture such as a fluorene/trichlorethane solution is sprayed on the model surface. As the wind-tunnel flow is started, the mixture is sublimed away from the model surface. The sublimation rate varies depending on the shear at the model surface. Since the surface shear stress is dramatically different for laminar and turbulent flows, two contrasting regions of different color result. The line of demarkation between the two regions is the transition location.

Surface streamlines. The principal method of visualizing surface streamlines in transonic wind tunnels is by oil flows. A mixture such as titanium dioxide and oil is painted on the model prior to testing. After the flow stabilizes, photographs are taken.

A typical oil flow on an advanced Lockheed transport wing at off-design conditions is shown in Fig. 38. Note that the separation lines and shock waves show up clearly in the oil flow.

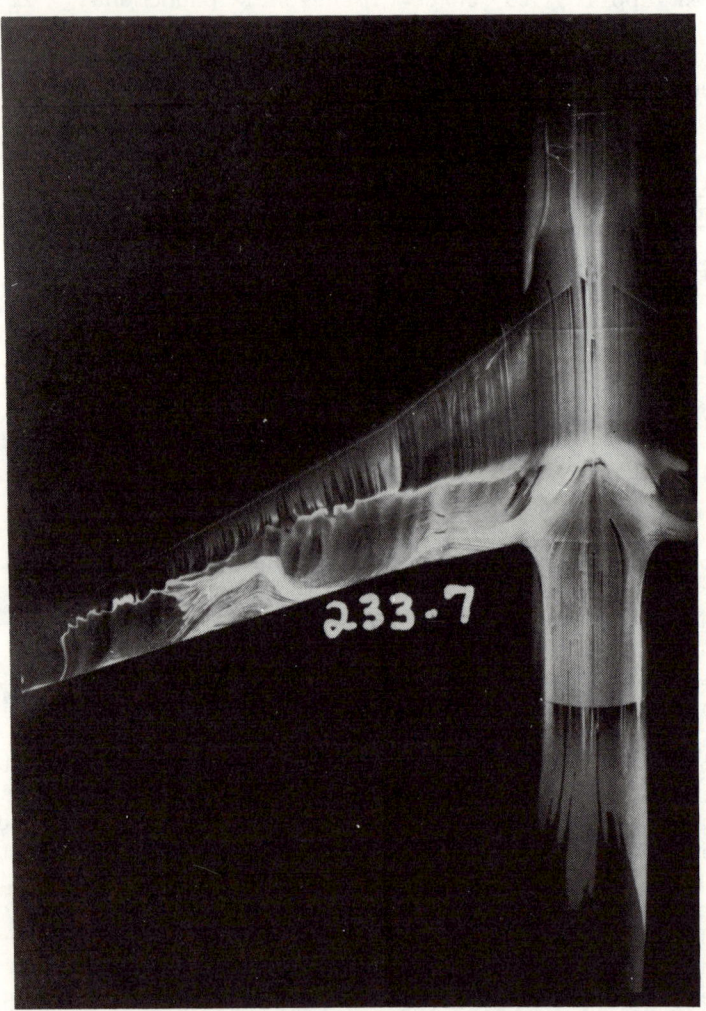

Fig. 38 Flow visualization on transport wing using oil flow.

A more innovative approach developed by ONERA uses a kerosene-base fluid with paint colorant that issues from orifices in the model surface. Using different colors, the flow streamlines and the location of shock waves and separation regions can be established. This technique permits the observation of more than one stabilized flow condition during a tunnel run.

A new technique for visualizing surface streamlines called "fluorescent minitufts" has recently been developed by Crowder et al.[25] The basis of the method is the use of extremely thin nylon monofilament for the minitufts, a process of attaching them to the model surface with small drops of lacquer-type adhesive, and the use of fluorescense photography for recording the visual data.

Figure 39 shows the appearance of fluorescence minitufts on a Boeing empennage research model. Approximately 1,050 tufts were applied to the starboard side of the vertical tail and rudder and 800 tufts were applied to the lower surface of the horizontal tail and rudder. These minitufts were made from nylon monofilament with a diameter of 0.0019 in. and a length of 1.5 in.

Inspection of Fig. 39 indicates the flow streamlines and vortical flows present on the empennage. An obvious

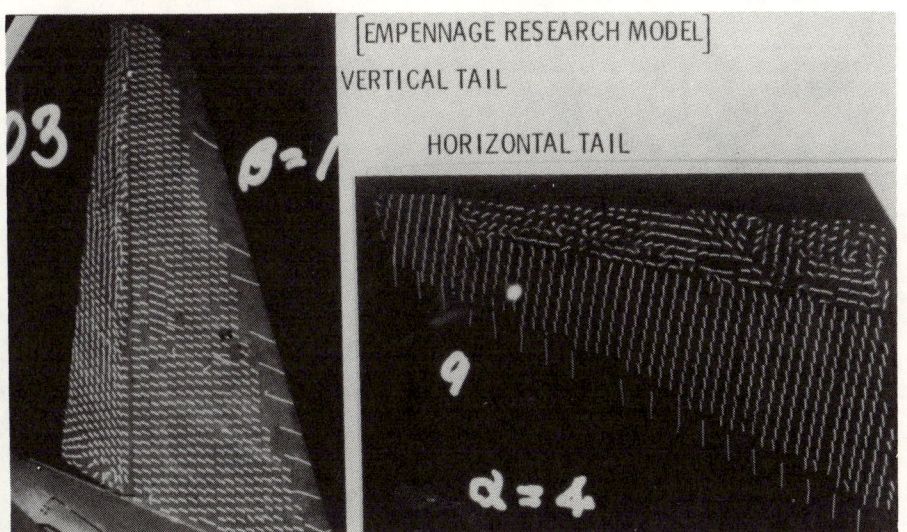

Fig. 39 Flow visualization on transport empennage using fluorescent minitufts.

advantage of the fluorescense minituft method is that flow visualization can be obtained simultaneously with other measurements. In Ref. 25, results at transonic speeds indicate that drag data on a transport model are changed by approximately 2 drag counts when tufts are used with the tufts-on data showing less drag.

Flowfield visualization. Many of the flowfield phenomena around an aircraft are not speed or Reynolds number dependent. One such phenomena is vortex flows which are present on maneuvering fighter aircraft or on cargo transport fuselage afterbodies. These flows can be visualized with different colors effectively and inexpensively using a water tunnel.

A water tunnel photograph of the Northrop F-18 fighter at an angle of attack of 35 deg is shown in Fig. 40.[26] This photograph obtained in the Northrop water tunnel depicts forebody and strake vortices, vortex interaction, and vortex breakdown using various color dyes.

A new technique has been developed by Crowder et al.[25] for flowfield surveys around aircraft at transonic speeds called Photo-Graphical Flow-Field Survey Technique. This method allows total pressure isobar patterns around the aircraft to be mapped and displayed graphically where different colors correspond to different total pressure levels.

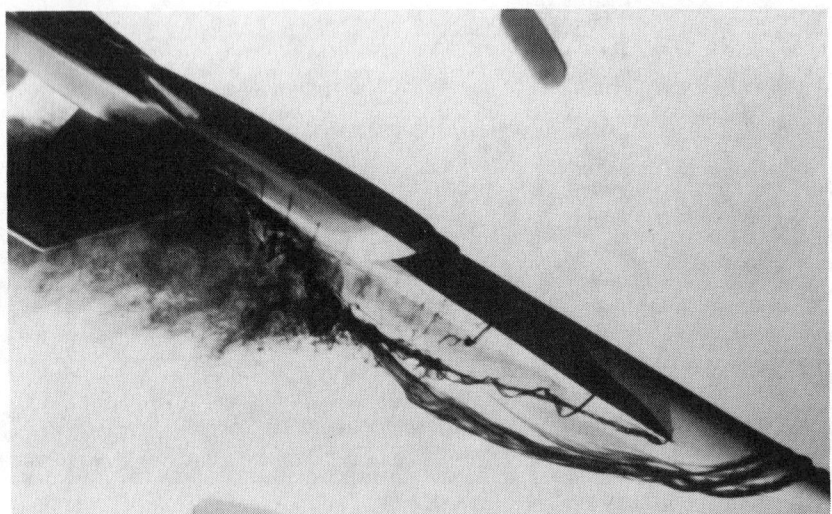

Fig. 40 Flow visualization on F-18 in Northrop water tunnel.

Fig. 41 Flow visualization of transport wake using photographical flowfield survey technique.

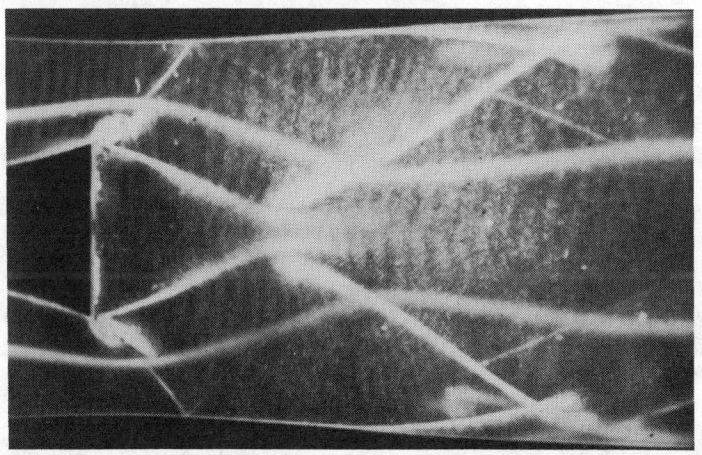

Fig. 42 Flow visualization of wake from wedge-shaped plug with rounded leading edge using simultaneous smokeline and Schlieren.

The photographical flowfield survey technique uses a pressure probe to transverse the flowfield. The data are transmitted off-board the transverse mechanism by a multicolored signal light projected onto Polaroid film by a camera lens. Figure 41 illustrates the use of the technique to display the total pressure isobars in a wing-body wake of a transport airplane model.[25] The vortex nature of the flow is quite evident.

Throughout the years, one of the most important techniques of flowfield visualization has been the Schlieren system. A Schlieren photograph which illustrates the shock waves in a transonic wake behind a wedge shaped plug with rounded leading edge[27] is presented in Fig. 42.

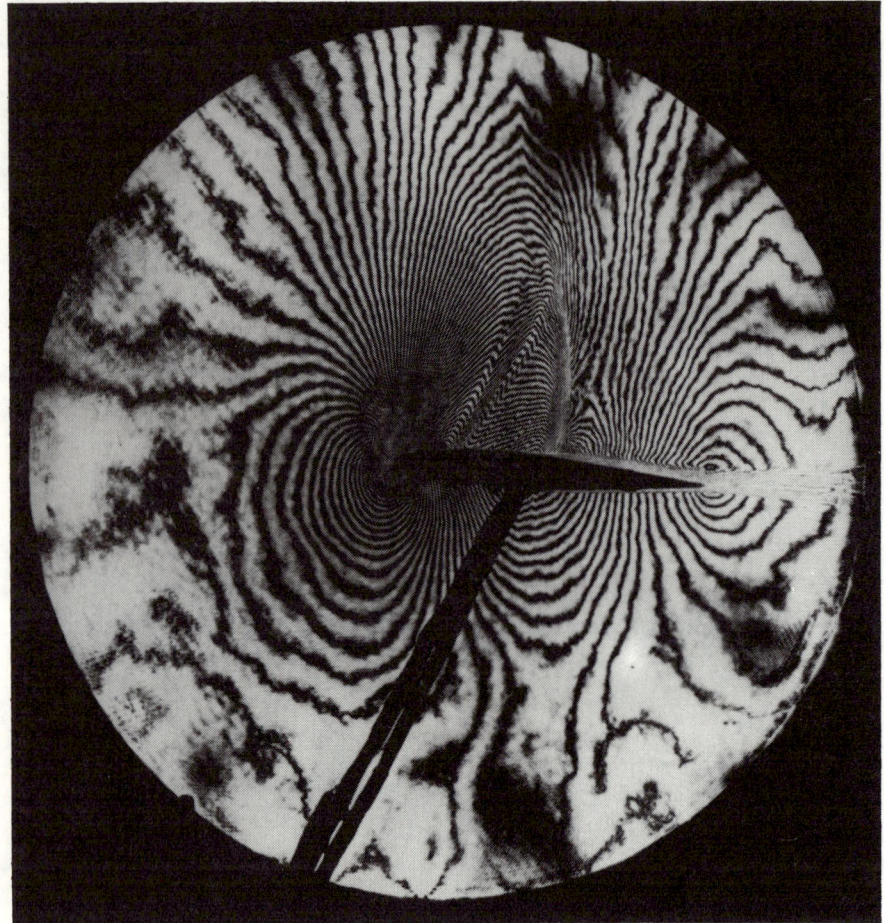

Fig. 43 Flow visualization of flowfield on NACA 64A010 airfoil using dual-plate holography.

Superimposed on the Schlieren photograph in Fig. 42 are smokelines depicting the path of the streamlines through the flow. Smokelines are used for flow visualization extensively at low speeds; however, as shown by Mueller[27] smokelines can also be used to advantage at transonic speeds.

In Fig. 43 an interferogram is shown for transonic flow over a NACA 64A010 airfoil, in the NASA Ames 2-Foot Wind Tunnel.[28] This particular flow visualization technique is called holographic interferometry. The fringe pattern illustrates the flowfield density variation, shock waves, and boundary layer around the airfoil. From the measurements of the density field, Mach number contours can be quantified.

Flow Measurement Instrumentation

Standard Instrumentation

Standard instrumentation is instrumentation which is generally used in most experiments and includes internal strain-gauge balances and scanivalve pressure transducers.

Typical photographs of this instrumentation are presented in Figs. 44 and 45. The balance is used to measure forces and moments on the model. The scanivalve pressure transducer is used in conjunction with pressure tubes installed in the model surface to measure the static pressure on the model surface.

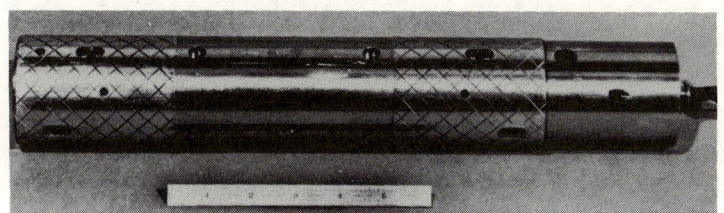

Fig. 44   Internal six-component strain-gauge balance.

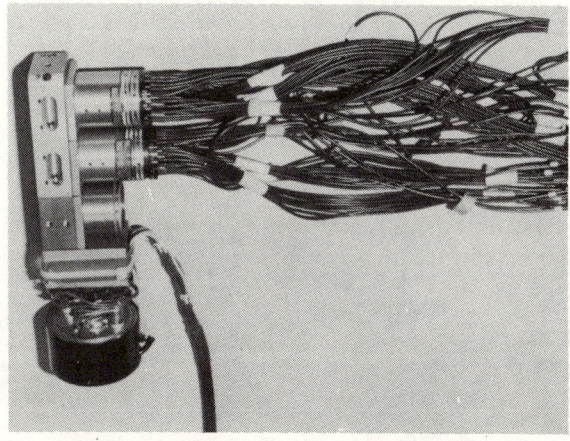

Fig. 45   Scanivalve pressure measurement system.

The use of the above instrumentation is well advanced in transonic testing. The principal problem in the use of this instrumentation arises when the balance or transducer measurement range is large in comparison to the actual measurements. When this occurs, considerable inaccuracies are introduced into the data. The obvious solution is to design the experiment so that the instrumentation selected is compatible with the expected measurements.

Specialized Instrumentation

For many of the types of models discussed earlier (Fig. 14-17), specialized instrumentation is required to obtain the needed data. Typical instrumentation that falls into this category are: pressure probes and rakes, hot wire probes, hot film probes, skin-friction measurement devices, and laser velocimeters.

Pressure probes are used to measure static and dynamic pressures in the flowfield and in the model boundary layer. A traversing pressure probe capable of measuring two velocity components in the airfoil boundary layer is shown in Fig. 14.

A photograph of a standard DISA hot wire probe is shown in Fig. 46. Hot wires have been successfully used at transonic speeds to measure flowfield and boundary-layer flows by Hortsmann et al.[29] In Ref. 29, the hot wires were backed with epoxy to improve the sturdiness of the wire at transonic speeds.

Specialized hot film probes have been developed by Roos[30] to measure shock wave oscillations at transonic speeds. The hot film probe is shown in Fig. 47 inserted in the flowfield around a supercritical airfoil.

Skin friction measurement devices are used to measure the skin friction of the surface directly. A variety of

Fig. 46 Hot wire probe.

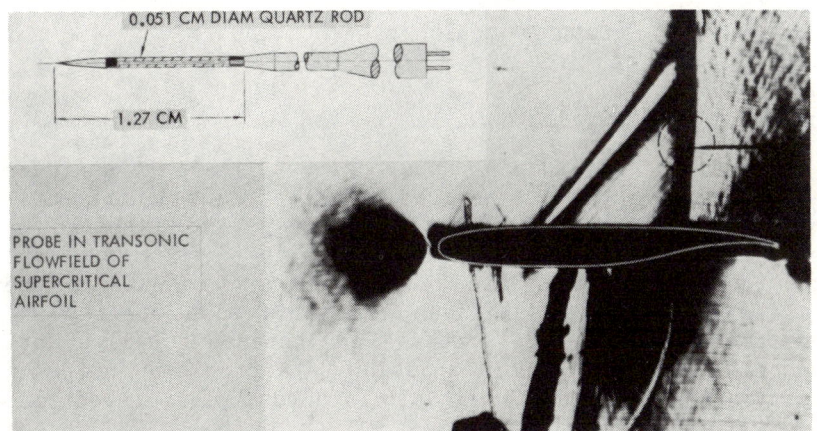

Fig. 47 Hot film probe used to locate unsteady shock wave movement.

methods such as Preston tubes and Stanton tubes have historically been used for this purpose.

The instrumentation for making flowfield and boundary-layer measurements, that have been described above, have one common problem: they interfere with the flow. Obviously, to minimize their interference the size of the probes relative to the model size must be kept as small as practical.

A new technique for specialized measurements, whose use is rapidly increasing, is the laser velocimeter. Currently, lasers are being used extensively to measure flow properties in high speed jet flows and in aerodynamic flows. Problems currently restricting the general use of lasers for flowfield measurement are:

1) Size of measurement volume (restricts use for boundary layers).

2) Flow seeding techniques.

3) Relative vibration between laser beam and model.

4) Measurement time is long compared to run time of blowdown tunnels.

Substantial work is currently underway in government, industry, and universities to find solutions to these problems.

Data Reduction/Presentation

On-Line Data Reduction

In the performance of an experiment, the test engineer is at a disadvantage if he does not have reduced data until the test is over. If practical, every effort should be made to include on-line data reduction in the design of the experiment. On-line data reduction greatly enhances the ability of the test engineer to obtain high quality and more useful experimental data. For instance, on-line data reduction aids in trouble shooting data problems, helps in decision making, and allows judicious test point selection.

Data Presentation

If large quantities of data are obtained in an experiment, faster ways of preparing the data for analysis are needed. One method is on-line plotting. Another is to utilize computer graphics to assist in data presentation. Obviously, if the engineer can readily display and analyze his experimental data, a more thorough and faster analysis can be made.

Concluding Remarks

With the growing demand for aerodynamic information to design new aircraft, the quality and quantity of experimental data produced for each wind-tunnel test hour must be increased. This will require aerodynamic engineers to have a good understanding of aerodynamic testing technology in order to maximize the amount of useful information produced during an experiment and to knowledgeably use the data.

This article has attempted to contribute to the general understanding of aerodynamic technology by presenting a state-of-the-art review of experimental testing at transonic speeds with particular emphasis on problems that arise in experimental testing and ways that are currently used to minimize or solve the problem. Specifically, problems/solutions were discussed that deal with wind-tunnel wall interference, models and support systems, Reynolds number simulation, aerodynamic flow visualization, and aerodynamic flow measurements.

References

[1] Harper, C. W., "Prospects in Aeronautics Research and Development," *Journal of Aircraft*, Vol. 5, No. 5, 1968, pp. 417-426.

[2] Blackwell, J. A., Jr. and Pounds, G. A., "Wind-Tunnel Wall Interference Effects on a Supercritical Airfoil at Transonic Speeds," *Journal of Aircraft*, Vol. 14, No. 10, Oct. 1977, pp. 929-935.

[3] Goethert, B. H., *Transonic Wind Tunnel Testing*, Pergamon Press, New York, 1961.

[4] Garner, H. C., Rogers, E. W. E., Acum, W. E. A., and Maskell, E. C., "Subsonic Wind Tunnel Wall Corrections," AGARDograph 109, 1966.

[5] Blackwell, J. A., Jr., "Wind Tunnel Blockage Correction for Two-Dimensional Transonic Flow," *Journal of Aircraft*, Vol. 16, No. 4, April 1979, pp. 256-263.

[6] Bauer, F., Garabedian, P., Korn, D., and Jameson, A., "Supercritical Wing Sections II," *Lecture Notes in Economics and Mathematical Systems*, Vol. 108, Springer-Verlag, New York, 1975.

[7] Morky, M. and Ohman, L. H., "Comparison of Methods for Two-Dimensional Wind Tunnel Interference Corrections From Wall Pressure Measurements," 52nd Meeting of Supersonic Tunnel Association, Notre Dame, Ind., Sept. 1979.

[8] Morky, M., Peake, D. J., and Bowker, A. J., "Wall Interference on Two-Dimensional Supercritical Airfoils, Using Wall Pressure Measurements to Determine the Porosity Factors for Tunnel Floor and Ceiling," NAE LR-575, 1974.

[9] Strahara, S. S. and Spreiter, J. R., "A Transonic Wind Tunnel Interference Assessment - Axisymmetric Flows," AIAA Paper 79-0208, New Orleans, La., Jan. 1979.

[10] Hinson, B. L. and Burdges, K. P., "An Evaluation of Three-Dimensional Transonic Codes Using New Correlation - Tailored Test Data," AIAA Paper 80-0003, Pasadena, California, Jan. 1980.

[11] Ballhaus, W. F., Bailey, F. R., and Frick, J., "Improved Computational Treatment of Transonic Flow About Swept Wings," NASA CP-2001, Nov. 1976.

[12] Vidal, R. J. and Erickson, J. C., Jr., "Experiments on Supercritical Flows in a Self-Correcting Wind Tunnel," AIAA Paper 78-788, San Diego, Calif., April 1978.

[13] Goodyer, M. J. and Wolf, S. W. D., "The Development of a Self-Streamlining Flexible Walled Transonic Test Section," AIAA Paper 80-0440-CP, Colorado Springs, Col., March 1980.

[14] Burdges, K. P., "Experimental Measurements of Shock/Boundary Layer Interaction on a Supercritical Airfoil," AIAA Paper 79-1499, Williamsburg, Va., July 1979.

[15] Jacobs, P. F., Flechner, S. G. and Monotoya, L. C., "Effects of Winglets on a First Generation Transport Wing," Vol. 1 Longitudinal Aerodynamic Characteristics of a Semispan Model at Subsonic Speeds, NASA TND 8473, June 1977.

[16] Bavitz, P. C., "Configuration Development of Advanced Fighters," AFWAL TR-80-3142, Nov. 1980.

[17] Blackerby, W. T. and Cahill, J. F., "High Reynolds Number Tests of a C-141A Aircraft Semispan Model to Investigate Shock-Induced Separation," NASA CR-2604, Oct. 1975.

[18] MacWilkinson, D. G., Blackerby, W. T. and Paterson, J. H., "Correlation of Full Scale Drag Predictions With Flight Measurements on the C-141A Aircraft," Vol. 2. Wind Tunnel Test and Basic Data, NASA CR-2334, June 1973.

[19] Smith, R., Private Communication, Jan. 1981.

[20] Loving, D. L. "Wind-Tunnel - Flight Correlation of Shock-Induced Separated Flow," NASA TND-3580, 1966.

[21] Blackwell, J. A., Jr., "Scale Effects on Supercritical Airfoils," 11th Congress of ICAS, Sept. 1978.

[22] Blackwell, J. A., Jr., "Preliminary Study of Effects of Reynolds Number and Boundary-Layer Transition Location on Shock-Induced Separation," NASA TND-5003, Jan. 1969.

[23] Cahill, J. E., Treon, S. L., and Hofstettler, W. R., "Feasibility of Testing a Large Chord, Swept-Panel Model to Determine Wing Shock Location at Flight Reynolds Numbers," AGARD Proceedings No. 83, April 1971.

[24] Howell, R. R. and McKinney, L. W., "The U.S. 2.5-Meter Cryogenic High Reynolds Number Tunnel," High Reynolds Number Research Workshop, NASA Langley Research Center, Oct. 1976.

[25] Crowder, J. P., Hill, E. G. and Pond, C. R., "Selected Wind Tunnel Testing Developments at the Boeing Aerodynamics Laboratory," AIAA Paper 80-0458 CP, Colorado Springs, Col., March 1980.

[26] Erickson, G. E., "Water Tunnel Flow Visualization: Insight Into Complex Three-Dimensional Flowfields," Journal of Aircraft, Vol. 17, No. 9, Sept. 1980, pp. 656-662.

[27] Mueller, T. J., "On the Historical Development of Apparatus and Techniques for Smoke Visualization of Subsonic and Supersonic Flows," AIAA Paper 80-0420-CP, Colorado Springs, Col., March 1980.

[28] Johnson, D. A. and Bachalo, W. D., "Transonic Flow Past a Symmetrical Airfoil - Inviscid and Turbulent Flow Properties," AIAA Journal, Vol. 18, No. 1, Jan. 1980, pp. 16-24.

[29] Horstmann, C. C. and Rose, W. C., "Hot-Wire Anemometry in Transonic Flow," AIAA Journal, Vol. 15, March 1977, pp. 395-401.

[30] Roos, F. W., "Hot-Film Probe Technique for Monitoring Shock Wave Oscillations," AIAA Paper 79-0331, New Orleans, La., Jan. 1979.

Chapter V.

Potential Equation Methods for Transonic Flow Prediction

David Nixon* and G. David Kerlick†
Nielsen Engineering & Research, Inc., Mountain View, Calif.

1. Introduction

One of the main difficulties facing the aircraft designer is the prediction of the aerodynamic loads at transonic speeds. The difficulty is caused by the fact that, in the transonic regime, even the most primitive representation of the aerodynamics must be described by a nonlinear equation or set of equations. This is in contrast to the subsonic or supersonic flow regimes where an adequate prediction of the aerodynamic loads can be obtained using linear theory. It is also the case that aerodynamic prediction in the transonic regime is of extreme importance, since it is in this speed range that most civil aircraft cruise and most military aircraft maneuver. Because of the difficulties of predicting the loading and the common occurrence of transonic flight the transonic regime is probably the most critical flow regime for present day aircraft.

A typical transonic flow has a subsonic freestream from which the flow accelerates over the wing to supersonic speeds; the necessary deceleration from supersonic speed to the freestream velocity usually, but not always, takes place through a shock wave. A wing which has such a supersonic domain but decelerates to the freestream value continuously, that is, no shock wave, is called a supercritical wing. The importance of having a shock-free wing is that the wave drag penalty associated with shock waves

---

Presented at the Transonic Perspective Symposium, NASA/Ames Research Center, Moffett Field, Calif., Feb. 18-20, 1981. Copyright © American Institute of Aeronautics and Astronautics, Inc., 1981. All rights reserved.
*Manager, Computational Fluid Dynamics Department.
†Research Scientist, Computational Fluid Dynamics Department.

is avoided. It should be noted that according to the theorems of Morawetz,[1,2] a supercritical shock-free flow cannot be simply perturbed to give another shock-free flow. This means that the shock-free condition is an isolated case. Generally the flow will have a shock wave.

In steady flow theory there are two main problems. First, one must predict the essentially inviscid phenomena of a mixed subsonic/supersonic flow with shock waves. Second, one must predict the viscous interaction, either by coupling a boundary-layer theory to an inviscid theory or by solving a viscous subset of the Navier-Stokes equations. Early steady flow prediction methods are typified for two-dimensional flow by the techniques of Murman et al.,[3] and for three-dimensional flow by Bailey and Ballhaus,[4] who solve a small disturbance approximation to the full potential equation; Jameson,[5] and Jameson and Caughey,[6] solved the full potential equation for two and three dimensions, respectively. Techniques which solve more complex equations than the potential formulation are given by Magnus and Yoshihara[7] who solve the Euler equations for two-dimensional flow and Deiwert[8] and Steger[9] who solve the "thin-layer" approximation to the Navier-Stokes equations for two-dimensional flow. Steger and Pulliam[10] have solved the "thin-layer" Navier-Stokes equations for three-dimensional flow. It should be emphasized that the above references are mentioned only as being typical of available solution procedures; they are not intended to form a comprehensive survey of the subject.

A typical transonic flow with a subsonic freestream Mach number contains a supersonic zone, bounded by the sonic line and a shock wave, through which the flow decelerates to subsonic flow. A sketch of such a flow is shown in Fig. 1. Mathematically, this means that the flow must be described by a nonlinear equation of mixed elliptic/hyperbolic type, since a subsonic flow is described by an elliptic equation

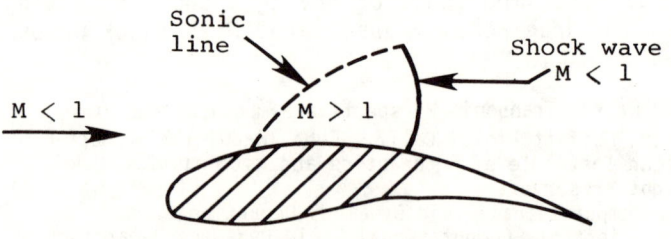

Fig. 1 Sketch of a typical transonic flow.

and a supersonic flow by a hyperbolic equation. The
boundary between the hyperbolic and elliptic regions must
be found as part of the solution. The difficulty is that
the mathematical theory of mixed equations is not as well
advanced as, say, that for purely elliptic equations.
Because of this, much of the early work on transonic flow
theory concentrated on solving a simple mixed-type equation,
for example, Tricomi's equation. The technique used most
often was the hodograph method which interchanges the role
of dependent and independent variables; this procedure
linearizes the basic equation but at a cost of considerably
complicating the boundary conditions. An excellent review
of this work is given by Ferrari and Tricomi.[11]

The complexity of the equations that need to be solved
depends on the flow phenomena in question. Perhaps the
most complex flow is one when the wing is at a high angle
of attack and the upper surface of the wing has shock waves
which are sufficiently strong to separate the flow,
probably without reattachment. In addition, the flow may
not be steady. This is a very extreme form of flow pattern,
but it may be encountered in a severe maneuver of a fighter
or other military aircraft. If it is necessary to examine
this type of motion, then there is little alternative to
solving the Navier-Stokes equations with a realistic model
for the turbulent flow, especially in the separated region.
At the other extreme the wing may have a flow pattern with
relatively weak shock waves, the shocks being sufficiently
weak that the flow does not separate. There are no significant viscous effects. This type of flow can be predicted
by solving the transonic small disturbance equation, which
is the simplest equation that can describe a typical
transonic flow. In between these two extremes exists a
range of flow phenomena that are described by equations of
varying complexity. The present paper is concerned with
the prediction of fairly well behaved flows, that is, those
that can be adequately modeled by potential theory, with
perhaps the addition of a boundary-layer model. The basic
numerical techniques are described together with typical
grid generation methods. Examples illustrating the application of these potential flow techniques to practical
problems are also given.

2. Basic Equations for Transonic Flow

The Navier-Stokes equations are generally accepted as
the basic equations governing most fluid dynamic phenomena
of interest to aerodynamicists. The equations are capable

of representing mathematically the physical phenomena encountered in transonic flows, including mixed subsonic-supersonic flow, shock waves, boundary layers, and separation. They also apply to turbulence, a random, dissipative, three-dimensional phenomenon that involves many characteristic length and time scales. Since present computer speed and capacity do not permit resolution of all these scales for practical aerodynamic problems, some type of averaging must be used. The commonly used Reynolds time-averaged procedure averages the equations over a time interval that is long compared with turbulent eddy fluctuations but small compared with macroscopic flowfield changes. This process introduces new terms, called "Reynolds stresses," that represent time-averaged transport of turbulent momentum and energy. Thus, there are more unknowns in the averaged equations than there are equations. This is commonly referred to as the closure problem. The process of expressing the Reynolds stress terms in terms of empirical functions and constants or transport equations is referred to as turbulence modeling.

The unsteady, compressible Navier-Stokes equations for a perfect gas are a set of equations representing the conservation of the physical quantities of mass, momentum, and energy. In addition there are two equations of state. The complete set of equations are as follows

$$(\partial \rho/\partial t) + \partial(\rho u_j)/\partial x_j = 0 \quad \text{(mass)} \tag{1}$$

$$\partial(\rho u_i)/\partial t + \partial(\rho u_i u_j)/\partial x_j = -\partial p/\partial x_i + \partial \tau_{ij}/\partial x_j$$

$$i,j = 1,2,3 \quad \text{(momentum)} \tag{2}$$

$$\partial(\rho h)/\partial t + \partial(\rho h u_j)/\partial x_j = \partial p/\partial t + u_j(\partial p/\partial x_j)$$

$$+ \tau_{ij}(\partial u_i/\partial x_j) - (\partial q_i/\partial x_j) \quad \text{(energy)} \tag{3}$$

where ρ is the density, u_i the velocity vector, p the pressure, and h is the specific enthalpy; τ_{ij} is the stress tensor and q_i is the heat flux vector. The stress tensor, τ_{ij}, and the heat flux vector q_i are given by

$$\tau_{ij} = \lambda \delta_{ij}(\partial u_\ell/\partial x_\ell) + \mu[(\partial u_i/\partial x_j) + (\partial u_j/\partial x_i)] \tag{4}$$

$$q_j = k(\partial T/\partial x_j) \tag{5}$$

where δ_{ij} is the Kronecker delta, $\lambda = -2\mu/3$ is the bulk viscosity and μ is the dynamic viscosity; T is the temperature and k is the thermal conductivity. The equations of state for a perfect gas are

$$p = \rho RT \qquad (6a)$$

and

$$h = c_p T \qquad (6b)$$

where R is the gas constant and c_p is the specific heat at constant pressure.

It is not possible to solve these equations in closed form, at least for practical problems, and some form of numerical solution must be attempted. Unfortunately, in a turbulent flow, the length and time scales of the turbulent eddies vary enormously and it is only practicable to solve the Navier-Stokes equations in an averaged sense. The usual form of averaging leads to the Reynolds averaged equations. Thus, the general variable $f_i(x_i,t)$ is written as

$$f_i(x_i,t) = \overline{f}(x_i) + f'(x_i,t) \qquad (7)$$

where the bar denotes a time average of $f_i(x_i,t)$ and is defined at a time t_o by

$$\overline{f}_i(x_i) = \lim_{\Delta t \to \infty} \frac{1}{\Delta t} \int_{t_o}^{t_o + \Delta t} f(x_i,t) dt \qquad (8)$$

where the time Δt is large relative to the time scale of turbulent fluctuations. Introducing this concept into the conservation equations gives, for example, the modified momentum equation

$$(\partial/\partial t)(\rho\bar{u}_i + \overline{\rho'u_i'}) + (\partial/\partial x_j)(\bar{\rho}\bar{u}_i\bar{u}_j + \bar{u}_i\overline{\rho'u_j'} + \bar{u}_j\overline{\rho'u_i'}$$
$$+ \bar{\rho}\,\overline{u_i'u_j'} + \overline{\rho u_i'u_j'}) = -(\partial\bar{p}/\partial x_j) + (\partial/\partial x_j)\bar{\tau}_{ij} \qquad (9)$$

The disadvantage of some sort of averaging is now apparent since the process has introduced additional correlations such as $\overline{\rho'u_i'u_j'}$ terms which must be found as part of the solution. In short, therefore, there are more unknowns than equations and additional equations must be derived by some form of turbulence modeling. The science, or art, of turbulence modeling has been studied for many years and there are numerous modeling schemes, ranging from simple algebraic methods to those involving one or more additional differential equations. Apart from schemes like "large eddy simulation,"[12] these turbulence modeling techniques relate the correlations to stress-like terms which have the effect of reducing the Navier-Stokes equations to a form identical to Eqs. (1-5) but for the mean quantities $\bar{\rho}$, $\bar{u}_j$, $\bar{p}$, etc. and with the coefficients of viscosity and thermal conductivity augmented by an "eddy viscosity" or "eddy-conductivity." These additions to the viscosity and thermal conductivity coefficients may be found algebraically, like the theory described by Cebeci[13] or Baldwin and Lomax.[14] They may also be given by more complex relations such as those described by Reynolds and Cebeci.[15]

The Navier-Stokes equations will describe any type of flow over an airplane in any speed range; however, the problem of turbulence modeling can restrict the application to practical problems, depending on the modeling scheme used. At present, the necessary use of the algebraic model restricts the applicable flows to those with little or no separation, although separated flows have been investigated.[16]

A slight simplification of the Navier-Stokes equations is the "thin-layer" approximation.[14] In this approximation the viscous terms in the streamwise, or near streamwise direction, are neglected. The rationale behind this approximation is that the streamwise diffusion terms are small compared to the normal diffusion terms. A second important reason is that because of limitations on computer storage it is not possible to provide sufficient grid points to resolve the viscous terms adequately in all directions and only those in a thin layer close to the surface are resolved. The "thin-layer" equations are similar to the more classic boundary-layer equations with the main exception that the normal momentum equation is retained. This prevents the occurrence of the usual boundary-layer singularity at a separation point.

It is sometimes the case that the viscous effects may be small compared to the inviscid effects in a flow, for example, in a well behaved attached flow. It is then possible to neglect the viscous terms entirely; it is also assumed at this stage that the time-dependent behavior can be neglected. This approximation leads to the steady Euler equations

$$\partial(\rho u_j)/\partial x_j = 0 \quad \text{(mass conservation)} \quad (10)$$

$\partial(\rho u_i u_j)/\partial x_j = -\partial p/\partial x_i$ (momentum conservation) (11)

$\partial(\rho h u_j)/\partial x_j = u_j(\partial p/\partial x_j)$ (energy conservation) (12)

It is assumed that the gas does not conduct heat. The Euler equations will represent inviscid rotational flows in all speed ranges. However, the set of Eqs. (10-12) is not yet complete since the neglect of the viscous terms in the Navier-Stokes equations leads to some nonuniqueness of the solutions. First, from the second law of thermodynamics, entropy cannot decrease; in the Navier-Stokes equations this condition is enforced by the viscous terms. This results in the impossibility of "expansion shock waves." However, in the Euler equations, it is possible to have expansion shock waves and therefore these equations must be solved subject to an "entropy condition" which eliminates the unrealistic solutions.

The second point concerns the trailing edge of an airfoil. At a sharp trailing edge the local flow is controlled by the action of the viscous stresses; this determines the circulation around the airfoil and hence the lift. In an inviscid flow the behavior at the trailing edge is no longer controlled and a multiplicity of solutions can be found. In order to model the real viscous flow accurately, therefore, an additional condition, the Kutta-Joukowski condition, is imposed to give a physically realistic flow. For solutions of the Euler equations several forms of the Kutta-Joukowski condition can be applied, the most usual being continuity of pressure across the wake and the velocity component normal to the wake being equated to zero.

The Euler equations can be used to represent a flow which is always attached and in which the inviscid effects dominate the viscous effects. The flow may be rotational.

In an inviscid fluid the only mechanism for producing entropy changes is through the presence of a shock wave. It may be shown[17] that the entropy rise through a shock wave is proportional to

$$(M^2 - 1)^{3/2}$$

where M is the local Mach number ahead of the shock, whereas other flow variable changes are of the order of $(M^2 - 1)^{1/2}$. It then follows that an approximation to the Euler equations is inherent in the assumption of isentropic flow, since if the shock wave is sufficiently weak such that

$$(M^2 - 1)^{3/2} \ll (M^2 - 1)^{1/2} \qquad (13)$$

then the entropy production is a higher-order effect. Furthermore, if the flow at infinity is uniform, then by Crocco's theorem the assumption of isentropy leads directly to the assumption of irrotational flow. Thus, in two-dimensional flow

$$(\partial u_i/\partial x_j) = (\partial u_j/\partial x_i) \qquad j \neq i \qquad (14)$$

The main advantage of assuming irrotational flow is that Eq. (14) suggests the introduction of a velocity potential, Φ, such that

$$u_i = (\partial \Phi/\partial x_i) \qquad (15)$$

The introduction of two additional assumptions requires that two of the Euler equations be redundant (actually they are incompatible with the new system). The question then arises as to which equations to retain. It has been found that a transonic flow is most sensitive to errors in the conservation of mass equation. This is due to the singular behavior of the density at sonic conditions. Hence, it is usually assumed that mass must be conserved.

The density in the conservation of mass equation can be found from Bernoulli's equation for isentropic flow; thus

$$(\rho/\rho_\infty) = \{1 + [(\gamma - 1)/2]M_\infty^2[(1 - u_i u_i)/q^2]\}^{\frac{1}{\gamma-1}} \qquad (16)$$

where M_∞ is the freestream Mach number. The pressure is found by using the isentropic relation

$$p/p_\infty = (\rho/\rho_\infty)^\gamma \qquad (17)$$

In Eqs. (16) and (17), γ is the ratio of specific heats, q denotes a total velocity, and the subscript ∞ denotes free-stream values. Hence, the governing equations for a potential flow are Eq. (10) and Eqs. (15-17). As in the Euler equations, an entropy condition and the Kutta-Joukowski condition must be applied.

Equation (16) can be combined with Eqs. (10) and (15) to give the quasilinear form of the potential equation

$$u_i u_j (\partial u_i/\partial x_j) = a^2 (\partial u_k/\partial x_k) \qquad (18)$$

where a is the local speed of sound and is given by

$$a/a_\infty = 1 + [(\gamma - 1)/2]M_\infty^2[1 - (u_i u_i/q_\infty^2)] \quad (19)$$

The potential equation is a relatively simple equation but its numerical solution for transonic flow requires that the implementation of the boundary conditions be of equal accuracy. This means that the tangency boundary condition should be applied on the actual airfoil surface rather than on a mean surface. For finite-difference solutions this usually requires the development of a body-conforming finite-difference grid which adds to the overall complexity of the problem. Consequently, a simple form of the potential equation, the small disturbance equation, can be derived.

It is assumed that the deviations of the flow from the freestream values are small compared to the freestream. Thus, if U, V, and W are the velocity components in the coordinate directions, x,y,z, respectively, and x is the freestream direction and z is the spanwise variable, then

$$\begin{aligned} U &= U_\infty + u = U_\infty(1 + \phi_x) \\ V &= v = U_\infty \phi_y \\ W &= w = U_\infty \phi_z \end{aligned} \quad (20)$$

If these equations are combined with Bernoulli's equation, and if second-order terms in ϕ_y and ϕ_z, and third-order terms in ϕ_x are neglected, then

$$\rho/\rho_\infty \approx 1 - M_\infty^2 \phi_x - (\phi_x^2/2) M_\infty^2[1 - (2 - \gamma)M_\infty^2] \quad (21)$$

If Eqs. (20) and (21) are combined with the conservation of mass equation, and if all second-order terms in the perturbation velocity potential ϕ are neglected except those in the x derivatives, then the resulting equation is

$$\{1 - M_\infty^2 - M_\infty^2[3 + M_\infty^2(\gamma - 2)]\phi_x\}\phi_{xx} + \phi_{yy} + \phi_{zz} = 0 \quad (22)$$

Equation (22) is the transonic small disturbance equation.

The transonic small disturbance equation is formally valid as $M_\infty \to 1$. Thus, in Eq. (22) the term in the square brackets may be replaced by $(\gamma+1)$ and thus becomes

$$[1 - M_\infty^2 - (\gamma+1)M_\infty^n \phi_x]\phi_{xx} + \phi_{yy} + \phi_{zz} = 0 \qquad (23)$$

where n is a fairly arbitrary exponent and is usually determined by comparison of solutions of the above equation to more exact solutions. Equation (23) is sometimes referred to as the classic small disturbance equation.

It was found by Lomax, et al.[18] that this version of the small disturbance equation does not give adequate solution of swept shock waves and hence proposed the inclusion of certain higher-order terms in the potential equation to give the modified or extended small disturbance equation

$$[(1 - M^2)\phi_x - (\gamma+1)M_\infty^n \phi_x^2/2 + (\gamma-3)M_\infty^2 \phi_z^2/2]_x$$
$$+ [\phi_z - (\gamma-1)M_\infty^2 \phi_x \phi_z]_z + (\phi_y)_y = 0 \qquad (24)$$

The small disturbance equation is the simplest equation that can describe a transonic flow.

In order to solve the basic differential equations, boundary conditions are required. In addition to the "entropy" and Kutta-Joukowsky conditions noted earlier, the tangency condition of no flow through a solid body and a zero load condition on a wake must be imposed. Finally, a far-field boundary condition is imposed which requires that, apart from the wake, flow disturbances vanish an infinite distance from the body.

3. Some Basic Concepts

3.1 Weak Solutions

In an earlier section it has been emphasized that an accurate representation of shock waves is of great importance in unsteady transonic flow predictions. Consequently, it is instructive to examine shock wave representation in a finite difference scheme.

Consider the partial differential equation

$$\partial U/\partial x + \partial E/\partial y + \partial F/\partial z = 0 \qquad (25)$$

where, if U is a vector of the flow variables, then E and F are vectors which are functions of U. Equation (25) is in divergence or conservation form. For the equations of gas-dynamics this equation can represent the conservation of the physical quantities mass, momentum, and energy. If the equations are hyperbolic, then in general the solution will have either shock waves or contact discontinuities. In this case the derivatives in Eq. (25) are infinite across the discontinuity and the equations are only piecewise valid. In a numerical solution there are two options. Either all discontinuities can be fitted into the solution and Eq. (25) can be used in a piecewise sense, or some means can be found to generalize the conservation law, Eq. (25), so that a solution procedure is valid across the discontinuities. The latter procedure leads to the concept of weak solutions.[19]

If $W(x,y,z,t)$ is a smooth test vector field of the same order as U and which vanishes for $|x| + |y| + |z|$ large enough, then a function $U(x,y,z)$ is a weak solution of Eq. (25) if it satisfies the integral relation

$$\int_{-\infty}^{\infty}\int_{-\infty}^{\infty}\int_{-\infty}^{\infty} [U(\partial W/\partial x) + E(\partial W/\partial y) + F(\partial W/\partial z)] dx\, dy\, dz$$

$$+ \int_{-\infty}^{\infty}\int_{-\infty}^{\infty} W(0,y,z)U(0,y,z) dy\, dz = 0 \qquad (26)$$

where the values in the second integral are boundary values. Note that Eq. (26) does not contain derivatives of U, E, and F. It can be easily shown that a function $U(x,y,z)$ which is a solution of Eq. (26) and has continuous first derivatives, is a solution of Eq. (25). If the solution to Eq. (25) does not have continuous first derivatives, then across the surface of discontinuity, $f(x,y,z)$, the weak solution gives

$$[U]_-^+ f_x + [E]_-^+ f_y + [F]_-^+ f_z = 0 \qquad (27)$$

where $[\ ]_-^+$ is defined by the relation

$$[A]_-^+ = A^+ - A^-$$

where A^+ and A^- are the values just ahead of and behind the surface of discontinuity, respectively. For the Euler equations the jump relation, Eq. (27), is identical to the Rankine-Hugoniot shock relations if Eq. (27) represents the conservation of the physical quantities, mass, momentum, and energy.

Numerical schemes are approximations to differential equations and if a device such as an artificial viscosity is used to stabilize the scheme then the approximate solution has continuous first derivatives. It can be shown[19] that a converged solution of Eq. (25) using a finite-difference scheme satisfies Eq. (26) and is therefore a weak solution of Eq. (25). Hence, a weak solution is a solution of the modified differential equation. If the artificial viscosity could be gradually reduced to zero then the weak solution will admit discontinuities which satisfy Eq. (27). This is the basis of "shock capturing," where a discontinuous solution is approximated by a continuous solution, which in the limit of zero mesh size, would give a solution with a discontinuity given by Eq. (27).

Generally, weak solutions are not unique and the preferred weak solution should be obtained by a specific limiting process, such as the added artificial dissipation going to zero. The mechanism that leads to the preferred solution must satisfy the laws of thermodynamics, even though an approximate set of equations, such as the Euler equations, can admit solutions which do not satisfy all the laws of thermodynamics.

If the equations are written in conservation form, it should be noted that the finite differencing should also be in conservation form to obtain a consistent solution.

Consider now a simple first-order accurate finite difference scheme of the form

$$(f_x)_{i-1/2} = (\Delta x)^{-1}(f_i - f_{i-1}) + (\Delta x/2)f_{xx}$$

$$\equiv D_x f + (\Delta x/2)f_{xx} \qquad (28)$$

where D_x denotes the difference. The $(\Delta x/2)f_{xx}$ term is the leading term of the truncation error. If the derivatives in Eq. (25) are replaced by the differences, then to second order, the difference equation

$$D_x U + D_y E + D_z F = 0 \qquad (29)$$

is equivalent to the differential equation

$$(\partial/\partial x)[U + (\Delta x/2)U_x] + (\partial/\partial y)[E + (\Delta y/2)E_y]$$
$$+ (\partial/\partial z)[F + (\Delta z/2)F_z] = 0 \qquad (30)$$

The jump relations of this equivalent differential equation are

$$[U + (\Delta x/2)U_x]_-^+ f_x + [E + (\Delta y/2)E_y]_-^+ f_y$$
$$+ (F + (\Delta z/2)F_z]_-^+ f_z = 0 \qquad (31)$$

If a stable algorithm can be found to solve Eq. (29) for a vanishingly small mesh size then the equivalent equations, Eqs. (30) and (31) reduce to Eqs. (25) and (27). Such a difference scheme is termed consistent.

It should also be noted that if the truncation error terms in Eq. (30) are taken to the right-hand side then they resemble the viscous terms in the Navier-Stokes equations; this is called the artificial viscosity of the scheme.

If a nonconservative equation is considered, such as

$$A(\partial U/\partial x) + B(\partial E/\partial y) + C(\partial F/\partial z) = 0 \qquad (32)$$

where A,B,C are functions of U, then the exact form of the weak jump relations is not easily apparent. In practice the jump relation is not unique and depends on the mesh size as, for example, in Fig. 2. Thus, conservative equations and difference relations should always be used.

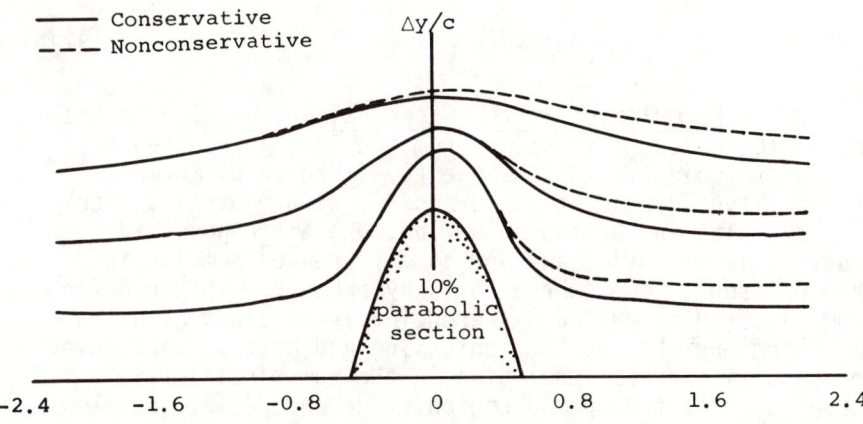

Fig. 2 Displacement of streamline predicted by conservative algorithms.

3.2 Numerical Methods

Now we turn to the discussion of some finite-difference schemes. In order to examine the basic ideas it is advantageous to initially consider the two-dimensional version of the potential formulation.

If K is the similarity constant $(1 - M_\infty^2)/\tau^{2/3}$, where τ is the thickness/chord ratio of the airfoil, then a simple version of the small disturbance equation in two dimensions developed from Eq. (23) can be written as

$$A\phi_{xx} + \phi_{yy} = 0 \tag{33}$$

where

$$A = K - (\gamma+1)\phi_x \tag{34}$$

Note that in Eq. (33) the y-axis has been scaled by the factor $\tau^{1/3}$ and ϕ is the disturbance potential, scaled by the factor $\tau^{-2/3}$. The tangency boundary condition is transferred to the x-axis in the small disturbance formulation. Thus

$$\phi_y = (df/dx) \text{ at } y = 0 \tag{35}$$

where f(x) denotes the body surface. If the small disturbance equation is written in the conservation form

$$(\partial/\partial x)[K\phi_x - (\gamma+1)\phi_x^2/2] + \phi_{yy} = 0 \tag{36}$$

then the corresponding weak jump condition is

$$[\phi_y]_-^+ - (dy/dx)[K\phi_x - (\gamma+1)\phi_x^2/2]_-^+ = 0 \tag{37}$$

3.2.1 Formulation of the Numerical Method.

The general method to be described stems from the idea introduced by Murman and Cole,[20] and subsequently improved by Murman,[21] of using type dependent differencing, with central difference formulas in the subsonic zone, where the governing equation is elliptic, and upwind difference formulas in the supersonic zone, where it is hyperbolic. The transfer from centered to upwind differencing is obtained by using a switching operator at the sonic line and at the shock wave. The resulting directional bias in the numerical scheme corresponds to the upwind region of dependence of the flow in the supersonic zone. If we consider the transonic flow past a profile with fore and aft symmetry such as an ellipse, the desired solution of the potential flow equation is not symmetric. Instead it exhibits a smooth acceleration over the front half of the profile followed by a discontinuous compression through a shock wave. In the absence of a directional bias in the numerical scheme, the fore and aft symmetry would be preserved in any solution which would be obtained, resulting in the appearance of improper discontinuities. For example, see Fig. 3.

The general method of constructing a difference approximation to a conservation law of the form

$$f_x + g_y = 0$$

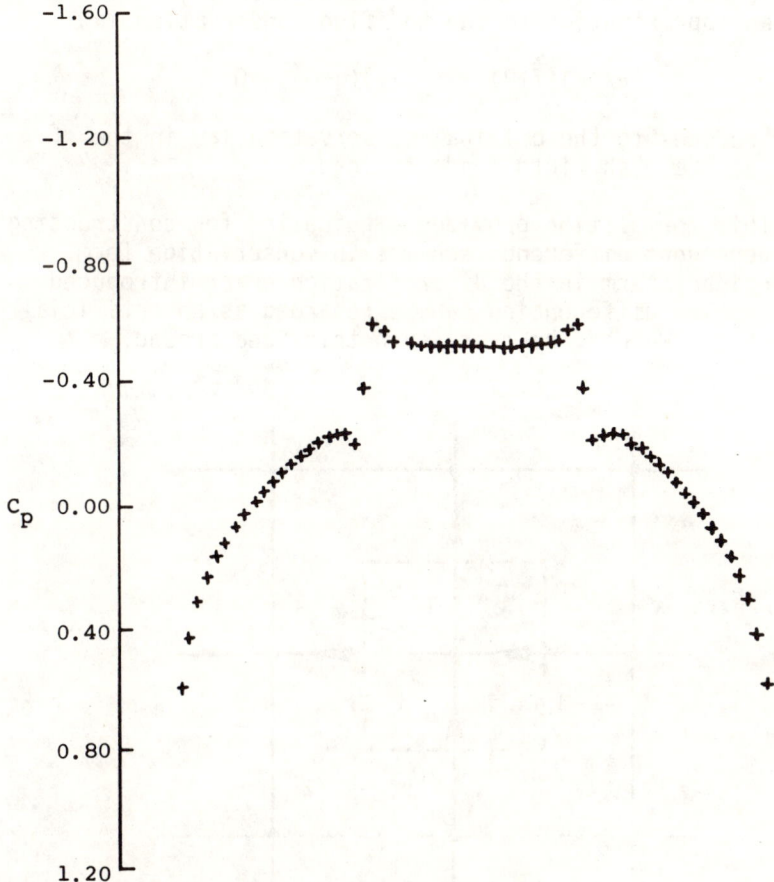

Fig. 3 Effect of violating the entropy conditions: 10% thick biconvex airfoil.

is to preserve the flux balance in each cell, as illustrated in Fig. 4. This leads to a scheme of the form

$$(\Delta x)^{-1}(F_{i+1/2,j} - F_{i-1/2,j}) + (\Delta y)^{-1}(G_{i,j+1/2} - G_{i,j-1/2})$$

where F and G should converge to f and g in the limit as the mesh width tends to zero. If shock waves may be present in the flow, then, in the constructions of a conservative difference approximation, the artificial viscosity must be in conservative form, as discussed in Section 3.1. Thus the artificial viscosity is of the form

$$(\partial P/\partial x) + (\partial Q/\partial y)$$

where Q and Q are of order Δx. The difference scheme is then an approximation to the modified conservation law

$$(\partial/\partial x)(f+P) + (\partial/\partial y)(g+Q) = 0$$

which reduces to the original conservation law in the limit as the mesh width tends to zero.

This formulation provides a guideline for constructing type-dependent difference schemes in conservation form. The dominant term in the discretization error introduced by an upwind differencing can be regarded as an artificial viscosity. We can, however, turn this idea around.

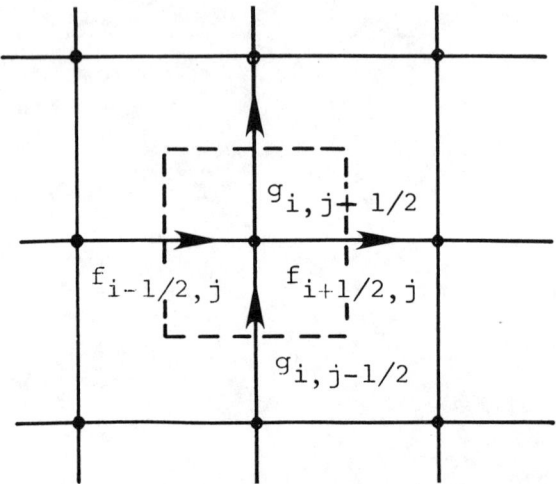

Fig. 4 Flux balance for difference scheme in conservation form.

Instead of using a switch in the difference scheme to introduce an artificial viscosity, we can explicitly add an artificial viscosity which produces an upwind bias in the difference scheme at supersonic points. Suppose that we have a central difference approximation to the differential equation in conservation form. Then the conservation form will be preserved as long as the added viscosity is also in conservation form. The effect of the viscosity is simply to alter the conserved quantities by terms proportional to the mesh width Δx which vanish in the limit as the mesh width approaches zero, with the result that the proper jump conditions must be satisfied. By including a switching function in the viscosity to make it vanish in the subsonic zone we can continue to obtain the sharp representation of shock waves which results from switching the difference scheme.

There remains the problem of finding a convergent iterative scheme for solving the nonlinear difference equations which result from the discretization. Suppose that in the (n+1) cycle the residual R_{ij} at the point $i \Delta x$, $j \Delta y$ is evaluated by inserting the result $\phi_{ij}^{(n)}$ of the nth cycle into the difference approximation. Then the correction

$$C_{ij} \equiv \phi_{ij}^{(n+1)} - \phi_{ij}^{(n)}$$

is calculated by solving an equation of the form

$$NC + \sigma R = 0 \qquad (38)$$

where N is a discrete linear operator, and σ is a scaling function. In a relaxation method N is restricted to be a lower triangular or block triangular matrix so that the elements of C can be determined sequentially. In the analysis of such a scheme it is helpful to introduce an analogy where the number of steps is proportional to an artificial time coordinate t. The vector R is an approximation to $L\phi$, where L is the operator appearing in the differential equation. If we consider C as representing $\Delta t\, \phi_t$, where t is the artificial time coordinate, and N is an approximation to a differential operator $(1/\Delta x)F$, then Eq. (38) is an approximation to

$$F\phi_t + \sigma(\Delta x/\Delta t)L\phi = 0 \qquad (39)$$

Thus we should choose N so that this process converges in time.

With this approach the formulation of a relaxation method for solving a transonic flow is reduced to three main steps:

1) Construct a central difference approximation to the differential equation.

2) Add a numerical viscosity to produce the desired directional bias in the hyperbolic region.

3) Add time-dependent terms to embed the steady-state equation in a convergent time-dependent process.

Methods constructed along these lines have proved extremely reliable. Their main shortcoming is a rather slow rate of convergence.

3.2.2 <u>The Murman Difference Scheme</u>. The basic ideas can be illustrated by considering the solution of the transonic small disturbance equation. The treatment of the small disturbance equation is simplified by the fact that its characteristics are locally symmetric about the x direction. Thus the desired directional bias can be introduced simply by switching to upwind differencing in the x direction at all supersonic points. To preserve the conservation form some care must be exercised in the method of switching. Let p_{ij} be a central difference approximation to the x derivatives at the point $i\,\Delta x$, $j\,\Delta y$:

$$p_{ij} = K(\Delta x)^{-2}[(\phi_{i+1,j} - \phi_{ij}) - (\phi_{ij} - \phi_{i-1,j})]$$

$$- (\gamma+1)(\Delta x)^{-3}[(\phi_{i+1,j} - \phi_{ij})^2 - (\phi_{ij} - \phi_{i-1,j})^2]/2$$

$$= A_{ij}(\Delta x)^{-2}[\phi_{i+1,j} - 2\phi_{ij} + \phi_{i-1,j}] \tag{40}$$

where

$$A_{ij} = K - (\gamma+1)(\Delta x)^{-1}(\phi_{i+1,j} - \phi_{i-1,j})/2 \tag{41}$$

Also let q_{ij} be a central difference approximation to ϕ_{yy}

$$q_{ij} = (\Delta y)^{-2}(\phi_{i,j+1} - 2\phi_{ij} + \phi_{i,j-1}) \tag{42}$$

Define a switching function μ with the value unity at supersonic points and zero at subsonic points

$$\mu_{ij} = \begin{cases} 0 & \text{if } A_{ij} > 0 \\ 1 & \text{if } A_{ij} < 0 \end{cases} \quad (43)$$

Then the original scheme of Murman and Cole[20] can be written as

$$p_{ij} + q_{ij} - \mu_{ij}(p_{ij} - p_{i-1,j}) = 0 \quad (44)$$

Let

$$P = \Delta x (\partial/\partial x)[K\phi_x - (\gamma+1)\phi_x^2/2]$$

$$= \Delta x \, A \, \phi_{xx}$$

where A is the nonlinear coefficient defined by Eq. (34). Then the added terms are an approximation to

$$-\mu(\partial P/\partial x) = -\mu\Delta x \left[A\phi_{xxx} - (\gamma+1)\phi_{xx}^2 \right]$$

This may be regarded as an artificial viscosity of order Δx which is added at all points of the supersonic zone. Since the coefficient $-A$ of ϕ_{xxx} is positive in the supersonic zone, it can be seen that the artificial viscosity includes a term similar to the viscous terms in the Navier-Stokes equation.

Since μ is not constant, the artificial viscosity is not in conservation form, with the result that the difference scheme does not satisfy the conditions stated earlier for the discrete approximation to converge to a weak solution satisfying the proper jump conditions. To correct this, all that is required is to recast the artificial viscosity in a divergence form as $(\partial/\partial x)(\mu P)$. This leads to Murman's fully conservative scheme,

$$p_{ij} + q_{ij} - \mu_{ij}p_{ij} + \mu_{i-1,j}p_{i-1,j} = 0 \quad (45)$$

At points where the flow enters and leaves the supersonic zone μ_{ij} and $\mu_{i-1,j}$ have different values, leading to special parabolic and shock point equations, respectively

$$q_{ij} = 0$$

and
$$p_{ij} + p_{i-1,j} + q_{ij} = 0$$

The nonlinear difference equations, Eqs. (40-43) and (44) or (45), may be solved by a generalization of the line relaxation method for elliptic equations. At each point we calculate the coefficient A_{ij} and the residual R_{ij} by substituting the result $\phi_{ij}^{(n)}$ of the previous cycle in the difference equations. Then we set

$$\phi_{ij}^{(n+1)} = \phi_{ij}^{(n)} + C_{ij}$$

where the correction C_{ij} is determined by solving the linear equations

$$(\Delta y)^{-2}(C_{i,j+1} - 2C_{ij} + C_{i,j-1})$$
$$+ (1 - \mu_{ij})A_{ij}(\Delta x)^{-2}[-(2/\omega)(C_{ij} + C_{i-1,j})]$$
$$+ \mu_{i-1,j} A_{i-1,j} (\Delta x)^{-2}(C_{ij} - 2C_{i-1,j} + C_{i-2,j}) + R_{ij} = 0 \quad (46)$$

on each successive vertical line. In these equations ω is the over-relaxation factor for subsonic points with a value in the range $1 < \omega < 2$. In a typical line relaxation scheme for an elliptic equation, provisional values $\tilde{\phi}_{ij}$ are determined for the line $x = i\Delta x$ by solving the difference equations with the latest available values $\phi_{i-1,j}^{(n+1)}$ and $\phi_{i+1,j}^{(n)}$ inserted at points on the adjacent lines. Then new values $\phi_{ij}^{(n+1)}$ are determined by the formula

$$\phi_{ij}^{(n+1)} = \phi_{ij}^{(n)} + \omega\left[\tilde{\phi}_{ij} - \phi_{ij}^{(n)}\right]$$

By eliminating $\tilde{\phi}_{ij}$ we can write the difference equations in terms of $\phi_{ij}^{(n+1)}$ and $\phi_{ij}^{(n)}$. Then it can be seen that ϕ_{yy} would be represented by

$$\frac{1}{\omega}\delta_y^2\phi^{(n+1)} + (1 - 1/\omega)\delta_y^2\phi^{(n)}$$

in such a process, where δ_y^2 denotes the second central difference operator. The appropriate procedure for treating the upwind difference formulas in the supersonic zone, however, is to march in the flow directions, so that the values $\phi_{ij}^{(n+1)}$ on each new column can be calculated from the values $\phi_{i-1,j}^{(n+1)}$ and $\phi_{i-2,j}^{(n+1)}$ already determined on the previous columns. This implies that ϕ_{yy} should be represented by $\delta_y^2 \phi^{(n+1)}$ in the supersonic zone, leading to a discontinuity at the sonic line. The correction formula Eq. (46) is derived by modifying this process to remove this discontinuity. New values $\phi_{ij}^{(n+1)}$ are used instead of provisional values $\tilde{\phi}_{ij}$ to evaluate ϕ_{yy} at both supersonic and subsonic points. At supersonic points ϕ_{xx} is also evaluated from $\phi_{ij}^{(n+1)}$ and $\phi_{ij}^{(n)}$ equivalent to $\tilde{\phi}_{ij}$. In the subsonic zone the scheme acts like a line relaxation scheme, with a comparable rate of convergence. In the supersonic zone it is equivalent to a marching scheme, once the coefficients A_{ij} have been evaluated. Since the supersonic difference scheme is implicit, no limit is imposed on the step length Δx as A_{ij} approaches zero near the sonic line.

3.2.3 <u>Solution of the Exact Potential Flow Equation</u>. It is less easy to construct difference approximations to the potential flow equation with a correct directional bias because the upwind direction is not known in advance. If, however, the supersonic flow is confined to a bubble above the profile, it may be possible to use a coordinate system in which the x coordinate is more or less aligned with the flow in the supersonic zone. For this purpose we use a conformal mapping to make the profile coincide with a x coordinate line (see Section 4). In the following discussion the quasilinear form of the potential equation, Eq. (18) is written in the form

$$(a^2-u^2)\phi_{xx} - 2uv\phi_{xy} + (a^2-u^2)\phi_{yy} = 0 \qquad (47)$$

A simple difference approximation to the quasilinear form can then be constructed in the following manner: The velocity components u and v are evaluated throughout the flow field by central difference formulas, and the speed of sound is determined by Eq. (19). Then at subsonic points we use central difference formulas for ϕ_{xx}, ϕ_{xy}, and ϕ_{yy}, while at supersonic points we switch to upwind difference formulas for ϕ_{xx} and ϕ_{xy}. The upwind difference formulas can be regarded as approximations to $\phi_{xx} - \Delta x \phi_{xxx}$ and $\phi_{xy} - (\Delta x/2)\phi_{xxy}$. Thus they introduce an artificial viscosity

$$\Delta x \left[(u^2-a^2)\phi_{xxx} + uv\phi_{xxy} \right] = \Delta x \left[(u^2-a^2)u_{xx} + uvv_{xx} \right]$$

When the flow is not perfectly aligned with the x coordinate there exist supersonic points at which

$$u^2 < a^2 < u^2 + v^2$$

One characteristic lies ahead of the y coordinate line at such a point, so that the difference scheme does not have the correct domain of dependence (see Fig. 5). Also the artificial viscosity $\Delta x(u^2 - a^2)$ introduced by the upwind difference formula for ϕ_{xx} is then negative.

The treatment of flows with large supersonic zones in a curvilinear coordinate system suited to the geometry

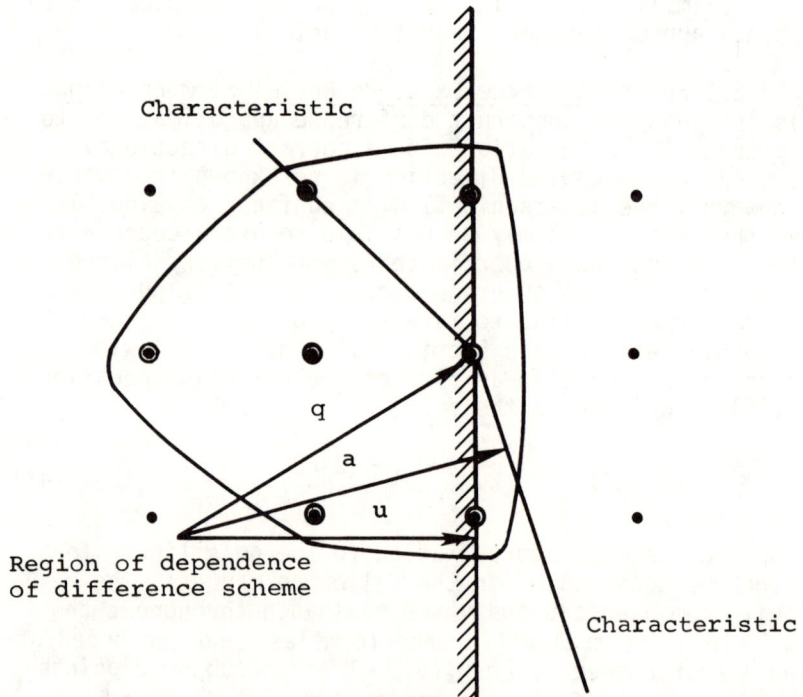

Fig. 5 Simple difference scheme.

of the problem requires the use of a more elaborate difference scheme, in which the direction of upwind differencing is independent of the coordinate system, and is instead rotated to conform with the local flow direction.[6,22,23] To illustrate the construction of such a scheme consider the potential flow equation, Eq. (47) in Cartesian coordinates. The required rotation of the upwind differencing at any particular point can be accomplished by introducing an auxiliary Cartesian coordinate system which is locally aligned with the flow at that point. If s and n denote the local streamwise and normal directions, then Eq. (47) becomes

$$(a^2 - q^2)\phi_{ss} + a^2 \phi_{nn} = 0 \qquad (48)$$

Since u/q and v/q are the local direction cosines, ϕ_{ss} and ϕ_{nn} can be expressed in the original coordinate system as

$$\phi_{ss} = q^{-2}(u^2 \phi_{xx} + 2uv\phi_{xy} + v^2 \phi_{yy}) \qquad (49)$$

and

$$\phi_{nn} = q^{-2}(v^2 \phi_{xx} - 2uv\phi_{xy} + u^2 \phi_{yy}) \qquad (50)$$

Then at subsonic points central difference formulas are used for both ϕ_{xx} and ϕ_{nn}. At supersonic points central difference formulas are used for ϕ_{nn}, but upwind formulas are used for the second derivative contribution to ϕ_{ss}, as illustrated in Fig. 6. At a supersonic point at which u > 0 and v > 0, for example, ϕ_{ss} is constructed from the formulas

$$\phi_{xx} = (\Delta x)^{-2}(\phi_{ij} - 2\phi_{i-1,j} + \phi_{i-2,j})$$

$$\phi_{xy} = (\Delta x \Delta y)^{-1}(\phi_{ij} - \phi_{i-1,j} - \phi_{i,j-1} + \phi_{j-1,j-1}) \qquad (51)$$

$$\phi_{yy} = (\Delta y)^{-2}(\phi_{ij} - 2\phi_{i,j-1} + \phi_{i,j-2})$$

It can be seen that the scheme reduces to a form similar to the scheme of Murman and Cole for the small disturbance equation if either u = 0 or v = 0. The upwind difference formulas can be regarded as approximations to

$$\phi_{xx} - \Delta x \phi_{xxx}$$

$$\phi_{xy} - (\Delta x/2)\phi_{xxy} - (\Delta y/2)\phi_{xyy}$$

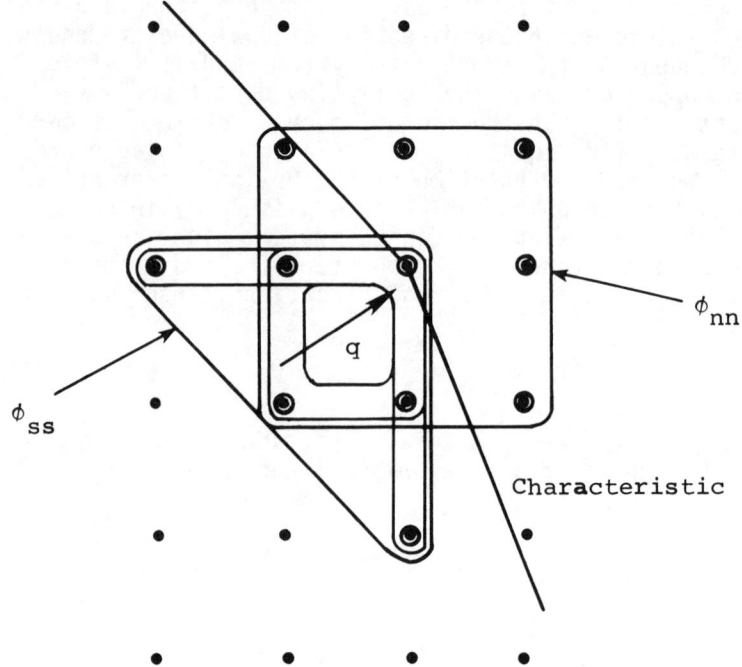

Fig. 6 Rotated difference scheme.

and
$$\phi_{yy} - \Delta y \phi_{yyy}$$

Thus at supersonic points the scheme introduces an effective artificial viscosity

$$(1 - a^2/q^2)[\Delta x(u^2 u_{xx} + uv v_{xx}) + \Delta y(uv u_{yy} + v^2 v_{yy})] \qquad (52)$$

which is symmetric in x and y.

In the construction of a discrete approximation to the conservation form of the potential equation, Eqs. (10) and (16), it is convenient to accomplish the switch to upwind differencing by the explicit addition of an artificial viscosity in the manner proposed in Section 3.2.1. Thus, we solve an equation of the form

$$S_{ij} + T_{ij} = 0 \qquad (53)$$

where S_{ij} is a central difference approximation to the left-hand side of Eq. (10), and T_{ij} is the artificial viscosity, which is constructed as an approximation to an expression in divergence form $\partial P/\partial x + \partial Q/\partial y$, where P and Q are appropriate quantities with a magnitude proportional to the mesh width. The central difference approximation is constructed in the natural manner as

$$S_{ij} = (\Delta x)^{-1}[(\rho u)_{i+1/2,j} - (\rho u)_{i-1/2,j}]$$
$$+ (\Delta y)^{-1}[(\rho v)_{i,j+1/2} - (\rho v)_{i,j-1/2}] \quad (54)$$

The treatment of flows which are not well aligned with the coordinate system requires the use of a difference scheme in which the upwind bias conforms to the local flow direction. The desired bias can be obtained by modeling the added terms T_{ij} on the artificial viscosity of the rotated difference scheme for the quasilinear form described in the previous section. Since Eq. (10) is equivalent to Eq. (18) multiplied by ρ/a^2, P and Q should be chosen so that $\partial P/\partial x + \partial Q/\partial y$ contain terms similar to Eq. (52) multipled by ρ/a^2. The following scheme has proved successful:[24] Let μ be a switching function which vanishes in the subsonic zone,

$$\mu = \max[0, 1-a^2/q^2] \quad (55)$$

Then P and Q are defined as approximations to

$$-\mu\left[(1-\varepsilon)|u| \Delta x \rho_x + \varepsilon u \Delta x^2 \rho_{xx}\right] \quad (56)$$

and

$$-\mu\left[(1-\varepsilon)|v| \Delta y \rho_y + \varepsilon v \Delta y^2 \rho_{yy}\right] \quad (57)$$

respectively, where the parameter ε controls the accuracy. If $\varepsilon = 0$, the scheme is first-order accurate, and at a supersonic point where $u > 0$, $v > 0$, P then approximates

$$\Delta x(\rho/a^2)(1-a^2/q^2)(u^2 u_x + uv v_x) \quad (58)$$

When this formula and the corresponding formula for Q are inserted in $\partial P/\partial x + \partial Q/\partial y$, it can be verified that the terms containing the highest derivatives of ϕ are the same as those in Eq. (52) multiplied by ρ/a^2. In the construction of P and Q the derivatives of ρ are represented by upwind difference formulas. Thus the formula for the viscosity finally becomes

$$T_{ij} = -(\Delta x)^{-1}(P_{i+1/2,j} - P_{i-1/2,j})$$
$$- (\Delta y)^{-1}(Q_{i,j+1/2} - Q_{i,j-1/2}) \quad (59)$$

where if $u_{i+1/2,j} > 0$,

$$P_{i+1/2,j} = u_{i+1/2,j}\mu_{ij}\left[\rho_{i+1/2,j} - \rho_{i-1/2,j}\right.$$
$$\left. - \varepsilon(\rho_{i-1/2,j} - \rho_{i-3/2,j})\right] \quad (60)$$

and if $u_{i+1/2,j} < 0$,

$$P_{i+1/2,j} = u_{i+1/2,j}\mu_{i+1,j}\left[\rho_{i+1/2,j} - \rho_{i+3/2,j}\right.$$
$$\left. - \varepsilon(\rho_{i+3/2,j} - \rho_{i+5/2,j})\right] \quad (60)$$

while $Q_{i,j+1/2}$ is defined by a similar formula.

Eqs. (54) and (60) call for the evaluation of the velocity components and density at the midpoint of each mesh interval. The precise method by which this is accomplished has been found to have little influence on the result.

3.2.4 <u>Approximate Factorization</u>. A faster procedure than successive line over-relaxation (SLOR) is obtained when the operator N in Eq. (38) of Section 3.2.1 is factored so that each of the factors N_1, N_2 have a simple, easily invertible form and so that their product approximates the differential operator L. Thus we write

$$N_2 N_1 C + \sigma R = 0 \quad (61)$$

Here again $R_{ij} = L\phi_{ij}^n$ is the residual and $C_{ij}^n = \phi_{ij}^{n+1} - \phi_{ij}^n$ is the correction.

The most commonly used approximate factorization is the AF2 scheme of Ballhaus and Steger[25] which takes the following two-step form for the TSD equation.[26]

$$[\alpha - (1-\varepsilon_j)A_j\vec{\delta}_x - \varepsilon_{j-1}\overleftarrow{\delta}_x]\tilde{\phi}_{ij}$$

$$= \alpha\{[(1-\varepsilon_j)A_j\vec{\delta}_x + \varepsilon_{j-1}A_{j-1}\overleftarrow{\delta}_x]\overleftarrow{\delta}_x + \delta_{yy}\}\phi^n \quad (62a)$$

$$[\alpha\vec{\delta}_x - \delta_y^2]C_{ij}^n = \tilde{\phi}_{ij} \quad (62b)$$

Here $A_j = K - (\gamma+1)\delta_x\phi$ is a difference approximation to Eq. (34), and $\tilde{\phi}_{ij}$ is an intermediate value stored by the code. The difference operators $\vec{\delta}_x, \overleftarrow{\delta}_x$, and δ_x are forward, backward, and central first x-differences, respectively, and δ_y^2 is a second-order central second y-difference. The "acceleration parameter" $\alpha = (\Delta t)^{-1}$ represents frequency in artificial time. In practice a sequence of α's is chosen to suppress errors of different spatial frequency. The choice of this sequence has usually been obtained by numerical optimization. Although the optimal choice can be expected to depend on the class of problems one is attempting to solve, AF2 has been found by numerical experiment to be up to an order of magnitude faster than SLOR for most problems.

3.2.5 <u>Finite-Volume Method</u>. For three-dimensional flows the generation of the finite-difference grid becomes a far from trivial task and, consequently, some consideration must be given to developing an algorithm that allows the maximum grid flexibility. The two-dimensional algorithms presented in the previous section have been extended to three dimensions. The small disturbance extensions have been undertaken by Bailey and Ballhaus[4] and by Ballhaus, Bailey, and Frick;[27] extensions to three dimensions of the full-potential algorithm have been undertaken by Jameson and Caughey.[28] Extensions to complex geometries of the small disturbance method are conceptually fairly easy, as demonstrated by Boppe.[29] However, because the small disturbance assumptions are invalid in regions with large gradients, there are practical limitations to small disturbance theory, as illustrated by Hinson, et al.[30] It is necessary, therefore, to focus attention on a full-potential equation method which gives a more accurate treatment of geometries with large curvature gradients. One outcome is the finite-volume algorithm of Jameson and Caughey.[6]

The finite-volume algorithm assumes that the six-faced volume elements comprising the mesh in the physical space can be transformed to cubes in the computational space. The mapping to each cube is assumed to be local so that

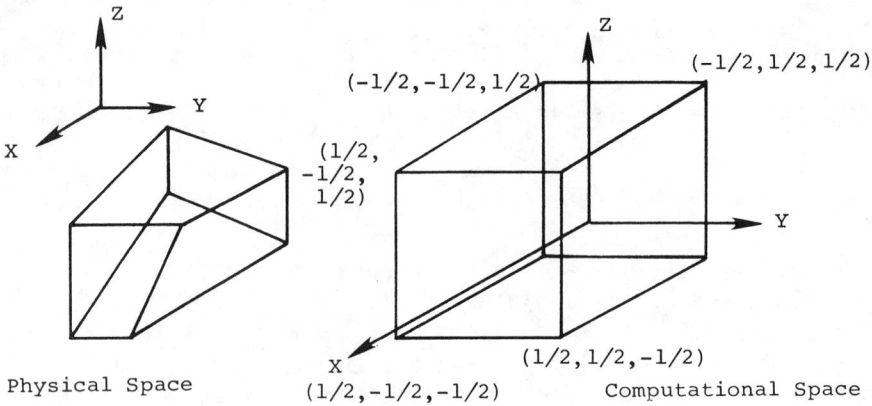

Fig. 7 Mapping from physical space to computational space.

transformations can be based on the physical values at the vertices of the volume elements. The location of the vertices (or mesh points) in physical space may be determined by any suitable procedure, and two specific examples are given in following sections. The mapped cubes have trilinear variations of coordinates ranging from -1/2 to 1/2 (Fig. 7), and the potential is assumed to vary trilinearly within each cell. With the coordinate variation assumption, the corresponding points in the physical space can be located from points in the computational space by the local trilinear mapping formula,

$$x = 8 \sum_{k=1}^{8} x_k \left(\frac{1}{4} + X_k X\right)\left(\frac{1}{4} + Y_k Y\right)\left(\frac{1}{4} + Z_k Z\right) \quad (63)$$

where X_k, Y_k, and Z_k are the mapped vertices of the cubes ($\pm 1/2$) and the x_k terms are the corresponding physical values. There are equivalent formulas for y, z, and ϕ, the velocity potential. With this mapping, continuity of x, y, z and ϕ is preserved at the cell boundaries. The mapping also allows derivatives of the transformation and potential to be evaluated anywere in the cell.

The flow equation that we wish to solve is the conservation relation:

$$\partial(\rho u)/\partial x + \partial(\rho v)/\partial y + \partial(\rho w)/\partial z = \partial(\rho u_i)/\partial x^i = 0 \quad (64)$$

The finite-volume algorithm is a conservative differencing scheme which satisfies the above equation using the cubical

TRANSONIC POTENTIAL METHODS

cells in the computational space. Density is computed from the isentropic relation, Eq. (16).

The first step in the procedure is to determine the governing equation, Eq. (64) in computational space. The result is

$$\partial(\rho J U^i)/\partial X^i = 0$$

where X^i are the transformed coordinates [X, Y, and Z in Eq. (63)], U^i are the contravariant velocity components, and J is the determinant of the transformation matrix $\tilde{J}$ with the elements $\partial x^i/\partial X^j$. The contravariant velocity is defined by

$$U^i = g^{ij}(\partial \phi/\partial x^j) \equiv (\tilde{J}^T \tilde{J})^{-1}(\partial \phi/\partial x^j)$$

A differencing algorithm which conserves $\rho J U^i$ on the cubical cells is derived by creating a set of secondary cells whose vertices lie at the centers of the primary cubical cells. The flux quantity $\rho J U^i$ is evaluated at the center of each primary cell (the vertices of the secondary cell, Fig. 8). The flux computed at the corner is assumed to be constant over that portion of the secondary cell face that lies within the primary cell. If the global mapping is sufficiently smooth to allow a Taylor series expansion of the physical coordinates in terms of

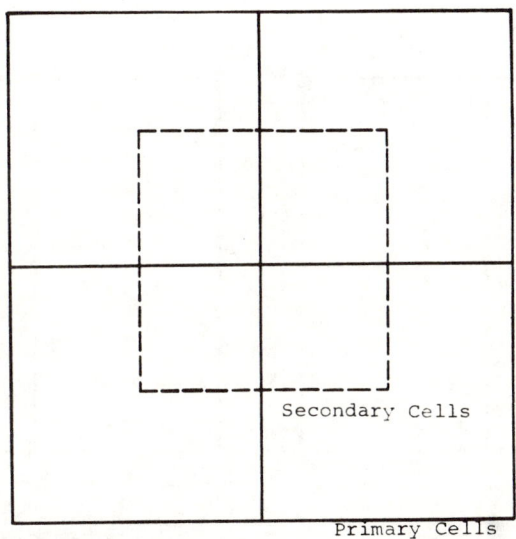

Fig. 8 Primary and secondary cell in computational plane.

the computational coordinates, then the local linear truncation error terms for the flux will cancel and the flux conservation formulas will be accurate to the second order.

With this approach a problem arises in that the difference operator decouples odd and even points as shown in Fig. 9. This results in a homogeneous solution where ϕ can be +1 at odd points and -1 at even points. This problem is overcome by displacing the flux evaluation point away from the vertices by adding a higher-order correction term. This displacement recouples the odd and even points and eliminates the homogeneous solution. For the simple case of the flux being given by ϕ_x, the displacement relation used by Jameson and Caughey[6] is

$$\phi_x = \phi_{x_0} + \varepsilon \phi_{xy_0} \qquad (65)$$

where the subscript o represents the center of the primary cell and ε can vary from 0 to 1/2 where the cell height is assumed to be 1. In regions where the flow is supersonic, upwind differencing is employed. This is accomplished by adding terms to the conservation equation which produce an upwind bias, as discussed in Section 3.2.3. The terms are selected such that the proper domain of dependence is used in the differencing. The effect of this is to produce a

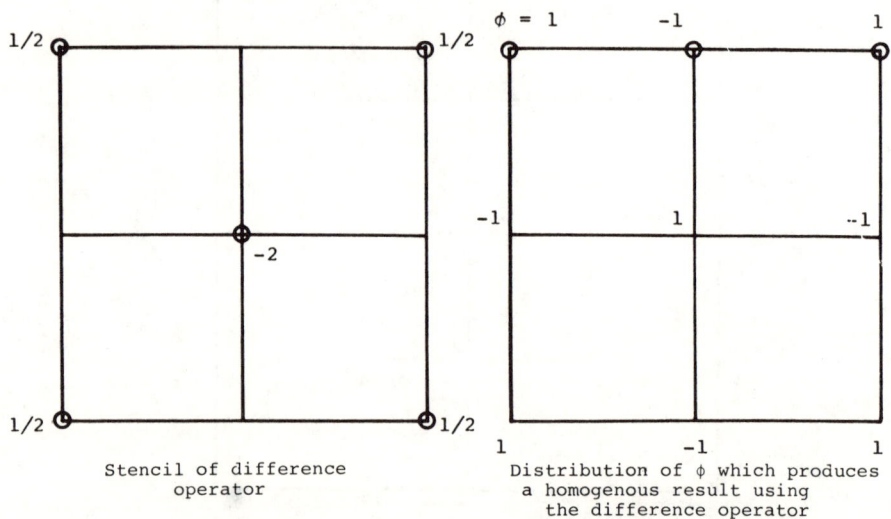

Stencil of difference operator

Distribution of ϕ which produces a homogenous result using the difference operator

Fig. 9 Decoupled solution arising from difference operator.

rotated difference operator of the form

$$\partial\phi/\partial s = (U^i/q_c)(\partial\phi/\partial X^i) \qquad (66)$$

where s is the streamwise direction, q_c is the contravariant velocity, and the first-order difference operators $\partial/\partial X^i$ are chosen to be in the upwind direction. The terms added to the flux equation are

$$\rho^i = -\mu J |U^i| \Delta X^i \rho_{X^i} \qquad (67)$$

where μ is a switching function

$$\mu = \max[0, (1-a^2/q^2)] \qquad (68)$$

and q/a is the local Mach number. The presence of these terms has the effect of adding artificial viscosity to the solution. This does require, however, that the mesh be smooth in the supersonic zone or the effect of the higher-order derivatives associated with the artificial viscosity will cause the solution to give erroneous results.

The last terms which have to be added to the equation are timelike derivatives which have the effect of embedding the steady-state equation in an artificial time-dependent equation. The final equation that is solved is a discrete approximation to

$$(\partial/\partial X^i)(\rho J U^i + P^i) = \alpha\phi_{XT} + \beta\phi_{YT} + \gamma\phi_{ZT} + \delta\phi_T \qquad (69)$$

where the P^i are the upwind biasing terms in the supersonic zones; α, β, and γ are chosen to make the flow direction timelike as in the steady state, and $\delta\phi_T$ is a damping factor.

The complete numerical scheme is outlined below.

1) Evaluate the contravariant velocity components and densities at the centers of the primary cells.

2) Satisfy continuity on the secondary cells using the flux values calculated in step 1 plus the recoupling terms.

3) Add artificial viscosity in the supersonic zones to produce an upwind bias and enforce the entropy condition.

4) Add the time-dependent terms to embed the steady-state equation in a convergent, time-dependent process which evolves to the solution.

The main difficulty associated with developing a computer code based on the finite finite-volume algorithm is that of generating a grid system and incorporating boundary conditions. A desirable grid is one which conforms to all the solid boundaries. Boundary conforming grids provide an accurate and convenient means of specifying boundary conditions. They also can be very efficient in that the grid density can be readily controlled at the boundaries where the gradients of the flow parameters can vary most rapidly.

Since the finite-volume method only requires sets of coordinates corresponding to the corner points of the cubic computational cells, there is no need to have a single mapping function to generate the grid. The procedure chosen is one that uses a sequence of rather simple transformations. The overall mapping is required to be smooth so that the higher-order terms in the transformations do not cause numerical instabilities, particularly in the vicinity of shocks.

3.2.5 <u>Multiple Grid Techniques</u>. The computation time required for a full three-dimensional calculation is prohibitive for routine calculations at the moment and in the continual research for ever faster techniques the multiple grid method emerges as a possible candidate for future development.

The multiple grid method was first proposed by Federenko[31] who realized that it should be possible to accelerate an iterative scheme for solving difference equations by calculating corrections for the fine grid equations on a sequence of successively coarser grids. This idea was subsequently analyzed by Bakhvalov[32] and then extended and applied to a variety of problems by Brandt.[33] It has recently been proved under rather general assumptions by Nicolaides[34] and Hackbusch[35] that the number of operations required to solve the equations arising from a finite-element or a finite-difference approximation to an elliptic problem by a multiple grid method is directly proportional to the number of unknowns.

There is less experience of the use of multiple grid methods for nonelliptic problems. The first demonstration of the use of a multiple grid method for a transonic flow problem was by South and Brandt[36] who solved the transonic small disturbance equation for a nonlifting flow and observed a high rate of convergence. Difficulties were experienced, however, by South and Brandt in the treatment of lifting flows and in calculations on nonuniform and curvilinear meshes. There was a tendency to produce an oscillating sonic line, and for the calculations to enter a variety of limit cycles between several grids. These difficulties appeared to be due to insufficient smoothing of the errors on fine grids before passing to coarser grids, and South and Brandt were able to obtain convergence for a wider range of cases by using multiple line relaxation sweeps in different directions.[37]

In the other applications these difficulties have been attacked by combining the multiple grid method with a generalized alternating direction method, suitable for transonic flows, as the smoothing algorithm.

The case which will be considered is that of two-dimensional transonic flow past an airfoil using the potential flow approximation discussed in Section 2. The potential flow equation is given by Eqs. (10) and (16), namely

$$\partial(\rho u)/\partial x + \partial(\rho v)/\partial y = 0$$

Consider the solution of the equation

$$L^h u = f \qquad (70)$$

by a relaxation method, where L^h approximates a linear differential operator L on a grid with a spacing proportional to the parameter h. Let U be an approximation to the solution, and let V be the correction to U such that $U + V$ satisfies Eq. (70). Then the basis of the multiple grid method is to replace Eq. (70) by

$$L^{2h} V + I_h^{2h} L^h U = f \qquad (71)$$

where L^{2h} is the same approximation of L on a grid in which the spacing has been doubled, and I_h^{2h} is an operator which transfers to each grid point of the coarse grid the residual $L^h U - f$ of the coincident point of the fine mesh, or alternatively a weighted average of the residuals at neighboring points. After solution of Eq. (71), the approximation on the fine grid is updated by interpolating the correction calculated on the coarse grid to the fine grid, so that U is replaced by

$$U^{new} = U + I_{2h}^{h} V \qquad (72)$$

where I_{2h}^{h} is an interpolation operator. Equation (71) can in turn be solved by introducing an approximation on a yet coarser grid, so that a multiple sequence of grids may be used, leading to a rapid solution procedure. Multigrid methods are faster for two reasons. First, the number of operations required for a relaxation sweep on one of the coarse grids is much smaller than the number required on the fine grid. Second, the rate of convergence is faster on a coarse grid, reflecting the fact that corrections can be propagated from one end of the grid to the other in a small number of steps.

To extend this idea to nonlinear equations, Eq. (71) may be reorganized by adding and subtracting the current solution U to give

$$L^{2h}(U + V) + I_{h}^{2h} L^{h} U - L^{2h} U = f$$

or

$$L^{2h} \bar{U} = \bar{f} \qquad (73)$$

where $\bar{U}$ is the improved estimate of the solution to be determined on the coarse grid, and $\bar{f}$ is an appropriately modified right-hand side,

$$\bar{f} = f + L^{2h} u - I_{h}^{2h} L^{h} U \qquad (74)$$

The updating formula, Eq. (72), now becomes

$$U^{new} = U + I_{2h}^{h}(\bar{U} - U) \qquad (75)$$

This avoids the need to introduce a special perturbation operator to represent the correction Eq. (71).

The success of the multiple grid method generally depends on the use of a relaxation algorithm which rapidly reduces the high-frequency components of error on any given grid, because on a coarser grid these components cannot be distinguished from low-frequency components. This aliasing process will cause improper corrections to be computed on coarse grids, and can prevent convergence.

It turns out that point and line relaxation schemes do not necessarily provide the required smoothing of all high-frequency components of error on a nonuniform or curvilinear mesh. To illustrate this, consider the model problem

$$a\phi_{xx} + b\phi_{yy} = 0 \qquad (76)$$

with positive coefficients a and b. Let δ_x^2 and δ_y^2 denote second-difference operators in the x and y directions, and suppose that the difference approximation has the form

$$L\phi \equiv (A\delta_x^2 + B\delta_y^2)\phi = 0 \qquad (77)$$

where if Δx and Δy are the mesh widths,

$$A = a(\delta x)^{-2} \quad B = b(\delta y)^{-2} \qquad (78)$$

In this analysis we use an alternating direction method as the smoothing algorithm. Consider the model Eq. (76) and suppose that the correction $\delta\phi$ is calculated by the equation

$$(\alpha - A\delta_x^2)(\alpha - B\delta_y^2)\delta\phi = \omega\alpha L\phi \qquad (79)$$

where α is a parameter to be chosen, ω is an over-relaxation factor, and the residual $L\phi$ is calculated using the result of the previous iteration.

In the case of transonic flow we have to allow for change from elliptic to hyperbolic type as the flow becomes locally supersonic. In the model problem this corresponds to one of the coefficients, a, becoming negative. The classical alternating direction scheme then leads to an ill-posed problem which admits oscillatory solutions which are undamped in time and grow in the x direction.

The following generalized alternating direction scheme is therefore used. Let the scalar parameter α in Eq. (79) be replaced by a difference operator

$$S \equiv \alpha_0 + \alpha_1\delta_x^- + \alpha_2\delta_y^- \qquad (80)$$

where δ_x^- and δ_y^- denote the one-sided difference operators in the x and y directions. This yields the scheme

$$(S-A\delta_x^2)(S-B\delta_y^2)\delta\phi = \omega S L\phi \qquad (81)$$

in which the residual $L\phi$ is differenced by the operator S. The corresponding time-dependent equation is now a hyperbolic equation of the form

$$\beta_0 \phi_t + \beta_1 \phi_{xt} + \beta_2 \phi_{yt} = a\phi_{xx} + b\phi_{yy}$$

where the coefficients β_0, β_1, β_2 depend on the parameters α_0, α_1, α_2.

The following simple strategy has been found to be effective. Begin each cycle on the fine gird. The alternating direction iteration is performed once on each grid until the coarsest grid is reached. Then it is performed once on each grid going back up to the second finest grid, and the cycle terminates with the interpolation of the correction from the second finest grid to the fine grid. It is convenient to measure the work in units representing the work required to perform an iteration of the alternating direction scheme on the fine grid. Since each grid has 1/4 as many cells as the next finer grid, the work required to perform each cycle is

$$1 + 2\left(\frac{1}{4} + \frac{1}{16} + \frac{1}{64} \ldots\right) \leq 1\frac{2}{3} \text{ units}$$

plus the overhead of computing and transmitting residuals from one grid to the next, and interpolating the corrections.

It is the usual practice to accelerate the alternating direction scheme Eq. (79) by using a sequence of values of the parameter α designed to give rapid damping of the error components in a series of frequency bands. The multigrid alternating direction method economizes the work required by passing to the coarse grids to treat the error components in the low-frequency bands. If a sequence of six parameters were used to treat six frequency bands, for example, the work required to complete one cycle through the parameters would be six units, whereas the work required to perform the alternating direction iterations of a multigrid cycle with six grids would be less than 1 2/3 units with the present stategy. A more complete discussion of this approach is given in Ref. 38.

4. Grid Generation

Before any of the algorithms described in the previous section can be used a finite-difference grid must be constructed. In the case of the two-dimensional small distur-

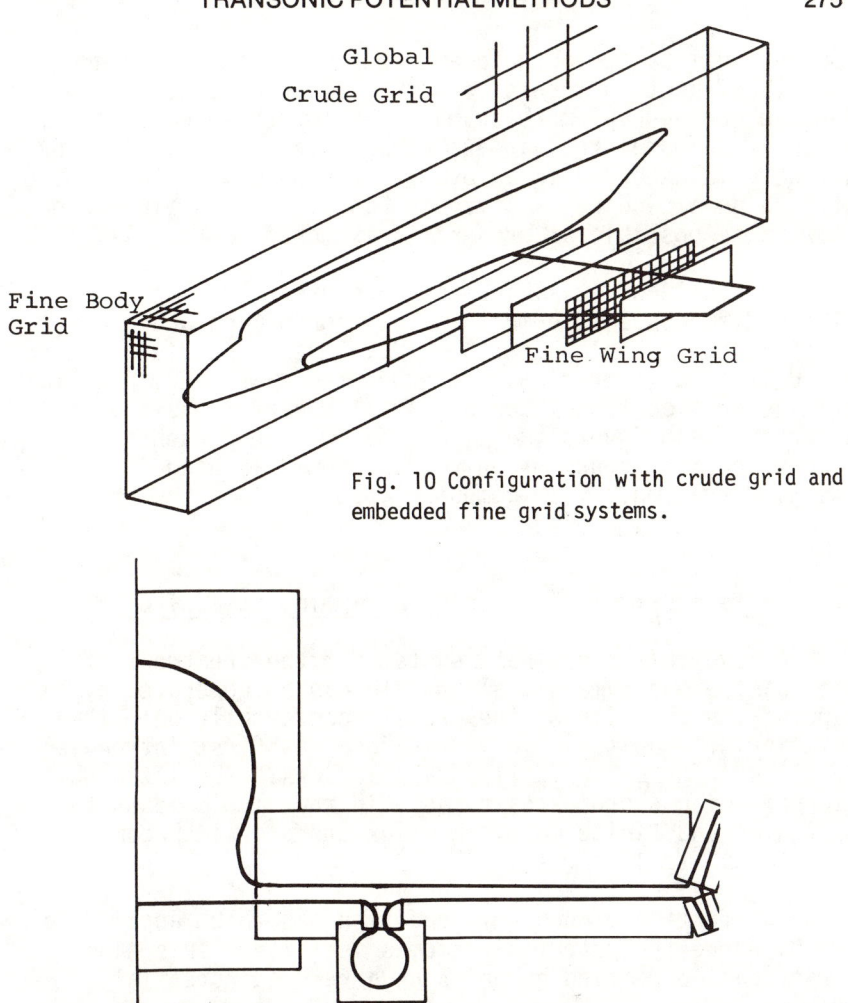

Fig. 10 Configuration with crude grid and embedded fine grid systems.

Fig. 11 Future grid component build-up capability for complex aircraft configurations.

bance equation a simple Cartesian grid is used. Points can be clustered near the airfoil and near the leading and trailing edges where flow gradients are large. For the three-dimensional small disturbance codes a sheared Cartesian grid was used initially; the shearing aligned a coordinate line with the wing leading and trailing edges. Later developments by Boppe,[29] however, concentrated on the embedding of a fine grid in an overall coarse grid. The basic idea is to concentrate grid points near the body and to model the far field by a coarse grid. This type of embedding avoids problems associated with stretching a conventional Cartesian grid and models the far field adequately without an excessive number of grid points. A further and perhaps more dramatic use of these embedded grids is to model the various components of an airplane by a fine grid embedded in an overall coarse grid. An example of such an embedding is given in Fig. 10 and an impression of what is possible at present is sketched in Fig. 11.

For full potential equation computer codes the finite-difference grid must be body conforming and a specialized technology has been developed recently in the field of grid generation. Although there are very powerful grid generation techniques now available, for example the numerical technique of Refs. 39 and 40, the present discussion will consider only those commonly used in existing transonic flow computer codes.

4.1 Airfoil Calculations Using a Mapping to a Circle

A favorable coordinate system for the treatment of a flow past a two-dimensional profile can be generated by mapping the exterior of the profile conformally onto the interior of a unit circle. This idea was first introduced by Sells[41] for subsonic flow calculations. The introduction of polar coordinates r and θ in the circle leads to a regular and finite mesh, in which the profile becomes the coordinate line $r = 1$.

The far-field boundary condition has to be applied to $r = 0$, where the potential becomes infinite. This singularity can be removed by defining a reduced potential

$$G = \phi - r^{-1}\cos(\theta+\alpha) + E(\theta+\alpha) \qquad (82)$$

where $2\pi E$ is the circulation, and α is the angle of attack. Then G is finite and single valued. The modulus of the mapping function also becomes finite at $r = 0$, and the use of finite-difference formulas to represent derivatives of quantities depending on the mapping function can lead to large errors. The conservation of mass equation, Eq. (10), becomes

$$(\partial/\partial\theta)(\rho Hu/r) + r(\partial/\partial r)(\rho Hv/r) = 0 \qquad (83)$$

where H is the modulus of the derivative of the transformation to the exterior of the circle, and u and v are the

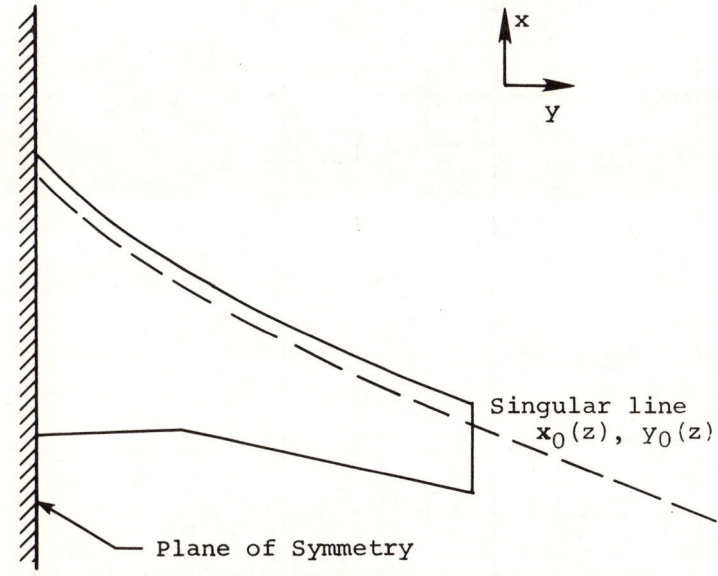

Fig. 12 Configuration of swept wing.

velocity components in the θ and r directions,

$$u = H^{-1}[r(G_\theta - E) - \sin(\theta + \alpha)]$$
$$v = H^{-1}[r^2 G_r - \cos(\theta + \alpha)] \qquad (84)$$

The circulation constant E is determined by the Kutta condition, which requires the velocity to be finite at the trailing edge, where $H = 0$. Thus ϕ_θ must also vanish giving

$$E = G_\theta - \sin\alpha \text{ at } r = 1, \theta = 0.$$

4.2 Sheared Parabolic Grid for a Finite Wing

The construction of a satisfactory curvilinear coordinate system to suit the geometry of the configuration is one of the most difficult aspects of the three-dimensional problem. Here nonorthogonal coordinates will be generated by a sequence of elementary transformations.[23] First we introduce parabolic coordinates in planes containing the wing section by the complex square root transformation

$$X_1 + iY_1 = \left\{x - x_0(z) + i[y - y_0(z)]\right\}^{1/2}, \quad Z_1 = z \qquad (85)$$

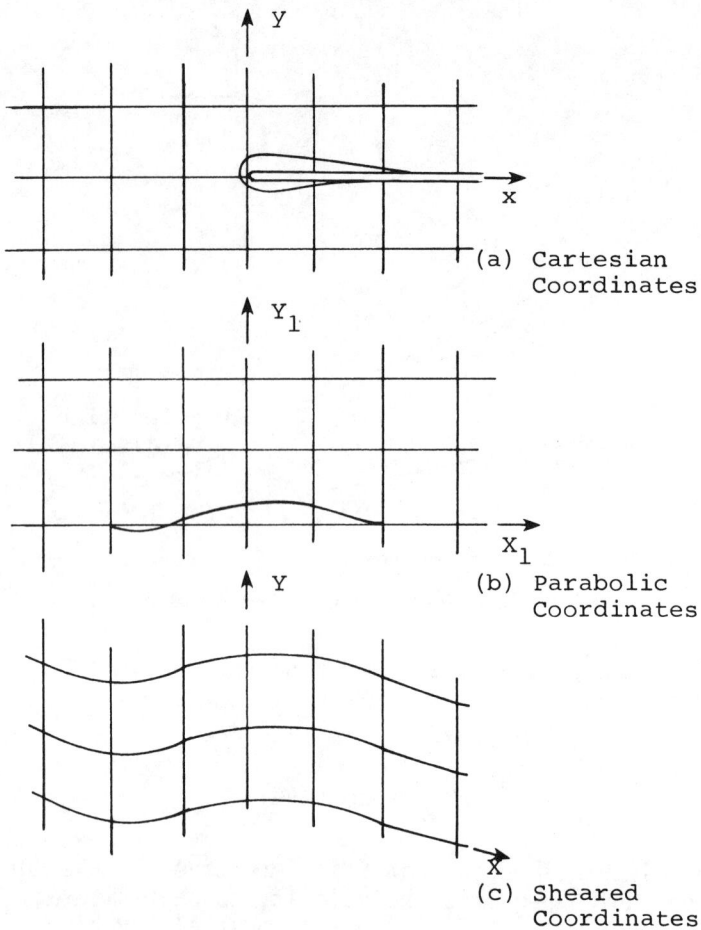

Fig. 13 Construction of coordinate system for swept wing calculation.

where z is the spanwise coordinate, and x_0 and y_0 define a singular line of the coordinate system located just inside the leading edge (see Figs. 12 and 13). The effect of this transformation is to unwrap the wing to form a shallow bump

$$Y_1 = S(X_1, Z_1) \tag{86}$$

Then we use a shearing transformation

$$X = X_1 \;,\; Y = Y_1 - S(X_1, Z_1) \;,\; Z = Z_1 \tag{87}$$

to map the wing surface to a coordinate surface. Finally, in order to obtain a finite computational domain X, Y, and Z are replaced by stretched coordinates $\bar{X}$, $\bar{Y}$, and $\bar{Z}$. The stretching frequently used is to set $X = \bar{X}$ in an inner domain $-\bar{X}_m \leq \bar{X} \leq \bar{X}_m$, and to set

$$X = \bar{X}_m + (\bar{X}-\bar{X}_m)\{1 - [(\bar{X}-\bar{X}_m)/(1-\bar{X}_m)]^2\}^{-\alpha}$$

when $\bar{X} > \bar{X}_m$, with a corresponding formula when $\bar{X} < \bar{X}_m$, so that $X = +\infty$ when $\bar{X} = +1$. Typically the exponent α has the value 1/2. Similar stretchings are used for Y and Z.

The vortex sheet is assumed to coincide with the cut behind the singular line which is opened up by the square root transformation Eq. (85). Thus a jump Γ is introduced in the potential between corresponding points representing the two sides of the vortex sheet. A complication is caused by the continuation of the cut beyond the wing tip. Points on the two sides of the cut must be identified as the same point in the physical space. Also a special form of the equations must be used at points lying on the singular line beyond the wing.

4.3 Sheared Parabolic Grid for a Wing-Body Combination

In this case the body cross section[6] is approximated by a circle which is then conformally transformed to a slit. Since the body cross section is not usually circular the result of this transformation is a "wavy" slit. This is transformed to a plane slit by a simple shearing. By this transformation a wing-body configuration is transformed to a wing-on-a-wall configuration which can then be treated using the ideas in the previous subsection.

4.4 Body-Oriented Cylindrical Coordinate System

This technique[42] uses a sequence of transformations to generate a boundary-conforming coordinate grid for an arbitrary wing-fuselage combination. A typical example consists of a swept, tapered wing of arbitrary planform, dihedral and section shape mounted upon a finite fuselage of varying cross-sectional area and shape. Let x,y,z be Cartesian coordinates in the streamwise, vertical, and spanwise directions, repectively; the geometry and solution are assumed to be symmetric about the plane $z = 0$. We introduce polar coordinates r and θ in the crossflow planes, defined by

$$r = (y^2 + x^2)^{1/2}$$
$$\theta = \arctan(y/z) \qquad (89)$$

In each crossflow plane, the fuselage surface is defined by $r = R_f(x,\theta)$, and a normalized radial coordinate $\bar{r}$ is defined by

$$\bar{r} = [r - R_f(x,\theta)]/[R_t - R_f(x,\theta)] \qquad (90)$$

where R_t is the radius of the cylindrical surface passing through the wing tip. The wing sweep and dihedral are normalized by referencing the coordinates in each surface of constant $\bar{r}$ to the location of a singular line $x_s(\bar{r})$, $\theta_s(\bar{r})$ passing just inside the leading edge of each wing section according to

$$\bar{x} = [x - x_s(\bar{r})]/c(\bar{r}) + \log 2$$
$$\bar{\theta} = 2\{\theta - (1 - 4\theta^2/\pi^2)\theta_s(\bar{r})/[1 - 4\theta_s^2(\bar{r})/\pi^2]\} \qquad (91)$$

where $c(\bar{r})$ is the local chord length.

In a surface of constant $\bar{r}$ intersecting the wing surface (i.e., for $\bar{r} \leq 1$), the geometry looks like that in Fig. 14a. The conformal transformation

$$\bar{x} + i\bar{\theta} = \log[1 - \cosh(\xi + i\eta)] \qquad (92)$$

maps these surfaces to the geometry shown in Fig. 14b. The width of the strip $S(\xi,\bar{r})$ defining the wing surface is a slowly varying function if the location of the singular line has been carefully chosen. Thus the introduction of

$$X = \xi$$
$$Y = \eta/S(\xi,\bar{r}) \qquad (93)$$
$$Z = \bar{r}$$

results in a nearly orthogonal coordinate system in the planes Z = constant.

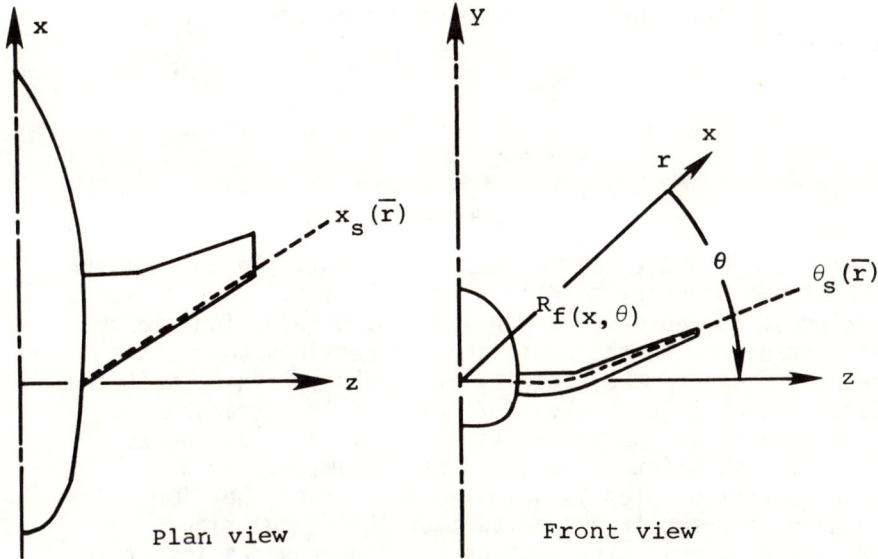

Fig. 14a Coordinate system for wing-body.

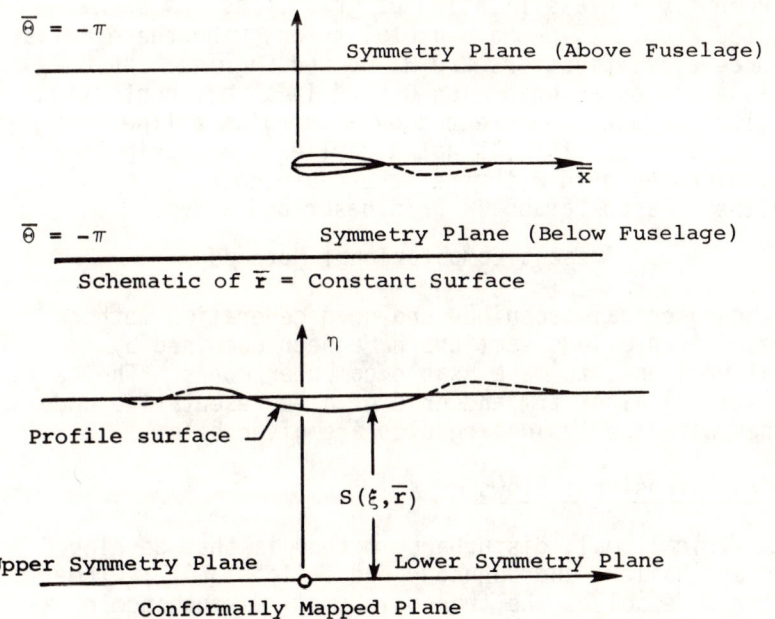

Fig. 14b Transformation sequence for wing-body.

A Cartesian grid is introduced into the X,Y,Z domain for

$$-X_{lim} \leq X \leq X_{lim}$$
$$0 \leq Y \leq 1 \qquad (94)$$
$$0 \leq Z \leq Z_{lim}$$

where X_{lim} and Z_{lim} are chosen sufficiently large that the freestream boundary condition can be specified on these surfaces. Suitable one-dimensional stretching functions are introduced in the X, Y, and Z directions to cluster mesh surfaces in the vicinity of the body. This stretched Cartesian grid in the X, Y, Z space is then transformed back to give the Cartesian (i.e., x, y, z) coordinates of each mesh point. The computation time required for the mesh generation step is a small fraction of the time required for the transonic part of the calculation. On a 122,880 cell grid, the mesh generation requires less time than that for 4 relaxation sweeps when the fuselage cross-section is defined by 17 Fourier components at each of 21 axial stations.

Recently a classification of grid types has been attempted and it is perhaps useful to describe these grids The three classifications are H, O, and C grids. An H grid is a Cartesian grid. An O grid is a body conforming grid with the wake represented by a coordinate line, for example the grids of Sells noted earlier. A C grid is a body-conforming grid with the wake represented by a cut, as in the sheared parabolic grid described above.

5. Present Computational Methods

The numerical technique and grid generation methods discussed in previous sections have been combined by several engineers to make usable computer codes. The basic computational algorithm and grid of a representative code together with the typical results are given below.

5.1 Two-Dimensional Flows

A typical small disturbance method is that developed by Murman, Bailey, and Johnson[3] and called TSFOIL. The computer code solves the transonic small disturbance equation with either the conservative or nonconservative algorithm discussed in Section 3.2.2. The grid is Cartesian with the clustering near the leading and trailing edge of

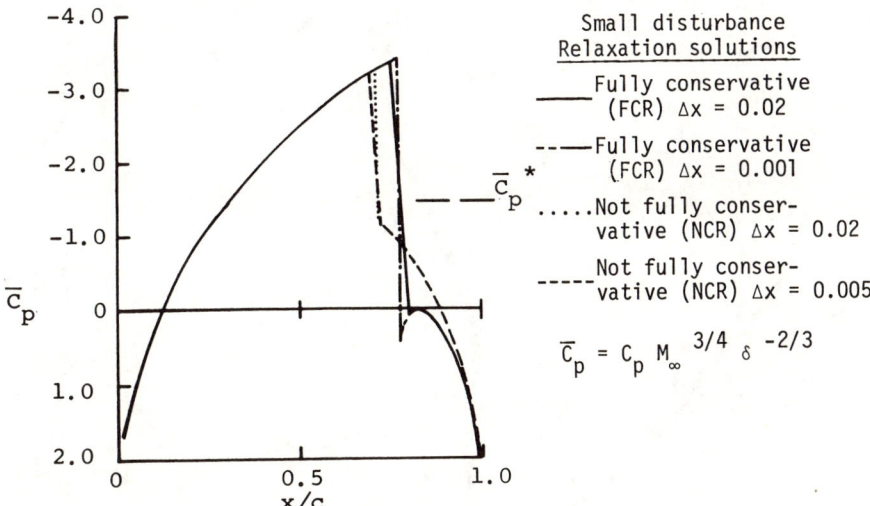

Fig. 15 Comparison of FCR and NCR solutions for the similarity surface pressure distribution on a parabolic ARC airfoil with $\delta = 0.06$ and $M_\infty = 0.872$.

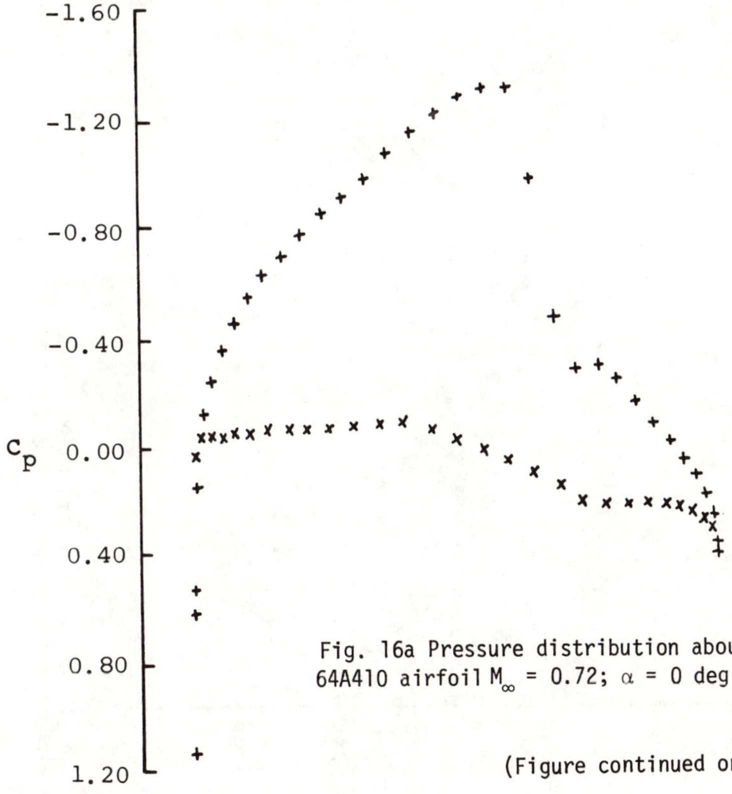

Fig. 16a Pressure distribution abour a NACA 64A410 airfoil $M_\infty = 0.72$; $\alpha = 0$ deg 64×16 grid.

(Figure continued on next page)

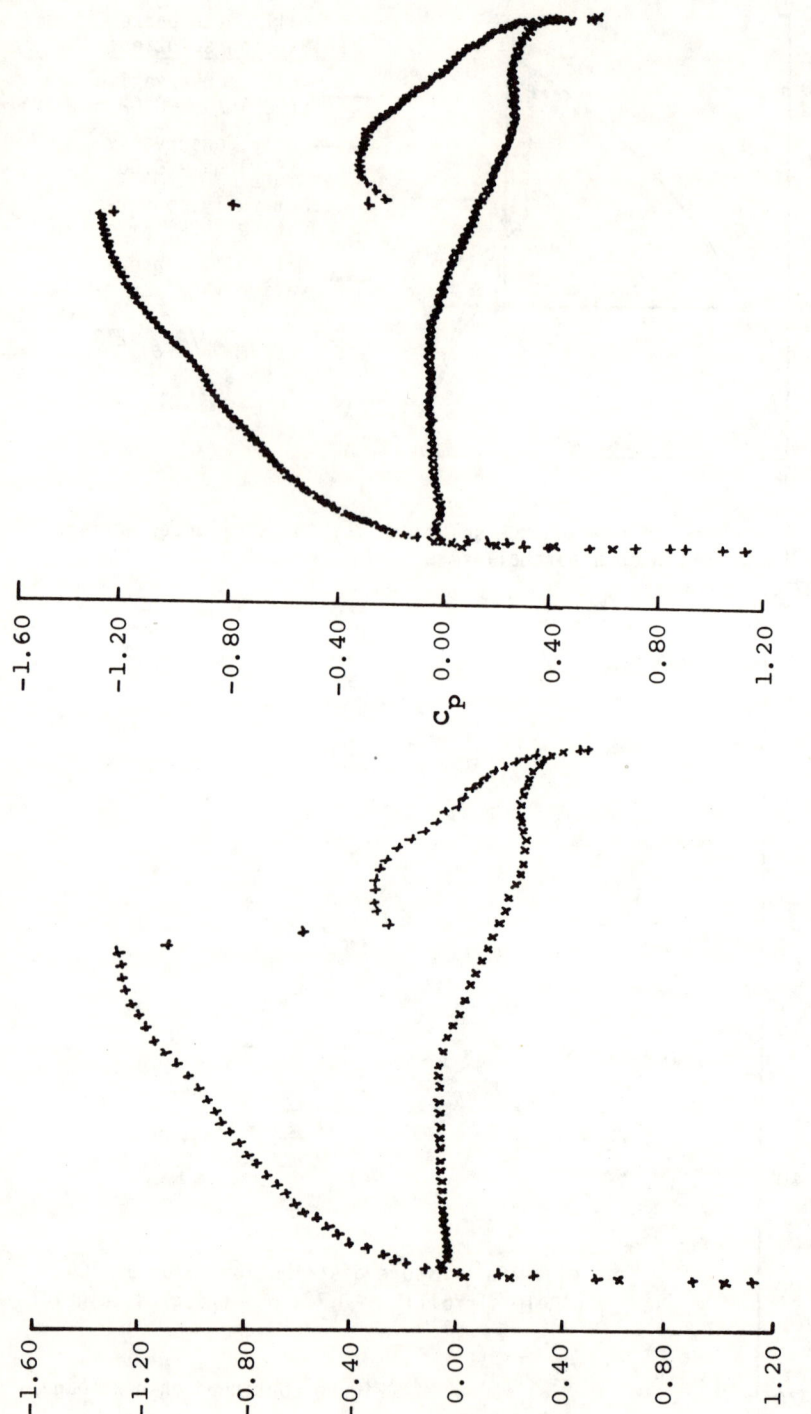

Fig. 16b Pressure distribution around a NACA 64A410 airfoil; $M_\infty = 0.72$; $\alpha = 0$ deg 128×32 grid.

Fig. 16c Pressure distribution around a NACA 64A410 airfoil; $M_\infty = 0.72$; $\alpha = 0$ deg 256×64 grid.

the airfoil. The algorithm uses successive line over-relaxation (SLOR) in its iteration procedure.

A typical result is shown in Fig. 15 where both the nonconservative and conservative solutions are compared. The airfoil is a 6% thick parabolic arc airfoil with $M_\infty = 0.87$.

A typical computer code that solves the full potential equation is FLO-6, written by Jameson, which uses SLOR iteration with the rotated difference scheme described in Section 3.2.3. The computational grid is obtained by transforming the exterior flowfield to the interior of a unit circle as described in Section 4.1. The flow around a NACA 64A 410 airfoil at $M_\infty = 0.72$ and zero angle of attack is shown in Fig. 16. A series of solutions for an increasing number of grid points is shown. Engineering accuracy is obtained on the 128 x 32 mesh. An example of a potential flow calculation coupled with the boundary-layer model of Melnik et al.[43] is shown in Fig. 17.

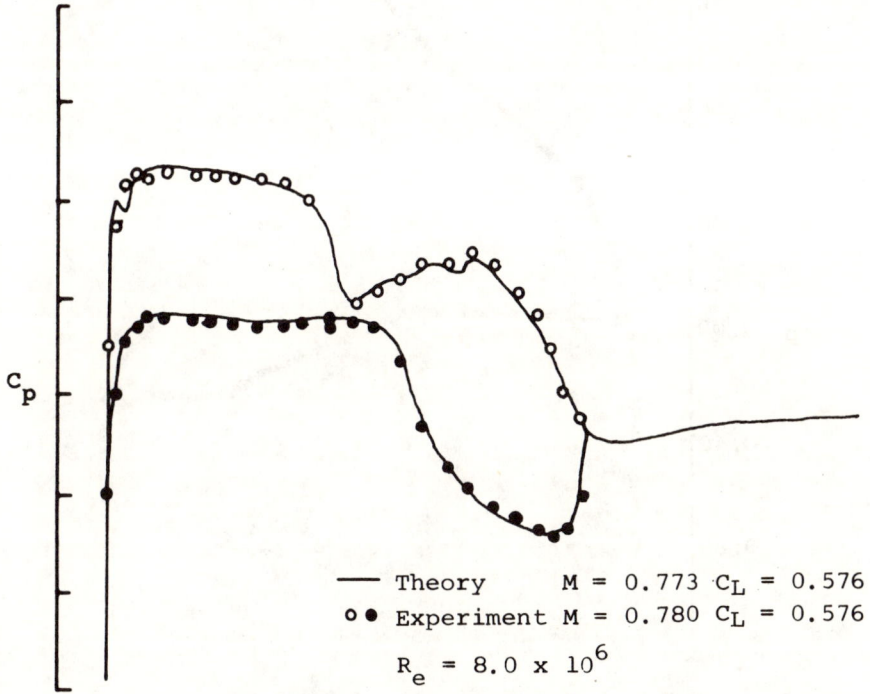

Fig. 17 Pressure distribution on a Whitcomb airfoil.

The faster AF2 algorithm described in Section 3.2.4 is used to solve the conservative full potential equation in the program TAIR written by Holst et al.[44,45] A body-fitted computational grid is generated by a numerical mapping procedure for an arbitrary airfoil. The convergence of the solution is monitored by internal logic, and the acceleration parameters are updated when necessary to speed up convergence or suppress divergence. A factor of 4-10 in convergence speed is claimed over SLOR.[44]

The last of the two-dimensional flow methods mentioned here is the multigrid method discussed in Section 3.2.5; the circle grid mapping of Section 4.1 is used. The flow around a NACA 64A 410 airfoil at M_∞ = 0.72 and zero angle

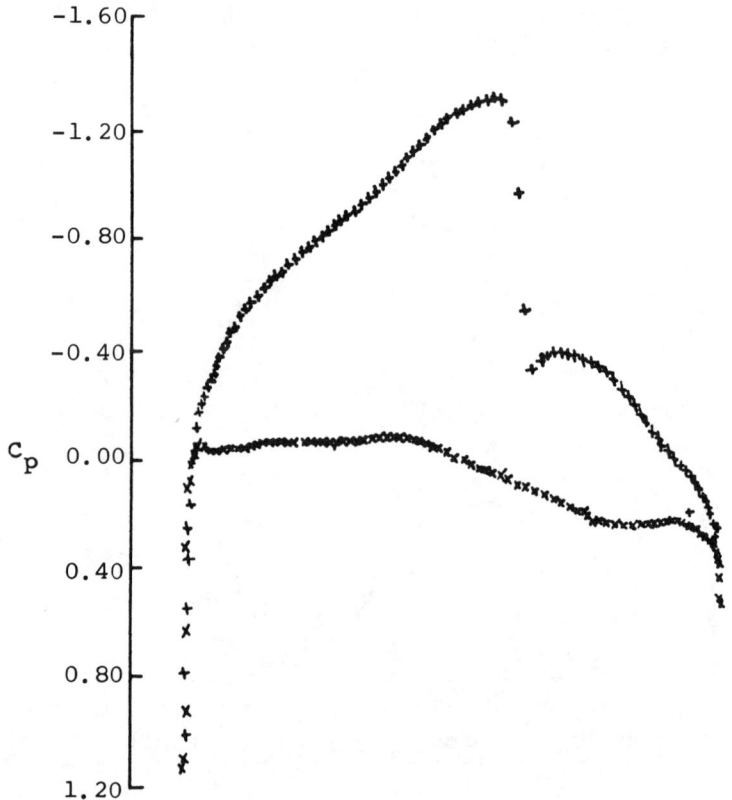

Fig. 18 Multigrid solution for the pressure distribution around a NACA 64A410 airfoil; M_∞ = 0.72, α = 0 deg 192×32 grid.

of attack is shown in Fig. 18. Five grids are used in the multigrid calculations.

5.2 Three-Dimensional Flows

The earliest of the three-dimensional flow calculation methods is that of Bailey and Ballhaus[4] which used a straightforward extension of two-dimensional Murman-Cole algorithm with a simple sheared Cartesian grid. In recent years the modified version of this code, developed primarily by Boppe,[27] has proved more popular. This modified version uses the embedded fine grid/coarse grid technique discussed in Section 4. A typical example for the flow around a wing-body configuration, sketched in Fig. 19, is shown in Fig. 20. Experimental pressure distributions are shown for comparison. There is a tendency for the predicted pressure to overexpand at the leading edge. This is due to the inaccurate treatment of the wing leading edge.

One of the first of the three-dimensional full potential equation methods to appear in the literature is that of Jameson and Caughey[26] (FLO-22) which used a nonconservative algorithm based on the two-dimensional technique described in Section 3.2.3. The grid used is the sheared parabolic grid described in Section 4.2. The flow around the ONERA M6 wing at M_∞ = 0.84 and 3 deg angle of attack is shown in Fig. 21. Results of coupling this code with a strip boundary layer are shown in Fig. 22. In both cases

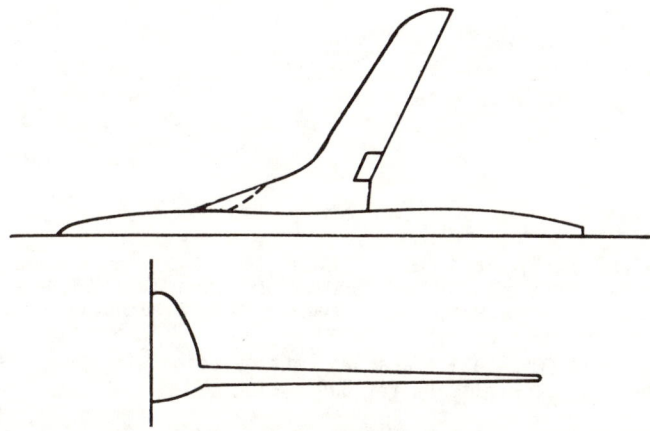

Fig. 19 Supercritical wing/area-ruled fuselage transport configuration (NASA TM X-3431) 57 deg glove leading edge sweep used for analysis (actual glove sweep 72 deg).

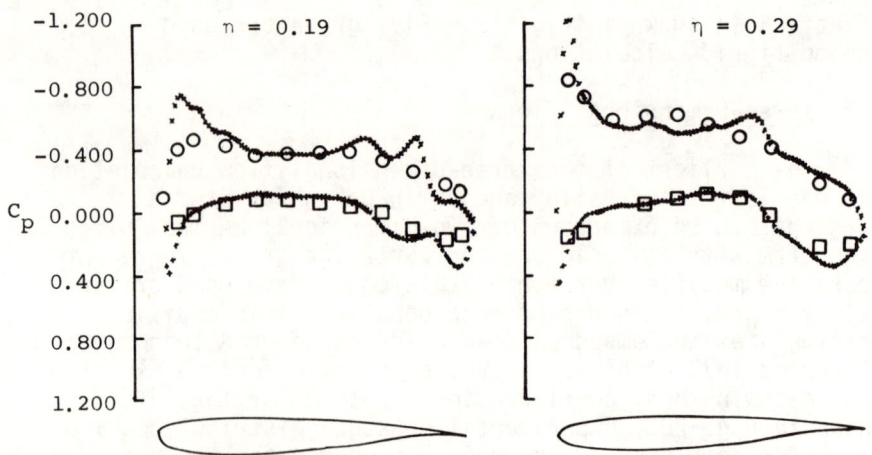

Fig. 20a NASA TM X-3431 transport configuration wing pressure distribution. Correlation for basic wing.

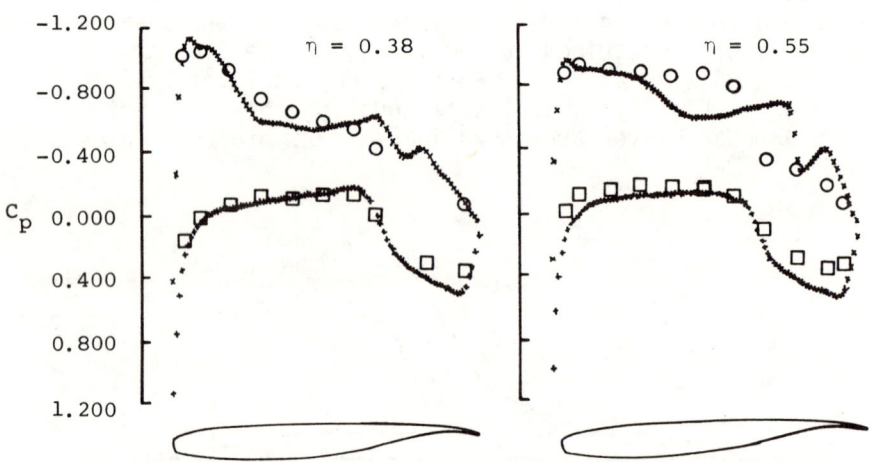

Fig. 20b NASA TM X-3431 transport configuration wing pressure distribution. Correlation for basic wing.

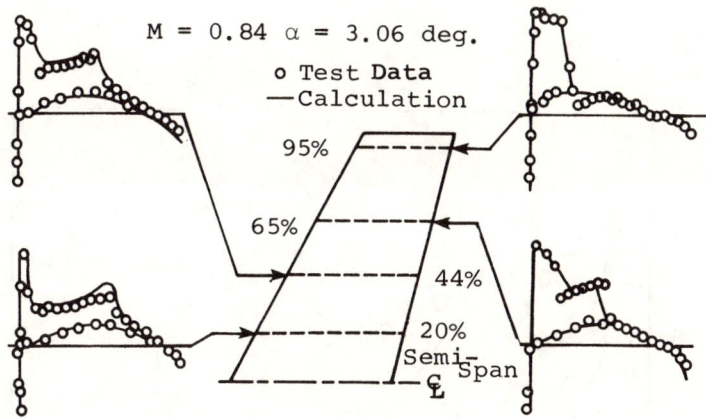

Fig. 21 Comparison of FLO-22 results with experiment for Onera wing M-6; test conducted at $R_{\bar{c}} = 18 \times 10^6$.

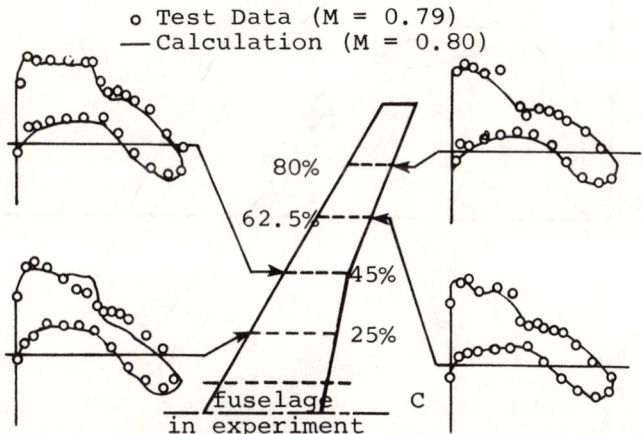

Fig. 22 Comparison of FLO-22 results with experiment for NASA supercritical wing; test conducted at $R_{\bar{c}} = 2.4 \times 10^6$.

the experimental data is shown and it may be seen that the agreement between the predicted and experimental pressure distributions is quite good.

The most recent of the Jameson and Caughey methods uses the finite-volume[6] technique for wing-body configurations. These use the conservative finite-volume algorithm described in Section 3.2.4 with either the grid described in Section 4.2 (FLO-28 code) or the grid described in Section 4.3 (FLO-30 code). A typical result obtained by the FLO-28 code is shown in Fig. 23. The example is the

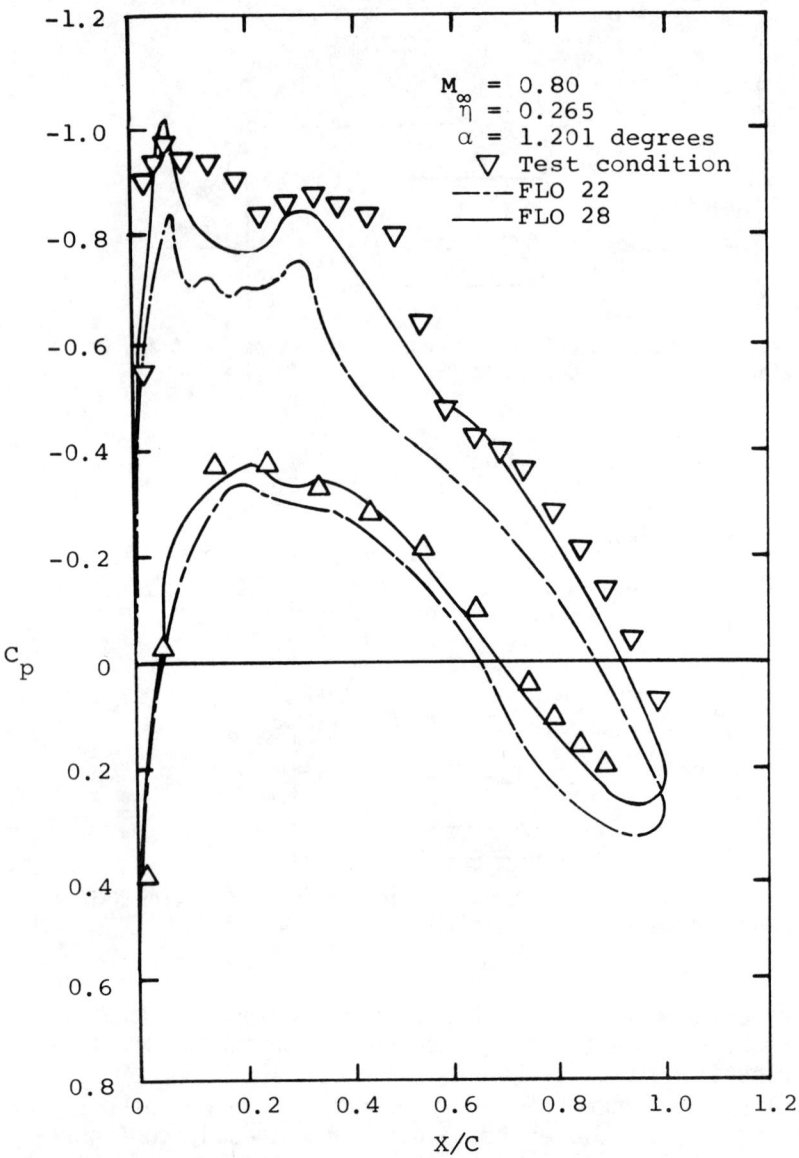

Fig. 23a Pressure distribution around wing-body combination fuselage cross section.

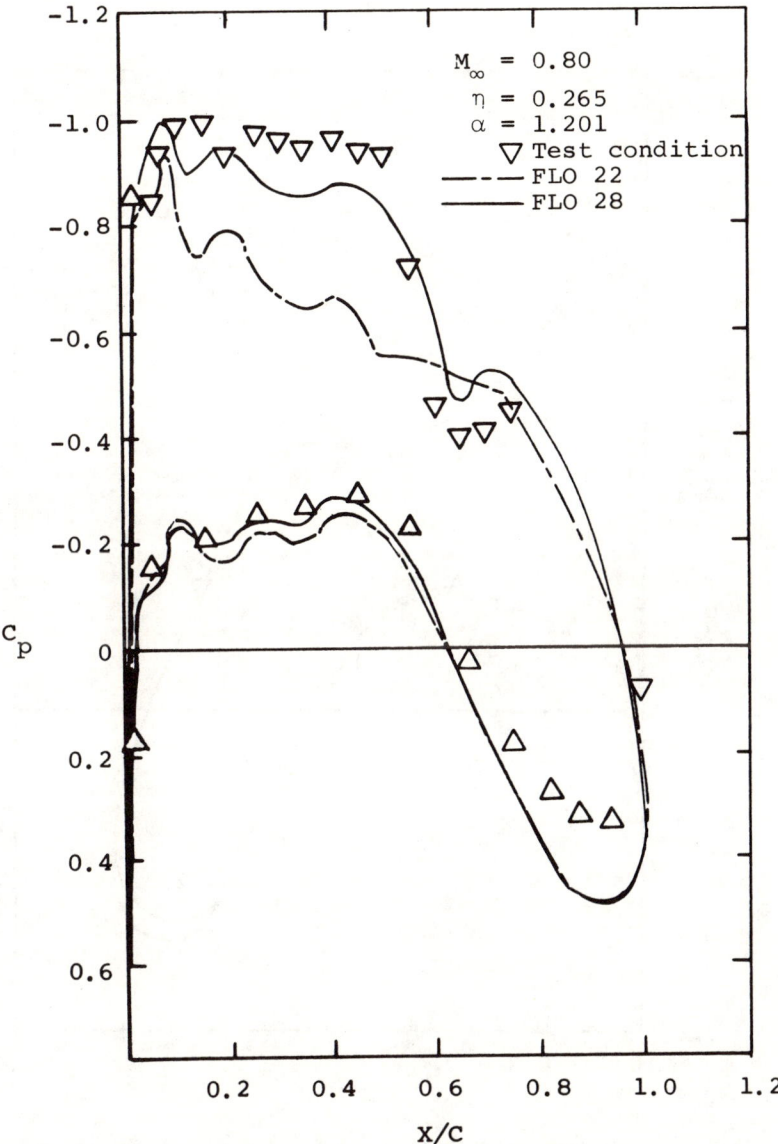

Fig. 23b Pressure distribution around wing-body combination fuselage cross section.

(Figure continued on next page)

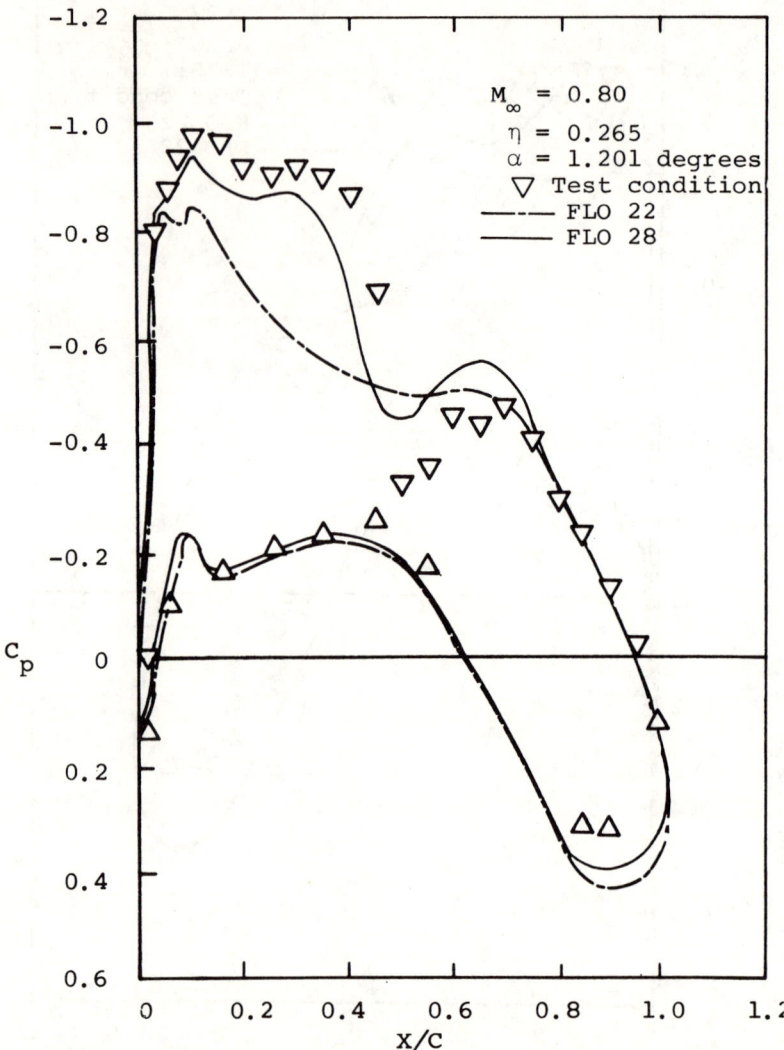

Fig. 23c Pressure distribution around wing-body combination fuselage cross section

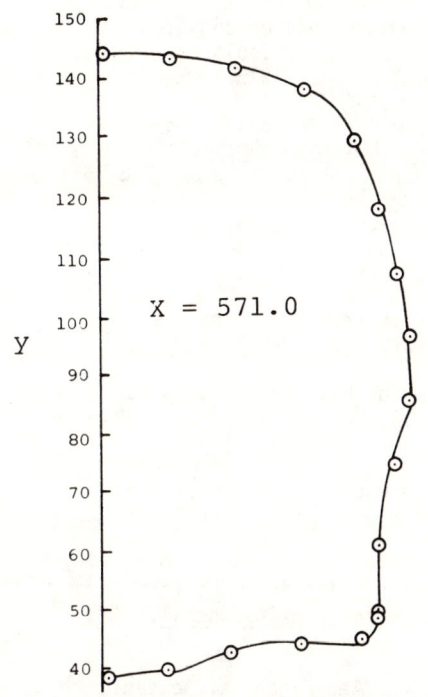

Fig. 24 Fuselage cross section.

wing-body configuration shown in Fig. 24. It may be seen that the agreement with the experiments is adequate.

Acknowledgment

The authors would like to thank Kathy Woerner, Bernard Halliwell, and Carol Sevilla for their careful preparation of the manuscript.

References

[1] Morawetz, C. S., "On the Non-Existence of Continuous Transonic Flow Past Profiles I," Communications on Pure and Applied Mathematics, Vol. 9, No. 1, Feb. 1956, pp. 45-68.

[2] Morawetz, C. S., "On the Non-Existence of Continuous Transonic Flow Past Profiles II," Communications on Pure and Applied Mathematics, Vol. 10, No. 1, Feb. 1957, pp. 107-131.

[3] Murman, E. M., Bailey, F. R., and Johnson, M. J., "TSFOIL - A Computer Code for 2D Transonic Calculations Including Wind Tunnel Wall Effects and Wave Drag Evaluations," NASA SP 347, 1975.

[4] Bailey, F. R. and Ballhaus, W. F., "Comparisons of Computed and Experimental Pressures for Transonic Flows About Isolated Wings and Wing Fuselage Configurations," NASA SP 347, 1975.

[5] Jameson, A., "Transonic Potential Flow Calculations Using Conservation Form," *Proceedings of the AIAA 2nd Computational Fluid Dynamics Conference*, Hartford, Conn., June 1975, pp. 148-161.

[6] Jameson, A. and Caughey, D. A., "A Finite Volume Method for Transonic Potential Flow Calculations." AIAA Paper No. 77-635, June 1977, Albuquerque, N. Mex.

[7] Magnus, R. and Yoshihara, H., "Inviscid Transonic Flow Over Airfoils," AIAA Paper No. 70-47, January 1970, New York, N. Y.

[8] Deiwert, G. S., "Numerical Simulation of High Reynolds Number Transonic Flows," *AIAA Journal*, Vol. 13, No. 10, 1975, pp. 1354-1359.

[9] Steger, J. L., "Implicit Finite Difference Simulation of Flow About Arbitrary Two-Dimensional Geometries", *AIAA Journal*, Vol. 16, No. 7, 1978, pp. 679-686.

[10] Steger, J. L. and Pulliam, T. M., "Implicit Finite Difference Simulations of Three Dimensional Geometries," *AIAA Journal*, Vol. 18, No. 2, 1980, pp. 159-167.

[11] Ferrari, C. and Tricomi, F., *Transonic Aerodynamics*, Academic Press, New York, 1968, pp. 21-104.

[12] Ferziger, J. H, "Large Eddy Numerical Simulations of Turbulent Flows," *AIAA Journal*, Vol. 15, No. 9, 1977, pp 1261-1267.

[13] Cebeci, T., "Calculation of Compressible Turbulent Boundary Layers with Heat and Mass Transfer," AIAA Paper No. 70-741, June 29-July 1, 1970, Los Angeles, Calif.

[14] Baldwin, B. S. and Lomax, H., "Thin Approximation and Algebraic Model for Separated Turbulent Flows," AIAA Paper No. 78-257, January 1978, Huntsville, Ala.

[15] Reynolds, W. C. and Cebeci, T, "Calculation of Turbulent Flows," *Turbulence*, edited by Peter Bradshaw, Springer-Verlag, New York, 1978, pp. 193-229.

[16] Levy, L. L, "An Experimental and Computational Investigation of the Steady and Unsteady Transonic Flow Field About an Airfoil in a Solid Wall Test Channel," AIAA Paper No. 77-678, June 1977, Albuquerque, N. Mex.

[17] Liepmann, H. W. and Roshko, A., *Elements of Gas Dynamics*, Wiley, New York, 1957, pp. 60-61.

[18] Lomax, H., Bailey, F. R., Ballhaus, W. F., "On the Numerical Simulation of Three Dimensional Transonic Flow with Application to the C141 Wing," NASA TN D-6933, 1973.

[19] Lax, P. D., "Weak Solutions of Nonlinear Hyperbolic Equations and Their Numerical Computation," Communications on Pure and Applied Mathematics, Vol. 7, No. 1, Feb. 1954, pp. 159-193.

[20] Murman, E. M. and Cole, J. D., "Calculation of Plane Steady Transonic Flows," AIAA Journal, Vol. 9, 1971, pp. 114-121.

[21] Murman, E. M., "Analysis of Embedded Shock Waves Calculated by Relaxation Methods," AIAA Journal, Vol. 12, 1974, pp. 626-633.

[22] South, J. C. and Jameson, A., "Relaxation Solutions for Inviscid Axisymmetric Flow Over Blunt or Pointed Bodies," Proceedings of the AIAA Conference on Computational Fluid Dynamics, Palm Springs, Calif., July 1973, pp. 8-17.

[23] Albone, C. M., "A Finite Difference Method for Computing Supercritical Flows in an Arbitrary Coordinate System," RAE Tech. Rept. 74090, 1974.

[24] Jameson, A., "Numerical Computation of Transonic Flows with Shock Waves," Symposium Transsonicum II, Springer-Verlag, New York, 1976, pp. 384-414.

[25] Ballhaus, W. F. and Steger, J. L., "Implicit Approximate Factorization Schemes for the Low Frequency Transonic Equation," NASA TM X-73082, 1975.

[26] Ballhaus, W. F., Jameson, A., and Albert, J., "Implicit Approximate Factorization Schemes for Steady Transonic Flow Problems," AIAA Journal, Vol. 16, No. 6, June 1978, pp. 573-579.

[27] Ballhaus, W. F., Bailey, F. R., and Frick, J., "Improved Computational Treatment of Transonic Flow About Swept Wings," Advances in Engineering Science, Vol. 4, NASA CP 2001, 1976, pp. 1311-1321.

[28] Jameson, A. and Caughey, D. A., "Numerical Calculation of the Transonic Flow Past a Swept Wing," ERDA Research & Development Rept. L00-3077-140, 1977.

[29] Boppe, C., "Transonic Flowfield Analysis for Wing Fuselage Configurations," NASA CR 3243, 1980.

[30] Hinson, B. L. and Burdges, K. P., "An Evaluation of Three Dimensional Transonic Codes Using New Correlation Tailored Data," AIAA Paper No. 80-0003, Jan. 1980, Pasadena, Calif.

[31] Federenko, R. P, "The Speed of Convergence of One Iterative Process," USSR Computational Mathematics and Mathematical Physics, Vol. 4, No. 3, 1964, pp. 227-235.

[32] Bakhvalov, N. S., "On the Convergence of a Relaxation Method with Natural Constraints on the Elliptic Operator," USSR Computational Mathematics and Mathematical Physics, Vol. 6, No. 5, 1966, pp. 101-135.

[33]Brandt, A., "Multi-Level Adaptive Solution to Boundary Value Problems," Mathematics of Computation, Vol. 31, No. 138, April 1977, pp. 333-390.

[34]Nicholaides, R. A., "On the ℓ^2 Convergence of an Algorithm for Solving Finite Element Systems," Mathematics of Computation, Vol. 31, No. 14, Oct. 1977, pp. 892-906.

[35]Hackbusch, W., "Convergence of Multi-Grid Iterations Applied to Difference Equations," Köln University Mathematics Institute Report 79-5, April 1979.

[36]South, J. C. and Brandt, A., "Application of a Multi-Level Grid Method to Transonic Flow Calculations," Transonic Flow Problems in Turbomachinery, edited by T. C. Adamson and M. F. Platzer, Hemisphere, Washington, D. C., 1977, pp. 180-207.

[37]South, J. C. and Brandt, A., "The Multi-Grid Method: Fast Relaxation for Transonic Flows," presentation at 13th Annual Meeting of Society of Engineering Science, Hampton, Va., Nov. 1976.

[38]Jameson, A., "Acceleration of Transonic Potential Flow Calculations on Arbitrary Meshes by the Multiple Grid Method," AIAA Paper No. 79-1458, July 1979, Williamsburg, Va.

[39]Thompson, J. F., Thames, F. C., and Mastin, C. M, "Automatic Numerical Generation of Body-Fitted Curvilinear Coordinate System for Field Containing Any Number of Arbitrary Two-Dimensional Bodies," Journal of Computational Physics, Vol. 15, No. 3, July 1974, pp. 299-319.

[40]Steger, J. L. and Sorenson, R. L., "Automatic Mesh-Point Clustering Near a Boundary in Grid Generation with Elliptic Partial Differential Equations," Journal of Computational Physics, Vol. 33, No. 3, 1979, pp. 405-410.

[41]Sells, C. C. L, "Plane Subcritical Flow Past a Lifting Airfoil," Proceedings of the Royal Society, Vol. 308A, No. 1494, 1969, pp. 377-401.

[42]Caughey, D. A. and Jameson, A., "Numerical Calculation of Transonic Potential Flow about Wing/Body Combinations," AIAA Journal, Vol. 17, No. 2, 1979, pp. 175-181.

[43]Melnik, R. E., Chow, R., and Mead, H. R., "Theory of Viscous Transonic Flow Over Airfoils at High Reynolds Number," AIAA Paper No. 77-680, June 1977, Albuquerque, N. Mex.

[44]Holst, T. L. and Ballhaus, W. F., "Fast, Conservative Schemes for the Full Potential Equation Applied to Transonic Flows," AIAA Journal, Vol. 17, No. 2, Feb. 1979, pp. 145-152.

[45]Dougherty, F. C., Holst, T. L., Grundy, K. L., and Thomas, S. D., "TAIR - A Transonic Airfoil Analysis Computer Code," NASA TM 81-296, May 1981.

Chapter VI.

Reynolds Averaged Navier-Stokes Computations of Transonic Flows—the State-of-the-Art

Unmeel Mehta* and Harvard Lomax†
NASA/Ames Research Center, Moffett Field, Calif.

Introduction

During the past five years, numerous pioneering archival publications have appeared that have presented computer solutions of the mass-weighted, time-averaged Navier-Stokes equations[1] for transonic problems pertinent to the aircraft industry. These solutions have been pathfinders of developments that could evolve into a major new technological capability, namely the computational Navier-Stokes technology, for the aircraft industry. So far these simulations have demonstrated that computational techniques and computer capabilities have advanced to the point where it is possible to solve forms of the Navier-Stokes equations for transonic research problems. At present there are two major shortcomings of the technology: limited computer speed and memory, and difficulties in turbulence modeling and in computation of complex three-dimensional geometries. These limitations and difficulties are the pacing items of the continuing developments, although the one item that will most likely turn out to be the most crucial to the progress of this technology is turbulence modeling. The objective of this paper is to discuss the state-of-the-art of this technology and suggest possible future areas of research.

At present, the viscous transonic flow research is conducted by either a zonal viscous-inviscid interaction procedure or a global Navier-Stokes procedure. For the state-of-the-art of viscous-inviscid interaction procedures, refer to the proceedings of an AGARD Symposium on

---

Presented at the Transonic Perspective Symposium, NASA/Ames Research Center, Moffett Field, Calif., Feb. 18-20, 1981. This paper is declared a work of the U.S. Government and therefore is in the public domain.
*Research Scientist.
†Senior Scientist.

"Computation of Viscous-Inviscid Interactions."[2] These procedures have achieved some success but most either predict poorly or fail when faced with flow separation. The procedure of Le Balleur[3] for small separated regions appears promising. There does not appear to be a single one of these procedures which gives acceptable results under a wide range of conditions. However, these procedures are being further developed, and in those cases where they can be trusted, they should be computationally cheaper to use than a global Navier-Stokes calculation. Presumably both the viscous-inviscid interaction procedures and the global Navier-Stokes approach will contribute to the understanding of various transonic flow phenomena and in providing insight for developing efficient numerical methods.

This paper discusses some of the flow conditions for which the Navier-Stokes equations appear to be required. There are three different types of interaction of a shock wave with a boundary layer: 1) shock/boundary-layer interaction with no separation, 2) shock-induced turbulent separation with immediate reattachment (we refer to this as a shock-induced separation bubble), and 3) shock-induced turbulent separation without reattachment. The shock-induced separation is caused by a strong shock wave. A proper treatment of interaction of this shock with a boundary layer requires the Navier-Stokes equations, at least locally.[4]

Shock waves that terminate in the vicinity of boundary layers are seldom steady, particularly on transonic wings and control surfaces. In some cases, the shock/boundary-layer interactions are observed to oscillate periodically with relatively large amplitudes.[5] These fluctuations can cause stalling, buffeting, flutter, and control surface buzz. The first two phenomena arise at large angles of attack when the upper-surface separation of the boundary layer extends from the shock wave to the trailing edge and beyond. The last two phenomena are manifested when the separated boundary layer experiences lateral oscillations in the wake. A different type of transonic flow problem is recently reported by McCroskey et al.[6] They report transonic flow near the leading edge for freestream Mach numbers as low as 0.2 on an oscillating airfoil. This flow is characterized by a small supersonic bubble with or without shock waves. At Mach numbers between 0.3 and 0.5, the airfoil may experience shock-induced leading edge stall.[6]

There are primarily two motivations for understanding separated flows: 1) controlling and minimizing the effects of separation when it is an undesirable feature, and 2) organizing separation so that it constitutes a natural way of improving aerodynamic performance. The latter occurs, for example, in three dimensions where strakes are used to create streamwise vortices that increase performance at cruise and climb conditions. It appears that aircraft designers are not so much worried about incipient or microscopic separation bubbles of small extent as they are about a boundary layer failing to reattach before the trailing edge. If that happens, it may cause, depending on its severity, stall and buffet, pitch-up motion, and possible degradation of lateral stability.

When the boundary-layer assumptions are almost valid through a small separated region which is not caused by a shock wave, it is possible to determine, using the boundary-layer equations, the main effects of the separation with an integral method, and the quantitative structure of the separated region with a differential method. But when the separation region is not small, this approach fails, and the Navier-Stokes equations are required.

In computational aerodynamics, both the physics and numerics are equally important. Physics is involved in selecting the appropriate governing equations and formulating suitable initial and boundary conditions. Numerics, on the other hand, deals with generating a grid system; devising stable, accurate, and efficient approximating schemes for solving the differential equations along with the initial and boundary conditions; and actually carrying out the solution procedure. All of the processes are important, and they all affect the accuracy of the solution. For the purposes being discussed here, the accuracy required of the solution is determined by the practical requirements of the aircraft industry. If this solution fulfills these requirements, then it is accurate enough. The above processes dealing with physics and numerics for the Navier-Stokes equations constitute the Navier-Stokes technology.

At present, computer simulations of transonic flowfields are usually validated by comparison with experiments which are very often in themselves simulations of computational flow constraints. This reliance on experiment results principally from the fact that the effects of turbulence must be modeled and the models are essentially empirical. In addition, this reliance results when a numerical solution, for all practical purposes, is not shown to be independent of the discretization errors. It is usually not possible to show the extent to which a large-scale, numerical simulation is affected by these errors. On the other hand, the validity of an experiment is, more often than not, questionable. Generally, a quantitative assessment of effects of any known deficiencies in the data is lacking. Rarely are the initial and boundary conditions completely documented. There is usually a minimum rather than a comprehensive set of data.

With the general shortcomings of both numerical and experimental simulations as background, this paper discusses the state-of-the-art of predictive Navier-Stokes technology dealing with the processes discussed in the Introduction to this paper and presents some computed simulations of transonic flows.

Governing Equations

Navier-Stokes Equations

The continuum, compressible fluid mechanics is described by the classical Navier-Stokes equations, properly modified to take into account variations in density and temperature, along with equations governing conservation of mass and energy and an equation of state, taken from equilibrium thermodynamics. This system is referred to here simply as the Navier-Stokes equations. It shall be assumed that solutions of this system, subject to appropriate initial and boundary conditions, do exist and are unique. However, only local existence theorems in two- and three-dimensional problems have been established,[7] and the Cauchy problem for a perfect polytropic gas in three dimensions is solvable "in the large" provided the initial data are close to constants.[8] In short, the mathematical analysis of the above system is far from complete.

In the Navier-Stokes equations, the assumptions concerning the stress tensor and the heat-flux vector exclude rarefaction shocks without specifically assuming the second law of thermodynamics. Therefore, the entropy condition[9] need not be satisfied by a numerical method for these equations. The effect of viscosity and heat conductivity develops a continuous transition through a shock wave. In the transonic flow regime, these equations are valid through this wave which is, however, quite thin if its intensity is strong enough. For example, at a Mach number of 1.05 and Reynolds number of 10^7, the shock thickness in air is almost the same as the thickness of the linear sublayer of a turbulent boundary layer on a smooth flat surface. The latter thickness corresponds to about $y^+ \sim 5$, where y^+ is the Reynolds number based on the friction velocity at the surface and a length scale of turbulence. At lower Mach numbers the shock is even thicker. A shock wave with such a small thickness is not usually resolved in current transonic simulations. (Likewise, the contact discontinuity is not resolved.) Instead, it is considered to be a discontinuity, the location of which is part of the solution procedure. However, its thickness may not be small when it begins to interact with a viscous boundary layer and can even lose its identity as it penetrates into the viscous region.

Reynolds Averaged Navier-Stokes Equations with Mass-Weighted Variables

In the study of turbulence by means of the Navier-Stokes equations, it is usual to use some form of averaging. For example, Monin and Yaglom[10] present a general space-time averaging procedure for functions $\bar{f}(x,t)$ given by the equation

$$\langle \bar{f}(\underline{x},t) \rangle = \iint_{-\infty}^{\infty} \bar{f}(\underline{x} - \underline{\zeta}, t - \tau) g(\underline{\zeta},\tau) d\underline{\zeta}\, d\tau \qquad (1)$$

Here, the overbar and the underscore indicate an instantaneous value and a vector field, respectively. The non-negative weighting function, g, satisfies the normalizing condition

$$\iint_{-\infty}^{\infty} g(\underline{\zeta},\tau) d\underline{\zeta}\, d\tau = 1 \qquad (2)$$

The choice of this weighting function determines the significance of the averaged quantities. For example, if g is a constant over some time interval T and zero outside of it, and the dependence on $\underline{\zeta}$ is a Dirac delta function, $\langle \bar{f}(x,t) \rangle$ is referred to a time-averaged quantity. In unsteady flows, the interval T must be large compared to the periods characteristic of time scales that cannot be resolved computationally, but small compared to the period of resolvable flow motion.

The system of equations determined by applying the above time-averaging procedure constrained with the Reynolds conditions[10] gives rise to Reynolds averaged Navier-Stokes equations. For compressible fluids, these equations contain second-order moments, such as $\langle \rho'u' \rangle$, and a third-order moment $\langle \rho'u'u' \rangle$, due to fluctuations in the fluid density.[11] Here, the prime denotes fluctuating quantity. Therefore, for these fluids, instead of time-averaged flow quantities, mass-weighted time-averaged quantities are preferable. For example, the mass-weighted velocity u_i equals to $\langle \bar{\rho}\bar{u}_i \rangle / \langle \bar{\rho} \rangle$. This averaging procedure eliminates the above moments from the averaged Navier-Stokes equations but it does not remove density fluctuations from turbulence. This procedure is first used in the study of atmospheric turbulence by Hesselberg.[12] A comprehensive discussion of this procedure for compressible turbulent flows is presented by Favre[13] and by Cebeci and Smith.[14] Henceforth, the equations resulting from this type of averaging are simply called the Reynolds averaged Navier-Stokes equations. These equations, without external forces, may be written in dimensional form as

$$\dot{\rho} + (\rho u_i)_{,i} = 0 \tag{3}$$

$$(\rho \dot{u}_i) + (\rho u_j u_i)_{,j} = -p_{,i} + (\rho \sigma_{ij})_{,j} \tag{4}$$

$$(\rho \dot{h}) + (\rho u_j h)_{,j} = \dot{p} + u_j p_{,j} + \rho \sigma_{ij} u_{i,j} - (\rho q_j)_{,j} \tag{5}$$

Here, ρ and p are, respectively, time-averaged mass density and pressure; u_i and h are, respectively, mass-weighted mean velocity and enthalpy. The Cartesian tensor summation convection is used. The overdot indicates a partial derivative with respect to time, and subscripts after commas denote partial differentiation. Further, the symbols σ_{ij} and q_j, respectively, represent the specific time-averaged total shear stress and heat flux as follows:

$$\sigma_{ij} = 2\nu[S_{ij} - (1/3)u_{k,k}\delta_{ij}] - R_{ij} \tag{6}$$

$$q_j = (\nu/P_{r_L})h_{,j} + \langle \bar{\rho} u_j' h' \rangle / \rho \tag{7}$$

where ν is the kinematic viscosity. These include contributions of both the molecular and turbulent transport. The mean rate of strain tensor S_{ij} and the Reynolds stress tensor $-\rho R_{ij}$ are given by

$$S_{ij} = (u_{i,j} + u_{j,i})/2 \tag{8}$$

$$R_{ij} = \langle \bar{\rho} u_i' u_j' \rangle / \rho \tag{9}$$

The above equations are identical to the equations used to determine laminar flows, except for the Reynolds stress tensor and turbulent heat-flux vector. In addition, these equations essentially exhibit a term by term correspondence with those for the incompressible fluids. This correspondence permits extension of the large body of experience existing with modeling turbulence for constant density flows to transonic flows, provided turbulence structure in both these flows is closely the same. Interpretation of Morkovin's hypothesis[15] suggests that this is the case for boundary layers and wakes at freestream Mach numbers less than about 5 and of jets at Mach numbers less than about 1.5.[16] This hypothesis states that the effects of density fluctuations on turbulence are negligible when the root-mean-square density fluctuation is small compared with the absolute density. Transport equation turbulence models, which are discussed below in terms of mass-weighted, time-averaged variables, contain additional terms due to compressibility effects. These terms are negligible according to the above hypothesis in the transonic regime.

Modeling of Turbulence

There are two approaches for turbulence modeling: the first-order approach in which the Reynolds stress tensor is modeled, and the second-order approach in which this tensor is determined from the Navier-Stokes equations. In the former approach, one forms the equations for the first-order quantities, such as mean velocities, and models the second-order quantities that appear in them. See Eqs. (4) and (10). In the latter approach, equations are formed for the first- and second-order quantities (u_i and R_{ij}), and the third-order terms are modeled. These equations may be simplified to yield algebraic stress models, which still require differential equations, both for the turbulent kinetic energy and energy dissipation.[17]

In transonic, turbulent flow simulations, the first-order approach is almost always used, and it forms the basis for the so-called zero-equation (algebraic), one-equation, and two-equation models. In practice, the actual form of these models and the manner of applying them generally differ in detail from investigator to investigator. General definitions and characteristics of these models are available from Launder and Spalding,[18] Cebeci and Smith,[14] Reynolds,[19] Reynolds and Cebeci,[20] Rubesin,[21] Launder,[22] and Rodi.[23] (Simulation of transition is not considered in this state-of-the-art review.)

Some zero-equation models are based on the Prandtl mixing length hypothesis. But other first-order turbulence models are based on the "Newtonian" assumption, and they are, therefore, eddy viscosity models. Boussinesq's eddy viscosity concept[24] is based on an analogy with the gradient diffusion mechanism of the kinetic theory of gases. Methods based on this concept are also known as eddy diffusivity or gradient transport methods. Corrsin[25] has presented limitations of gradient transport models. In these methods, the eddy viscosity, ν_T, is assumed to be a scalar and is defined by a Newtonian constitutive equation of the form

$$R_{ij} = (1/3)v^2\delta_{ij} - 2\nu_T[S_{ij} - (1/3)u_{k,k}\delta_{ij}] \qquad (10)$$

Here, $v^2 = R_{ii}$ is the turbulent kinetic energy. The v^2 term may be absorbed in p. This relation restricts R_{ij} and S_{ij} to the same principal axes, which is not true in general. It is possible to modify this relation in order to remove this restriction.[26] Algebraic models relate ν_T directly to Reynolds averaged quantities. Both one- and two-equation models contain a partial differential equation for turbulent kinetic energy, which defines a turbulence velocity scale, v. One-equation models use a prescribed, empirical length scale distribution, and two-equation models use an additional partial differential equation to define a turbulence length scale, ℓ. A combination of the turbulence velocity and length scale determines the value of the eddy viscosity

$$\nu_T = cv\ell \qquad (11)$$

where c is a constant. Models using partial differential equations for turbulent quantities are also called transport equation models.

In the eddy conductivity concept, the transport of heat due to the time-averaged product of fluctuating enthalpy and fluctuating velocity is modeled. It is assumed that the turbulent heat flux follows a law similar to Fourier's law. Further, it is generally assumed that dynamic eddy viscosity, μ_T, and turbulent thermal conductivity have the same functional relationship with temperature. Although the turbulent Prandtl number varies across the boundary layer, it is commonly considered to be a constant, and it is usually taken to be 0.9 for air. Apparently, more complex modeling of the turbulent heat flux than this has yet to be attempted in transonic simulations.

There are many zero-equation models. As an example, a model used by Baldwin and Lomax[27] for attached, separated, and wake flows is briefly outlined below. This model is patterned after that of Cebeci.[28] The turbulent boundary layer is regarded as a composite layer consisting of inner and outer regions. In each region, the distributions of ν and ℓ are prescribed by two different empirical expressions. For example, in the log-law region, ℓ is proportional to the boundary-layer thickness. The proportionality of ℓ to y is extended into the viscous sublayer with a damping function suggested by Van Driest.[29] In the outer region, the vorticity is used to define the boundary-layer thickness.

In the inner layer, $0 \le y \le y_c$, the expressions for ν and ℓ are

$$(\nu)_{inner} = \ell |\Omega| \qquad (12)$$

and

$$(\ell)_{inner} = \alpha_1 y [1 - \exp(-y \sqrt{|\sigma_{12}|\rho|}_w / 26 \nu_w)] \qquad (13)$$

with $c = 1.0$ in Eq. (11). Here, $\alpha_1 = 0.4$, Ω is the vorticity, and subscript w indicates wall values.

In the outer region, $y > y_c$, the expressions for ν and ℓ used by the Baldwin-Lomax (B-L) model are

$$(\nu)_{outer} = \begin{cases} L_{max} \\ \text{or} \\ 0.25 U_{dif}^2 / L_{max} \end{cases} \text{the smaller} \qquad (14)$$

and

$$(\ell)_{outer} = y_{max} C_{BL} \alpha_2 \qquad (15)$$

with c the Clauser constant equal to 0.0168 in Eq. (11). The quantity U_{dif} is the difference between maximum and minimum absolute velocity, the value of C_{BL} is 1.6, and the Klebanoff intermittency factor, α_2, is given by

$$\alpha_2 = [1 + 5.5(0.3 y/y_{max})^6]^{-1} \qquad (16)$$

The quantities y_{max} and L_{max} are determined from

$$L(y) = y |\Omega| [1 - \exp(-y \sqrt{|\sigma_{12}|\rho|}_w / 26 \nu_w)] \qquad (17)$$

The above exponential term is negligible in the outer part of the boundary layer. In wakes, it is set to zero. The quantity L_{max} is the maximum value of $L(y)$ that occurs in this equation, and y_{max} is the value of y at which it occurs.

The region of validity of the inner and outer scales is determined by y_c. It is the smallest value of y at which values of inner and outer eddy viscosity are the same. The value of α_1 in the inner region and of c in the outer region are assumed to be universal constants for $R_\theta > 5000$, where R_θ is based on the momentum thickness. At lower Reynolds numbers, they are functions of Reynolds number.

As an example of a two-equation model, the Wilcox-Rubesin (W-R) model[30] is presented in Eqs. (18) and (19). This model is an extension of the model developed by Wilcox and Traci,[31] which evolved from the model formulated by Saffman and Wilcox[32] and that by Saffman.[33] In the earlier models, the term determining the rate of production of kinetic energy was inconsistent with that in a stress equation formulation. The present model removes this inconsistency. In this model, the turbulent kinetic energy and the specific energy dissipation are given by

$$(\rho \dot{v}^2) + (\rho u_j v^2)_{,j}$$
$$= 2\rho\sigma_{ij}u_{i,j} - \beta_1\rho\omega v^2 + [(\mu + \beta_2\nu_T)v^2_{,j}]_{,j} \quad (18)$$

$$(\rho \dot{\omega}^2) + (\rho u_j \omega^2)_{,j}$$
$$= \beta_3(\omega^2/v^2)\rho\sigma_{ij}u_{i,j} - [\beta_4 + \beta_5(\ell_{,k})^2]\rho\omega^3$$
$$+ [(\mu + \beta_5\nu_T)\omega^2_{,j}]_{,j} \quad (19)$$

where the length scale is defined by

$$\ell = v/\omega \quad (20)$$

The eddy viscosity is computed from Eq. (11) and the constitutive Eq. (10) is used to provide R_{ij}. Wilcox and Rubesin[30] recommend following values of the constants in the above model

$$\beta_1 = 0.09, \quad \beta_2 = \beta_5 = 0.5, \quad \beta_4 = 0.15,$$

$$\beta_6 = 1/11, \quad \beta_7 = 10/9$$

$$c = [1 - (1 - \beta_6^2)\exp(-Re_T/2)]/2$$

$$\beta_3 = \beta_7[1 - (1 - \beta_6^2)\exp(-Re_T/4)]/c$$

The turbulence Reynolds number is calculated as

$$Re_T = v\ell/\nu$$

The Reynolds averaged Navier-Stokes equations along with turbulence model equations constitute the governing equations of the Navier-Stokes technology.

Conservation Law Forms

The Reynolds averaged Navier-Stokes equations, as presented in Eqs. (3-5), are not in a form generally suitable for simulations of flowfields around aerodynamic shapes. For such shapes, surface-oriented coordinates are preferred. Furthermore, the choice of dependent variables made for these equations is not the only choice available.

For unsteady flows, Moretti[34] recommends using the velocity components, pressure (actually ln p) and entropy for the dependent variables. This is motivated by the fact that, in inviscid flows, there are two types of surfaces across which flow quantities can be discontinuous: characteristic surfaces and stream surfaces. Across the characteristic surfaces, pressure and velocities are discontinuous, but not entropy. In contrast, across stream surfaces, entropy is discontinuous, but not pressure. Any other thermodynamic parameters, such as energy or density, are discontinuous across both the surfaces. The above recommendation does not, however, lead to the conservative law form[9,35,36] of the governing equations. If this is not crucial, then the above variables may be appropriate. Another choice is to use density, energy, and the contravariant components of the velocity vector. This leads to a nondivergence form of the equations. It is possible to put these equations in the divergence or conservative law form.[37,38] A third choice is to use the Cartesian components of velocity and conservative variables. It is this third choice written in the conservative law form that follows.

The conservative law form of the Navier-Stokes equations in conservative variables facilitates capturing of discontinuities and maintenance of global conservation of fluxes. The importance of these issues is decided by applications and acceptable error bounds. As indicated in our previous discussion of Navier-Stokes equations, shock waves and contact regions are treated as physical discontinuities during flow simulations. The above conservative form of these differential equations avoids fictitious sources along these discontinuities. Further, there is a weak solution of these equations, in the absence of differentiability, across the discontinuities. In principle, these theoretical results facilitate capturing of discontinuities. Whether these theoretical advantages are maintained or not in simulations depends upon the numerical scheme along with the grid system which will be discussed in the next two sections of this paper. Likewise, the issue of maintaining global conservation of fluxes depends upon the numerical scheme. Analytical integration of a convective flux term with respect to an independent variable yields a difference between boundary values of the flux. Construction of a differencing scheme that preserves this conservation property is relatively easy if the conservative law form is used to begin with. Sometimes it is possible to formulate a differencing scheme which conserves fluxes, starting with the nonconservative form. However, it is really the discrete form that governs the conservation of fluxes and not the differential form. Global conservation of fluxes does not automatically assure that the discontinuities are captured correctly. Conservation errors are analogous to truncation errors. As long as conservation errors remain bounded and do not affect acceptable accuracy, it is immaterial whether or not the governing equations are in the conservative law form.

The above considerations concerning conservation of fluxes also apply to transport equations for turbulence modeling. These equations are not in the conservative law form [e.g., Eqs. (18) and (19)], albeit the Reynolds stress equations are based on conservation laws, namely the Navier-Stokes equations. If the transport models were formulated in the conservation law form, then similar numerical treatment is possible of all the governing equations of Navier-Stokes technology.

When the Navier-Stokes equations in the conservative law form are transformed from the Cartesian coordinates to arbitrary curvilinear coordinates, the resulting equations are not automatically in the conservative law form, although they can be made to be so.[39] This is done in order to facilitate global conservation of fluxes. Theoretically, however, this form is not necessary for obtaining the weak solution as $(\nu + \nu_T) \to 0$ and for avoiding fictitious sources along discontinuities, provided the metric coefficients multiplying the transformed derivatives and their first derivatives with respect to the new independent variables are continuous. This can be readily demonstrated following Lax,[40] but the derivation is not given here. Further, it can be shown that the shock speed in the Cartesian coordinates and the curvilinear coordinates differ by a factor containing the metric coefficients.

Reynolds Averaged Navier-Stokes Equations in Curvilinear Coordinates

The Reynolds averaged Navier-Stokes equations in conservative Cartesian variables are presented in arbitrary curvilinear coordinates (ξ, τ). In nondimensional form, these equations can be written

$$\frac{\partial Q}{\partial \tau} + \sum_{i=1}^{d} \frac{\partial C_i}{\partial \xi_i} = \frac{1}{Re} \sum_{i=1}^{d} \frac{\partial V_i}{\partial \xi_i} \tag{21}$$

where d is the number of dimensions, and Q, C, and V are vectors

$$Q = \mathcal{J}[\rho, \rho u_1, \ldots, \rho u_d, e]^T$$

$$C_i = Q \mathcal{U}_i + p \Phi_i$$

$$V_i = \sum_{j=1}^{d} R_j \frac{\partial \xi_i}{\partial x_j}$$

$$\mathcal{U}_i = \frac{\partial \xi_i}{\partial t} + \sum_{j=1}^{d} u_j \frac{\partial \xi_i}{\partial x_j}$$

$$\Phi_i = \mathcal{D} \sum_{j=1}^{d} \frac{\partial \xi_i}{\partial x_j} [0, \delta_{j1}, \ldots, \delta_{jd}, u_j]^T$$

$$R_i = \left[0, \sigma_{i1}, \ldots, \sigma_{id}, \sum_{j=1}^{d} u_j \sigma_{ij} + q_i \right]^T$$

$$\sigma_{ij} = (\mu + \mu_T) \sum_{k=1}^{d} \left(-\frac{2}{3} \sum_{\ell=1}^{d} \frac{\partial \xi_\ell}{\partial x_k} \frac{\partial u_k}{\partial x_\ell} \delta_{ij} \right.$$

$$\left. + \frac{\partial \xi_k}{\partial x_j} \frac{\partial u_i}{\partial \xi_k} + \frac{\partial \xi_k}{\partial x_i} \frac{\partial u_j}{\partial \xi_k}\right) \quad (22)$$

$$q_i = \frac{\gamma}{Pr}\left(\mu + \frac{Pr}{Pr_T}\mu_T\right) \sum_{j=1}^{d} \frac{\partial \xi_j}{\partial x_i} \frac{\partial e_I}{\partial x_j}$$

$$e_I = \frac{e}{\rho} - \sum_{k=1}^{d} \frac{u_k u_k}{2}$$

$$p = (\gamma - 1)\rho e_I$$

$$\frac{\partial \xi_i}{\partial t} = -\sum_{j=1}^{d} \frac{\partial \xi_i}{\partial x_j} \frac{\partial x_j}{\partial \tau}$$

$$\mathcal{D} = \frac{\partial(x_1, \ldots, x_d)}{\partial(\xi_1, \ldots, \xi_d)}$$

$$\frac{\partial \xi_i}{\partial x_j} = \frac{1}{\mathcal{D}} \frac{\partial(x_{j+1}, x_{j+2})}{\partial(\xi_{i+1}, \xi_{i+2})}$$

In the previous expression, subscripts (i, i+1, i+2) and (j, j+1, j+2) vary in a cyclic order, (1, 2, 3), (2, 3, 1), etc. The Stokes hypothesis, $(3\lambda + 2\mu = 0)$, of local thermodynamic equilibrium has been used, and total energy-per-unit volume and the internal energy-per-unit mass are represented by e and e_I, respectively.

The second-order, thin shear layer approximation neglects in Eq. (21) all streamwise, spanwise, and cross derivatives of the viscous as well as turbulence stress terms. The momentum equation in the direction away from the surface (ξ_2 direction) is retained. If this were also neglected, we have the first-order, thin shear layer approximation which is analogous to the classical boundary-layer approximation. Investigators using the second-order thin shear layer approximation justify it on the basis that the neglected terms in the complete equations are not computed correctly with the available grid resolution anyway, so why keep them. This approximation is, however, valid only for "small" separation bubbles and for "weak" shock/boundary-layer interactions. This approximation applied to Eq. (21) leads to the following equation

$$\frac{\partial Q}{\partial \tau} + \sum_{i=1}^{d} \frac{\partial C_i}{\partial \xi_i} = \frac{\mathcal{D}}{Re} \sum_{i=1}^{d} \frac{\partial \xi_2}{\partial x_i} \frac{\partial R_i^T}{\partial \xi_2} \qquad (23)$$

with

$$\sigma_{ij}^T = (\mu + \mu_T)\left(-\frac{2}{3}\sum_{\ell=1}^{d} \frac{\partial \xi_2}{\partial x_\ell} \frac{\partial u_\ell}{\partial \xi_2} \delta_{ij} + \frac{\partial \xi_2}{\partial x_j} \frac{\partial u_i}{\partial \xi_2} + \frac{\partial \xi_2}{\partial x_i} \frac{\partial u_j}{\partial \xi_2}\right) \qquad (24)$$

$$q_i^T = \frac{\gamma}{pr}\left(\mu + \frac{Pr}{Pr_T}\mu_T\right) \frac{\partial \xi_2}{\partial x_i} \frac{\partial e_I}{\partial \xi_2}$$

Instead of Eq. (23), some investigators (e.g., Steger[41] and Pullium and Steger[42]) use, for convenience, the following equation

$$\frac{\partial Q}{\partial \tau} + \sum_{i=1}^{d} \frac{\partial C_i}{\partial \xi_i} = \frac{1}{Re} \sum_{i=1}^{d} \frac{\partial}{\partial x_2}\left(\mathcal{D} R_i^T \frac{\partial \xi_2}{\partial x_i}\right) \qquad (25)$$

Boundary Conditions

Boundary conditions for the above governing equations are determined by mathematics and physics. The mathematical character of these equations dictates the number and type of these conditions that determine, along with initial conditions, the well-posedness of a problem using these equations. Further, this mathematical character is determined by the theory of characteristics. Theoretical analyses of two kinds are available: one based on the classical energy method (e.g., Elvius and Sundström[43]) which follows the earlier work of Serrin,[44] and the other on the normal mode concept.[45] Most of the work is done for the compressible Eulerian and shallow water equations. A few recent studies deal with the compressible Navier-Stokes equations (e.g., Oliger and Sundström[46] and Gustafsson and Sundström[47]). These studies consider both number and a possible set of admissible forms of the boundary conditions. At present, such studies serve as a guide rather than as a useful tool in practical transonic simulations. A theoretical study of the well-posedness of any problem using the governing equations of the Navier-Stokes technology has yet to be done. The boundary conditions discussed below are based on both the mathematical character of the equations and physical considerations. They are not based on the analytical procedures mentioned above.

The mathematical character of the system represented by the linearized form of Eq. (21) is incompletely parabolic[48] or parabolic-hyperbolic. Without the time derivative, it is elliptic-hyperbolic. The system given by Eq. (23) or (25) is incompletely hyperbolic or hyperbolic-parabolic. This system is parabolic only in (ξ_2 - t) plane. The global character of these systems remains the same even if the local character may be, for instance, purely hyperbolic. Therefore, the boundary conditions are determined by the global character of these systems.

First the boundary conditions for the system represented by Eq. (21) will be discussed. Consider each equation of this system separately from the others as an equation determining Q_i; the other Q's in this equation are assumed to be known quantities. Here Q_i is a component of vector Q. The mass conservation equation requires one boundary condition in each coordinate direction for Q_1. The second derivative of Q_2 in Q_2-momentum equation requires two boundary conditions in each coordinate direction. Likewise, two boundary conditions are required for the remaining Q's. This means that if $\partial \mathcal{R}$ is the boundary of computational region $\mathcal{R}$, then everywhere on $\partial \mathcal{R}$ conditions specifying $Q_2, \ldots, Q_{d+1}$ are required; and on a part of $\partial \mathcal{R}$, a condition specifying Q_1 is needed. These considerations determine the number of boundary conditions on $\partial \mathcal{R}$. The type of the boundary condition for a Q_i in any direction is determined by the highest derivative of this Q_i in that direction. The boundary condition should be one order lower than the highest derivative. This constraint yields boundary conditions which are either Dirichlet, Neumann, or mixed type.

The above heuristic considerations help formulate boundary conditions based on physics. A set of these conditions for Eq. (21) are presented below. In external flow problems, two kinds of boundaries arise: rigid wall boundaries, $\partial \mathcal{R}_W$, and open boundaries, $\partial \mathcal{R}_O$. The rigid wall constrains the flowfield along $\partial \mathcal{R}_W$. This physical constraint is relatively easy to formulate and convert into computational boundary conditions. Open boundaries do not provide a material constraint, and hence appropriate conditions are not obvious.

The rigid wall boundary provides velocity and temperature conditions on $\partial \mathcal{R}_w$. The behavior of a real gas at ordinary conditions (Knudsen numbers less than 10^{-2}) is accurately described by the no-slip and no-temperature jump conditions. These are the only two physical conditions available. (In contrast, for inviscid flows there is only one physical boundary condition, namely, no flow normal to the rigid walls. Further, for a flow past an airfoil, a condition is required at or near the trailing edge of the airfoil, if the inviscid streamline leaving the airfoil is to correspond with that in viscous flow.) Considering the case of impermeable walls, the no-slip condition translates into vanishing contravariant velocity components, $\mathcal{U}_i = 0$. Further, the temperature condition gives either a Dirichlet or a Neumann condition for the total energy.

The mass conservation equation contains the material derivative of Q_1. Consequently, on $\partial \mathcal{R}$, Q_1 changes if its previous history is known, otherwise, a condition on Q_1 must be specified. This means that if fluid is on $\partial \mathcal{R}$ or inside $\mathcal{R}$, Q_1 is determined by the mass conservation equation. But if fluid enters $\mathcal{R}$ by crossing $\partial \mathcal{R}$, Q_1 must be specified. Therefore, Q_1 cannot be specified on $\partial \mathcal{R}_w$, and it must be calculated from its material derivative. When this recourse leads to numerical difficulties, a new governing equation is formulated by appropriately combining the momentum equations to form the normal derivative of pressure. After expressing pressure in terms of Q's (equation of state), we have

$$\sum_{\ell=1}^{d} \sum_{j=1}^{d} \frac{\partial \xi_\ell}{\partial x_j} \frac{\partial \xi_2}{\partial x_j} \frac{\partial p}{\partial \xi_\ell} = \rho \left(\frac{\partial}{\partial \tau} \frac{\partial \xi_2}{\partial t} + \sum_{j=1}^{d} u_j \frac{\partial}{\partial \tau} \frac{\partial \xi_2}{\partial x_j} \right)$$
$$+ \frac{1}{Re} \left(\frac{\partial \xi_2}{\partial x_1} \sum_{i=1}^{d} \frac{\partial V_i}{\partial \xi_i} + \frac{\partial \xi_2}{\partial x_2} \sum_{i=1}^{d} \frac{\partial V_i}{\partial \xi_i} + \frac{\partial \xi_2}{\partial x_3} \sum_{i=1}^{3} \frac{\partial V_i}{\partial \xi_i} \right) \quad (26)$$

The left-hand side of the above expression simplifies for orthogonal curvilinear coordinates. As Eq. (26) is derived from the momentum equations, and as it replaces the mass conservation equation, it is not a boundary condition on

Q_1. This equation is subjected to the no-slip condition when it is used. The above viscous terms vanish when the first-order thin shear layer approximation is valid; and they can be neglected only when the second-order thin shear layer approximation is valid.

These conditions are also valid for internal flow problems. However, when simulations of the external flow problems include wind-tunnel wall effects, one alternative is to use the no-slip condition. Another alternative is not to compute the wall boundary layers. In this case, obviously walls cannot be considered as open boundaries if they interfere with the flowfield around an aerodynamic body from that observed in free flight. This is the situation to present transonic wind tunnels. An ideal situation is to measure all required flow quantities just outside the wind-tunnel wall boundary layers and use these values as boundary conditions. Probably the next best avenue is to measure only pressure, again perhaps just outside the wall boundary layers, and then consider the boundary formed by pressure measurement locations as an open boundary. Another approach is to contour the wind-tunnel walls, such that they coincide with streamlines in free-flight conditions. The slip boundary condition is enforced along these contoured walls. This is restrictive, because in unsteady flows these free-flight streamlines, at a short distance from the body, can be time dependent. Instead of these alternatives, the adaptive-wall wind tunnels (see for instance, Sears[49]) could allow the use of the free-flight boundary conditions.

The inflow, outflow, and tangent flow open boundaries require different treatments. The above discussion dealing with the material derivative of Q_1 shows that on an inflow boundary, Q_1 must be specified. On outflow, Q_1 is determined from the mass conservation equation; and on tangent flow boundary Eq. (26) is used.

For external flow problems, boundary conditions are available at infinity, but not at finite distances. If the inflow boundary is at, say, about ten times the characteristic length of an aerodynamic body, then the influence of the body at that distance should be negligible, and therefore, it is possible to use the conditions at infinity as inflow boundary conditions. This leads to the specification of remaining Q's.

The main difficulty in specifying the outflow boundary conditions across a wake of a body is that the boundary values are part of the solution and hence not known a priori. However, we do know something about the outflow boundary. There are no physical boundary layers, and flowfield is inviscid for all practical purposes. This suggests that the boundary conditions must not introduce any boundary layer. This requirement is also valid for the open tangent flow boundary. Further, on the part of the outflow boundary which cuts the wake region, the flow is rotational. During passage of vortices or "eddies" through this boundary, pressure values on this boundary vary. The variation depends upon the strength of these vortices. In addition, extrapolation along curvilinear coordinates, when the Reynolds averaged Navier-Stokes equations are in the conservation law form as in Eq. (21), may introduce errors because of one or more of the relations between metric coefficients are not satisfied. This situation is analogous to that between Eq. (23) and Eq. (25). Because of these reasons, this paper presents here a possible set of conditions on the open outflow and tangent flow boundaries. Simply stated, the Euler equations are considered as boundary conditions for the Navier-Stokes equations. In other words, the viscous and heat conduction terms are neglected on the outflow and tangent flow boundaries. This approach was applied to the incompressible Navier-Stokes equations by Mehta and Lavan[50] and Mehta.[51] The above conditions satisfy the type constraint on the boundary conditions as required by the mathematical character of the system represented by Eq. (21). When the wind-tunnel flows are simulated with open boundaries, as discussed above, the outflow condition on Q_1 may be replaced by the measured pressure values.

For the system represented by Eqs. (23) or (25), again heuristic arguments are used for determining the number of boundary conditions. The above rigid wall boundary conditions are applicable to the equations of this system. However, the open boundaries for these equations require a different treatment. In ξ_2 direction, the above considerations are valid. But in other coordinate directions, the system represented by these equations is hyperbolic. Therefore, the direction of flow of information dictates the boundary conditions. The local characteristics or eigenvalues determine the number and the admissible forms of boundary conditions. For a hyperbolic system, the eigenvalues are real. The number of negative eigenvalues with distinct eigenvectors determines the number of boundary conditions. This number is the same as the number of inward characteristics into $\mathcal{R}$. In other words, if inflow is supersonic in the "hyperbolic" directions, then all Q's must be specified, otherwise one less specification is required. On outflow boundaries (in these directions), if the flow is supersonic, then nothing can be specified; and if it is subsonic, one condition is required.

As it is indicated in the section Numerical Methods, in practice mainly numerical methods without artificial viscosity (page 330, Ref. 36) require extraneous diffusion, which is provided by the addition of higher-order even power derivatives in each of the original differential equations. (This extraneous diffusion is often called "numerical dissipation.") These added terms remove unresolved high frequencies, but they are supposed to be constrained so as not to alter the resolved frequencies beyond the error bounds for these frequencies. One point of view is that these terms do not change the mathematical character of the original equation because they are a part of the truncation errors of a numerical method. (See page 331, Ref. 36.) These terms disappear when $\Delta\xi \to 0$. An alternative point of view is to consider these terms as part of the original differential equations, since $\Delta\xi$ is never equal to zero. In this case, these terms do not change the character of the original parabolic equations. Also, they do not change the global character of the original hyperbolic equations, provided they do not introduce any boundary layers at the boundaries. This is achieved by not adding these terms either on the boundaries or next to the boundaries in ξ_2 direction. This avoids additional boundary conditions for both parabolic and hyperbolic equations. These terms may form interior "boundary layers" such as captured (smeared) shocks. In this case, the "additional boundary conditions" for these terms are automatically provided by the appropriate neighboring, interior flow quantities.

Computational Grids

A computational grid system is a necessary part of any numerical solution based on a finite-difference, a finite-volume, or a finite-element method. The selection of a grid system is based primarily on the requirement for accuracy in the final solution. Secondary considerations are the effect on computational efficiency of the solution algorithm using available computer architecture, and finally the ease of grid generation. These concepts are discussed below.

Accuracy Requirements

Accuracy requirements are determined by the application of the numerical solutions of governing equations along with initial and boundary conditions. If the solutions serve the purpose for which they were intended, then the accuracy requirements are satisfied for that particular application. These requirements vary with purposes of applications and frequently tend to be subjective. Unlike the accuracy requirements, the discretization (truncation) errors are independent of both purposes of applications and subjectiveness. Therefore, in the discussion that follows, the accuracy constraints are not quantified, and the emphasis is placed on the discretization errors.

Simulations of flow regions, throughout which the scales of motion are essentially the same in all directions, are probably best carried out by equispaced Cartesian meshes. In this case, the evaluation of mesh errors on the solution is completely determined by the size of the single-space interval. On the other hand, a flow-field with a surface along which there is a turbulent boundary layer, is generally computed using a highly "stretched" mesh "normal" to the surface. This mesh is very fine near the surface and usually is constructed to increase exponentially in the direction "normal" to it. In this case, the errors of the solution are much more intimately tied to the grid structure, and the evaluation of errors is not simple. The situation is again not easy to evaluate when attempts are made to align and cluster meshes with shock waves, the position of which are not known a priori. The relationship between solution errors and grid choice is very important to the evaluation of transonic viscous flow simulations. It is still in the stage of development and our comments here are based on limited information.

It seems to be generally accepted that one of the coordinate families should lie along any surface that is generating a viscous boundary layer. Usually, at least 15 grid points spaced nonuniformly from the surface to the "outer edge" of the layer are required for even marginal resolution. This is accomplished most conveniently by using a body-oriented system. This further facilitates application of surface-boundary conditions. It is reasonable to expect that the accuracy is best when grid lines leaving the surface are normal to it, although this does not appear to be crucial.

It also appears to be generally accepted that one of the coordinate families should be made to lie along a shock, if this is possible. This is often quite possible for bow shocks which interface with a completely known freestream flowfield. For interior shocks, this is much more difficult; and turbulent, transonic Navier-Stokes simulations have been, so far, done with shock-capturing techniques rather than shock-fitting ones. This is primarily due to the fact that the latter methods introduce algebraic and data management complexities in the viscous interaction regions. In contrast, the shock-capturing techniques do introduce errors, and these errors depend upon the grid system and the choice of numerical method (see Numerical Techniques for Computing Shocks). When one of the coordinate families is not aligned with a shock, these techniques tend to thicken the shock-wave region. This thickening may modify the shock/boundary-layer interaction phenomena. Therefore, one of the principal weaknesses in evaluating the errors and reliability of the present Reynolds averaged Navier-Stokes codes is estimating the effect of the grid system on the shock strength, location, and thickness. A systematic and dependable study of this issue would be most welcome.

The discrete governing equations for flow simulations around complicated aerodynamic geometries involve the following geometrical quantities, depending upon the numerical methodology: in case of finite-difference methods, there are metric coefficients and the Jacobian of topological transformations; when there are finite-volume methods, we have lengths or surface areas, and areas or volumes; and the finite-element methods contain shape factors. All these geometrical quantities are obviously grid dependent and they appear along with physical quantities in the overall numerical process. Clearly, it is the combination of physical and geometrical quantities that appear in the difference formulas that should be accurately resolved in order for the simulated flow to be a useful solution of the governing system. This suggests that geometrical quantities require proper representation as do the physical quantities. The standard technique of refining the mesh is usually not available, because computational resources do not permit it. A necessary but far from sufficient condition is to make sure that a given grid, along with a numerical scheme, maintains the freestream if the freestream boundary conditions are applied.

Methods for Improving Flow Simulation Accuracy

One requirement of accurate solutions is that they be, for all practical purposes, independent of the grid system. So far, this has not been systematically demonstrated for turbulent, transonic simulations. The generally accepted practice of indicating the order of truncation error of a numerical method does not quantify the discretization errors. Although quantification of these errors is difficult, it is possible to determine their effects through grid refinement studies. On the other hand, minimization of these errors may be achieved by a proper choice of both the numerical method and the grid system. Usually, there is more freedom in choosing the grid system than in choosing the accuracy of the numerical method. Further, the choice of the grid system is determined by a priori knowledge about the solution. Most of this knowledge is available in terms of generalities rather than specifics. For example, surface boundary layers are always resolved with the help of some stretching function near the known surface. But without the specific information, such as the magnitude and location of gradients in the flowfield, the grid system employed can often be wasteful and not satisfactorily concentrated on those regions where a better resolution is desirable.

For a better utilization of grid point resources, there is a growing interest in solution adaptive grid systems. In a moving finite-element method, which allows both nodal amplitudes and nodal positions to move continuously with time, nodes generally move automatically to those regions where they are most needed.[52] In finite-difference methods, there are currently two basic strategies. The first strategy involves tracking a fluid property, such as the density gradient, and inserting or regridding so that finely spaced grid points are in the immediate vicinity of that selected property (for instance Dwyer et al.[53] and Kovenya and Yanenko[54]). The second strategy is to minimize the leading term or terms of the modified equations that determine the order of the truncation error of a numerical method (e.g., Pierson and Kutler,[55] and Rai and Anderson[56]).

So far, the adaptive grid techniques have been primarily applied in one- and two-dimensional Burgers' equations, and for a two-dimensional heat equation. Extension of these techniques to the Navier-Stokes equations for turbulent, transonic simulations is a difficult undertaking. Questions, such as what flow variables to monitor, which truncation errors to minimize, whether all flow variables and/or truncation errors should be considered simultaneously, and which parts of the flow domain require special checking, need to be resolved. Of course, how to best adapt the grid system is a major research effort. These issues become much more involved when there are unsteady shock/boundary-layer interactions. The obvious payoff of solution adaptive grid systems is in terms of efficient use of computer resources.

Management of Grid Systems

The secondary criteria for selection of a grid system, mentioned in the beginning of this section, deals with the care of the grid system and the associated data base. Implicit numerical algorithms for both finite-difference and finite-volume methods are more efficient when based on grids with ordered discretizations (see Effect of Grid Topologies on Computational Efficiency), and zonal methods and finite-element methods do not necessarily produce well-ordered data bases. The real importance of well-ordered data bases occurs in studies involving three-dimensional spaces, and there is very little information available in this area.

Methods for Generating Grid Systems

At the present time, the property of providing a body-fitted grid system for external aerodynamic problems is automatically satisfied in two dimensions by all currently popular curvilinear grid generating schemes. But appropriate interior grid systems for each class of topological geometries require trial-and-error manipulations of different variables in these schemes. (See, for instance, Sorenson.[57]) Based on methodology, there are two types of grid generating schemes, algebraic and differential. An algebraic grid generation scheme is a direct approach. It may be further classified into a conformal transformation procedure and a nonconformal transformation procedure. A description of conformal transformations for computational aerodynamics is given, for example, by Sells,[58] Ives,[59] and Moretti.[60] (A conformal transformation may be defined either by an analytical function or by two Laplace equations resulting from the fact that the real and imaginary parts of an analytical function are harmonic. The procedure based on the latter definition does not require a separate discussion.) Some of the nonconformal procedures are the parametric multisurface transformations,[61,62] transfinite interpolations,[63] and the isoparametric mappings.[64] (Note: Eiseman has not used the adjective "parametric.") On the other hand, a differential grid generation scheme is an indirect approach. This again may be further categorized as that based on a hyperbolic differential system and on an elliptic differential system. A hyperbolic procedure was first presented by Barfield,[65] and then it was extended and analyzed by Starius.[66] Recently, Steger and Chaussee[67] have modified Starius' procedure. Thompson, Thames, and Mastin[68] exposed the elliptic procedure to the computational aerodynamic community by extending, in particular, the work of Barfield,[65] Godunov and Prokopov,[69] and that of Amsden and Hirt.[70]

When a boundary of a flowfield can be mapped with an analytical function, when the resulting distribution of boundary grid points is nearly satisfactory, and when the interior grid distribution is less of a concern, conformal transformations are the best: They give rise to simple geometrical mapping quantities, and it is easy to assemble a grid system with them. Furthermore, they provide exact values of geometrical quantities. These transformations, however, cannot be extended to three dimensions, but they can be used in two-dimensional cross sections of a three-dimensional flowfield.

The hyperbolic transformation procedures give, in two dimensions, orthogonal curvilinear grid systems. With these procedures, it is not automatically possible to control either the location of the outer boundary or the distribution of points on it. Therefore, they cannot be used directly for internal flow problems or for patching different grid systems. Further, their application and usefulness in three dimensions remain to be demonstrated. On the other hand, the elliptic transformation procedures have been extended to three dimensions (e.g., Mastin and Thompson[71] and Lee et al.[72]). Under these conditions, the possibility and ease of a reasonable grid control is still to be demonstrated. The elliptic procedures require more computational time than the other procedures. These procedures, generally, assemble nonorthogonal grid systems. They allow some flexibility and consequently control in the nature of grid system at the boundaries. But they do not allow local grid control without affecting the entire grid system because of the ellipticity of the grid generating differential system. The solutions of both hyperbolic and elliptic grid generating differential procedures are constrained by the accuracy requirements just as are the solutions of the flow governing equations.

Unlike the elliptic procedures, the parametric multisurface transformation procedures allow local grid control. By the very nature of these procedures, they provide more flexibility than the other procedures, which results in precise grid control. However, they require a more complex specification of generating variables than the others.

Sometimes it is possible to choose the type of grid pattern. For instance, turbulent transonic and inviscid transonic (or transonic viscous-inviscid interaction) simulations past an airfoil are almost always conducted, respectively, with the "C" and the "O" grid. The "H" grid may also be used. This introduces a geometrical singularity, if the two halves of a vertical line in the "H," one below and one above the horizontal line, meet at an angle other than 180 deg, as in ">—|." This requires a special treatment. Use of the "C" grid avoids the difficulty of the mesh singularity. However, as it is usually programmed, it does not make efficient use of mesh points in the region behind the trailing edge. This is also the case for the "H" grid. Most currently available Reynolds averaged Navier-Stokes codes with "C" grids have been used with a number of grid points ranging from 45 to 85 on the airfoil surface. (In contrast, current inviscid transonic simulations are generally conducted with "O" grids that use about 160 grid points on the airfoil surface.) Of the three grids as they are usually programmed, the "O" grid gives the best airfoil resolution for the same number of grid points. However, its use can create numerical difficulties at a sharp trailing edge.

An important problem that is beginning to emerge with the availability of more powerful computers is the generation of a grid system around a complete aircraft. Recently, there have been some attempts at generating a grid system around some parts of an aircraft. Lee and Rubbert[73] and Lee et al.[72] have explored the possibility and presented some ramifications of constructing a grid system for three-dimensional configurations such as a wing-body-nacelle shape. The computational domain is divided into a multiple set of rectangular blocks. An elliptic grid generating scheme is used within each block. With this approach, there are two shortcomings. It introduces geometrical singularities in the transformed domain where there were none to begin with and the grid control in the physical domain is poor, particularly across block boundaries and along the trailing edge of the wing. The above investigators have also considered a single-block around a wing-body configuration. In this case, geometrical singularities become regular in the transformed domain. Eriksson[63] has used an algebraic scheme for the same configuration. The resulting "C" grid pattern around both the leading edge of the wing and the wing tip appears to be acceptable. Moretti[60] has shown how to assemble a grid system in cross-sectional planes of a fuselage and arrow wing configuration using conformal transformations. Complex three-dimensional geometries are first rendered quasi-two-dimensional, then two-dimensional grid generating techniques are applied.

The complexity of generating suitable three-dimensional grid systems is somewhat analogous to problems in the design and manufacturing. In these disciplines, computational geometry, that is the computer representation, analysis, and synthesis of shape information,[74] has been invaluable. It has given rise to the fields of computer aided design (CAD) and computer aided manufacturing (CAM). In CAD and CAM, parametric transformation procedures are used to describe a single surface.[75] In a grid system, multiple surfaces are defined, and the constraints placed on these surfaces are much more severe than on a single surface. The roots of parametric multisurface transformation procedures appear to be in computational geometry. In three dimensions, complexity of generating grid systems, and difficulties in visualizing a grid system during and after it has been generated, call for using interactive graphics, just as in CAD and CAM.

Component Adaptive Grid Systems

In the numerical simulation of three-dimensional flows, each component of an aircraft has its own "natural" grid system, which is usually not "natural" for the other components. Consequently, different grid systems, each suitable for a particular component, are constructed. This leads to the concept of component adaptive grid systems, also referred to as the zonal grid approach. These different systems must, of course, interact. This is accomplished by embedding one type of grid into another (e.g., Atta[76]) or by some other form of patching neighboring regions (Forcey et al.,[64] Eiseman and Smith,[77] and

Lee and Rubbert[73]). In any form of grid patching, the region of interaction between the different grid systems requires special consideration, for instance, maintenance of global conservation and consistent accuracies. Perhaps the most important problem in the practical use of zonal grid systems is their effect on the numerical stability of the solution process.

Numerical Methods

Two Crucial Nonlinear Convective Phenomena

In order to clarify the discussion presented below, it is useful to develop a concept that can be used to relate physical and numerical phenomena. We search for some form of scale in both time and space that is common to both phenomena, and find an excellent candidate in the frequency content of a harmonic analysis made of the physical variables with reference to either time or space. The physical side of this concept can range from the very "natural" (in experimental studies of isotropic turbulence) to the rather "contrived" (in the harmonic analysis of a discontinuity). On the numerical side, these frequencies form part of the exact solution to certain model linear problems with periodic boundary conditions, but are only loosely related to the eigensystem of most difference equations actually being solved. Nevertheless, the association of frequency with scale is a very convenient concept when discussing some of the broader aspects of the numerical simulation of fluid flow.

The Euler equations model an unsteady flow that can contain a discontinuous solution referred to as a shock wave, or simply as a shock. For the Navier-Stokes equations, shock waves are not, strictly speaking, discontinuous, their thickness being of the same order as the thickness of the linear sublayer in a turbulent boundary layer (see Navier-Stokes Equations). The spectral analysis of a variable having a discontinuity, or an abrupt jump that is "nearly" discontinuous, is presented in Fig. 1. Notice that all, or "nearly" all, of the high-frequency terms have finite amplitude. In the theme of the previous paragraph, all or nearly all scales are present. This has an important influence on the construction of numerical methods used to compute flows with embedded shocks.

This paper discusses flows that have significant regions of turbulence and separation. Flows with attached

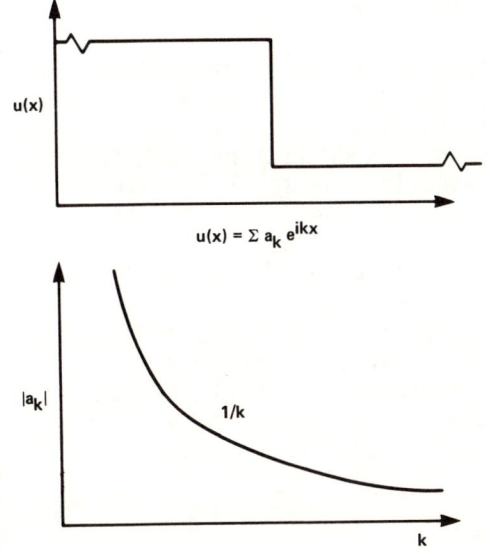

Fig. 1 Spectral analysis of a discontinuous function.

boundary layers can be computed using the methods we are discussing, but they usually can also be calculated by simpler and less expensive methods. Although the vorticity that is essential for the production of turbulence is generated by the viscous properties of the fluid-surface interface and curved shock wave, turbulence itself is generated away from the surface and caused by the nonlinear interactions of the convection terms in the Euler equations, the same terms responsible for the generation of shock waves. For the points relative to this paper, the most illuminating aspect of turbulent flow lies in the spectral representation of its inertial range shown in Fig. 2. This gives the amplitude of the turbulent kinetic energy associated with each harmonic in a spectral analysis of a typical high Reynolds number turbulent flow.[7,8] Notice that the scales of both axes in the figure are logarithmic, that almost all of the energy is carried in the low wave numbers, and that molecular dissipation is limited to the relatively high wave numbers where the energy content is low. The flow represented by the high-energy, low-frequency region is referred to as large-scale or large-eddy motion, and the flow represented by the opposite end as low-scale or small-eddy motion. The results shown in Fig. 2, referred to as the energy cascade, greatly influence models used to approximate the effects of turbulence.

Numerical Techniques for Computing Shocks

As stated previously in Accuracy Requirements, there are two common approaches used when devising numerical methods for calculating flowfields with shock waves: shock fitting and shock capturing. Shock-fitting methods employ some kind of test for detecting the shock location, and then treat the shock as a local discontinuity across which the Rankine-Hugoniot relations must be satisfied. Shock-fitting methods are probably to be preferred where they can be generated by reliable and efficient codes. They eliminate the need for conservation law forms of the governing equations (which has certain simplifying attractions), and they produce sharp discontinuities at the jump location. They are quite popular for computing many flows that can be modeled by the inviscid Euler equations,

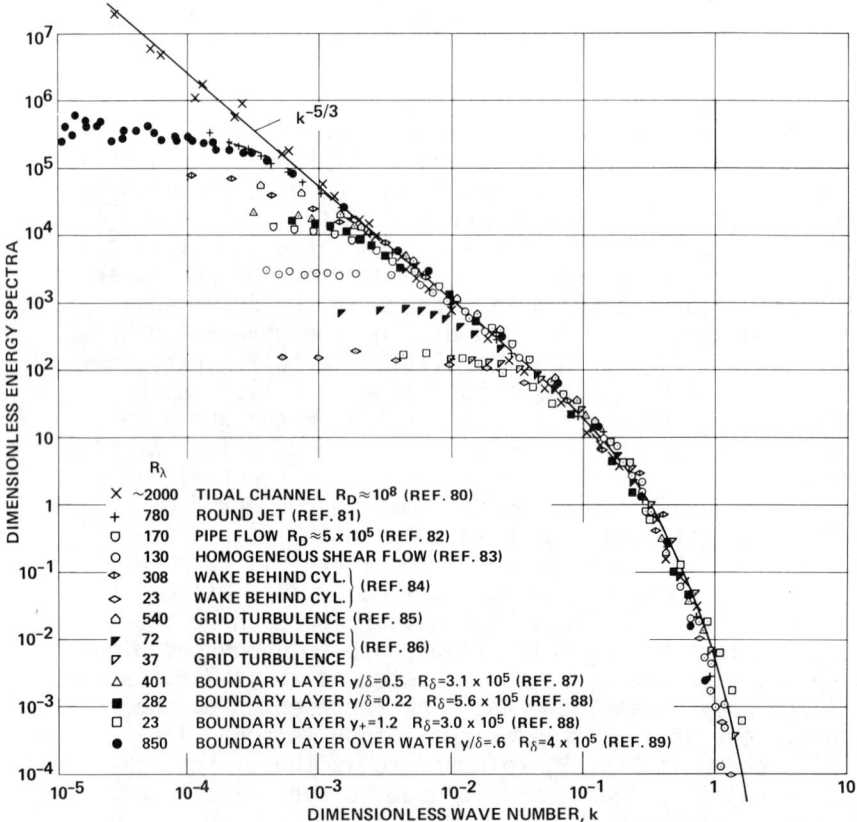

Fig. 2 Streamwise turbulent energy spectra for various turbulent flows.[79]

especially where the flowfield is supersonic, see, for instance, Kutler[90] and De Neef and Moretti.[91] However, the flows of interest in this report can have strong shock/boundary-layer interaction, and the effect of viscosity must be included in this region. Further, we are interested in the flows that contain three-dimensional and oblique shocks. Shock fitting under these conditions can become extremely difficult, and our remaining attention is limited to shock-capturing methods.

The point of a shock-capturing technique is that the shock forms and moves about in a mesh, while some kind of analytic connection is maintained between the flows on the two sides of the wave front. This does not mean that the shock-capturing methods cannot have built-in logical tests that try to isolate the shock location. Often they do, and they make use of the test results to make local adjustments to the differencing scheme to improve its capturing capability. Still, by definition, a shock-capturing numerical method connects the dependent variables on the two sides of the wave.

An immediate consequence of shock capturing relates to the spectral structure of a discontinuous function shown in Fig. 1. Since the capturing technique is based on some kind of numerical continuity across the shock, the harmonic analysis can be used to represent the result. It is well known that a finite grid can only support a finite number of frequencies in a discrete Fourier series. For example, an equispaced grid of M points can accurately accommodate $k = M/2$ harmonics of the form e^{ikx}. Frequencies higher than k reappear as lower frequencies, a property referred to as aliasing. In an unsteady flow with a moving shock, these higher frequencies are constantly being generated by the nonlinear convective interaction. For example, the product of the waves $e^{ikx} e^{i\ell x}$ brought about by terms such as $u \partial_x v$, produces two harmonics, one having a lower frequency proportional to $k - \ell$, and the other having a higher frequency proportional to $k + \ell$. This behavior can be verified in numerical simulations by observing how a simulated shock constantly tries to steepen. A linear discontinuity shows no such tendency. The situation just described can be summarized as follows:

1) Any discrete grid system can accurately support only a limited number of low frequencies. If higher frequencies are placed on it, they appear as amplitudes of low-order terms.

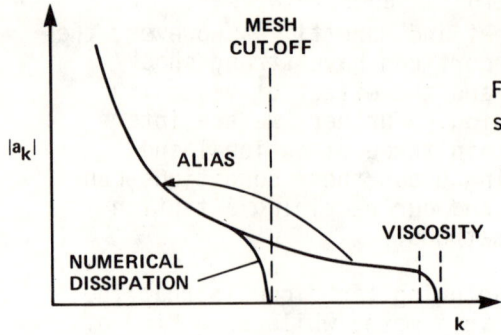

Fig. 3 Numerical dissipation of subgrid amplitudes.

2) Convective nonlinear interactions are constantly cascading low frequencies to higher ones.

The numerical difficulty brought about by this situation in the case of shocks is illustrated in Fig. 3. The frequencies to the right of the mesh cutoff line are referred to as subgrid frequencies. If their production is permitted, they must alias back into the low-frequency range causing numerical error. This error can be severe enough to cause numerical instability. The standard way to cope with the subgrid scale generation is to include in the computing process some form of numerical dissipation which removes the subgrid terms before any significant part of them cross the cutoff boundary. This is an arbitrary, numerical, error control procedure that has nothing to do with any physical dissipation which occurs at much higher frequencies.

The practical implementation of adding the numerical dissipation of the subgrid takes many forms. The process can be "hidden" in the differencing scheme. Such is the case for the various Lax-Wendroff types where the actual dissipative mechanism, which is provided by the fourth and higher even-order derivatives, is uncovered by inspecting the modified partial differential equation (e.g., Warming and Hyett[92] and Lerat[93]). Upwind space differencing schemes have the same property, which is again revealed by inspecting the modified partial differential equation. Central differencing schemes for the first derivative of a space term are well known to be nondissipative, so when these are used in shock-capturing algorithms, higher-order dissipation terms are deliberately added to the computations.[41,94,95,96,97] From the arguments presented here, they are no better or worse than the forms which have no overt dissipation. All numerical schemes that capture shock waves with satisfactory accuracy have some numerical error, and its quantification is usually subjective and problem dependent. This situation can be attenuated to a certain extent by mesh clustering, but is usually worse for Navier-Stokes codes than it is for potential codes, simply because the meshes for Navier-Stokes computations are usually coarser.

The preceding discussion does not present the only valid point of view for assessing shock-capturing techniques. An alternative point of view is based on the theory of characteristics in supersonic flows. For example, the usual justification of upwind differencing in locally supersonic regions is not based on dissipation but on the fact that upwind differencing can be made to approximate a local method of characteristics. The Lax-Wendroff methods also tend to approximate a local method of characteristics. In both cases, for all one-dimensional linear convective problems, a discontinuity can be solved exactly at a Courant number of one. Model problems, however, seldom occur in practical application. It is interesting to notice that central differenced first-derivative terms with deliberately added dissipation can be made to create a system that has the properties of upwind differencing.

A second consequence of using a shock-capturing method is to create the problem of insuring the proper location and strength of the shock as it moves about in the mesh. Lax[9,40] has shown that this can be suitably approximated if the difference equations are locally conservative. The most common way of enforcing this condition is to cast the governing partial differential equations in conservation law form, and then make sure the difference scheme maintains this property. When such a technique is employed, a shock profile, represented, for example, by the pressure distribution, is "smeared" over a few mesh points, but, for many practical applications, the general position and strength are adequately represented. Many variations of shock-capturing methods exist which attempt to make the wave structure "crisper" and to eliminate overshooting of shock profiles. Our experience with numerical calculations which include boundary layer indicates that the details of shock smearing and overshoot are not of critical importance in determining the flow behavior along body surfaces. From this point of view, a wide variety of published methods are quite adequate for capturing shocks in Navier-Stokes codes.

Numerical Techniques for Computing Turbulence Effects

The problem of computing turbulence is much more difficult than that of capturing shocks. In fact, at the Reynolds numbers typical of transonic aerodynamic flows, no attempt is made to compute turbulence; rather, an effort is made to approximate the effects of turbulence. The reason is, as in the case of shock capturing, the incapability of numerically resolving the full range of scale. However, in the case of turbulence, the problem is much more severe, since the scale to be resolved extends in all three space directions as well as in time.

A plot of the longitudinal turbulence energy spectra for eight different types of flow is shown in Fig. 2. It is seen that energy dissipating eddies (large k) are apparently independent of both Reynolds number and type of flow. Further, the form of the energy spectra in the inertial subrange at high Reynolds number conforms to the Kolmogoroff spectrum law ($k^{-5/3}$). This result is strictly experimental; no numerical simulation has yet produced real evidence of an inertial subrange in three dimensions. In order to accomplish this, a mesh needs to be provided that can support more than two orders of magnitude of frequency variation in all the three space dimensions. It is estimated that this will require a mesh with about $(1024)^3$ grid points. For an incompressible flow that contains all of the modes, the calculation would need a total storage of about 7×10^9 words using the most sophisticated numerical techniques.

The simple realities of computer resources force us to make one severe approximation and to accept one severe constraint in formulating our governing equations before the numerical methods can be considered. The approximation is the use of the Reynolds averaged equations discussed earlier. This eliminates the need to resolve the small eddy motion, but introduces the problem of closure. The constraint is to permit extreme coordinate distortion in only one direction. This permits us to approximate viscous effects normal to very thin layers, but, at high Reynolds numbers, in that direction only. Probably the most important result of all this is that the computational processes that finally emerge have the capability of qualitatively simulating flows with regions of separation and large-scale unsteady behavior. The crucial question, of course, is their reliability.

The role numerics plays in computing turbulence effects has yet to be discussed in this paper. Two quite different issues are involved: one is the manner in which the subgrid scales are accounted for, and the other is the manner in which the turbulence model is implemented. The subgrid scales are constantly being generated by the large-scale structure through the nonlinear wave interactions in the convective terms. The numerical control of the subgrid energy production is brought about by the addition of dissipation, either through the space derivative approximation or deliberately by additional terms. In either case the choice is arbitrary, except that it lies in the error band of the large-scale resolution, and that it prevents the accumulation of energy in the highest frequencies supported by the mesh. The role of this form of dissipation is often not clearly understood. It has nothing to do with physical viscosity at the scale that it is employed. Its detailed form is largely arbitrary, yet a solution would be physically incorrect if without it energy were to flow to subgrid levels and alias back into large-scale terms where it has no physical meaning. The numerical dissipation is essential to the numerical simulation of the effects of turbulence, but it is not, in conventional terminology, part of the turbulence model, see Fig. 4.

The second important role of numerics in Reynolds averaged codes lies in the detailed coding of the turbulence model. The analytic forms of several models were given earlier in the section Governing Equations. Unfortunately, these are not sufficient to describe the effect of a turbulence model on an actual calculation. The numerical effect of the complete model is the sum of all its parts, and this includes the grid clustering, the metric evaluations (see next section), the internal logic controlling the local evaluation of parameters such as mixing length, the choice of difference approximations, and the numerical dissipation. The "accuracy" of all this is difficult to evaluate since the conglomerate is the actual turbulence model and its fundamental basis is essentially empirical. The final judgment of the method is usually based on a comparison with some experiment, and the result may be good or bad depending on the choice of any one of the method constituents.

Many variations of turbulence models have been tried on transonic flows with turbulent boundary layers. How well these compare with transonic wind-tunnel experiments is discussed in the next major section.

Effect of Grid Choice on Numerical Stability

The basic reason for choosing a nonuniform grid is to improve the accuracy of a numerical solution for a given number of mesh points. There are two ways in which this is usually accomplished. One is to align, as closely as possible, a coordinate with a known or anticipated surface, such as a shock wave or body surface, in order to fit them more "naturally" into the mesh. The other is to cluster points in regions where there are rapid changes of gradients in order to reduce local truncation errors. As a corollary of the latter process, in order to conserve resources, points are often spread apart in regions where the curvature is small. Finite-difference, finite-volume, and finite-element methods all have these capabilities.

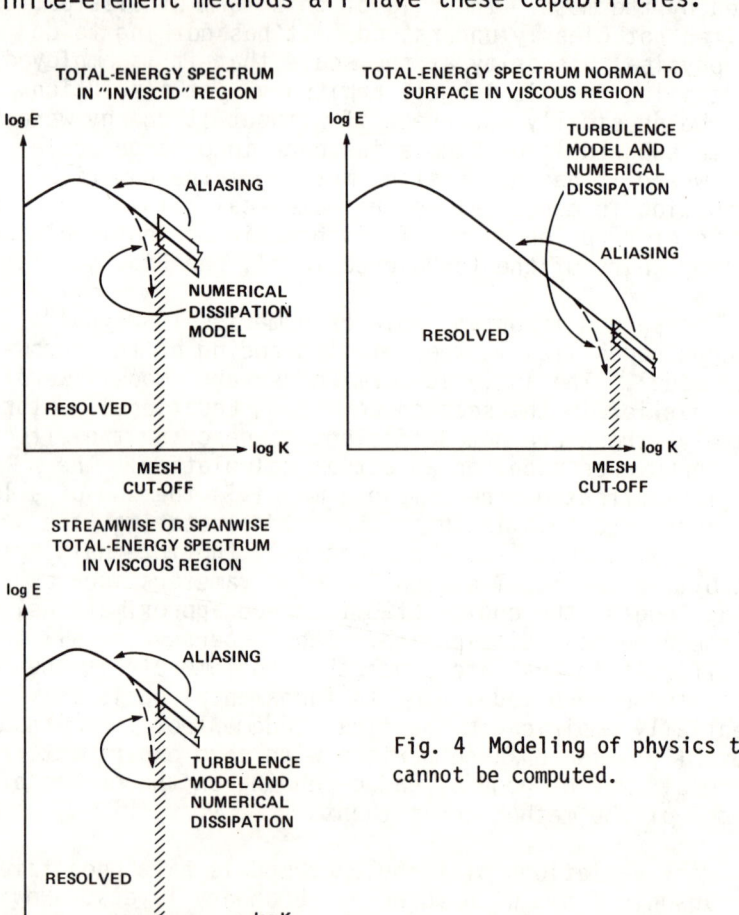

Fig. 4 Modeling of physics that cannot be computed.

The form chosen for a grid can have a profound effect on the solution process. By far the most important side effect of grid refinement on a numerical algorithm is its influence on numerical stability. As is well known, the time step of explicit methods is mainly bounded by the size of the space interval; this holds for nonequispaced as well as equispaced meshes. If a single time step is used for advancing the entire solution, an explicit method is generally limited by the smallest space interval in the mesh. This limitation can be extremely costly if the time step is forced to be very small compared to the time scales of motion that are of interest. In such cases, the algorithm is said to be stiff; if the stiffness is caused by the fineness of a space interval in the grid, the algorithm is said to be mesh stiff.

Codes using explicit numerical methods for the solution of the Reynolds averaged Navier-Stokes equations can be extremely mesh stiff when they are used to study flows with thin boundary layers. This occurs when the grid is very fine in the vicinity of the body in order to compute the viscous effects there. For example, a typical grid spacing normal to an airfoil surface can be in the order of 0.00001 chords for turbulent boundary-layer simulations at Reynolds numbers above 10^6. Grid point clustering around shocks and leading and trailing edges can also be the cause of mesh stiffness.

The most common way to avoid any form of stiffness is to use implicit, rather than explicit, algorithms. Almost all codes being used to analyze the compressible, Reynolds averaged Navier-Stokes equations have some parts that represent an implicit numerical technique. The use of such techniques involves the solution of coupled sets of simultaneous equations. In finite-difference codes, these simultaneous equations can usually be expressed as very sparse banded matrices that, in the great majority of cases, have tridiagonal structures. In fact, the numerical efficiency realizable from solving tridiagonal systems is so deeply embedded in finite-difference methods for the problems we are discussing that it has greatly influenced, and at present even limits, the choice of grid topologies. This is discussed in Effect of Grid Topologies on Computational Efficiency.

The Basic Difference Equations

The following is a brief evaluation of the finite-difference techniques currently being used to solve the types of problems in which we are interested — completely aside from any consideration of the turbulence model. It does not represent those formulations which compute the flow using different equations in different regions, such as an inviscid outer flow coupled with a boundary-layer calculation.

The difference equations used to solve the compressible Reynolds averaged Navier-Stokes equations take many forms and vary in accuracy, efficiency, and reliability. Throughout the 1970s, a variety of individual codes were developed and used to solve specific problems. The particular choice of problem was usually motivated by some experiment involving shock waves and turbulent boundary layers with varying amounts of separation. In numerical terminology, these codes represent methods that range from fully explicit (e.g., MacCormack[98]) to factored fully implicit (e.g., Briley and McDonald[97] and Beam and Warming[99]). The codes are usually written in terms of numerical operators which are applied in series to prescribed data bases. Thus, there may be a convection operator followed by a diffusion operator, or the algorithm may be "space split" so that a one-dimensional x operator is followed by a one-dimensional y operator to form the total x,y solution of a two-dimensional flow. These techniques are also referred to as factored forms. In some codes, certain of the factors represent explicit methods and others implicit ones.[100,101]

In general, at the present time, all of the codes used to solve the Reynolds averaged Navier-Stokes equations and the methods they represent, seem to have about the same potential for accuracy and efficiency of running time, although these can vary according to the capabilities of the individual coder. The numerical methods appear to be acceptable everywhere throughout the flowfield except possibly at the boundaries, a matter which is again an individual responsibility. The codes are generally at least first-order accurate in time. For high Reynolds numbers ($>10^6$), they require about 45 min running time on a CDC 7600 to reach a steady state, if one exists. This estimate is for codes that are at least partially implicit. It varies, of course, depending upon the number of grid points, the Mach and Reynolds number, and the angle of attack. If the codes are fully explicit the running times can be much longer.

Effect of Grid Topologies on Computational Efficiency

For the points to be made in this discussion our basic equation can be expressed in the form

$$d\vec{Q}/dt = A\vec{Q} - \vec{f} \qquad (27)$$

where A is a very large and very sparse nonlinear matrix that represents some combination of the flux Jacobian, the grid construction, and the space differencing. If the grid is chosen so that the physical space is mapped into a computational space that forms the insides of a rectangular box, and the boundary conditions are mapped onto the sides of the box, the matrix A becomes banded for most common choices of finite-difference schemes. The typical form of A for second-order finite-difference schemes is shown in Eq. (28) for a three-dimensional problem that is formulated in a computational box.

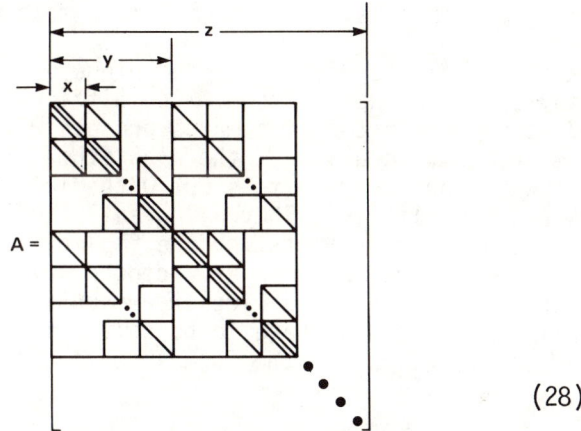

(28)

In this schematic structure, all matrix entries are zero except those represented by the diagonal lines. Each diagonal line represents a set of 5 × 5 block matrices each of which is composed of a local flux Jacobian. Suppose the mesh coordinates are represented by x, y, and z and there are a total of Mx, My, and Mz points in each coordinate direction. In the particular case shown in Eq. (28), the data vector Q is so arranged that the x data are closely packed, nearby y data skip blocks of x, and nearly z data skip blocks of y. Of course, this arrangement is arbitrary and, by permuting the data base, the variables in any one direction can be closely packed at the expense of the other two.

The steady-state solution of Eq. (27) can be viewed in two ways: 1) as the solution of the nonlinear system $Q = A^{-1}f$; 2) as the result of a converged time history of an unsteady process. The former requires the solution of a set of simultaneous equations having the form represented by A in Eq. (28) — which would have to be iterated because it is not linear. The latter would require the successive solution (with each time step) of a similar set of equations if the time-marching method were fully implicit.

Consider the prospect of carrying out either of these solution procedures. Although the matrix A is sparse and banded, notice that the half-bandwidth is $5 \times M_x \times M_y$ elements. A solution using simple Gaussian elimination would require about $(5 \times M_x \times M_y)(5 \times M_x \times M_y \times M_z)$ temporary storage locations to hold the information required to complete the backward sweep. This makes the solution of such a matrix by direct methods quite impractical on present day computers with even moderate mesh sizes.

A common finite-difference technique used in the unsteady approach that overcomes the difficulty just discussed is to factor the time march process without changing the order of accuracy of the algorithm. There are several ways for carrying this out with differing accuracies and stabilities. They all greatly reduce the temporary storage requirements for the implicit operation. Methods commonly referred to as factored fully implicit leads to a set of three matrices representing block tridiagonal equations that have to be solved in sequence. Each of the matrices has the form shown in Eq. (29).

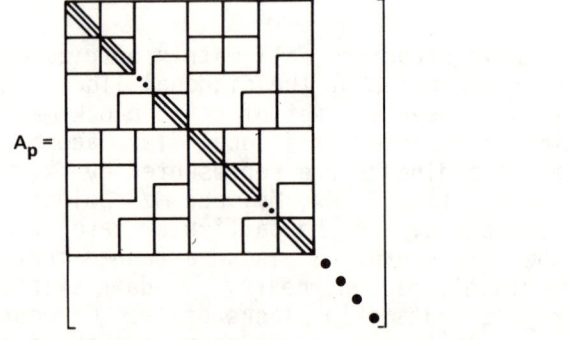

(29)

Notice that this time the matrix is formed by large uncoupled diagonal blocks each one of which is tridiagonal in subblocks of 5 × 5 matrices. In such cases, each large diagonal block can be solved independently and requires a temporary storage of only 5 × 5 × Mp words, where p represents x, y, or z.

The role of the topological box computational space in all of this is to provide the banded structure of the matrices in Eqs. (28) and (29). Zonal grids with interfaces, overlapping meshes, and other forms of nonregular grid structures lead to A matrices that are not banded and tend to deviate from the tridiagonal structure. This can greatly increase the complexity of the computational algorithm or drive it to explicit (or even numerically unstable) forms. In either case, efficiency and code reliability can suffer. Many forms of the finite-element approach will lead to the same difficulties for the same reason. The problem of generalizing mesh structures beyond computational boxes and keeping the codes that use them computationally reliable and efficient is one of the most pressing problems in finite-difference developments in the 1980s.

Comparison between Experiments and Calculations of Turbulent Transonic Flows

The following material draws from the relatively young and limited body of computed results based on the Reynolds averaged Navier-Stokes equations for transonic flows with strong viscous-inviscid interactions. We have taken this material from publications only from NASA/Ames Research Center simply because most of the work in this area has been carried out at this institution.

First this paper will consider some typical computed boundary-layer profiles for an attached flow. For example, Fig. 5 shows a group of such profiles ahead of a shock wave on an 18% thick circular arc airfoil.[102] These are compared with the compressible form of the law of the wall. In the figure, u^+ represents u normalized with the friction velocity. A simple mixing length model, given by Launder and Spalding,[18] was used to describe the turbulent transport. All computed profiles have one grid point in the viscous sublayer. The log-law region is well represented by the grid point distribution. This is generally the case of presently available Reynolds averaged Navier-Stokes computations in attached boundary layers. Computed velocity values at $x/c = 0.675$ differ from the empirical log-law distribution because of flow separation just downstream of this location.

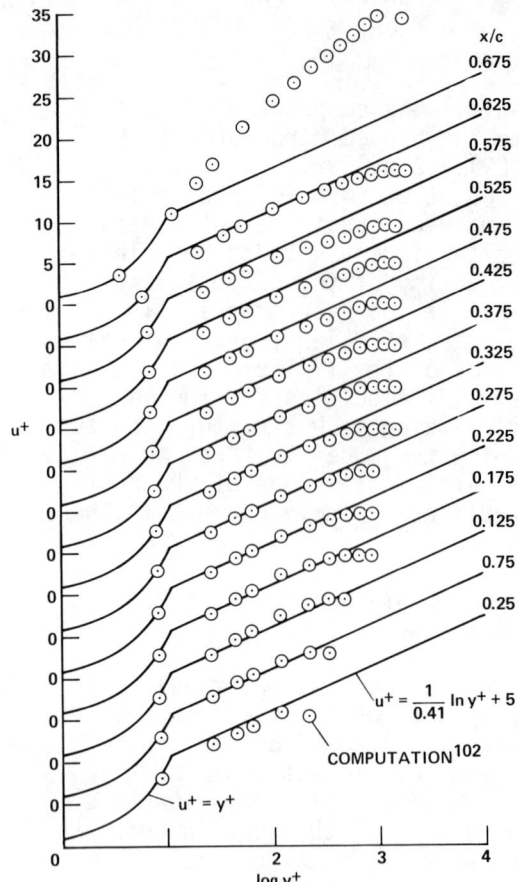

Fig. 5 Velocity profiles ahead of a shock wave.

For many practical uses, the turbulence modeling of attached turbulent boundary layers without shock-wave interaction is quite acceptable. The following discusses the status of the Navier-Stokes technology for turbulent, transonic simulations with emphasis on turbulence modeling for separated flows and on flow problems which are not feasible to solve with current viscous-inviscid interaction approaches. This discussion deals with representative simulations in which flowfields may be steady, unsteady, attached, or separated, both in two and three dimensions. Special problems such as "buffeting" flows, aileron buzz, and airfoil "stall" are considered. In addition, two types of turbulence simulations are presented,

NAVIER-STOKES COMPUTATIONS

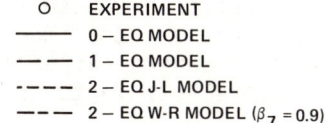

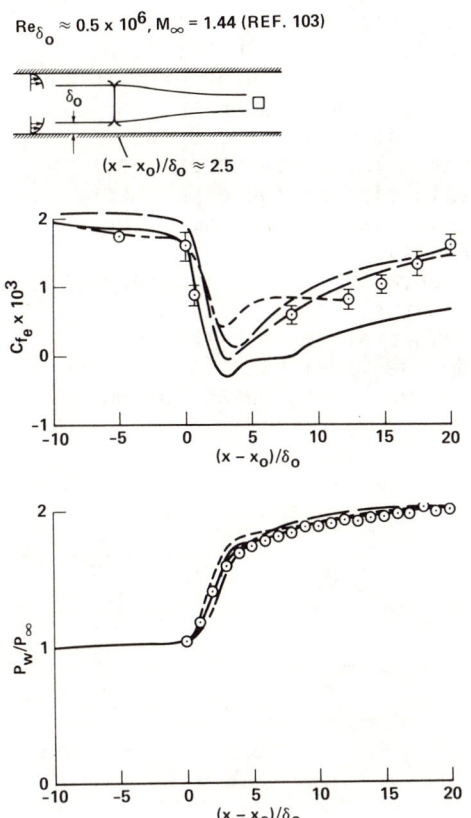

Fig. 6 Skin friction and surface-pressure distributions in axisymmetric shock/boundary-layer interaction.

one in which turbulence models are used in a predictive mode and the other where these models are used in a postdictive mode. This presentation is motivated in order to stimulate systematic questioning of what research directions are needed for accelerating advances in better predictions of separated turbulent flows in aerodynamic applications.

Axisymmetric Steady Flows

A computation of normal shock/boundary-layer interaction for an axisymmetric flow was carried out by Viegas and Horstman.[103] The tunnel geometry, experimental results, and several computations are shown in Fig. 6. This represents an attempt to compare the merits of four different types of turbulence models at Reynolds number (5.5×10^5) based on upstream boundary-layer thickness and low supersonic Mach number (1.44). The results for pressure distribution are essentially the same for all models. Differences are evident in calculation of skin friction, which depends on the slope of the boundary-layer profile at the surface. In fact, the particular algebraic model used showed a region of flow separation which did not appear in the other calculations. The above computors report that the most recent evaluation of the experiment indicates that flow does not separate. In light of the results shown in Fig. 7, a tentative conclusion can be drawn: This is probably representative of the accuracy one can expect from present forms of turbulence modeling and numerics. With regard to the algebraic model, the obvious question is: What details made the model used for Levy's results shown in Fig. 7 so superior to that used for the results in Fig. 6?

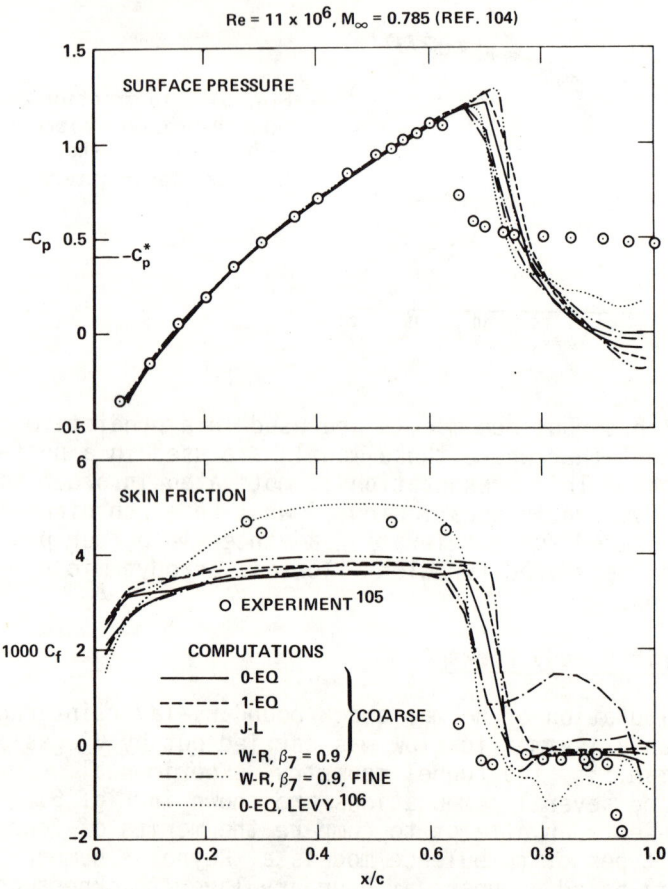

Fig. 7 Pressure and skin friction on 18% circular arc.

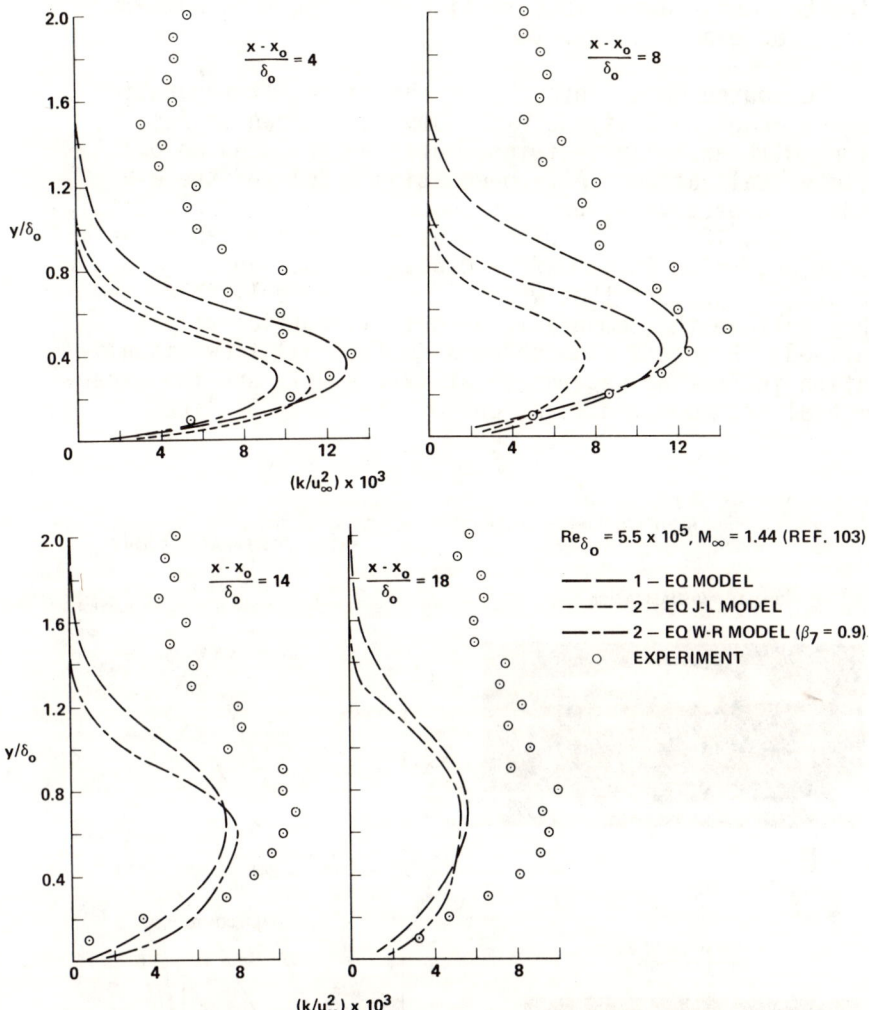

Fig. 8 Kinetic energy profiles in an axisymmetric flow.

As one looks into the details of more sensitive flow properties, one can anticipate further discrepancies. For example, the W-R model [Eqs. (18) and (19) with $\beta_7 = 0.9$] and the Jones-Launder (J-L) model[107] were used to compute the turbulent kinetic energy, $\overline{v^2}/2$. Measured and computed profiles of $\overline{v^2}/2$ are shown in Fig. 8 at various x locations downstream of the shock wave located at x_0. The measured energy was determined from a measurement of u_2' and with the assumption that $u_1' : u_2' : u_3' = 4 : 2 : 3$. This assumption was observed to be reasonable for equilibrium boundary-layer flows at high subsonic Mach numbers (see, for example, Ref. 108).

Computed Mach contours and the extent of separation region about an axisymmetric "bump" are shown in Fig. 9, along with an infinite fringe interferogram and an oil film visualization. A zero-equation model and the W-R model, respectively, predict shock locations approximately 0.16 and 0.12 chord lengths downstream of the experimental location, which is at $x/c \sim 0.66$. Johnson and Horstman[109] report that wall effects are negligible, and they believe the computational grid is sufficiently refined. Figure 9 also shows a surface oil flow visualization indicating separation at $x/c \sim 0.7$, and the experimental and computed locations of the $u_1 = 0$ line.

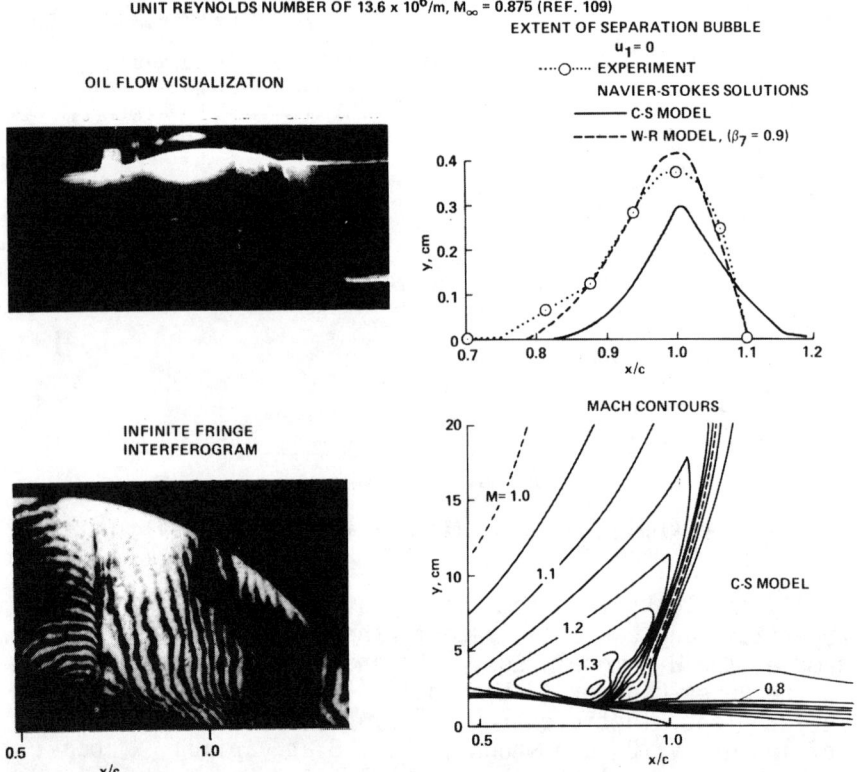

Fig. 9 Separation bubble and Mach contours over an axisymmetric "bump" model.

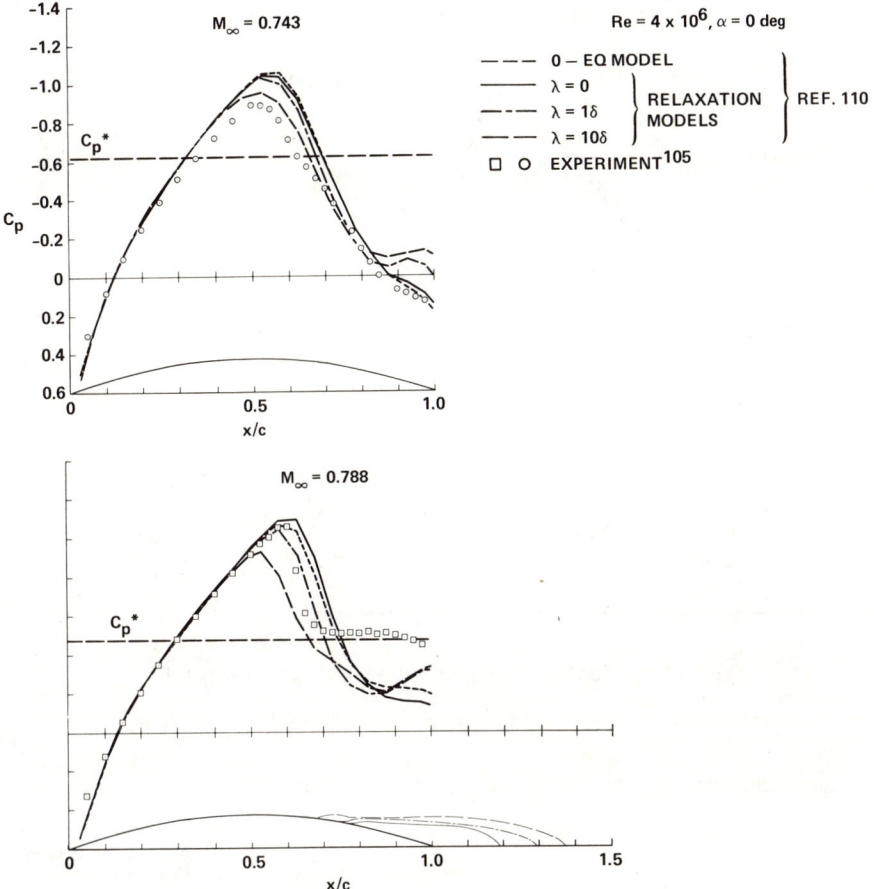

Fig. 10 Comparison of computed and experimental pressure distributions on 18% circular arc airfoil.

The amount of work expended on axisymmetric flows, both experimental and computational, is quite small compared to two-dimensional studies. Nevertheless, the above results are representative of the state-of-the-art of simulating such flows.

Two-Dimensional Steady Flows

Effect of four different algebraic eddy viscosity models on surface pressure distribution over an 18% circular arc airfoil is shown in Fig. 10. These models range from an unmodified boundary-layer, mixing length model to a streamwise relaxation model with three magnitudes of the

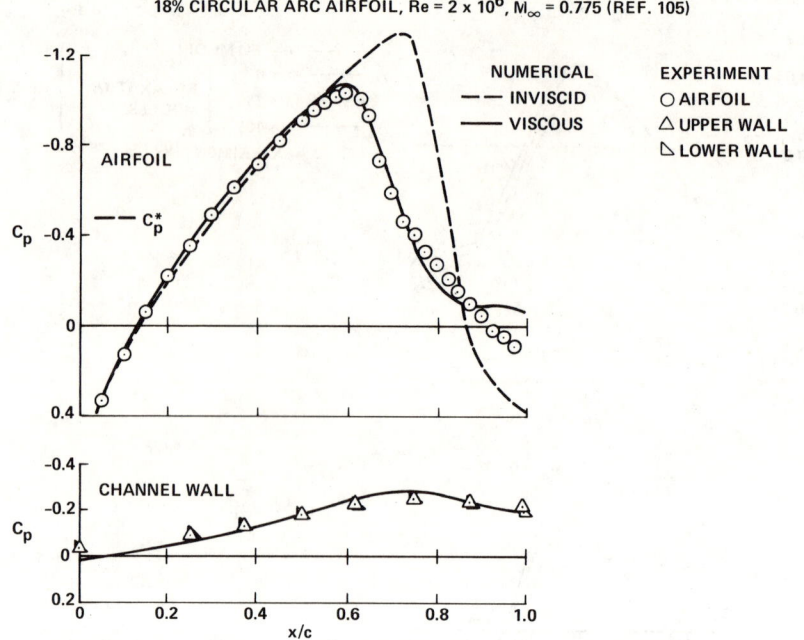

Fig. 11 Effects of viscosity at design conditions.

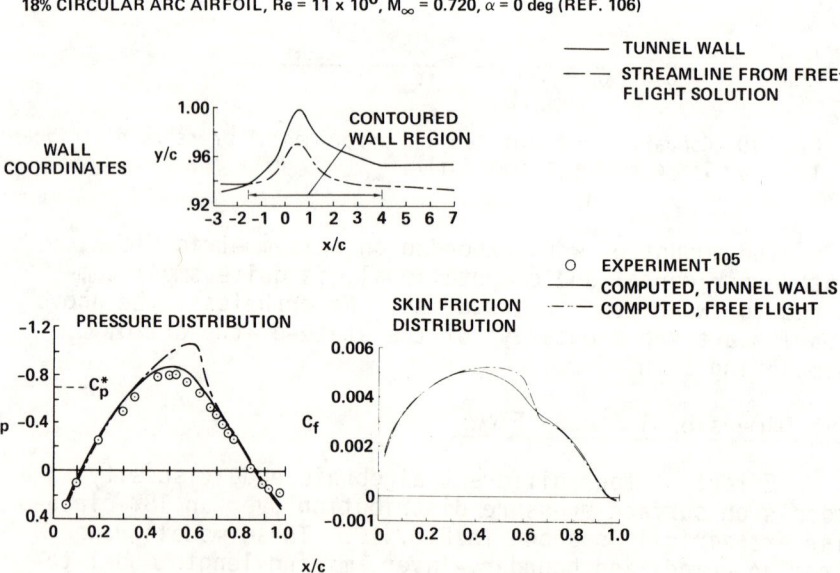

Fig. 12 Effect of tunnel wall boundaries at off-design conditions.

relaxation parameter λ.[110] Figure 10 shows the effect of this parameter on the extent of separation region for the high Mach number case. Except for the highly relaxed study, $\lambda = 10\delta$, the results are all about the same and show an agreement among themselves that can be expected of different forms of eddy viscosity turbulence models for flow with mild trailing edge separation.

The two cases shown in Fig. 10 have an interesting history. Consider first the results for the lower Mach number, 0.743. The experimental data came from a wind tunnel and the computations were made for free air. At the time the computations were made, they were considered to be acceptable because the upper and lower tunnel wall, which were at a distance of about a chord length from the model, had been contoured to match an inviscid free air calculation for a Mach number equal to 0.775. The contoured walls were diverged slightly to compensate for wall boundary-layer growth. The agreement between tunnel experiment and calculation under these conditions is shown in Fig. 11.[105] The effect of contouring for one Mach number and running for another is indicated in Fig. 12.[106] While it is not conclusive, it is reasonable to attribute most of the discrepancy between experiment and computation (excluding $\lambda = 10\delta$) in Fig. 10 to be due to an improper boundary condition on the upper surface of the computa-

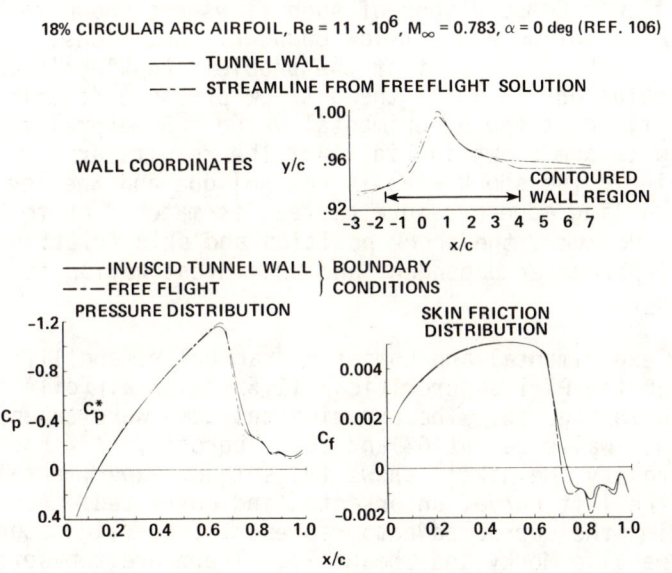

Fig. 13 Effect of tunnel wall boundaries near design conditions.

tional domain. The result for $\lambda = 10\delta$ is assumed to represent a bad model for the turbulent region.

The computed and experimental pressure distributions for $M_\infty = 0.788$ in Fig. 10 illustrate another possible source of trouble in making flow simulations. This is a case representing fairly steady (see the next subsection) shock-induced separation, where the pressure plateau behind the shock was very poorly estimated by all computations and the computed shock wave was nearly normal, instead of oblique as in the experiment. The effect of correcting the calculation by including the proper upper wall as a boundary condition made very little difference (Fig. 13). An effort to tie the discrepancies to the turbulence model was made by Coakley and Bergmann[104] with the second-order thin shear layer approximation. The results of this study are shown in Fig. 7. No essential difference in the result could be correlated with any of the forms of eddy viscosity models and mesh refinements that were tried. In fact, the zero-equation model result reported by Levy[106] was the closest to experiment both in pressure distribution and skin friction. However, we attach no significance to this fact insofar as any model can be considered as superior to the others.

The wind-tunnel results shown in Figs. 10, 11, 12, 13, and 7 were determined from measurements made in a channel flow. Computations of such flows are known to be sensitive to inflow and outflow boundary conditions. In order to check this aspect of the problem, Coakley[111] made some calculations in which the outflow pressure distribution was fixed at the experimental value. Some preliminary results are shown in Fig. 14. The results are encouraging. The shock wave is now oblique and the level of the trailing edge pressure plateau is matched quite closely. However, the shock position and skin friction are still parameter dependent and the investigation is continuing.

The experimental and computed drag polars and lift curves for the GK I supercritical 11.5% thick airfoil[112] are shown in Fig. 15. The experimental data were taken with tunnel walls set at 6% and 20.5% porosity.[115] For 20.5% porosity, Melnik[114] shows two sets of experimental data on the lift curve, uncorrected and corrected. According to him, the corrected data represent free-flight conditions (see also Morky and Ohman[116]). There are two sets of computed results. Deiwert[113] has solved the Reynolds

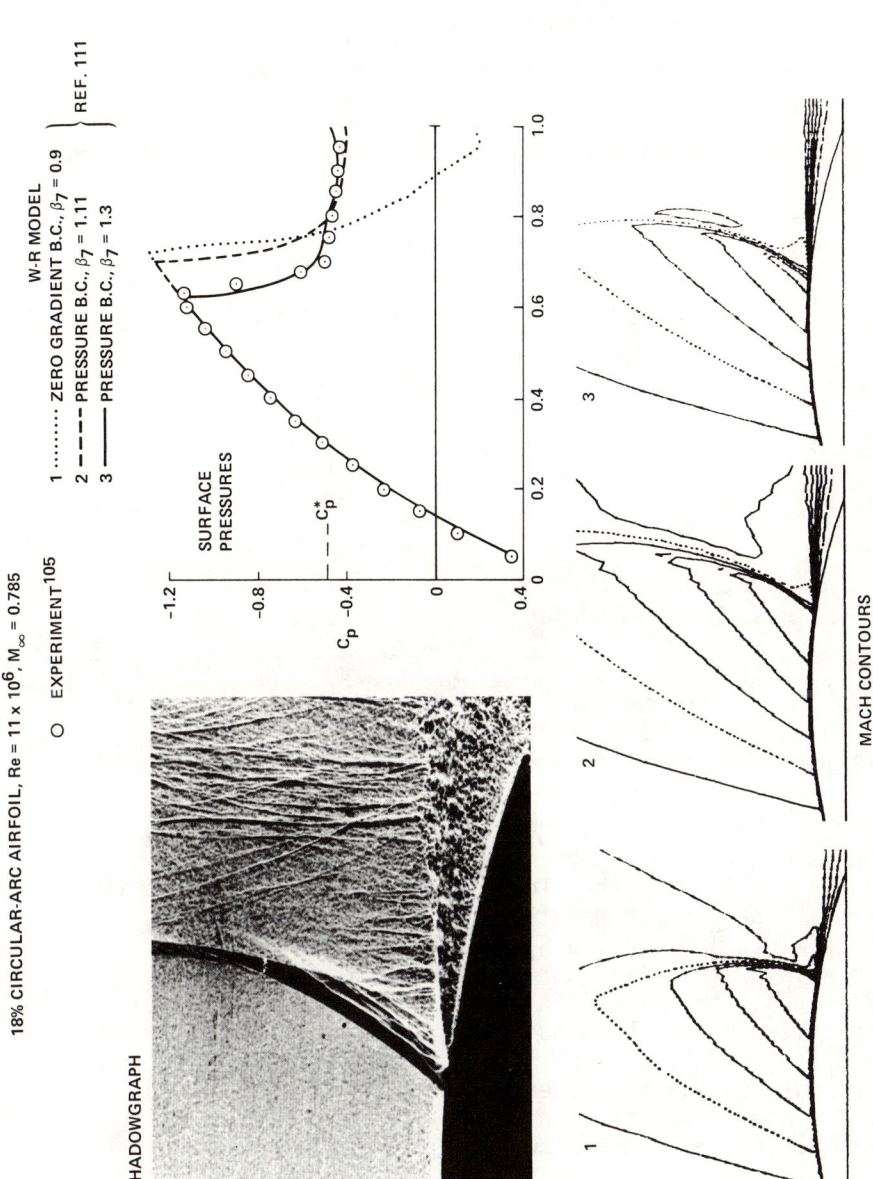

Fig. 14 Effect of downstream pressure condition and turbulence model on shock wave.

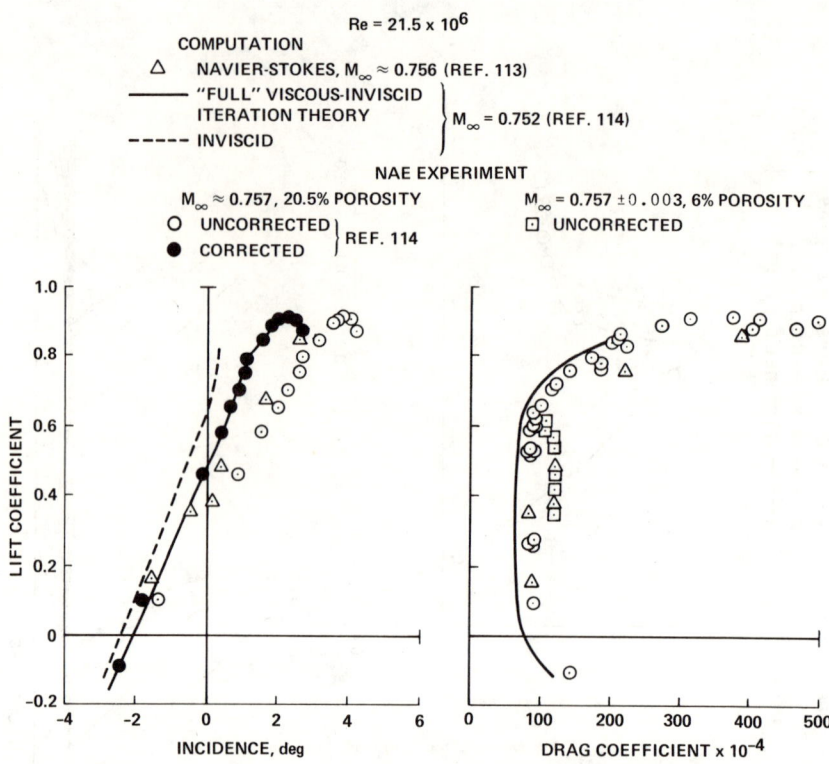

Fig. 15 Lift coefficients and drag polar for Garabedian-Korn I airfoil.

averaged Navier-Stokes equations with free-flight boundary conditions and an algebraic model without relaxation. Melnik[114] has used the "full" viscous-inviscid interaction theory. He has matched the lift coefficient with the experiments and applied a small Mach number shift of $M = -0.005$ to obtain agreement with the experimental shock position. The lift curve shows that both the viscous effects and the wind-tunnel interference effects are important. Drag values of both computations differ from the measured values. These computations again indicate that proper boundary conditions are required for taking into account wind-tunnel wall interference effects.

Two-Dimensional Unsteady Flows

An interesting set of experiments[105] and calculations[106] have been carried out for an 18% thick biconvex airfoil at zero-degree incidence. Both experiments and calculations showed a region of "buffeting" or self-excited, oscillating flow in the Mach number range between 0.72 and 0.79 for a Reynolds number around 11×10^6.

The experiment was conducted using a wind tunnel in which the upper and lower walls were contoured as mentioned in the preceding subsection. The calculations used slip flow boundary conditions along surfaces that matched these contours. The effect of turbulence was approximated by an algebraic eddy viscosity model similar to that used by Deiwert.[117] This model changed form in various regions bounded by the separation location, the location of reattachment of the separated streamline to the surface streamline, and the edge of the boundary layer. Unfortunately, the sensitivity of the solution to the model is an unknown.

Figures 16 to 19 show a comparison between the experiment and computed results. Figure 16 identifies the experimental Reynolds number and Mach number domains within which there are three distinctively different types of flow. The three types were reproduced by computations made at Mach numbers of 0.720, 0.754, and 0.783, and a chord Reynolds number of 11×10^6. At $M_\infty = 0.720$, the flow is steady and flow separation occurs near the trailing edge of the airfoil. At $M_\infty = 0.754$, there is unsteady

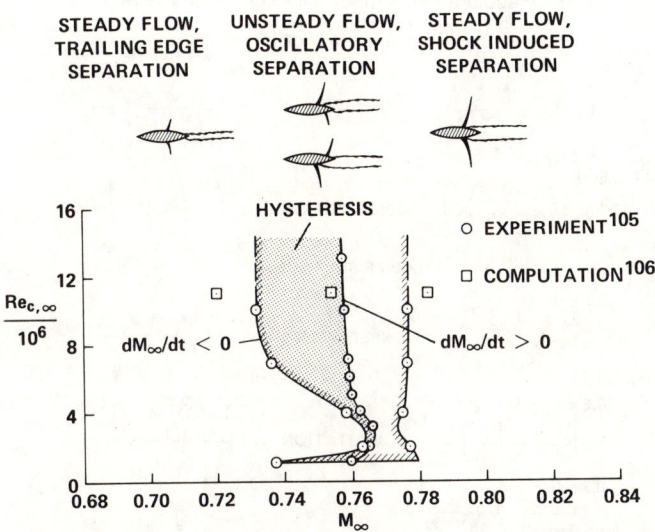

Fig. 16 Experimental flow domains for 18% circular arc airfoil.

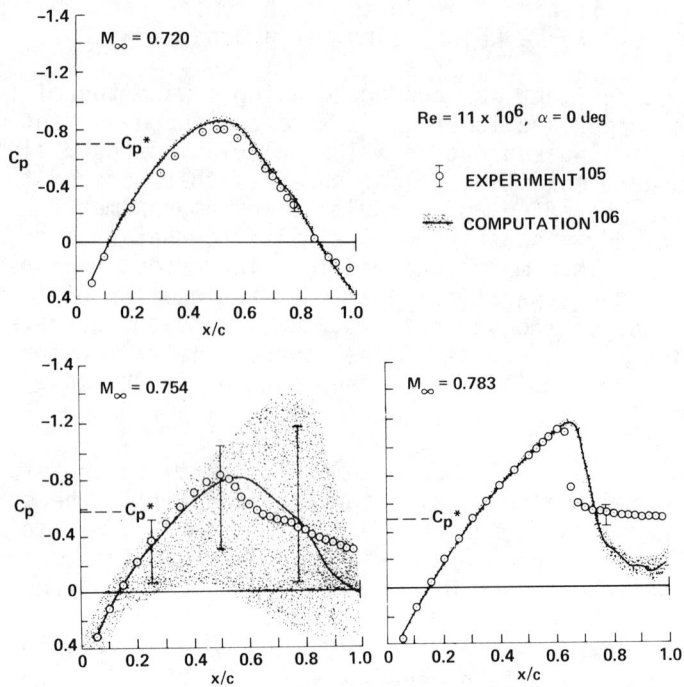

Fig. 17 Pressure distributions on 18% circular arc airfoil.

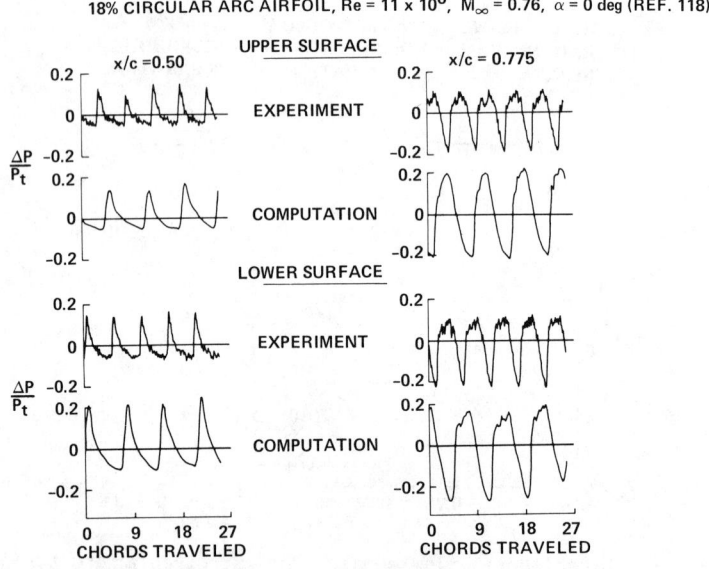

Fig. 18 "Buffeting" flow, surface pressure time histories.

periodic oscillation in shock-wave location and intensity; and the flow alternates between trailing edge and shock-induced separation and is quite different on the upper and lower surface at any given time. At $M_\infty = 0.783$, a shock wave induces boundary-layer separation at its base and the flow is relatively steady, except in the separated region.

Surface pressure comparison is demonstrated between computations and experiments for the above three different conditions in Fig. 17. The vertical bars on the experimental data represent maximum and minimum values of fluctuations about mean. The range of computed fluctuations about the mean computed values is denoted by the shaded area. The steady flow regions at $M_\infty = 0.720$ and 0.783 have been discussed in the previous subsection.

The unsteady flow at $M_\infty = 0.754$ is qualitatively very well predicted, but quantitative comparison is poor, except for the mean values of pressure over the forward half of the airfoil (Fig. 17). This is further supported by Fig. 18 which shows surface pressure time histories. Here, the instantaneous pressure oscillations are given about the mean pressure, normalized by the wind-tunnel total pressure. The computed and measured, reduced frequency of these oscillations are, respectively, 0.40 and 0.49. However, the amplitude of oscillations is quite different. For this case, the shock-wave shapes from shadowgraphs are compared with computed Mach number contours in Fig. 19 where the phase has been arbitrarily adjusted.[119] For another problem, namely, a 14% thick biconvex airfoil at $Re = 7 \times 10^6$ and $M_\infty = 0.83$, the computed unsteady lift forces and pitching moments are compared with those for $M_\infty = 0.85$ in Fig. 20.

It is not at all surprising that Reynolds averaged Navier-Stokes equations are capable of simulating unsteady flows when the computational time step is small compared to the period of resolvable flow motion which is of interest, but much larger than the high-frequency, small-scale fluctuations which have been averaged out of these equations (see earlier section, Governing Equations). The question of how high the resolvable frequency could be relative to the mean frequency of turbulence eddies is addressed by Chapman.[79]

Another unsteady phenomenon, this time associated with a moving boundary, is represented by the performance characteristics of the aileron of a P-80 (i.e., F-80) aircraft.

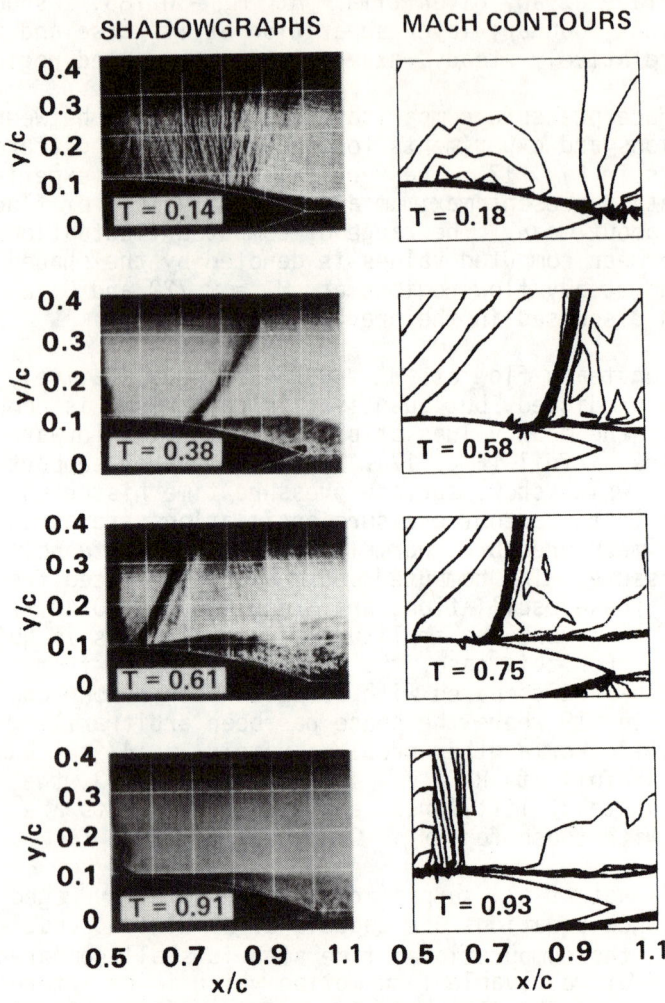

Fig. 19 Comparison of shadowgraphs and Mach contours.

This flow has been simulated by Steger and Bailey[122] using the algebraic eddy viscosity (B-L) model and the second-order thin shear layer approximation described in the section Governing Equations. The turbulence model was applied from the leading edge of the airfoil. The P-80 airfoil section is an NACA 65_1-213 with a = 0.5. The aileron buzz is a one-degree-of-freedom flutter problem.[123] The interrupted, inviscid shock-wave motion (e.g., Tijdeman and Seebass[124]) causes a phase shift in the response of the hinge moment to the aileron movement. In the experiment, at $M_\infty = 0.82$ and $\alpha = -1$ deg, the aileron, when freed at an angle near zero, would buzz with amplitude and frequency as indicated in Fig. 21. In the simulation, it would not buzz under these conditions. But if it was initially deflected to 4 deg, it would, on being released, buzz as shown in the figure. The computed and measured frequency are, respectively, 22.2 Hz and 21.2 Hz. Further, the computed and measured deflection of the aileron are, respectively, -1.1 ± 11.1 and -3 ± 9.2 deg. Similar calculations are made at different airfoil angles of attach to predict the measured buzz boundary.

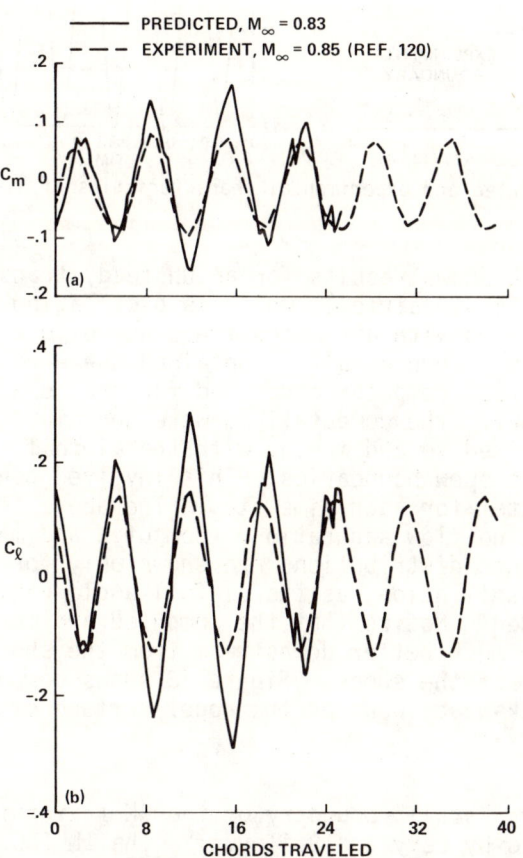

Fig. 20 Pitching moment and lift histories, 14% biconvex airfoil, $Re = 7 \times 10^6$, $\alpha = 0$ deg.[121]

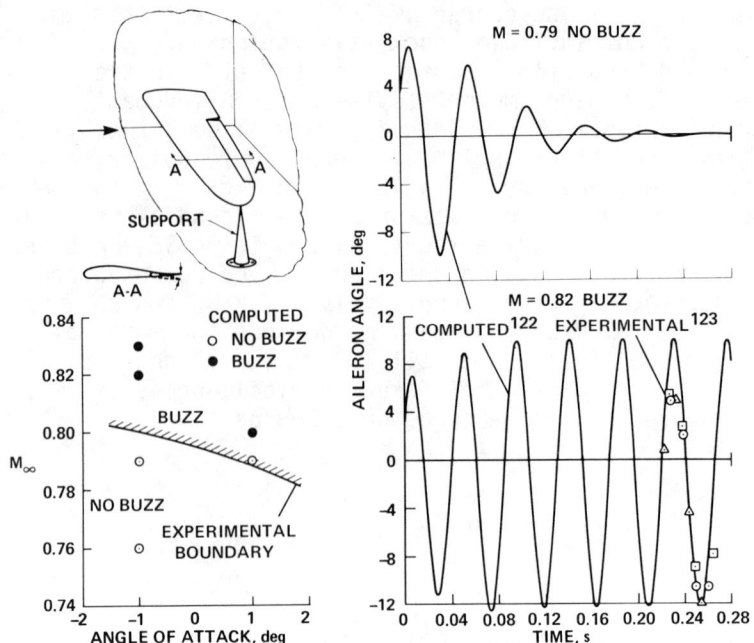

Fig. 21 Computed and experimental characteristics of aileron buzz.

Figure 22 shows results for an unsteady transonic flow over an NACA 64A010 airfoil, which is oscillating about its one-quarter chord with a reduced frequency of 0.2, based on one-half chord. Chyu et al.[125] obtained these results with the same CDC 7600 computer code used for the buzz study discussed above. The computations were done in a coordinate system fixed to and moving with the airfoil, but stationary at the open boundaries. This involved generation of a grid system for each time step. The above investigators report no flow separation. Computed and measured surface pressure distributions are shown only for one-half cycle of an oscillation, as the airfoil angle varies from 1 deg to -1 deg. Notice that the computed and measured results agree much better downstream from the shock wave than upstream of the shock. Figure 23 shows computed and measured shock-wave locus on the upper surface of the airfoil.

Recently, "stall" boundary of the GK I airfoil has been predicted by Levy and Bailey.[126] The ILLIAC IV computer code was the same as that used on the buzz study just discussed. Figure 24 shows computed and measured unsteady

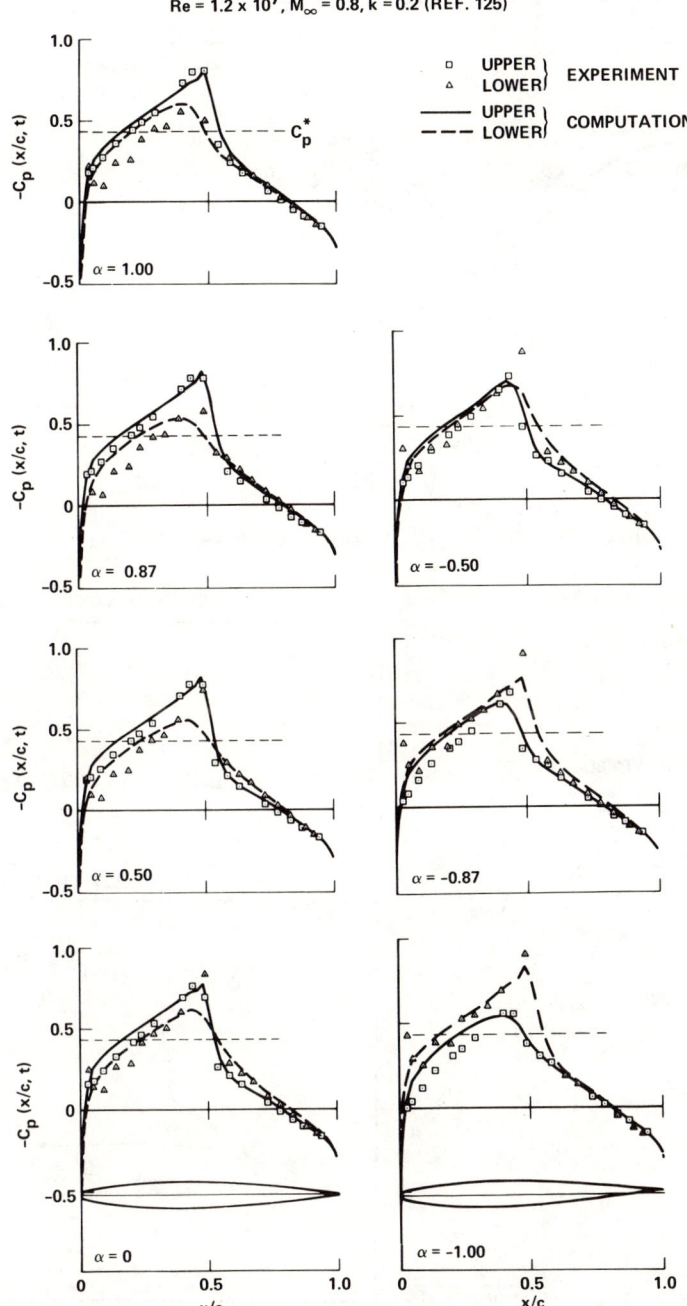

Fig. 22 Time histories of surface pressures on oscillating NACA 64A010 airfoil.

OSCILLATING NACA 64A010 AIRFOIL, Re = 1.2 x 10^7, M$_\infty$ = 0.8, k = 0.2

Fig. 23 Shock wave locus on upper surface.[125]

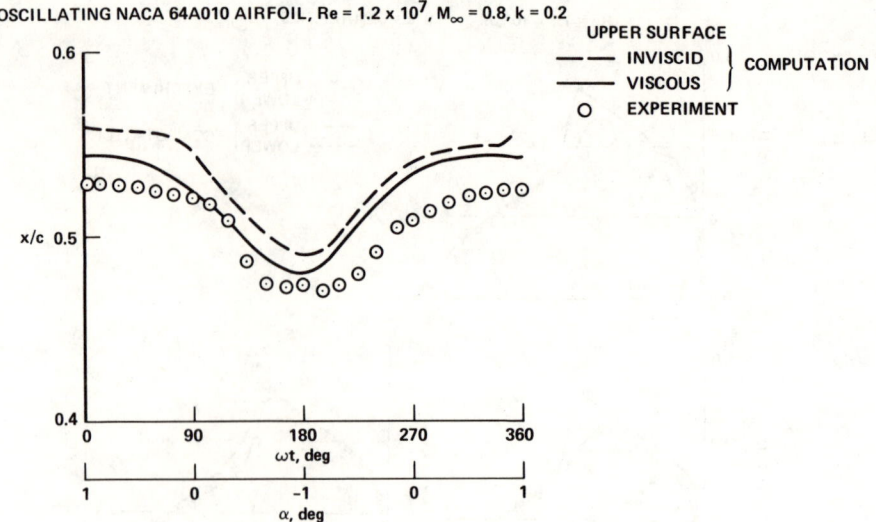

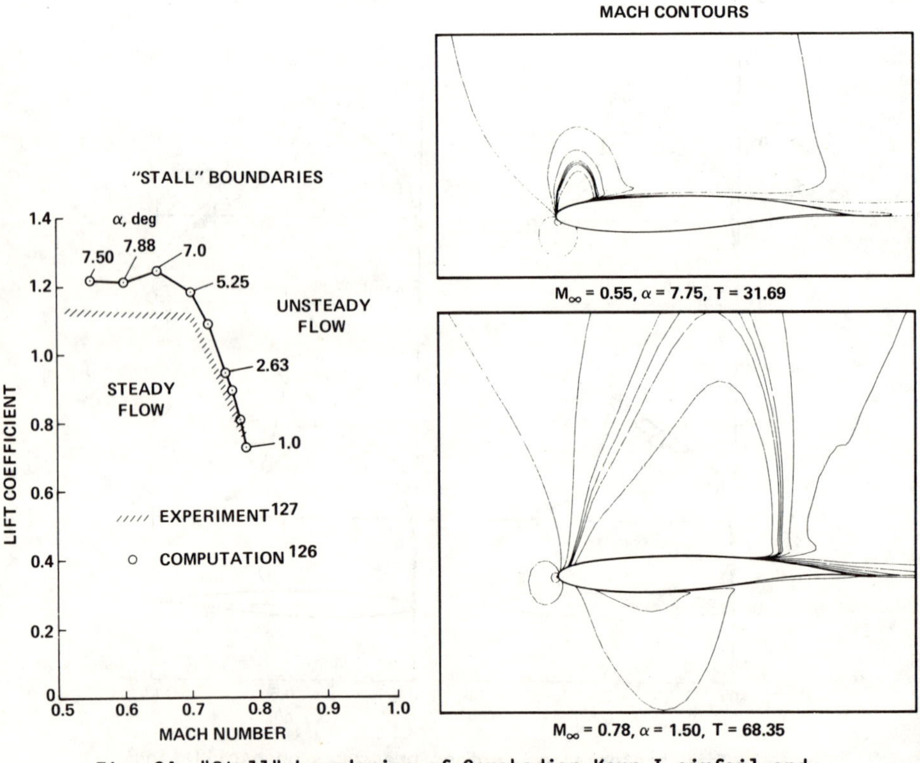

Fig. 24 "Stall" boundaries of Garabedian-Korn I airfoil and typical Mach contours at Re = 21 × 10^6.

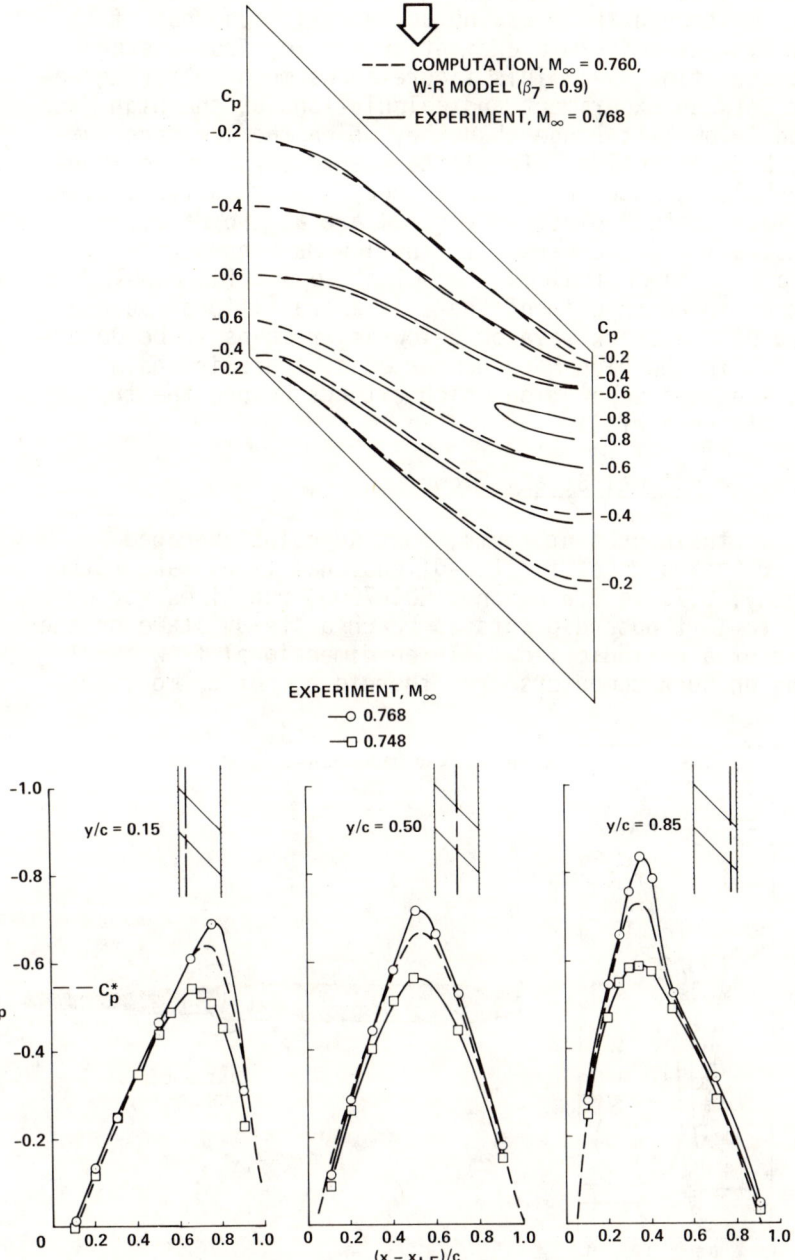

Fig. 25 Pressure distributions on a 45 deg swept wing.

flow boundaries and computed Mach contours. (Investigators of these boundaries have called them buffet boundaries, although there was no aeroelastic response of the airfoil to aerodynamic excitation arising from unsteady separated flow.[128]) This figure shows much better agreement between experiment and calculations at the high Mach number, low-lift range than they do on the low Mach number, high-lift side. The latter represents a case where a turbulence model has been pushed far beyond its limits. The Mach contour plots in Fig. 24 are at two different freestream Mach numbers. In the low Mach number case, there is a shock-induced, turbulent separation bubble. Whether in an experiment there is a transitional bubble ahead of the shock wave or below it, remains to be determined. In the high Mach number case, there is again shock-induced separation which extends beyond the trailing edge of the airfoil.

Three-Dimensional Steady Flows

In their present forms, most Reynolds averaged Navier-Stokes codes for two-dimensional flows take rather lengthy, 0.75 to 3.5 h on a CDC 7600, run times for grids of 4 to 10 thousand points to reach a steady state or the onset of a periodic flow. Three-dimensional flow simulations on such computers are, therefore, not common. On

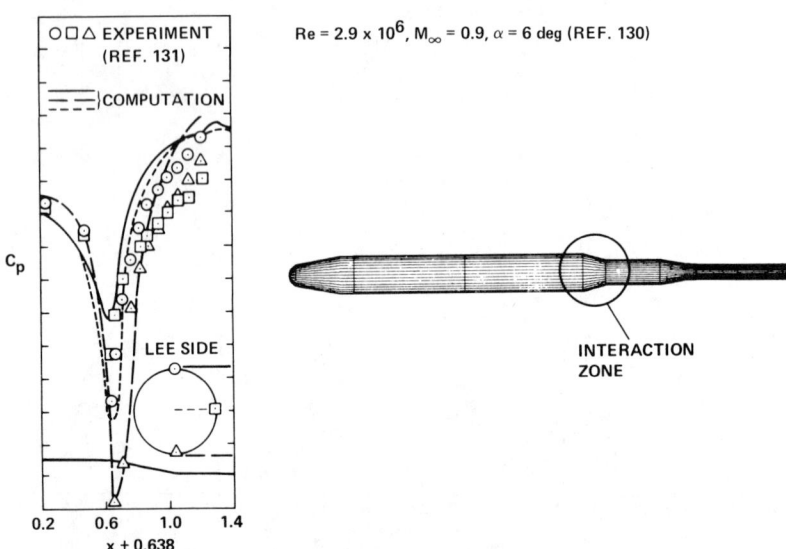

Fig. 26 Comparison of computed and experimental pressure distributions on a boattail afterbody.

the so-called class VI computers, such as the ILLIAC IV, however, some three-dimensional studies with moderate resolution are practical at a research level.

Surface pressure isobars for a subcritical unseparated flow over a 45 deg swept, 10% thick circular arc airfoil at a zero incidence and spanning a tunnel are shown in Fig. 25. The second-order thin shear layer approximation was used with the two-equation W-R turbulence model, and only the upper half of the flowfield was computed.

The computational and experimental Mach numbers are slightly different. Bertelrud et al.[129] have explained this difference between the Mach numbers as follows: the

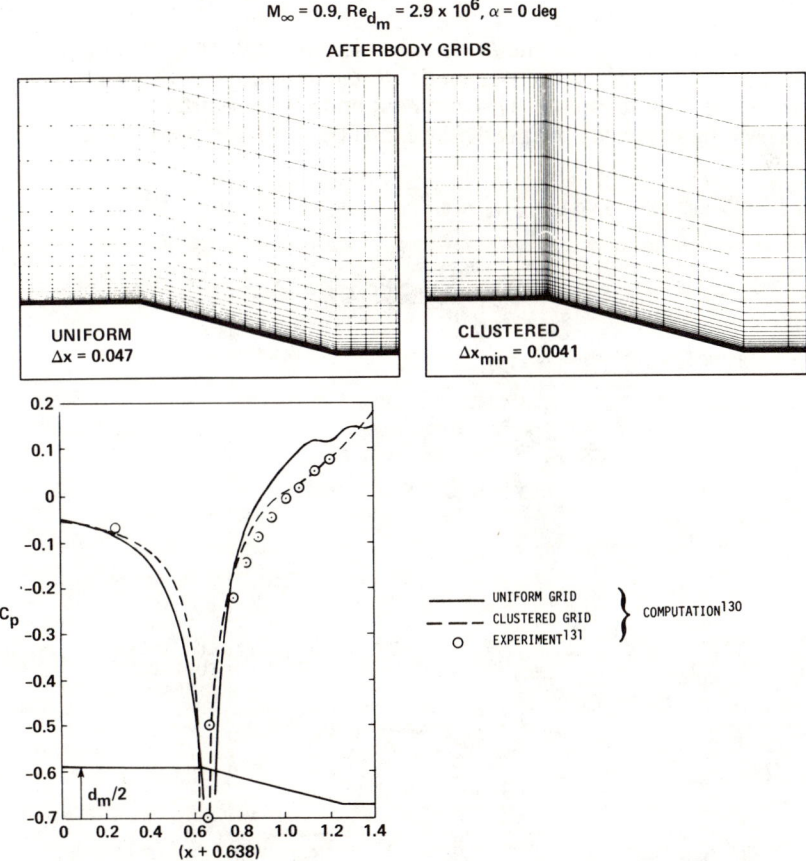

Fig. 27 Influence of grid near forebody/afterbody juncture on a boattail afterbody.

reference Mach number and pressure values for the experimental isobars were obtained at a location nearly one chord length ahead of the wing leading edge at the left wall of the channel. The computational boundary was located at 3.5 chord lengths ahead of same leading edge of the wing. Therefore, the computed state at the measuring location did not correspond to the measured state at that location. Figure 25 shows a comparison of measured and computed pressure distributions at three spanwise locations on the wing surface, and it gives the Mach number sensitivity.

Simulations of three-dimensional boattail afterbody flowfields have been obtained by Deiwert[130] with the second-order thin shear layer approximation and the same algebraic turbulence model used in the buzz study discussed above. In Fig. 26, surface pressure distributions are shown for a boattail model used by Shrewsbury.[131] The experimental data are shown in the insert by the triangles, squares, and circles corresponding to windward, lateral, and leeward positions. The corresponding computed results are shown by dashed, dotted, and solid lines. The junc-

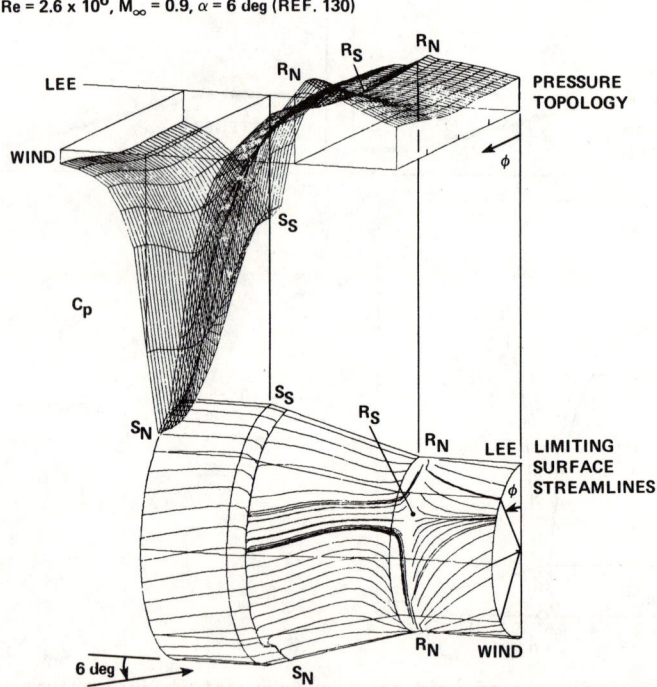

Fig. 28 Pressure topology and flow patterns on a boattail afterbody.

tion of the forebody and afterbody of the above boattail model is sharp. Deiwert[130] has reported some sensitivity of the computed results to the grid spacing in the vicinity of this junction (Fig. 27). Figure 28 shows computed results. The upper part is surface pressure topology and the lower one is a limiting surface flow pattern (surface shear directions) which approximates a surface oil flow pattern. The symbols S and R, respectively, stand for flow separation and flow reattachment, and the subscripts S and N, respectively, denote a saddle point and a node point. Downstream of the circumferential line S_S S_N, the flow is separated. Downstream of the circumferential line R_N R_S R_N, the flow is attached. The direction flow is from S_S to S_N and from R_N to R_S. Such details are available from present Navier-Stokes technology, and they are of considerable use towards a better understanding of complex flowfields and towards providing internal consistency checks for simulations.

Concluding Remarks

The Navier-Stokes technology is currently under vigorous development. It has opened new possibilities of simulating unsteady, separated, turbulent, compressible flows that were not accessible five years ago. In this paper we have presented the state-of-the-art, as we envision it. The primary utility of this technology is in applications where the present viscous-inviscid interaction computations fail, and this generally occurs in simulating separated flows that are nominally two-dimensional and unsteady, or three-dimensional steady or unsteady. There is little doubt that the Navier-Stokes technology can be of use to the aircraft designers and developers; the question is how much and when.

Over the past 10 years substantial advances have been made in computer speed and memory. Further, our ability to compute flows in rather complex geometries has greatly improved. It is becoming increasing clear that turbulence modeling in the regions of separation, which are not "small," is the weakest part of the Navier-Stokes technology. In fact, it is rapidly becoming the primary pacing item for Reynolds averaged Navier-Stokes, transonic flow simulations.

In the 1970s most of the work was done to prove the capability of simulating turbulent flows with mild separation. Many turbulence models were made to work on isolated experiments. In fact this approach was used to develop both the numerical techniques and a few empirical constants in the models. However, very little has been done to establish the reliability of a code, as distinct from a "model" when it is applied to a variety of experiments. The same turbulence "model" can give different results when used in different codes with the same or different numerical methods. This is due principally to lack of grid refinement studies.

Numerical simulations of the Reynolds averaged Navier-Stokes equations are, in general, predictive for attached boundary layers. Zero-equation models have been very useful in engineering analysis of these flows, but they must be interpreted with caution when used to approximate separated flows and flows with strong curvature effects. Simplicity of zero-equation models require more adjustment for separated flows; complex models, which contain more empirical constants, need less adjustment. From the results available at this time, however zero-equation models are judged, there is no clear evidence to show that one- or two-equation, first-order (eddy viscosity) models are much better.

One of the problems in constructing models for external separated flows is due to the fact that very little is known of the behavior of turbulence in such flows.[132,133] We do know, for instance, that in separated flows normal stresses are anisotropic and turbulence structure is not in equilibrium. Relaxation procedures and transport equations for turbulent scales can take into consideration some of the history effects, namely, the nonequilibrium nature of turbulence; but the Reynolds stress tensor is modeled to respond instantly to changes in mean strain field [Eq. (10)]. The first-order models, therefore, can be truly predictive only for flows in which turbulence is nearly in local equilibrium or for self-preserving flows. The second-order (stress equation) models are required for nonequilibrium flows. This is illustrated below.

Consider a distortion of a flowfield of fully developed, homogeneous turbulence by application of plane strain

(Fig. 29). This experiment acts as a test of turbulence models in separated flows when near surface effects are absent. The fluid is conditioned through screens, and it becomes parallel when it reaches the station where the constant rate of strain is applied. The subsequent straining of the fluid causes the initially nearly isotropic turbulence to become anisotropic. A measure of anisotropy is plotted as the ordinate, the lower portion of Fig. 29. At some distance downstream, the strain is removed and the fluid returns to parallel flow. The measurements of Tucker and Reynolds[134] are compared with computed results of Wilcox and Rubesin.[30] The computation with a second-order (Reynolds stress) model gives a better agreement with the measured values than that of the first-order (W-R) model. Although Wilcox and Rubesin modified Eq. (10) to remove the alignment of the Reynolds stress tensor and the mean rates of strain for the W-R model, the predicted return to isotropy is abrupt when the strain is removed. This kind of behavior is brought about by the shortcoming of the first-order models as explained.

The previous example illustrates two points. First, eddy viscosity models can probably never be completely predictive for separated flows. Some details of the structure will most certainly be lacking. More sophisticated models will pick up more of the details, but for stringent requirements they, too, may fail. The second point, and by far the most important one, is that it is probably possible to predict the gross behavior of a flow even when certain of the details are not well represented or even missing altogether. The most meaningful test of whether or not these calculations have useful information is whether or not they are used.

There are two schools of thought about modeling turbulence.[135] Some believe that under certain circumstances, rational second-order (or invariant) modeling can be developed for general computation procedures. They consider that this approach may at least provide a guide for the construction of the more empirical models. Others believe the structure of turbulence to be so complex that a search for universal closures is probably in vain. They believe that practical computations will require empirical techniques developed for particular flow topology. As for the current efforts in computing turbulence flows for industrial needs, Liepmann[136] has presented an adversely critical opinion.

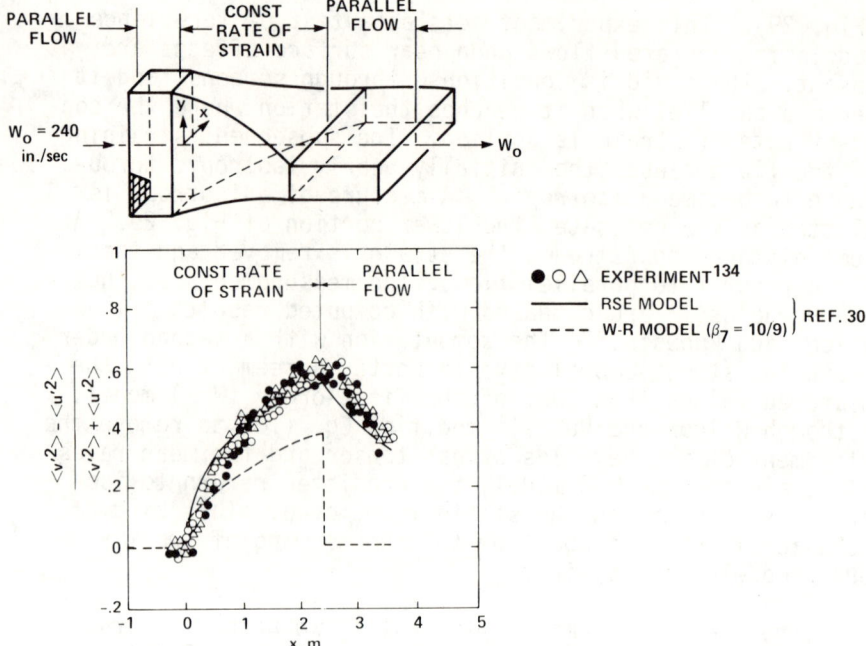

Fig. 29 Normally strained homogeneous flow, comparison of computed and experimental anisotropy.

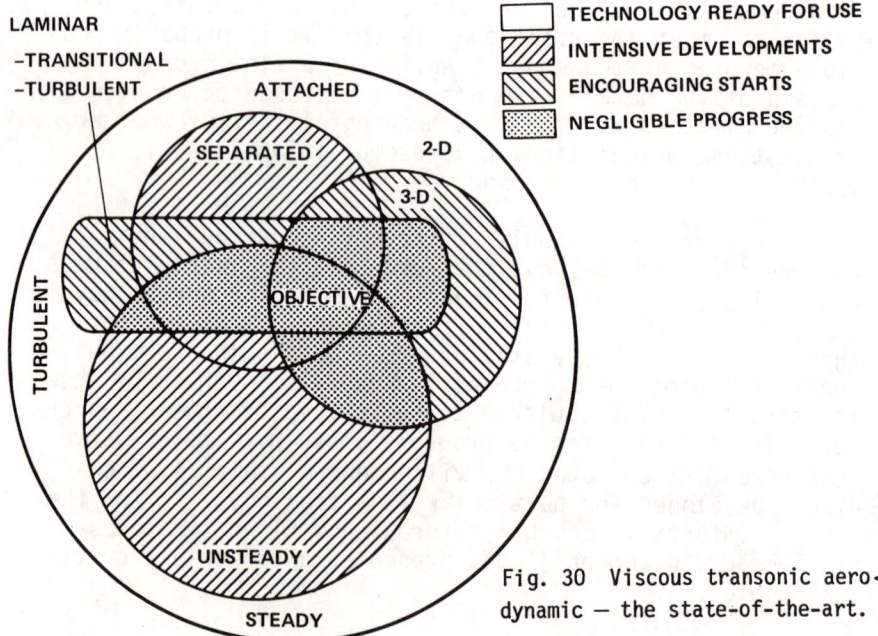

Fig. 30 Viscous transonic aerodynamic — the state-of-the-art.

There are probably five different parallel avenues of turbulent flow research: 1) different turbulence models are applied to the same geometrical flow problem in order to determine which one is the best; 2) the same model, without any change in its form or in its empirical constants, is applied to different geometrical flow problems so that its breadth of application can be determined; 3) for a given form of a model, a computer optimization is carried out to obtain the best set of model parameters relative to an available set of experiments; 4) for a specific flow problem, a determination of the range of flow parameters is carried out for which a given model with its empirical constants is valid; and 5) a model is developed for a particular flow problem based on a detailed experimental characterization of this flow. As demonstrated earlier, elements 1) and 2) are already being pursued; remaining avenues need to be pursued.

Computational aerodynamics is probably going to depend more on experimental inputs and checks and less on the solutions of the Navier-Stokes equations for developing turbulence models. Therefore, experimenters should be requested to document well the experiments they conduct during their quest for understanding turbulence in separated flows. Through that understanding, better turbulence models may result at least for these flows, and this should increase the utility of the Navier-Stokes technology. Further, the detailed documentation of experiments would facilitate use of experimental boundary conditions in computations. Both experimental and theoretical investigators need to work together to advance the state-of-the-art of turbulence models for separated flows. It is hoped that efforts will be devoted to extensive testing of these models on a variety of experiments without modifications to the basic coding. In addition, repeated grid-refinement studies are required to demonstrate that a turbulent numerical simulation tends to be independent of numerics.

In the 1980s, the complex three-dimensional geometries will require component adaptive methods. These procedures, along with limited availability of computed speed and memory, will guide the Navier-Stokes technology towards a viscous-inviscid interaction approach, which probably will consist of matching the Reynolds averaged Navier-Stokes solutions next to a body surface with either Euler or potential flow solutions away from the surface.

If the above efforts prove to work then not only capability but reliability would be established. At this point the Navier-Stokes technology will come of age.

In summary, the state-of-the-art of viscous transonic aerodynamics is presented in a Venn diagram illustrated in Fig. 30. At present, transonic, attached, two-dimensional, steady and fully turbulent flows can be routinely predicted. Extensive efforts are being made to predict both steady and unsteady, two-dimensional fully turbulent separated flows. Already promising starts have been made to simulate steady three-dimensional flows, either attached or separated. However, much remains to be done for laminar-transitional-turbulent flows. Further, there is negligible progress in meeting the final objective of predicting unsteady, three-dimensional, separated, and laminar-transitional-turbulent flows.

References

[1]Favre, A., "Equations des Gaz Turbulents Compressibles," Journal de Méchanique, Vol. 4, 1965, pp. 361-390.

[2]Computation of Viscous-Inviscid Interactions, AGARD CP-291, 1980.

[3]Le Balleur, J. C., "Computation of Flows Including Strong Viscous Interactions with Coupling Methods," AGARD CP-291, 1980.

[4]Melnik, R. E., "Turbulent Interactions on Airfoils at Transonic Speeds — Recent Developments," AGARD CP-291, 1980.

[5]Finke, K., "Unsteady Shock Wave Boundary Layer Interaction on Profiles in Transonic Flow," AGARD CP-168, 1975.

[6]McCroskey, W. J., McAlister, K. W., Carr, L. W., Pucci, S. L., Lambert, O., and Indergand, R. F., "Dynamic Stall on Advanced Airfoil Sections," Journal of American Helicopter Society, Vol. 26, 1981, pp. 40-50.

[7]Solonnikov, V. A. and Kazhikhov, A. V., "Existence Theorems for the Equations of Motion of a Compressible Viscous Fluid," Annual Review of Fluid Mechanics, Vol. 13, 1981, pp. 79-95.

[8]Matsumura, A. and Nishida, T., "The Initial Value Problem for the Equations of Motion of Viscous and Heat-Conductive Gases," Journal of Mathematics, Kyoto University, Vol. 20, 1980, pp. 67-104.

[9]Lax, P. D., "Hyperbolic Systems of Conservation Laws and the Mathematical Theory of Shock Waves," SIAM, 1973.

[10]Monin, A. S. and Yaglom, A. M., Statistical Fluid Mechanics of Turbulence, The MIT Press, Cambridge, Mass., Vol. 1, 1971, pp. 205-209.

[11] Van Driest, E. R., "Turbulent Boundary Layers in Compressible Fluids," *Journal of Aeronautical Sciences*, Vol. 18, 1951, pp. 145-160.

[12] Hesselberg, Th., "Die Gesetze der Ausgeglichene Atmosharischen Bewegungen," *Beiträge Physik freien Atmosphären*, Vol. 12, 1926, pp. 141-160.

[13] Favre, A., "Statistical Equations of Turbulent Gases, Problems of Hydrodynamics and Continuum Mechanics, SIAM," 1969.

[14] Cebeci, T. and Smith, A. M. O., *Analysis of Turbulent Boundary Layers*, Academic Press, Inc., New York, 1974.

[15] Morkovin, M. V., "Effects of Compressibility on Turbulence Flow," *The Mechanics of Turbulence*, ed. Favre, A., Gordon and Breach, New York, 1964.

[16] Bradshaw, P., "Compressible Turbulent Shear Layers," *Annual Review of Fluid Mechanics*, Vol. 9, 1977, pp. 33-54.

[17] Rodi, W., *Turbulence Models and Their Application in Hydraulics*, International Association for Hydraulic Research, Deft, The Netherlands, 1980.

[18] Launder, B. E. and Spalding, D. B., *Mathematical Models of Turbulence*, Academic Press, Inc., New York, 1972.

[19] Reynolds, W. C., "Computation of Turbulent Flows," *Annual Review of Fluid Mechanics*, Vol. 8, 1976, pp. 183-208.

[20] Reynolds, W. C. and Cebeci, T., "Calculation of Turbulent Flows," *Topics in Applied Physics*, Vol. 12, Turbulence, ed. Bradshaw, P., 1976, pp. 193-229.

[21] Rubesin, M. W., "Numerical Turbulence Modelling," AGARD LS-86, 1977.

[22] Launder, B. E., "Turbulence Transport Models for Numerical Computation of Complex Turbulent Flows," VKI Lecture Series 1980-3.

[23] Rodi, W., "Progress in Turbulence Modelling for Incompressible Flows," Paper No. 81-0045 presented at AIAA 19th Aerospace Sciences Meeting, St. Louis, Mo., Jan. 1981.

[24] Boussinesq, J., "Theorie de l'écoulement Tourbillant," Memoires Presentes par Divers Savants Sciences Mathématique et Physiques, Académie des Sciences, Paris, Vol. 23, 1877, p. 46.

[25] Corrsin, S., "Limitations of Gradient Transport Models in Random Walks and in Turbulence," *Advances in Geophysics*, Vol. 18A, 1974, pp. 25-60.

[26] Saffman, P. G., "Model Equations for Turbulent Shear Flow," *Studies in Applied Mathematics*, Vol. 53, 1974, pp. 17-34.

[27] Baldwin, B. S. and Lomax, H., "Thin Layer Approximation and Algebraic Model for Separated Turbulent Flows," Paper No. 78-257 presented at AIAA 16th Aerospace Sciences Meeting, Huntsville, Ala., Jan. 1978.

[28] Cebeci, T., "Calculation of Compressible Turbulent Boundary Layers with Heat and Mass Transfer," AIAA Journal, Vol. 9, 1971, pp. 1091-1097.

[29] Van Driest, E. R., "On Turbulent Flow Near a Wall," Journal of Aeronautical Sciences, Vol. 23, 1956, pp. 1007-1011.

[30] Wilcox, D. C. and Rubesin, M. W., "Progress in Turbulence Modeling for Complex Flow Fields Including Effects of Compressibility," NASA Technical Paper 1517, 1980.

[31] Wilcox, D. C. and Traci, R. M., "A Complete Model of Turbulence," Paper No. 76-351 presented at AIAA 9th Fluid and Plasma Dynamics Conference, San Diego, Calif., July 1976.

[32] Saffman, P. G. and Wilcox, D. C., "Turbulence-Model Predictions for Turbulent Boundary Layers," AIAA Journal, Vol. 12, 1974, pp. 541-546.

[33] Saffman, P. G., "A Model for Inhomogeneous Turbulent Flow," Proceedings of the Royal Society (London), Series A, Vol. 317, 1970, pp. 417-433.

[34] Moretti, G., "Numerical Analysis of Compressible Flow: An Introspective Survey," Paper No. 79-1510 presented at 12th Fluid and Plasma Dynamics Conference, Williamsburg, Va., July 1979.

[35] Lax, P. D., "Hyperbolic Systems of Conservation Laws II," Communications in Pure and Applied Mathematics, Vol. 10, 1957, pp. 537-566.

[36] Richtmyer, R. D. and Morton, K. W., Difference Methods for Initial-Value Problems, Interscience Publishers, New York, 1967.

[37] Vinokur, M., "Conservation Equations of Gasdynamics in Curvilinear Coordinate Systems," Journal of Computational Physics, Vol. 14, 1974, pp. 105-125.

[38] Eiseman, P. R. and Stone, A. P., "Conservation Laws of Fluid Dynamics — A Survey," SIAM Review, Vol. 22, 1980, pp. 12-27.

[39] Viviand, H., "Conservation Forms of Gas Dynamic Equations," La Recherche Aérospatiale, Vol. 1, 1974, pp. 153-158.

[40] Lax, P. D., "Weak Solutions of Nonlinear Hyperbolic Equations and Their Numerical Computation," Communications in Pure and Applied Mathematics, Vol. 7, 1954, pp. 159-193.

[41] Steger, J. L., "Implicit Finite-Difference Simulation of Flow About Arbitrary Geometries," AIAA Journal, Vol. 16, 1978, pp. 679-686.

[42]Pullium, T. H. and Steger, J. L., "Implicit Finite-Difference Simulations of Three-Dimensional Compressible Flow," AIAA Journal, Vol. 18, 1980, pp. 159-167.

[43]Elvius, T. and Sundström, A., "Computational Efficient Schemes and Boundary Conditions for a Fine-Mesh Baratropic Model Based on the Shallow-Water Equations," Tellus, Vol. 25, 1973, pp. 132-156.

[44]Serrin, J., "On the Uniqueness of Compressible Fluid Motions," Archives for Rational Mechanics and Analysis, Vol. 3, 1959, pp. 271-288.

[45]Kreiss, H. O., "Initial Boundary Value Problems for Hyperbolic Equations," Communications in Pure Applied Mathematics, Vol. 23, 1970, pp. 277-298.

[46]Oliger, J. and Sundström, A., "Theoretical and Practical Aspects of Some Initial Boundary Value Problems in Fluid Dynamics," SIAM Journal of Applied Mathematics, Vol. 35, 1978, pp. 419-446.

[47]Gustafsson, B. and Sundström, A., "Incompletely Parabolic Problems in Fluid Dynamics," SIAM Journal of Applied Mathematics, Vol. 35, 1978, pp. 343-357.

[48]Belov, Y. Y. and Yanenko, N. N., "Influence of Viscosity on the Smoothness of Solutions of Incompletely Parabolic Systems," Mathematical Notes, Academy of Sciences, USSR, Vol. 10, 1971, pp. 480-483.

[49]Sears, W. R., "On the Definition of Free-Stream Conditions in Wind-Tunnel Testing," Symposium on Numerical and Physical Aspects of Aerodynamics Flows, California State University, Long Beach, Calif., Jan. 1981.

[50]Mehta, U. and Lavan, Z., "Starting Vortex, Separation Bubbles, and Stall — A Numerical Study of Laminar Unsteady Flow Around an Airfoil," Journal of Fluid Mechanics, Vol. 67, 1975, pp. 227-256.

[51]Mehta, U., "Dynamic Stall of an Oscillating Airfoil," AGARD CP-227, 1977.

[52]Gelinas, R. J., Doss, S. K., and Miller, K., "The Moving Finite Element Method: Application to General Partial Differential Equations with Multiple Large Gradients," Journal of Computational Physics, Vol. 40, 1981, pp. 202-249.

[53]Dwyer, H. A., Raiszadeh, F., and Otey, G., "A Study of Reactive Diffusion Problems with Stiff Integrators and Adaptive Grids," Lecture Notes in Physics, Vol. 141, Seventh International Conference on Numerical Methods in Fluid Dynamics, 1981, pp. 170-175. (See also Dwyer, H. A., Kee, R. J., and Sanders, B. R., "Adaptive Grid Method for Problems in Fluid Mechanics and Heat Transfer," AIAA Journal, Vol. 18, 1980, pp. 1205-1212.)

[54]Kovenya, V. M. and Yanenko, N. N., "Numerical Method for the Viscous Gas Equations on Moving Grids," Computers and Fluids, Vol. 8, 1980, pp. 59-70.

[55]Pierson, B. L. and Kutler, P. K., "Optimal Nodal Point Distribution for Improved Accuracy in Computational Fluid Dynamics," AIAA Journal, Vol. 18, 1980, pp. 49-54.

[56]Rai, M. M. and Anderson, D. A., "Grid Evolution in Time Asymptotic Problems," NASA CP-2166, 1980, pp. 409-430.

[57]Sorenson, R. L., "A Computer Program to Generate Two-Dimensional Grids About Airfoils and Other Shapes By the Use of Poisson's Equation," NASA TM 81198, 1980.

[58]Sells, C. C. L., "Plane Subcritical Flow Past a Lifting Airfoil," Proceedings of the Royal Society (London), Series A, Vol. 308, 1968, pp. 377-401.

[59]Ives, D. C., "A Modern Look at Conformal Mapping Including Multiply Connected Regions," AIAA Journal, Vol. 14, 1976, pp. 1006-1111.

[60]Moretti, G., "Grid Generation Using Classical Techniques," NASA CP-2166, 1980, pp. 1-36.

[61]Eiseman, P. R., "A Coordinate System for a Viscous Transonic Cascade Analysis," Journal of Computational Physics, Vol. 26, 1978, pp. 307-338.

[62]Eiseman, P. R., "A Multi-Surface Method of Coordinate Generation," Journal of Computational Physics, Vol. 33, 1979, pp. 118-150.

[63]Eriksson, L. E., "Three-Dimensional Spline-Generated Coordinate Transformations for Grids Around Wing-Body Configurations," NASA CP-2166, 1980, pp. 253-264.

[64]Forcey, C. R., Edwards, M. G., and Carr, M. P., "An Investigation of Grid Patching Techniques," NASA CP-2166, 1980, pp. 265-294.

[65]Barfield, W. D., "An Optimal Mesh Generator for Lagrangian Hydrodynamic Calculations in Two Space Dimensions," Journal of Computational Physics, Vol. 6, 1970, pp. 417-429.

[66]Starius, G., "Constructing Orthogonal Curvilinear Meshes by Solving Initial Value Problems," Numerische Mathematik, Vol. 28, 1977, pp. 25-48.

[67]Steger, J. L. and Chaussee, D. S., "Generation of Body Fitted Coordinates Using Hyperbolic Partial Differential Equations," SIAM Journal of Scientific and Statistical Computing, Vol. 1, 1980, pp. 431-437.

[68] Thompson, J. F., Thames, F. C., and Mastin, C. W., "Automatic Numerical Generation of Body-Fitted Curvilinear Coordinate System for Field Containing Any Number of Arbitrary Two-Dimensional Bodies," Journal of Computational Physics, Vol. 15, 1974, pp. 299-319.

[69] Godunov, S. K. and Prokopov, G. P., "The Use of Moving Meshes in Gas-Dynamical Computations," USSR Computational Mathematics and Mathematical Physics, Vol. 12, 1972, p. 182.

[70] Amsden, A. A. and Hirt, C. W., "A Simple Scheme for Generating General Curvilinear Grids," Journal of Computational Physics, Vol. 11, 1973, pp. 348-359.

[71] Mastin, C. M. and Thompson, J. F., "Transformation of Three-Dimensional Regions onto Rectangular Regions by Elliptic Systems," Numerische Mathematik, Vol. 29, 1978, pp. 397-407.

[72] Lee, K. D., Huang, N. J. Y., and Rubbert, P. E., "Grid Generation for General Three-Dimensional Configurations," NASA CP-2166, 1980, pp. 355-366.

[73] Lee, K. D. and Rubbert, P. E., "Transonic Flow Computations using Grid Systems with Block Structure," Lecture Notes in Physics, Vol. 141, Seventh International Conference on Numerical Methods in Fluid Dynamics, 1981, pp. 266-271.

[74] Forrest, A. R., "Computational Geometry," Proceedings of the Royal Society (London), Series A, Vol. 321, 1971, pp. 187-195.

[75] Faux, I. D. and Pratt, M. J., Computational Geometry For Design and Manufacture, Ellis Horwood Limited, Halsted Press, 1979.

[76] Atta, E., "Component-Adaptive Grid Embedding," NASA CP-2166, 1980, pp. 157-174.

[77] Eiseman, P. R. and Smith, R. E., Jr., "Grid Generation Using Algebraic Techniques," NASA CP-2166, 1980, pp. 73-120.

[78] Tennekes, H. and Lumley, J. L., A First Course in Turbulence, The MIT Press, 1972.

[79] Chapman, D. R., "Computational Aerodynamics Development and Outlook," AIAA Journal, Vol. 17, 1979, pp. 1293-1313.

[80] Grant, H. L., Stewart, R. W., and Moillet, A., "Turbulence Spectra from Tidal Channel," Journal of Fluid Mechanics, Vol. 12, 1962, pp. 241-268.

[81] Gibson, M. M., "Spectra of Turbulence in a Round Jet," Journal of Fluid Mechanics, Vol. 15, 1963, pp. 161-173.

[82] Laufer, J., "The Structure of Turbulence in Fully Developed Pipe Flow," NACA Rept. 1174, 1954.

[83]Champagne, F. H., Harris, V. G., and Corrsin, S., "Experiments on Nearly Homogeneous Turbulent Shear Flow," Journal of Fluid Mechanics, Vol. 41, 1970, pp. 81-139.

[84]Uberoi, M. S. and Freymuth, P., "Spectra of Turbulence in Wakes Behind Circular Cylinders," Physics of Fluids, Vol. 12, 1969, pp. 1359-1363.

[85]Kistler, A. L. and Vrebalovich, T., "Grid Turbulence at Large Reynolds Numbers," Journal of Fluid Mechanics, Vol. 26, 1966, pp. 37-47.

[86]Compte-Bellot, G. and Corrsin, S., "Simple Eulerian Time Correlation at Full- and Narrow-Band Velocity Signals in Grid-Generated 'Isotropic' Turbulence," Journal of Fluid Mechanics, Vol. 48, 1971, pp. 273-337.

[87]Sanborn, V. A. and Marshall, R. D., "Local Isotropy in Wind Tunnel Turbulence," Colorado State University, Rept. CER 65 UAS-RDM71, 1965.

[88]Tieleman, H. W., "Viscous Region of Turbulent Boundary Layer," Colorado State University, Rept. CER 67-68HWT21, 1967.

[89]Coantic, M. and Favre, A., "Activities in, and Preliminary Results of, Air-Sea Interactions Research at I.M.S.T.," Advances in Geophysics, Vol. 18A, 1974, pp. 391-405.

[90]Kutler, P., "Computation of Three-dimensional, Inviscid Supersonic Flows," Lecture Notes in Physics, Vol. 41, 1974, pp. 287-374.

[91]De Neef, T. and Moretti, G., "Shock Fitting for Everybody," Computers and Fluids, Vol. 8, 1980, pp. 327-334.

[92]Warming, R. F. and Hyett, B. J., "The Modified Equation Approach to the Stability and Accuracy Analysis of Finite-Difference Methods," Journal of Computational Physics, Vol. 14, 1974, pp. 159-179.

[93]Lerat, A., "Numerical Shock Structure and Nonlinear Corrections for Difference Schemes in Conservation Forms," Lecture Notes in Physics, Vol. 90, 1979, pp. 345-351.

[94]Von Neumann, J. and Richtmyer, R. D., "A Method for Numerical Calculations of Hydrodynamical Shocks," Journal of Applied Physics, Vol. 21, 1950, p. 232.

[95]MacCormack, R. W. and Baldwin, B. S., "A Numerical Method for Solving the Navier-Stokes Equations with Application to Shock-Boundary Layer Interactions," AIAA Paper 75-1, 1975.

[96]Warming, R. W. and Beam, R. M., "Upwind Second-Order Difference Schemes and Applications in Aerodynamic Flows," AIAA Journal, Vol. 14, 1976, pp. 1241-1249.

[97]Briley, W. R. and McDonald, H., "Solution of the Multidimensional Compressible Navier-Stokes Equations by a Generalized Implicit Method," Journal of Computational Physics, Vol. 24, 1977, pp. 372-397.

[98]MacCormack, R. W., "The Effect of Viscosity in Hypervelocity Impact Cratering," Paper No. 69-354 presented at AIAA Hypervelocity Impact Conference, Cincinnati, Ohio, April 1969.

[99]Beam, R. and Warming, R. F., "An Implicit Factored Scheme for the Compressible Navier-Stokes Equations," AIAA Journal, Vol. 16, 1978, pp. 393-402.

[100]MacCormack, R. W., "An Efficient Explicit-Implicit-Characteristic Method for Solving the Compressible Navier-Stokes Equations," SIAM-AMS Proceedings, Vol. 11, 1978, pp. 130-155.

[101]Shang, J., "Implicit-Explicit Method for Solving the Navier-Stokes Equations," AIAA Journal, Vol. 16, 1978, pp. 496-502.

[102]Deiwert, G. S., "Numerical Simulation of High Reynolds Number Transonic Flows," AIAA Journal, Vol. 13, 1975, pp. 1354-1359.

[103]Viegas, J. R. and Horstman, C. C., "Comparison of Multiequation Turbulence Models for Several Separated Boundary-Layer Interaction Flows," AIAA Journal, Vol. 17, 1979, pp. 811-820.

[104]Coakley, T. J. and Bergmann, M. Y., "Effects of Turbulence Model Selection on the Prediction of Complex Aerodynamic Flows," Paper No. 79-0070 presented at AIAA 7th Aerospace Sciences Meeting, New Orleans, La., January 1979.

[105]McDevitt, J. B., Levy, L. L., Jr., and Deiwert, G. S., "Transonic Flow about a Thick Circular-Arc Airfoil," AIAA Journal, Vol. 14, 1976, pp. 606-613.

[106]Levy, L. L., Jr., "Experimental and Computational Steady and Unsteady Transonic Flows about a Thick Airfoil," AIAA Journal, Vol. 16, 1978, pp. 564-572.

[107]Jones, W. P. and Lauder, B. E., "The Prediction of Laminarization with a Two-Equation Model of Turbulence," International Journal of Heat Mass Transfer, Vol. 15, 1972, pp. 301-314.

[108]Archarya, M., "Effects of Compressibility on Boundary-Layer Turbulence," AIAA Journal, Vol. 15, 1977, pp. 303-304.

[109]Johnson, D. A., Horstman, C. C., and Bachalo, W. D., "A Comprehensive Comparison Between Experiment and Prediction for a Transonic Turbulent Separated Flow," to appear in AIAA Journal, 1982.

[110]Deiwert, G. S., "Computation of Separated Transonic Turbulent Flows," AIAA Journal, Vol. 14, 1976, pp. 735-740.

[111] Coakley, T. J., "Numerical Method for Gas Dynamics Combining Characteristic and Conservation Concepts," Paper No. 81-1257 presented at AIAA 14th Fluid and Plasma Dynamics Conference, Palo Alto, Ca., June 1981.

[112] Garabedian, P. R. and Korn, D. G.,"Numerical Design of Transonic Airfoils,"Numerical Solution of Partial Differential Equations, Academic Press, Inc., New York, Vol. 2, 1971, pp. 253-271.

[113] Deiwert, G. S., private communication, 1977.

[114] Melnik, R. E., "Recent Developments in a Boundary Layer Theory for Computing Viscous Flows Over Airfoils," USAF/FRG Data Exchange Meeting, Meersberg, Germany, BMVg-FBWT 79-31, 1979.

[115] Kacprzynski, J. J., Ohman, L. H., Garabedian, P. R., and Korn, D. G., "Analysis of the Flow Past a Shockless Lifting Airfoil in Design and Off-design Conditions," NRC, Aeronautical Report L-554, 1971. (See also Kacprzynski, J. J., "A Second Series of Wind Tunnel Tests of the Shockless Lifting Airfoil," No. 1. NRC/NAE Wind Tunnel Project Report 5x5/0062, 1972.)

[116] Morky, M. and Ohman, L. H., "Application of the Fast Fourier Transform to Two-Dimensional Wind Tunnel Wall Interference," Journal of Aircraft, Vol. 17, 1980, pp. 402-408.

[117] Deiwert, G. S., "On the Prediction of Viscous Phenomena in Transonic Flows," Transonic Flow Problems in Turbomachinery, eds. Adamson, T. C., Jr. and Platzer, M. F., Hemisphere Publishing Corp., New York, 1977, pp. 371-391.

[118] Seegmiller, H. L., Marvin, J. G., and Levy, L. L., Jr., "Steady and Unsteady Transonic Flows," AIAA Journal, Vol. 16, 1978, pp. 1261-1270.

[119] Marvin, J. G., Levy, L. L., Jr., and Seegmiller, H. L., "Turbulence Modelling for Unsteady Transonic Flows," AIAA Journal, Vol. 18, 1980, pp. 489-496.

[120] Mabey, D. G., Welsh, B. L., and Cripps, B., "Periodic Flows on a 14% Thick Biconvex Wing at Transonic Speeds," RAE Unpublished Report, 1980.

[121] Levy, L. L., Jr., "Predicted and Experimental Steady and Unsteady Transonic Flows about a Biconvex Airfoil," NASA TM 81262, 1981.

[122] Steger, J. L. and Bailey, H. E., "Calculation of Transonic Aileron Buzz," AIAA Journal, Vol. 18, 1980, pp. 249-255.

[123] Erikson, A. L. and Stephenson, J. D., "A Suggested Method of Analyzing for Transonic Flutter of Control Surfaces Based on Available Experimental Evidence," NACA RM A7F30, 1947.

[124] Tijdeman, H. and Seebass, R., "Transonic Flow Past Oscillating Airfoils," Annual Review of Fluid Mechanics, Vol. 12, 1980, pp. 181-222.

[125] Chyu, W. J., Davis, S. S., and Chang, K. S., "Calculations of Unsteady Transonic Flow over an Airfoil," AIAA Journal, Vol. 19, 1981, pp. 684-690.

[126] Levy, L. L., Jr. and Bailey, H. E., "Computation of Airfoil Buffet Boundaries," AIAA Journal, Vol. 19, 1981, 1488-1490.

[127] Ohman, L. H., Kacprzynski, J. J., and Brown, D., "Some Results from Tests in the NAE High Reynolds Number Two-Dimensional Test Facility on 'Shockless' and Other Airfoils," Canadian Aeronautical and Space Journal, Vol. 19, 1973, pp. 297-312.

[128] Fung, Y. C., An Introduction to the Theory of Aeroelasticity, John Wiley & Sons, Inc., New York, 1955. (See also: Jones, J. G., "Aircraft Dynamic Response Associated with Fluctuating Flow Fields," AGARD LS-74, 1975; and Mabey, D. G., "Prediction of Severity of Buffeting," AGARD LS-94, 1978.)

[129] Bertelrud, A., Bergmann, M. Y., and Coakley, T. J., "Experimental and Computational Study of Transonic Flow About Swept Wings," Paper No. 80-0005, presented at AIAA 18th Aerospace Sciences Meeting, Pasadena, Ca., January 1980.

[130] Deiwert, G. S., "Numerical Simulation of Three-Dimensional Boattail Afterbody Flow Field," AIAA Journal, Vol. 19, 1981, pp. 582-588.

[131] Shrewsbury, G. D., "Effect of Boattail Juncture Shape on Pressure Drag Coefficients of Isolated Afterbodies," NASA TM X-1517, 1968.

[132] Bradshaw, P., "Structure of Turbulence in Complex Flows," AGARD LS-94, 1978.

[133] Eaton, J. K. and Johnston, J. P., "A Review of Research on Subsonic Turbulent Flow Reattachment," AIAA Journal, Vol. 29, 1981, pp. 1093-1100.

[134] Tucker, H. J. and Reynolds, A. J., "The Distortion of Turbulence by Irrotational Plane Strain," Journal of Fluid Mechanics, Vol. 32, 1968, pp. 657-673.

[135] Lumley, J. L., "Computational Modeling of Turbulent Flows," Advances in Applied Mechanics, Vol. 18, 1978, pp. 123-176.

[136] Liepmann, H. W., "The Rise and Fall of Ideas in Turbulence," American Scientist, Vol. 67, 1979, pp. 221-228.

Chapter VII.

Transonic Design Using Computational Aerodynamics

M.E. Lores* and B.L. Hinson†
Lockheed-Georgia Company, Marietta, Ga.

Introduction

Numerous computational methods for analyzing transonic flows were developed during the 1970s. Although some methods involved the solution of time-averaged Navier-Stokes equations, only methods based upon separate solutions of a potential flow equation and a boundary-layer model have found widespread use in design applications in the aerospace industry. This situation exists for two basic reasons: 1) current computer limitations restrict the use of Navier-Stokes solution methods to only simple geometries, and 2) the attendant computer costs are large.

The first reason is clearly a valid justification for not using solution methods based on Navier-Stokes equations in aircraft design. However, the second reason warrants examination. Figure 1 shows the portions of the total design, development, production, and deployment costs of a typical military transport aircraft system that are associated with engineering man-hours and computer use. Also shown here are representative fleet life cycle cost savings that can be derived by reducing engineering man-hours and computer costs by 50% and by reducing drag by 10%.

These data show that major savings can be derived from improved design productivity and aircraft performance, while reductions in computer utilization do not significantly change the program costs. Consequently, in the development of computational design methods emphasis

---

Presented at the Transonic Perspective Symposium, NASA/Ames Research Center, Moffett Field, Calif., Feb. 18-20, 1981. Copyright © American Institute of Aeronautics and Astronautics, Inc., 1982. All rights reserved.
*Manager, Systems Engineering Department.
†Specialist Engineer, Propulsion and Acoustics Department.

should be placed on making the methods 1) easy to use and reliable, thus improving productivity; and 2) accurate, to permit the design of more efficient configurations.

This is not to say that computational efficiency is not important nor that it can be ignored in developing or evaluating new methods. Rather, the implication is that computational efficiency can be sacrificed to achieve increased design productivity or improved accuracy. Also implied is the fact that methods which are considered expensive to use, such as time-averaged Navier-Stokes methods, should be employed in aircraft design if they are needed to achieve accurate predictions of aircraft performance.

Achievement of aircraft design improvements offered by a new technology often requires the use of other new technologies. For example, high aspect ratio wings with aft-loaded airfoils offer a means of improving transonic aerodynamic cruise efficiency. However, because of their torsional and bending characteristics such wings present design challenges to structural engineers. New structural technologies such as composites and reliable structural design methods are required to avoid excessive increases in wing weight which could negate the improved aerodynamics of these wings. Consequently, aerodynamic and structural considerations must be taken into account simultaneously in

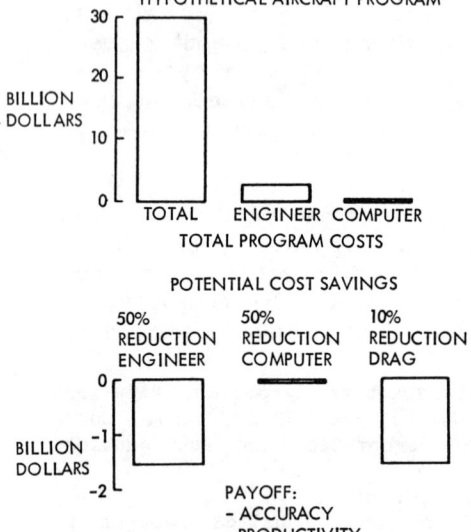

Fig. 1 Effect of computational methods on aircraft costs.

the design of high aspect ratio aft-loaded wings. Thus, new computational aerodynamic methods must be compatible with methods used by other engineering disciplines, and they must permit the easy inclusion of design constraints.

In this paper we will review typical transonic design methods, placing emphasis on their ease of use, accuracy, and versatility. We will then examine a recent case study involving the use of a promising technique, numerical optimization, in the cruise aerodynamic design of a wing for a transport aircraft.

Review of Design Methods

For the purposes of this paper, we will classify design methods as either inverse solution methods or numerical optimization methods. Inverse methods involve the specification of a desirable surface pressure distribution and the computation of the geometry which produces the pressures. Numerical optimization methods consist of the use of analysis methods with numerical minimization schemes to produce designs which minimize a user-specified design objective.

Inverse Solution Methods

In inverse methods, the value of the velocity potential on the surface is computed from the specified surface pressures to provide a Dirichlet boundary condition. Solution of a potential flow equation and subsequent integration of a tangential flow equation yields the desired geometry. One such method has been developed by Carlson[1]

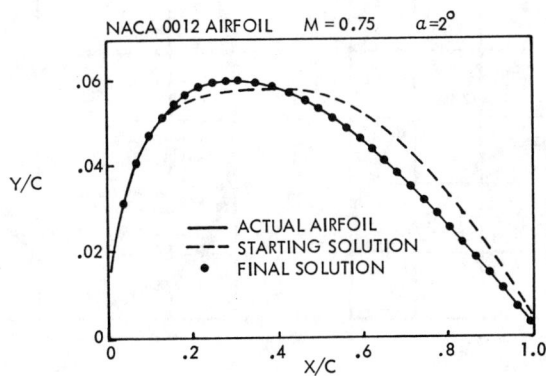

Fig. 2 Consistency of Carlson inverse code - inviscid example.

using both small disturbance and full potential two-dimensional potential flow equations, and a Cartesian coordinate system. A unique feature of Carlson's method is the use of a combined direct and inverse solution method. He specifies the leading edge geometry of the airfoil (typically the forward 10%) and computes the rest of the airfoil. By doing this the need to specify a leading edge stagnation condition is avoided, and a means of controlling the trailing edge thickness of the airfoil is provided.

The data shown in Fig. 2 demonstrate the self-consistency of Carlson's methods. Here, an inviscid analysis of a NACA 0012 airfoil was performed using Carlson's method, and the resulting pressures were input to compute the airfoil surface. The computed airfoil is in excellent agreement with the actual airfoil. Consistency in the case where an airfoil is generated using specified pressures and the resulting airfoil analyzed to reproduce the specified pressures can also be demonstrated.

Results from a more interesting example involving an aft-loaded airfoil are summarized in Fig. 3. On the left, the pressure distributions resulting from a viscous analysis of the airfoil are compared with experimental pressures. The computed pressures were then used to

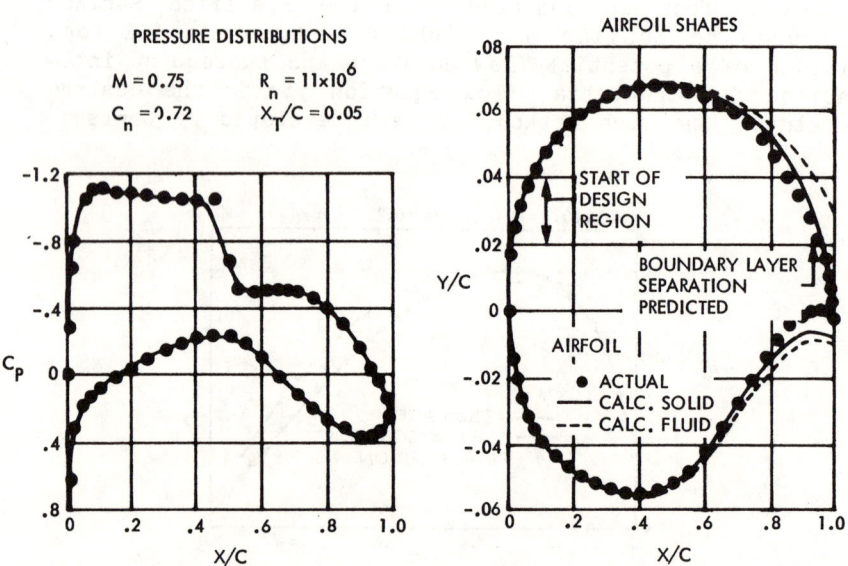

Fig. 3 Consistency of Carlson inverse code - viscous example.

generate a "fluid" airfoil geometry from which a boundary-layer displacement thickness was subtracted to produce a computed airfoil. Note that the agreement between the computed and the actual airfoil is not very good, especially in the cove region on the lower surface.

Since the inviscid study demonstrated consistency of the method while the viscous comparison did not, the weakness in this method seems to lie in the treatment of viscous effects.

A three-dimensional inverse method based on a modification of a NASA/Ames[2] extended small disturbance program has been developed by Shankar and his co-workers.[3] The method permits the inclusion of a fuselage. Reported results, taken from Ref. 3, are shown in Fig. 4. These data show a modest improvement in the pressure distribution for a conventional non-aft-loaded wing. Unlike Carlson's approach, this method requires the specification of a stagnation line.

Other inverse methods have been developed.[4-6] Unique features of the methods described in Refs. 4 and 5 are the adaptation of existing analysis programs to inverse solutions and the use of interactive graphics computations. In Ref. 6, Sobieski et al. describe a method for modifying wings to produce shock-free flow. A typical result taken from Ref. 6 is shown in Fig. 5.

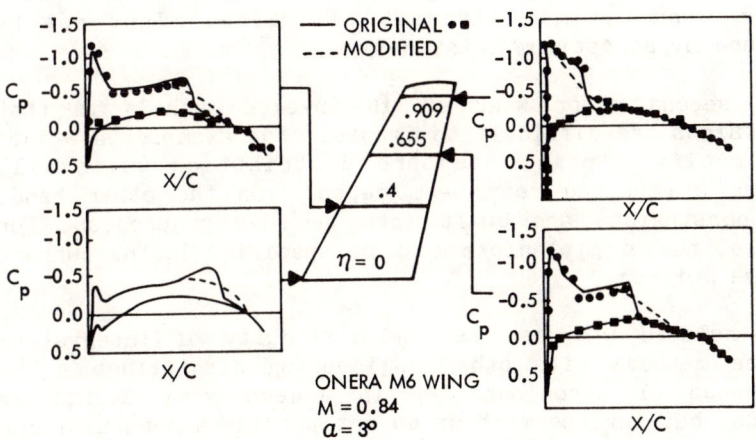

Fig. 4 Rockwell inverse example.

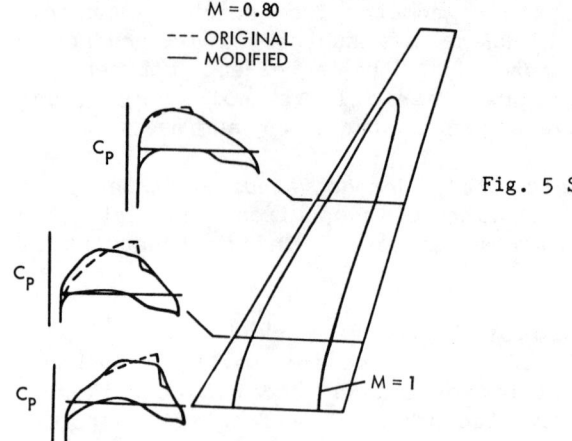

Fig. 5 Shock-free example.

Evaluation of Inverse Methods

The preceding discussion indicates that numerous inverse methods exist for airfoil and wing design. All of the methods successfully solve the problem formulations they are intended to solve. They require relatively little computer resources and they are easy to use.

The need to specify a desirable pressure distribution is a major weakness of these methods. Even with experienced designers, there is no assurance that the specified pressures will minimize drag, or weight, or whatever the aircraft design objective might be. Moreover, the need to specify design pressures implies a point design, and optimum aircraft efficiency might not be produced by an optimum point design.

A second major weakness in inverse methods is that constraints are difficult to impose. For example, how does one specify a prior pressure distribution which will provide a required rear beam depth? On the other hand, some constraints are built into the design process. For example, the wing planform must be specified in the current inverse methods.

A related weakness is the difficulty of integrating inverse methods with other engineering disciplines. The importance of aeroelasticity in modern wing design is obvious, but inverse methods do not provide a mechanism for combining aerodynamic and structural design to produce the best tradeoff between, for example, drag and weight.

Because of their relative simplicity and computational efficiency, coupled with these weaknesses, inverse methods seem best suited for initial wing design. Some other technique is then required to produce an optimized design.

Numerical Optimization

By combining them with numerical minimization schemes, analysis methods can be used to produce designs which are in some user specified sense optimized. The use of numerical optimization in transonic aerodynamic design was pioneered by Hicks[7,8] at NASA/Ames. Hicks was supported in this work by Vanderplaats who developed the numerical minimization computer code, CONMIN,[9] which is widely used. Because the currently reliable transonic analysis methods are restricted to isolated wings or simple wing bodies, transonic aerodynamic numerical optimization has been restricted to wing design.

Briefly, in numerical optimization the designer specifies a starting geometry which is usually based on previous experience, or is developed using airfoil design methods and sweep theory. The design objective and constraints are then specified. In principal, the design objective can be any parameter that can be reliably computed by the analysis method. The user must also specify the perturbations which can be made by the optimization process to minimize the objective while satisfying the constraints. These perturbations may be related to the geometry, flight conditions, or combinations thereof.

Design conditions must also be specified. Typically, a design Mach number and lift coefficient are specified, but this need not be the case: for example, design points involving more than one Mach number and lift coefficient are feasible.

The user specifies the parameter to be minimized and the optimization procedure is started. Optimization is accomplished by performing the following steps:

1) Each design variable is perturbed independently and the effect on the design objective and constraints is computed (gradient calculations).

2) Based on the results of Step 1, a search direction is computed which minimizes the design objective while satisfying constraints.

3) The design variables are incremented in the search direction, and objective and constraints are evaluated.

4) When the design objective can no longer be reduced in the current search direction, a new search direction is computed and the process is repeated.

Currently, each gradient calculation and each search step typically require a solution of the analysis program. Even though many of the solutions are computed by restarting from a previous converged solution, the computation times can be large. Numerical optimization is frequently criticized for this large computation time. However, as we previously noted, the attitude should perhaps be re-examined within the context of total aircraft development costs.

One of the primary advantages of numerical optimization is the freedom provided to the user in selecting a design objective; that is, the quantity to be minimized. Parameters such as drag, drag-to-lift ratio, the inverse of cruise efficiency (as reflected by drag divided by the product of Mach number times lift coefficient) have all been used as design objectives. Indeed, parameters such as operating costs can also be minimized if they can be reliably computed.

Unfortunately, this theoretical versatility has not been realized in practice. The primary reason for this failure is the inaccuracy of drag calculations. For example, the data shown in Figs. 6 and 7 compare measured and computed pressures and force and moment coefficients for two airfoils. These data are taken from Ref. 10 which evaluated[1] one of the most recent transonic airfoil analysis methods.[11] Figure 6 shows a case where the pressures are in good agreement, but a 12 count difference exists between measured and computed drag. On the other hand, the data in Fig. 7 reveal significant difference in pressures but only a 4 count drag difference.

We have experienced cases where the drag changes significantly as more iterations are performed from an apparently converged solution. Clearly, such uncertainty on drag calculations makes numerical optimization based on drag unreliable. Reference 12 discusses an example design where inaccurate calculations produced an erroneous design.

Because of the inaccuracies in drag calculations, most successful three-dimensional transonic designs done by

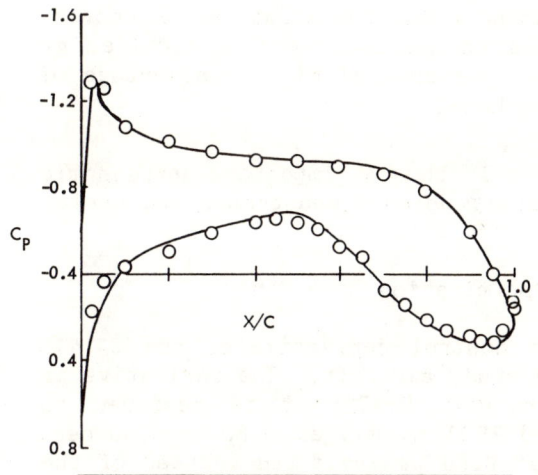

Fig. 6 Example of poor drag correlation with good pressure agreement.

LG4 - 612		
	Exp. Values	Grumfoil
$M =$	0.602	0.602
$R_c =$	7.16×10^6	7.16×10^6
$C_N =$	0.4614	0.4603
$\alpha =$	$2°$ (0.4°)	$0.7716°$
$C_d =$	0.01023	0.0090
$C_m =$	-0.098	-0.1071
$(x/c)_{tr} =$	0.075	0.075

numerical optimization involve the use of a design objective based on pressure distributions. Typically, the root-mean-square of the difference between specified and computed pressures is minimized. Thus, one of the important advantages that numerical optimization has over inverse methods is frequently lost.

Numerical optimization still retains three important advantages over inverse methods. First, since more than one analysis method can be used, multidiscipline optimization is feasible. For example, aerodynamic and structural analysis methods can be used to include static aeroelastic tailoring in wing design. Second, constraints are easily imposed. Finally, since improvements are historically first made in analysis codes, improved computational methods can be incorporated more rapidly in numerical optimization than in inverse solutions. Moreover, in an optimization design method, the basic design procedure does not have to be changed to take advantage of improved computational methods.

Numerical optimization is thus an important aerodynamic design method. A case study which demonstrates the use of numerical optimization is reviewed next.

Numerical Optimization Case Study

A twin-engine active control derivative of the C-141B was selected as the case study aircraft. The derivative is designated herein as the C-141B/AC2; it is designed to carry a 75,000 lb payload 3500 n. mi. at 0.80 Mach number. The cruise Mach number of 0.80 was selected instead of the 0.77 cruise Mach number of the C-141B to improve productivity and to provide a more challenging transonic design problem. Performance constraints include a field length of 7500 ft and an initial cruise altitude of 35,000 ft.

The design goal of this study was to significantly reduce the C-141B/AC2 empty weight (OWE) and fuel requirements from those of the C-141B.

Configuration Characteristics and Performance

The general arrangement of the C-141B/AC2 is shown in Fig. 8. Based on the technology levels used in sizing the aircraft, the C-141B/AC2 predicted OWE is approximately 75% of the C-141B, and the former aircraft requires approximately 44% of the fuel used by the latter airplane. These improvements are obtained using an aspect ratio 12 wing with a quarter-chord sweep of 25 deg and an average thickness-to-chord ratio of 0.109. As we will show, the use of the new design procedure produced a significantly thicker wing with only minor performance degradations.

Design Procedure

The transport wing design procedure is shown schematically in Fig. 9. The procedure is based upon the use of an isolated wing code for the transonic design, and the use of a more economical subsonic panel method which provides good geometric resolution to compute interference pressures. (Example: on lower wing surface due to gear pod.) Also included in the design procedure is a

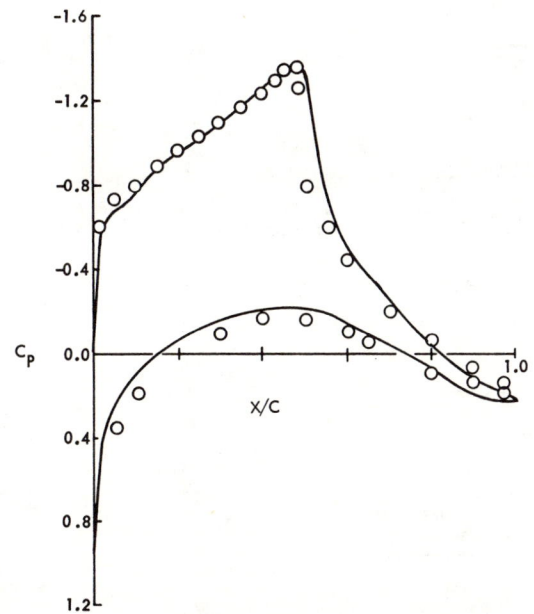

Fig. 7 Example of good drag correlation with poor pressure agreement.

NACA 65_1 - 213		
	Exp. Values	Grumfoil
M =	0.7058	0.7058
R_c =	2.91 x 10^6	2.91 x 10^6
C_N =	0.6062	0.5992
a =	4° (N.A.)	2.2362°
C_d =	0.0156	0.0160
C_m =	-0.0414	-0.0526
$(x/c)_{fr}$ =	0.05	0.05

systematic approach for selecting the starting wing geometry.

The key element in the procedure is the use of numerical optimization in the wing design. The wing design code was developed by linking CONMIN[9] with three-dimensional isolated wing analysis codes. Both an extended small disturbance code based on a program written by Bailey, Ballhaus, and Frick,[2] and Jameson's FLO-22 full potential equation program[13] were used in this study. The design objectives and constraints, and the permissible

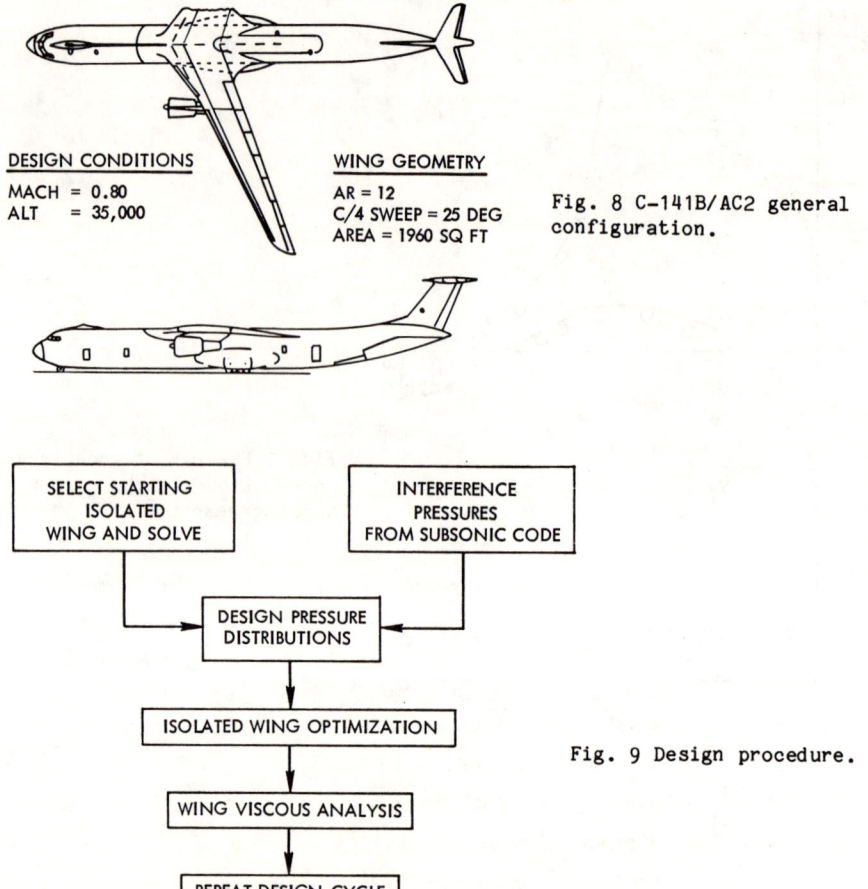

Fig. 8 C-141B/AC2 general configuration.

Fig. 9 Design procedure.

geometric perturbations (i.e., design variables) are detailed in the following paragraphs.

Design Objective and Constraints. To avoid the use of inaccurately calculated quantities such as drag in the optimization procedure, the design method was developed to permit the design of wings with specified chordwise pressure distributions. The capability of examining two pressure design objectives was provided. One design objective was the minimization of the RMS deviation between the target and actual pressures:

$$OBJ_1 = \left[\sum_N (C_P - C_{P_D})^2/N \right]^{1/2}$$

where N is the number of pressure coefficients, and C_{P_D} is the target pressure coefficient. The second objective considered was

$$OBJ_2 = \sum_N (C_P - C_{P_D}); \text{ constraint} = C_P < C_{P_D}$$

Notice the constraint is required to make the second objective meaningful.

Both objectives were tried in the design study, and the first objective proved to be superior. Consequently, OBJ_1 was used in the wing design.

Design Variables. Consistent with established wing geometry definition procedures, the wing geometry is determined by specifying the airfoil sections at various geometric control span stations and connecting these sections by linear loft elements. At each control station, the permissible surface perturbations are listed in Table 1. The magnitude of each of these 14 surface perturbation functions plus the section twist angle are the 15 design variables available for each surface at each geometric

Table 1 Transport wing design variables

$$V(1) = 3.89 \, (X/C)^{.25} (1 - X/C) e^{-20(X/C)}$$
$$V(2) = 10.68 \, (X/C)^{.5} (1 - X/C) e^{-20(X/C)}$$
$$V(j) = \sin^3 (\pi(X/C)^{r_j}), \, j = 3, 12$$
$$V(13) = (X/C)^8$$
$$V(14) = (X/C)^{20} e^a \sqrt{1 - (X/C)}$$
$$a = .5/(1 - (X/C)_M) - 20(X/C)_M$$
$$(X/C)_M = (X/C) \text{ for max camber}$$
$$V(15) = \text{TWIST}$$

	Sine deformations	
j	r_j	$(X/C)_{\text{Max def}}$
3	0.231	0.05
4	0.301	0.10
5	0.431	0.20
6	0.576	0.30
7	0.756	0.40
8	1.000	0.50
9	1.357	0.60
10	1.943	0.70
11	3.106	0.80
12	6.579	0.90

control span station. Thus, for a four control station wing, if all the surface perturbations were used and if all the sections were designed simultaneously, a total of 15 variables per surface per station x 2 surfaces x 4 stations = 120 design variables would be required.

Implementation. The simultaneous use of 120 design variables would result in an inordinately long computer run (greater than 10 h on a CDC 7600). Job turnaround time on such a run would be very long, and an error would have a catastrophic effect on computer budget. Consequently, wing design was accomplished in a series of steps. First, the upper surface was designed one section at a time proceeding from the root to tip. Next, the lower surface is similarly designed.

The optimization is done using the desired viscous pressure distribution. Consequently, the design procedure produces the "fluid" wing geometry (that is, the desired solid wing geometry plus the bounday-layer displacement thickness). The fluid wing is then analyzed and the entire process, or parts thereof, are repeated as required to produce the desired pressures.

Extraction of Solid Wing Geometry. The wing contours produced by the optimization include the boundary-layer displacement thickness, δ^*. The solid wing geometry is found by subtracting δ^* from the computed wing contours at each of the design stations. A conventional two-dimensional integral boundary-layer code[14] is used to compute δ^*.

We have found this step to be the most troublesome. Typically, even a well-designed aft-loaded airfoil will exhibit flow separation very near the trailing edge. Proven

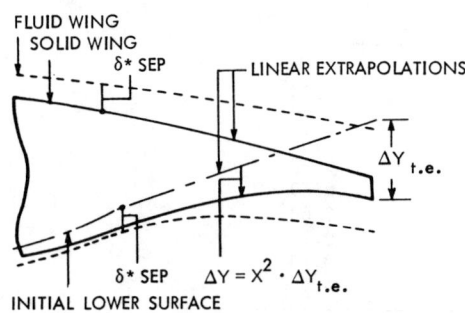

Fig. 10 Extraction of solid geometry.

three-dimensional separation prediction methods do not exist, and the validity of two-dimensional separation prediction methods is certainly questionable. Consequently, we would like to be able to compute the solid wing geometry from the fluid geometry even when our two-dimensional boundary-layer method predicts separation. This is not an easy task. Our approach is shown schematically in Fig. 10. On the upper surface we use a linear chordwise extrapolation at the solid wing geometry from the separation point to the trailing edge. On the lower surface we first linearly extrapolate and then use a parabolic wedge to control the trailing edge thickness.

The ad hoc approach is certainly very weak. The fact that we achieved what we consider to be a good wing design is somewhat surprising. An improved viscous flow analysis capability is surely needed if transonic wing design using numerical optimization is to become a reliable design tool.

Analysis of Optimized Wing. As the last step in the design procedure, the performance of the optimized wing is investigated using both full potential and extended small disturbance viscous transonic codes. The solid wing geometry is used in these calculations.

Configuration Design

The design procedure was used to design a new wing for the C-141B/AC2 configuration. The goal of this study was to improve the wing aerodynamics and at the same time increase the wing thickness for the Mach = 0.80, C_L = 0.60 design condition. Increased wing thickness was sought to increase the fuel volume, and to reduce wing weight.

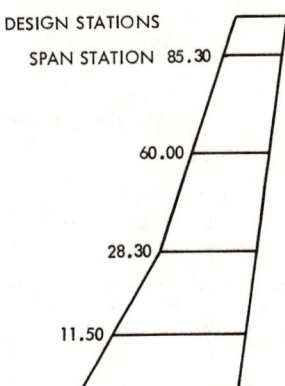

Fig. 11 Wing representation for design.

Current aft-loaded supercritical airfoil technology was used to determine the predicted C-141B/AC2 average wing thickness of 10.9%. Improvements in wing aerodynamic performance through the use of the new design procedure will be required to design a wing with increased thickness which operated efficiently at Mach = 0.80 and C_L = 0.60. The new design method will be shown to be a successful approach for obtaining the desired improvements in aerodynamic efficiency.

Numerical Optimization

The numerical optimization procedure previously described was used to determine the wing geometry which produces the desired wing pressures. The wing geometry was determined by designing the four wing control stations shown in Fig. 11 and using linear lofting to generate intermediate ordinates. Early numerical experiments showed that the 60% span design stations was needed because perturbations of the wing tip section were inadequate to

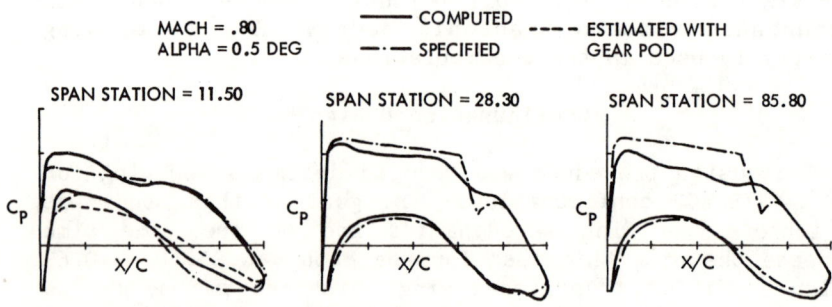

Fig. 12 ESD optimization pressures.

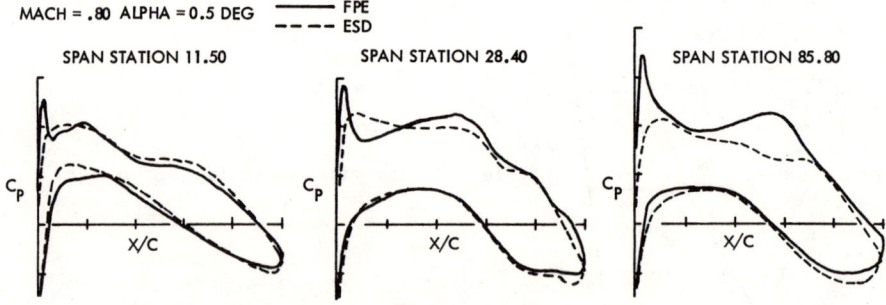

Fig. 13 Analysis of ESD wing.

control midsemispan and tip pressures simultaneously. A constant normalized section wing carrythrough was used.

Of note also in Fig. 11 is the specification of tip pressures near the 85% span station. This choice was made because of the relative inaccuracy of computed results near the wing tip.

Extended Small Disturbance Design. The Bailey-Ballhaus extended small disturbance program was used in the initial wing design. The final wing pressures are compared with the target pressures in Fig. 12. Also shown here is the agreement between target and computed pressures after the upper surface design of each span station. The agreement between target and computed pressures is fair.

Before continuing with the design process, the solid wing geometry was analyzed using viscous versions$_2$ of both full potential^{13} and extended small disturbance^2 codes. The results of these calculations are shown in Fig. 13. On the premise that the FPE results are correct, these data show that the ESD code mispredicts the wing leading edge flowfield, and this error causes complete disagreement between ESD and FPE results.

Full Potential Design. The failure of the ESD optimization to design accurately the wing leading edge made a second pass through the design procedure using a full potential equation analysis code necessary. The FPE optimization was done using the design variables and objective used in the ESD optimization. The target pressures were modified to produce a slightly weaker shock wave.

The span load distribution of the resulting wing was not satisfactory because the outboard section of the wing was too highly loaded, while the root section was unloaded. The lower surface of the root was modified using a simple Lockheed linear design method to increase the section loading at a small sacrifice in wing root thickness. The wing twist was also adjusted using a very efficient panel method to decrease the tip loading.

The resulting wing was analyzed using both the FPE and ESD viscous codes. The results are summarized in Fig. 14. The codes produce results in good agreement with one another, indicating that ESD methods yield accurate results if the leading edge is properly designed. The wing

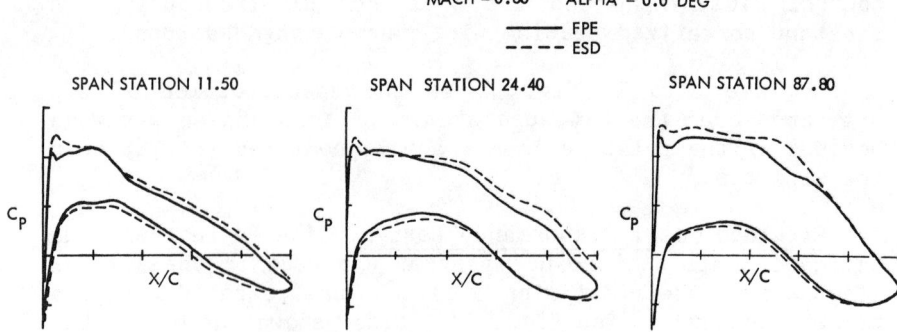

Fig. 14 Analysis of FPE wing.

pressures are quite satisfactory. There is no tendency for isobars to coalesce near the root trailing edge, nor is there a tendency for a leading edge shock wave to form. The desired midchord shock is weak (normal Mach number less than 1.16). Consequently, this FPE designed wing was selected as the final C-141B/AC2 wing design.

Wind-Tunnel Test

Test Facility

The design verification wind-tunnel tests were conducted in the Lockheed-Georgia Compressible Flow Wind Tunnel (CFWT). The tunnel is of the blowdown type, exhausting directly to the atmosphere.

The semispan configuration of the CFWT is shown in Fig. 15. The model is mounted on a five-component balance located in the floor. The balance and model rotate together on a turntable to vary angle of attack. A bleed duct is located 53.6 cm (21 in.) ahead of the balance centerline to remove the wind-tunnel boundary layer. The boundary-layer bleed system has an independent control valve and exhausts to atmosphere through a separate pipe system.

Models

An existing 0.0188 scale C-141 semispan model was used to obtain baseline data. The wing on this model is instrumented with 126 surface static pressure taps located at three span stations. The new C-141B/AC2 wing, Fig. 16, was machined from a solid billet of stainless steel. The

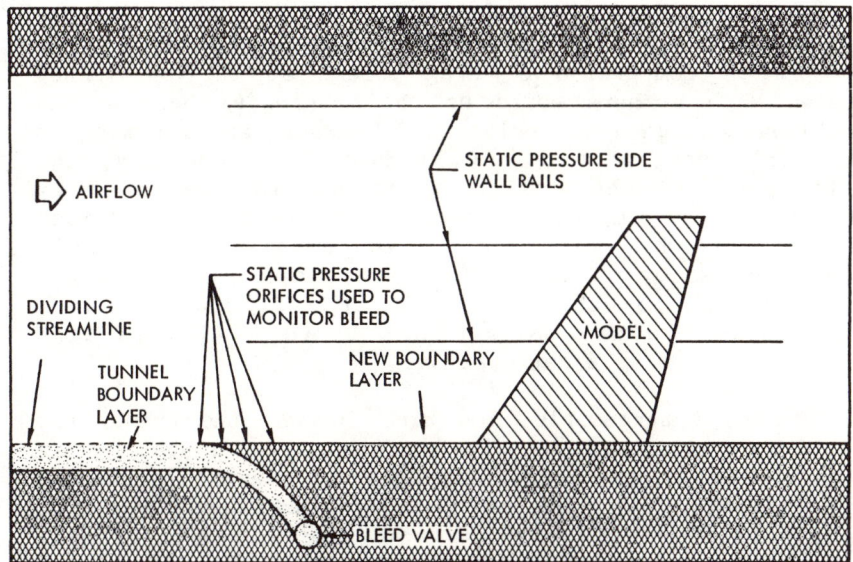

Fig. 15 Semispan test arrangement.

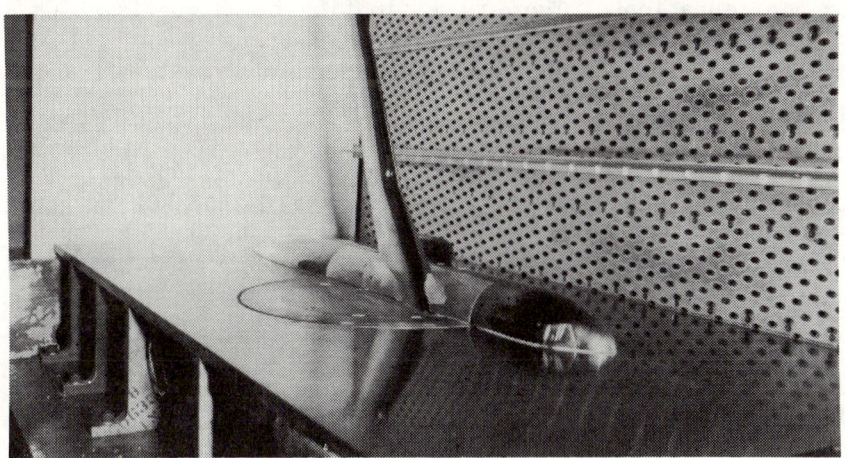

Fig. 16 C-141B/AC2 model installation.

wing has a total of 140 static pressure orifices located in chordwise rows at 5 spanwise stations.

Design Evaluation

Analysis of Test Data

The C-141B/AC2 design conditions and wing geometry are substantially different from the C-141. Consequently, a comparison of wing aerodynamics does not by itself provide a true measure of the effectiveness of the new wing. For example, a comparison of drag polars at the C-141B/AC2 design Mach numbers would not be meaningful because the C-141 was designed to cruise at 0.77 Mach, and its wing is well into drag rise at 0.80 Mach. Consequently, the efficiency of the design procedure is evaluated by comparison of complete airplane performance capabilities rather than by reference to incremental aerodynamic characteristics. To make these comparisons, flight aerodynamics for the C-141B/AC2 are extrapolated from wind-tunnel data using the known C-141 flight characteristics as a calibration.

In the Compressible Flow Wind Tunnel, the tunnel top and bottom walls are relatively close--approximately 3-1/2 mean chords--to the wing. Although a procedure for taking wall effects into account is under development,[15] time constraints precluded its use in this preliminary evaluation. Accordingly, the analysis method adopted herein is based on a comparison of uncorrected measured and estimated drag differences for the two wing fuselage gear pod configurations. The drag estimation technique is known to agree well with C-141 flight experience.

Aircraft Performance

The payload-range performance of the C-141B/AC2 is compared to that of the C-141B and the C-141B/AC2 target performance in Fig. 17. Of note is the better than

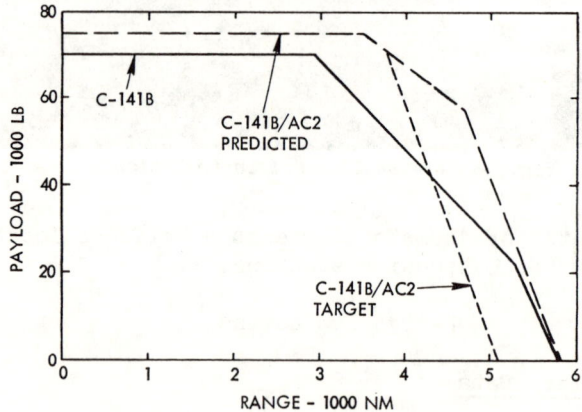

Fig. 17 Payload-range performance.

targeted ferry range of the C-141B/AC2 made possible by the thick wing resulting from the application of the new wing design method. The reduction in aircraft weights and fuel made possible by advanced technology are shown in Fig. 18. The predicted gross weight reduction is 2% less than the target value, while the block fuel decrease is 3% greater than targeted. The target empty weight reduction of 20% was achieved. Thus, with the exception of the 0.80 cruise Mach number, the study design objectives have been achieved.

Correlations

Correlations between computational results and experimental data have just been started. To date, a viscous version of FLO-22 developed by Henne has been used to generate solutions for both the designed wing and the measured manufactured wind-tunnel model. In the latter case, the wing spanwise twist was adjusted to simulate model deformation under load. The twist distribution was selected to provide the best match between computed and experimental pressures. Uncorrected test data were used in these early comparisons, and free air far field boundary conditions were used in the calculations.

Isolated wing calculated and measured lift, stability, and drag polar curves are compared in Fig. 19 for 0.80 Mach. These comparisons show that the use of measured model ordinates and adjusted twist improve the agreement between calculated and measured wing aerodynamics. The difference between computed and test zero lift angle of attack is in part due to wind-tunnel wall effects.

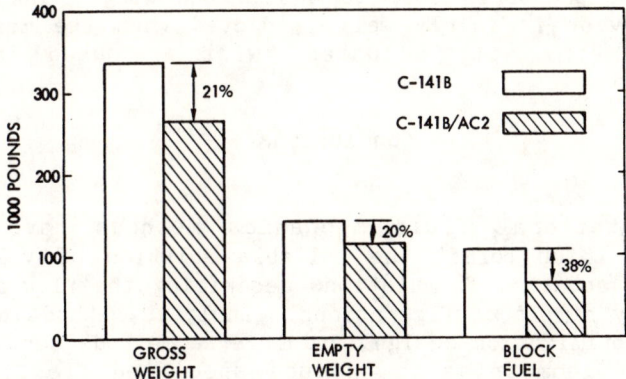

Fig. 18 Summary of aircraft parameters.

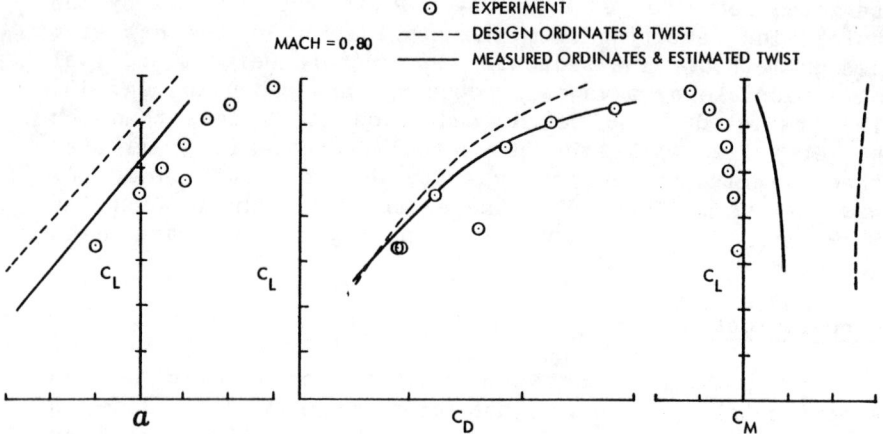

Fig. 19 FPE force correlation.

The pitching moment discrepancy can be explained by examination of the chordwise pressure distributions.

Computed and measured C-141B/AC2 isolated wing pressures are compared in Fig. 20. The use of measured ordinates and adjusted twist improves the correlation. However, the flow near the leading edge and the lower surface cove region are mispredicted. The discrepancy on the cove pressures is clearly due to flow separation which was not modeled. The differences near the leading edge have not yet been resolved.

Of particular note in the pressure distributions is the presence of a shock wave with an approximately constant sweep angle. This weak swept shock wave was specified in the target pressures used to design the wing. The shock wave behavior is fairly well predicted when the measured wing geometry with adjusted twist are used in the calculation.

Conclusions

Computational fluid mechanics methods have been developed which permit the reliable solution of velocity potential equation formulations describing the flow around relatively simple configurations. Analysis methods can be used in optimization design, and inverse methods permit the design of geometries to produce specified flowfields. Techniques such as the one described in Ref. 16 for treat-

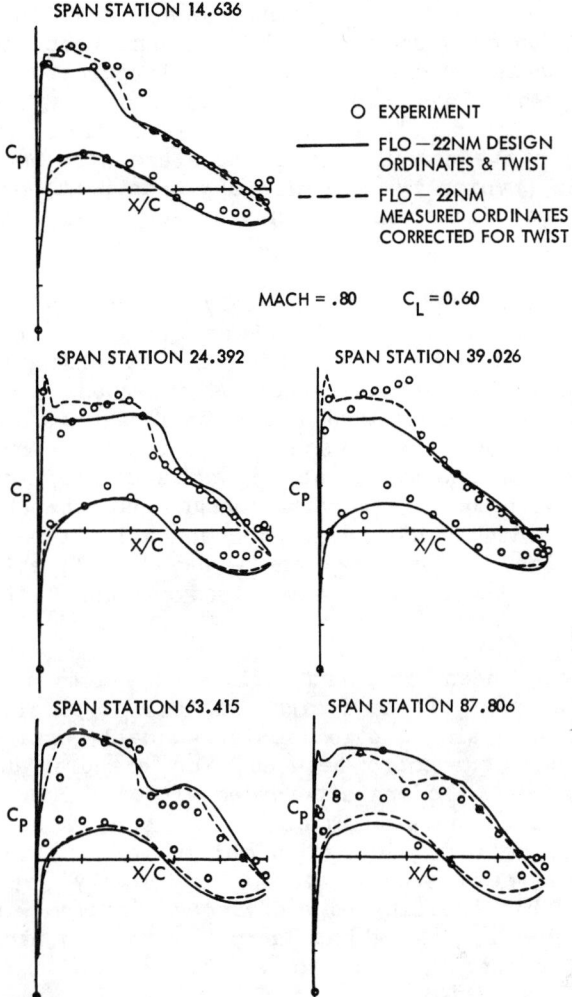

Fig. 20 Measured and computed C-141B/AC2 isolated wing pressures.

ing complex geometries are being developed. Consequently, the calculations of inviscid irrotational flows seem to be receiving adequate attention. Potential flow solution methods have been combined with boundary-layer analyses to simulate viscous flow solutions. This approximation is the major weakness in current computational design methods.

Most viscous analyses use two-dimensional boundary-layer methods and simplistic transition prediction techniques. The two principal deficiencies are the lack of reliable strong viscous interaction models to simulate shock wave/boundary-layer interactions, and inadequate treatment of thick viscous layers. In the computation of wing flows, these weaknesses result in inaccurate trailing edge flow predictions, and typically poor estimates of trailing edge pressure recovery. The situation is even more bleak for the calculation of fuselage afterbody flows where upsweep (typically on a military transport) or engine exhausts produce flows which cannot be accurately simulated using potential flow and boundary-layer models.

The deficiencies in viscous analyses are manifested in inaccurate calculation of integrated aerodynamic coefficients, in particular drag and pitching moment. Consequently configurations cannot be designed to minimize trimmed drag, an obvious weakness. As a result, surrogate design criteria such as pressure distributions are used. This in turn means that aerodynamicists who are very experienced with both aircraft design and computational methods are needed to use advanced computational methods in aircraft design. Such designers have profitably used computational methods to produce aerodynamically efficient wing designs.

Even experienced computational aerodynamists address only a limited part of the design problem. Typically, they concentrated on a single point design--usually cruise. They do not consider tradeoffs between, for example, drag and structural weight. Modern aft-loaded airfoils provide good transonic cruise aerodynamics. But they also impose stringent structural design problems associated with torsional stiffness and deflection of highly loaded but structurally thin trailing edge devices. Optimum wings can only be designed by taking both aerodynamics and structures into account.

A primary reason for the current restricted usage of computational aerodynamic methods is the cost of performing the calculations. In the past, a cost burden has been employed to control the usage of relatively limited and expensive computer resources. As the costs of computer resources continues to decrease, we should re-examine this practice. Indeed, we might reach a situation where computational resources can be treated like other company furnished equipment--in the extreme case like paper and pencils.

This observation implies that regardless of the extent that computers become "unburdened," we should not reject more computationally intensive viscous flow solution methods, if the methods are needed to achieve reliable computational design capability. We believe they are needed, because we cannot do truly effective aerodynamic designs until we can compute drag. It appears that we cannot compute drag using a weak interaction potential flow boundary-layer approximation. Therefore, we must face up to the need for some form of time-averaged Navier-Stokes equation solutions.

In summary, we believe that research in computational fluid mechanics should focus on: 1) improved accuracy; 2) configuration versatility; 3) multidiscipline design and analysis; and 4) efficiency—but not at the expense of the first three needs.

References

[1] Carlson, Leland A., Transonic Airfoil Design Using Cartesian Coordinates," NASA CR-2578, April 1976.

[2] Ballhaus, W. F., Bailey, F. R., and Frick, J., "Improved Computational Treatment of Transonic Flow About Swept Wings," NASA CP-2001, Nov. 1976.

[3] Shankar, V., Malmuth, N. D., and Cole, J. D., "Computational Transonic Design Procedure for Three-Dimensional Wings and Wing-Body Combinations," AIAA Paper 79-0344, Jan. 1979.

[4] Tranen, T. L., "A Rapid Computer Aided Transonic Airfoil Design Method," AIAA Paper 74-501, June 1974.

[5] Henne, P. A., "An Inverse Transonic Wing Design Method," AIAA Paper 80-330, Jan. 1980.

[6] Sobiesky, H., Fung, K. Y., and Seebass, A. R., "A New Method for Designing Shock-Free Transonic Configurations," AIAA Paper 78-1114, July 1978.

[7] Hicks, R. M. and Vanderplaats, G. N., "Design of Low-Speed Airfoils by Numerical Optimization," SAE Business Aircraft Meeting, Wichita, April 1975.

[8] Hicks, Raymond M. and Henne, Preston A., "Wing Design by Numerical Optimization," AIAA Paper 77-1247, 1977.

[9] Vanderplaats, Garret N., "CONMIN - A Fortran Program for Constrained Function Minimization," NASA TMX-62282, 1973.

[10] Ragab, S. and Lekoudis, S. G., "Validation of GRUMFOIL Computer Code," Lockheed IDC E-74-228-8, 1980.

[11] Melnik, R. C., Chow, R., and Mead, H. R., "Theory of Viscous Transonic Flow Over Airfoils at High Reynolds Number," AIAA Paper 77-680, 1977.

[12] Lores, M. E., Burdges, K. P., and Shrewsbury, G. D., "Analysis of a Theoretically Optimized Transonic Airfoil," NASA Contractor Report 3065, Nov. 1978.

[13] Jameson, Antony and Caughey, D. A., "Numerical Calculation of the Transonic Flow Past a Swept Wing," COO-3077-140, ERDA Math. & Comput. Lab., New York Univ., June 1977 (also available as NASA CR-153297).

[14] McNally, W. D., "FORTRAN Program for Calculating Compressible Laminar and Turbulent Boundary Layers in Arbitrary Pressure Gradients," NASA TND-5681, May 1970.

[15] Hinson, B. L. and Burdges, K. P., "An Evaluation of Three-Dimensional Transonic Codes Using New Correlation-Tailored Test Data," AIAA Paper No. 80-0003, Jan. 1980.

[16] Atta, E., "Component-Adaptive Grid Interfacing," AIAA Paper 81-0382, Jan. 1981.

Part II
Prediction Methods—Successes and Failures

In the first part of this book some prediction methods have been described from the point of view of the research scientist who develops such techniques. There is a considerable gap, however, between an algorithm that works for carefully selected research problems and those that work routinely in an aircraft design office. The following series of articles concern the application of the potential flow methods to real design problems. The authors have been encouraged to express their thoughts freely. Consequently, some problems in the present range of computer codes are discussed and in addition, some of the less technical constraints, such as insufficient budgets, that the designer faces in his work are noted. It is sometimes necessary for research scientists to realize that their favorite methods are being used in less than ideal circumstances.

The various potential codes mentioned in these articles are described briefly in the article by David Nixon and G. David Kerlick.

Chapter VIII.

Application of Computational Methods to Transonic Wing Design

I. C. Bhateley* and R. A. Cox[†]
General Dynamics Corporation, Fort Worth, Texas

Introduction

Transonic computational methods have finally matured to the stage where they can be effectively applied in the design cycle of both commercial and tactical aircraft. The analysis techniques have significant limitations in the type of governing equation they can solve, complexity of the geometry they can model, and the accuracy of the computed flowfield. Furthermore, computational speeds equivalent to a CDC 7600 are necessary to obtain valid solutions in less than an hour of central processor time.

Progress is also being made in developing design techniques by coupling analysis codes with optimizer codes and by solving direct inverse problems. These design methods require the expenditure of large amounts of computer time and are therfore used iteratively, with a man-in-the-loop examining the results intermittently.

This paper discusses some of the transonic computational methods that General Dynamics employs in refining aircraft wing designs. Available transonic methods and the design process are briefly discussed and then the application of these methods to two typical design problems is presented. The problems encountered during the exercise of

---

Presented at the Transonic Perspective Symposium, NASA/Ames Research Center, Moffett Field, Calif., Feb. 18-20, 1981. Copyright © 1981 by General Dynamics. Published by the American Institute of Aeronautics and Astronautics with permission.
*Engineering Specialist.
[†]Engineer.

these designs are also presented. Finally, recommendations for future work in this area are outlined.

Analysis Techniques

Some of the prediction methods that are publicly available are listed in Table 1. These methods are based on the solution of small disturbance or full potential equations using finite difference schemes. They utilize both conservative and nonconservative differencing techniques to obtain a solution for wing, wing-body, or wing-body-tail configurations.

In general, these codes model the real airplane geometry very crudely, as illustrated in Fig. 1, and most approaches are restricted to subsonic flow. Real flow effects are taken into account by applying viscous corrections to the geometry based on boundary-layer strip theory. A complete discussion of the theoretical methods listed in Table 1 is not intended in this paper since they are well documented in the cited references. However, the salient features of each approach as well as the basic differences between them are discussed.

Table 1 Available transonic codes

Code	Governing equations	Differencing technique	Geometry modeled	Viscous corrections	Mach no. range
Bailey-Ballhaus	modified small disturbance	CD & NCD	wing, body	yes	subsonic supersonic
Pandora-Boppe	modified small disturbance	NCD	wing, body, nacelles, tail, canards, winglets	yes	subsonic
Jameson FLO-22 Caughey	full potential	NCD	wing alone	no	subsonic
Jameson FLO-27 Caughey	full potential	CD	wing, body	yes	subsonic
Advanced Jameson codes FLO-28/30	full potential	CD	wing, improved body representation	no	subsonic

COMPUTATIONAL TRANSONIC WING DESIGN   407

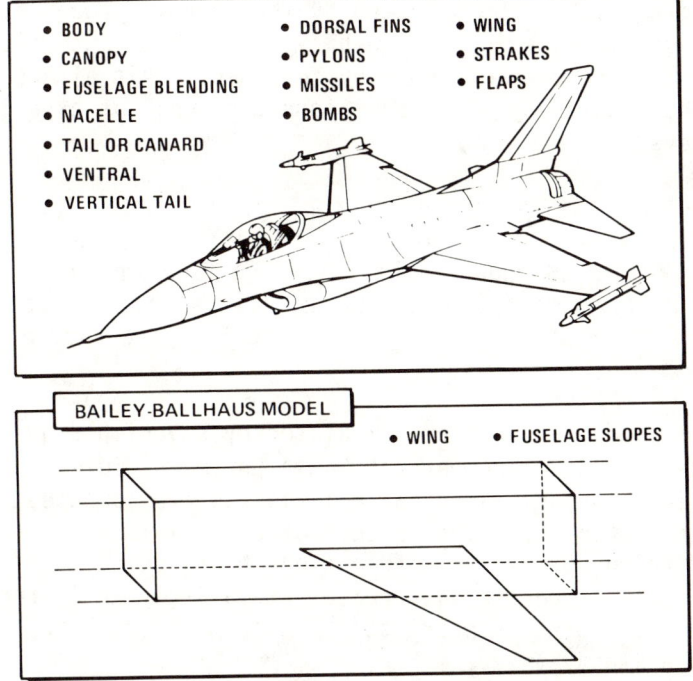

Fig. 1 Modelling limitations.

Bailey-Ballhaus Method

The Bailey-Ballhaus method, based on small disturbance theory, has evolved over a number of years, starting from work done by Bailey and Ballhaus[1] at NASA/Ames. Mason et al.[5] have compiled this work into an efficient user-oriented code. The basic small disturbance equation has been augmented[1] to allow a more accurate simulation of swept wing configurations. The current program can be used to compute three-dimensional transonic flows about wing-body-glove configurations. The program is now operational on both the NASA/Ames CDC 7600 and the NASA/Langley STAR computers.

The wing is approximated by applying linearized boundary conditions modified by Krupp-type scaling. The Murman-Cole[6] mixed finite difference approximations are used to generate a set of equations that are solved by a line-over-relaxation scheme on two-mesh systems using the Boppe[7] embedded mesh scheme. A conservative or nonconservative differencing scheme can be selected by the user. The coordinate transformations and the mesh systems can be internally generated by the code or input by the user. This allows usage of the program by the inexperienced as well as experienced users.

The fuselage is modeled by applying suitable flow inclination boundary conditions on a constant cross-section rectangular box as shown in Fig. 1. The flow inclinations are obtained by modifying the geometrical slopes in such a manner as to ensure correct global effect of the body on the flowfield.

The viscous effects are simulated by superimposing the boundary-layer displacement thickness on the airfoil ordinates. The boundary-layer characteristics are computed at each span station by a method developed for infinite swept wings, with the pressure distributions being calculated inviscidly. This process is repeated iteratively until the pressure distribtuions predicted by the inviscid theory converge to a steady value.

Pandora-Boppe Method

The Boppe[7-10] method is under development by Boppe of Grumman Aerospace Corporation under the sponsorship of NASA/Langley Research Center. The method is based on obtaining a solution of a modified small disturbance equation, which is slightly different from the equation used in the Bailey-Ballhaus code. The program can be used to compute three-dimensional transonic flowfields about arbitrary wing-body configurations. The latest version of the code allows one of the most accurate simulations of the fuselage. However, the fuselage analysis capability of this code was not exercised for this study. A pilot version of this code is operational at NASA/Langley.

The latest version of this program utilizes three grids to solve the wing-body problem. A global crude grid is utilized for the complete computational space with fine grid mesh systems embedded in regions very close to the wing and fuselage. A Cartesian grid is used for the crude and body mesh points, whereas a skewed grid aligned with the wing leading and trailing edges is used for the wing. The mesh points are internally generated in the pilot code. A non-conservative differencing scheme is employed to generate a set of equations that are solved by a SLOR scheme.

The Pandora code, also developed by Grumman under NASA/Langley sponsorship, is an extension of the Boppe methods. It uses embedded grids to model tails, canards, winglets, pylons, nacelles, etc. The results published[10] in general show very good agreement with experimental data.

Jameson-Caughey Nonconservative Code (FLO-22)

The Jameson-Caughey[11-13] FLO-22 code is a fully operational code for the analysis of three-dimensional wings in inviscid, adiabatic, and irrotational flow. Hicks[14] at NASA/Ames developed a wing design procedure by coupling this code with an automated optimization procedure. NASA/Langley[15] has implemented a vectorized version of this code on its STAR computer, which allows higher mesh density capability and faster execution times. The basic code and its derivatives have seen extensive use in transonic wing design.

The program is based on the solution of the full potential equation. A numerical sheared parabolic transformation is computed to transform a mesh with points concentrated near the leading edge in the physical space to a uniform mesh in the transformed space. The reduction in mesh points lying on the wing surface toward the wing tip of a tapered wing cause some loss of resolution in the computed results. The rotated differencing scheme is used to solve the transformed potential flow equation by use of standard SLOR techniques. The conservation condition is not satisfied when sonic velocity is exceeded anywhere in the flowfield.

Jameson-Caughey Finite-Volume Code (FLO-27)

The Jameson-Caughey FLO-27 code[16-18] is an improved version of the FLO-22 program. It is still in the pilot stage. The capability of analyzing a wing in the presence of a cylinderical fuselage is incorporated into the code. This program is operational both at NASA/Ames and Langley. The method is not as stable as FLO-22, and results cannot be computed for supersonic freestream Mach numbers.

The transformation used in the Jameson-Caughey FLO-27 procedure is similar to those in the Jameson FLO-22 code except that they are preceded by a Joukowsky transformation of the planes normal to the cylindrical fuselage axis. This reduces the fuselage to a slit in the transformed plane. The number of streamwise mesh points lying on the surface are constant across the span, giving improved resolution near the wing tips. The full potential equation is differenced in full conservation form. The resulting set of nonlinear equations are solved by standard SLOR techniques.

Advanced Jameson Codes (FLO-28/30)

The Jameson FLO-28 code was developed by Jameson under sponsorship of NASA/Ames. It incorporates the transformations necessary to model an arbitrarily shaped fuselage, but otherwise it is similar to FLO-27. The program is highly unstable and in our experience results cannot be computed for most configurations of interest at moderate lift coefficients and high Mach numbers.

The Jameson FLO-30 code is currently under development, also under the sponsorship of NASA/Ames. General Dynamics has not used this code to date.

Typical Results

Chordwise pressure distributions, computed for the F-16 configuration by some of the methods listed in Table 1, are compared with experimental data in Figs. 2 and 3. Additional comparisons can be seen in Ref. 19. The effect of conservative vs nonconservative differencing techniques is shown in Fig. 2. Although differencing schemes which conserve mass across shocks are more vigorous mathematically, the results predicted by methods using nonconservative differencing schemes seem to show better agreement with experiment. Nonconservative differencing schemes appear to simulate real flow effects more accurately in that it decreases the strength and steepness of the shock wave. Other points to be noted from this figure are that methods based on the solution of the small disturbance equation tend to under predict the leading edge suction pressures. Pressure predictions using linearized methods are also shown in this figure for comparison purposes.

The effect of including the fuselage in the analysis for the Bailey-Ballhaus and Jameson FLO-27 codes, is shown in Fig. 3. Converged solutions using the Jameson FLO-28 code could not be obtained. Including the body in the Jameson-Caughey FLO-27 code, represented as an infinite cylinder, made no appreciable or consistent change in the shock location. However, the computed pressures in the leading edge region for the wing-body case showed better correlation with experimental data than those calculated for the wing alone case. Including the body in the Bailey-Ballhaus code moved the shock upstream and better agreement with experiment and also improved the predicted suction pressures in the leading edge.

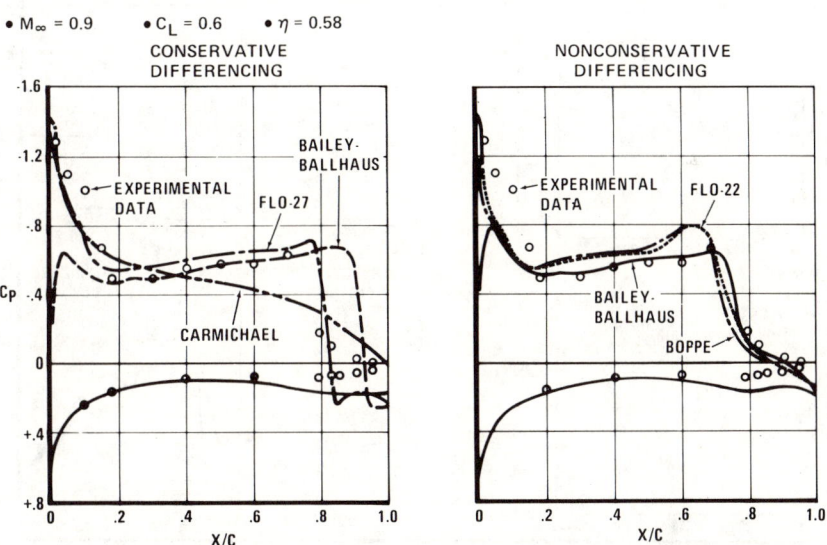

Fig. 2 Comparison of transonic code pressure predictions on F-16 inviscid wing alone.

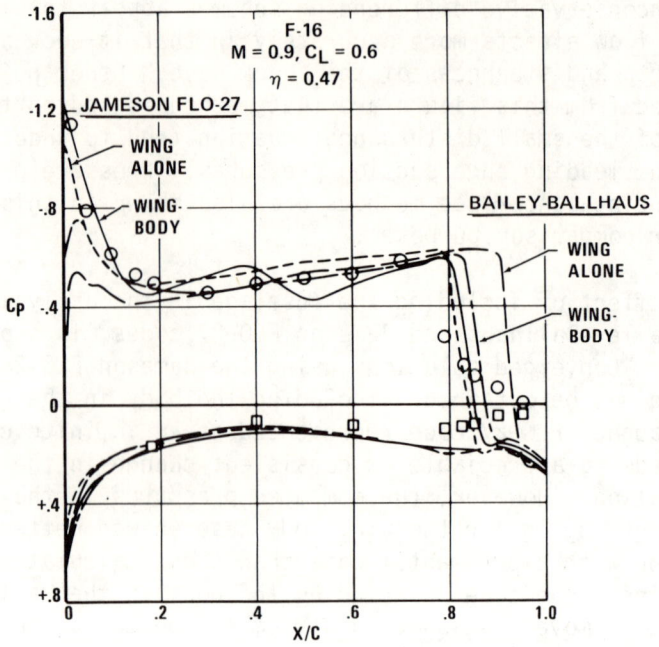

Fig. 3 Fuselage simulation in conservative transonic codes.

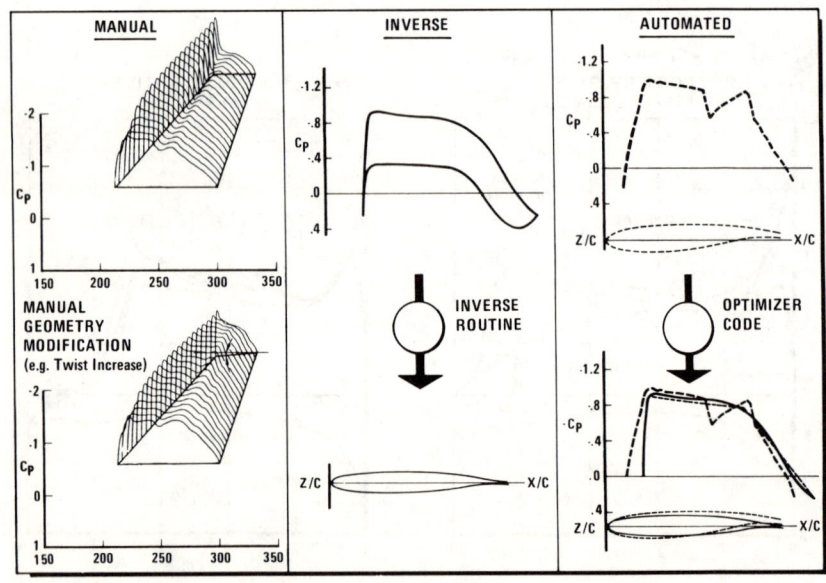

Fig. 4 Wing design refinement approaches.

Design Techniques

There are three basic techniques that can be used to refine aircraft wing designs (see Fig. 4). The first method is to use iteratively the analysis method and manually manipulate the configuration geometry until the desired design objectives are met. The second approach uses inverse design methods in which the geometry is generated corresponding to prescribed flowfield information. The last approach is the coupling of an analysis method with an automated optimization technique that systematically modifies the geometry to obtain the desired design objective without violating any constraints imposed on the system.

Several researchers, including General Dynamics, have been involved in developing coupled analysis-optimization codes. The approach used in a method developed by General Dynamics' Fort Worth Division is shown schematically in Fig. 5. Typically, the design process is carried out over one or two wing sections at a time to minimize computation time and to manually examine intermediate results during the optimization cycle. The drawback of this method is that the optimized wing cannot be defined by straight line segments joining sections at the root and the tip.

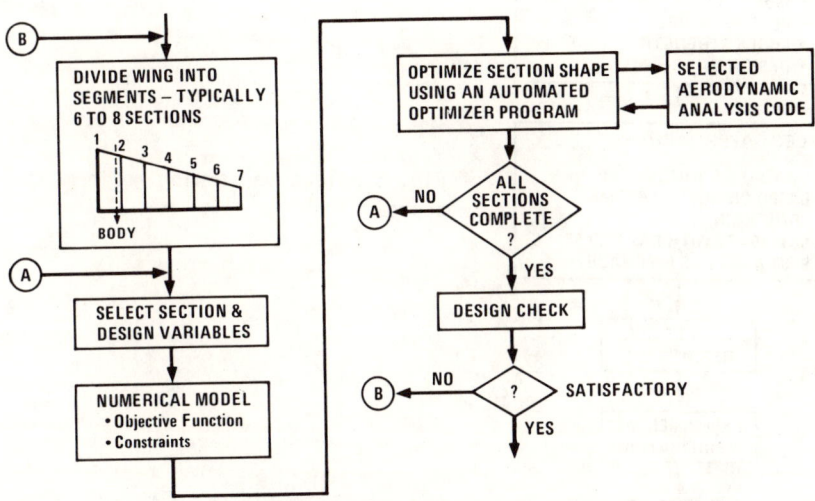

Fig. 5 Typical automated wing refinement approach.

Design Application

The method used to refine a wing for transonic conditions is shown schematically in Fig. 6. First a baseline analysis is carried out to reveal any shortcomings of the configuration at transonic Mach numbers. Based on this analysis, reasonable design goals are established. Next, a design approach is selected, using wind-tunnel data, if available, to calibrate the results predicted by the analysis method. Finally, the design is carried out and a configuration is defined for experimental verification.

The wing refinement approach has been carried out for a typical trainer configuration and has been initialized for an advanced fighter flap design.

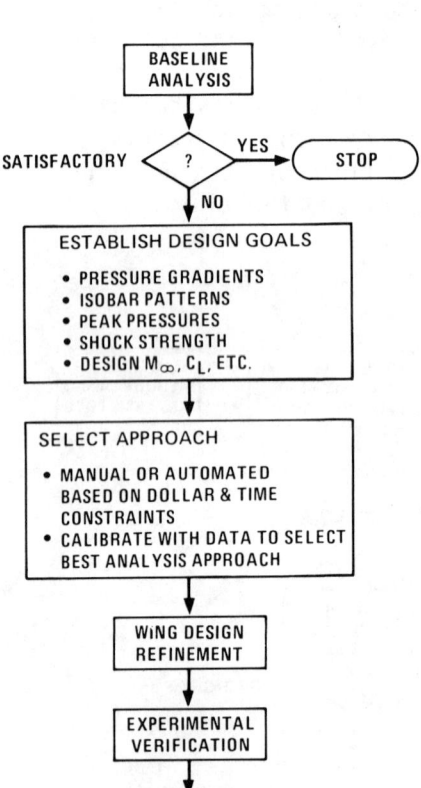

Fig. 6 Selection of wing refinement approach.

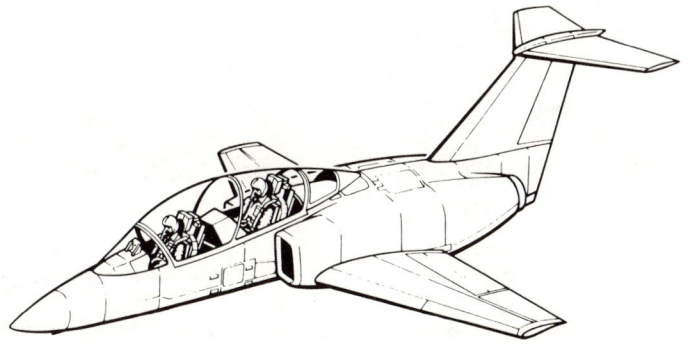

BASELINE

- AIRFOIL: ADVANCED SUPERCRITICAL
- L.E. SWEEP = 22.81°
- TAPER RATIO = 0.32
- t/c ROOT = 0.14
- t/c TIP = 0.10
- TWIST = 3°

TRANSONIC DESIGN GOALS

- DELAY DRAG RISE TO $M_\infty = 0.8$
- RETAIN LOW SPEED PERFORMANCE
- MAINTAIN TRIM DRAG LEVEL

Fig. 7 Application to typical trainer wing refinement.

Case I - Trainer Wing Refinement

The trainer configuration (shown in Fig. 7) was arrived at by means of wing geometry parametric trade studies using General Dynamics' Conceptual Design Synthesis Program. Based on the performance and structural requirements and limitations, the aircraft's wing area, span, sweep, and taper ratio were selected.

The streamwise airfoil section selected was a four-tenths camber NASA supercritical airfoil with a centerline thickness to chord ratio of 14% and a tip ratio of 10%.

A trainer aircraft encounters a wide variety of flight conditions when executing mission profiles and satisfying point performance requirements; however, time and funding limits the design analysis to only a few specific flight conditions. The flight condition that was found to be most crucial to mission performance consisted of an optimum cruise segment.

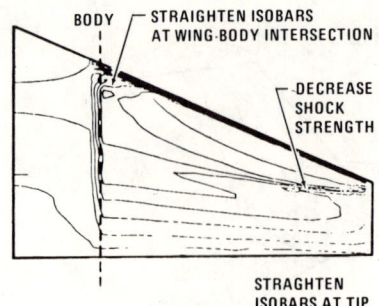

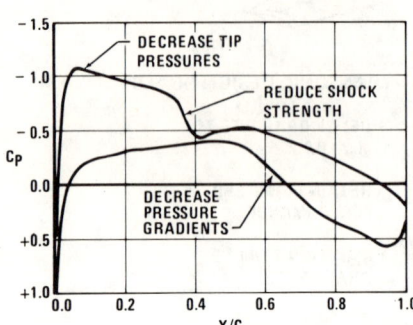

Fig. 8 Trainer baseline analysis.

Analysis of the Bailey-Ballhaus data (see Fig. 8) at a design point representative of this flight condition revealed a shock that unswept towards the tip. In addition, relatively high leading edge suction pressures, strong adverse pressure gradients on the lower surface, and a normalization of the wing isobars at the wing-body intersection were also observed.

The method employed for the trainer wing design consisted of defining an objective (or design) pressure distribution and calculating an error based on the difference between the design pressure distribution and the computed pressure distribution.

The three major goals of the wing design effort were 1) to decrease adverse pressure gradients to eliminate regions of separated flow, 2) to achieve shock-free flow at the design point with minimal detrimental effects at off-design conditions, and 3) to minimize three-dimensional effects on the wing by incorporating pressure distributions that are as similar as possible across the entire span. A design pressure distribution was derived, for the wing upper surface, in order to achieve these goals.

In keeping with a supercritical-type pressure distribution, a design pressure distribution was chosen with a pressure coefficient that remained constant over the majority of the upper surface of the two-dimensional airfoil section (see Fig. 9). The two-dimensional airfoil section refers to the section perpendicular to the percent chord line at the airfoil maximum thickness (38% for the baseline airfoil).

A sloped rooftop pressure distribution was incorporated in transforming the two-dimensional section pressures to the three-dimensional yawed wing according to the procedure of Lock and Bridgewater.[20] At the design point (M = 0.78/C_L = 0.46) a value of M_{normal} = 1.06 was necessary to achieve equivalent upper surface wing lifts between the baseline pressure distributions and the design pressure distributions. M_{normal} refers to the component of the upper surface Mach number normal to the local wing sweep (the sweep of a specific constant-percent chord line).

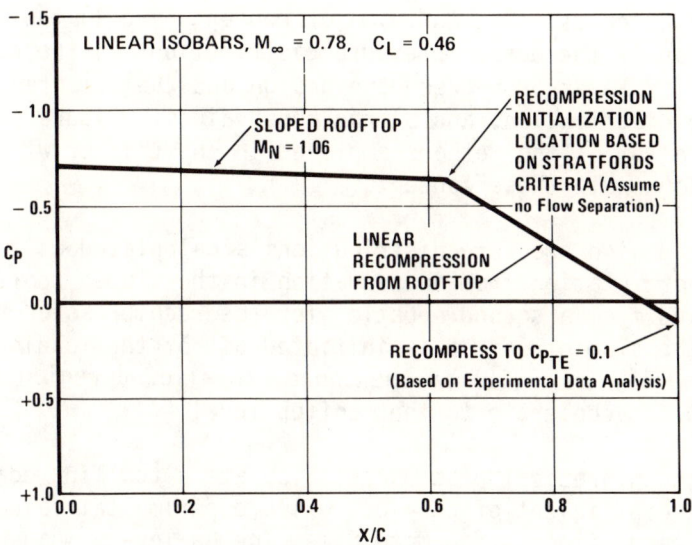

Fig. 9 Design objective.

The chordwise location of the initialization of recompression from the sloped rooftop pressure to the trailing edge pressure was determined on the basis of Stratford's criteria for a compressible turbulent boundary layer.[21] The location of the initiation of linear recompression from the rooftop to the trailing edge of a two-dimensional airfoil can be calculated from the freestream Reynolds number, Mach number, rooftop Mach number, coefficient of pressure at the airfoil trailing edge, and the boundary-layer transition location. The trailing edge pressure coefficient was chosen at $C_p = 0.1$. Stratford's criteria are based on two-dimensional theory; however, it was still utilized on the three-dimensional wing design because the design spanwise pressure distributions are constant along the span (approximating a two-dimensional flowfield).

The Bailey-Ballhaus code allows a great deal of versatility in its operation, and was chosen as the analysis code for the wing design. The code was coupled with the Vanderplaats optimization routine in order to allow specific chordwise modifications to several airfoil sections along a span.

Computational design requires numerical modelling of physical concepts. The three major concepts are 1) the definition of the design pressure distribution, 2) the calculation of an error between the design and the computed pressure distribution, and 3) the constraints. Figure 10 depicts a few of the methods employed in the trainer wing design to handle these concepts.

The design pressure distribution is calculated at a desired chord station from one equation in the sloped rooftop region and from a second equation for the recompression region. The sloped rooftop is initiated aft of the leading edge (usually 5% to 10% of the chord) to allow a region for the flow to accelerate to the rooftop level.

Most theoretical procedures calculate a trailing edge pressure coefficient of $C_p = 0.2$ while experimental values usually range from $C_p = 0.0$ to 0.1. The Bailey-Ballhaus theoretical pressures will diverge from the design pressures at the trailing edge, but ahead of the trailing edge the experimental and theoretical pressures should be equal. To account for these numerical inaccuracies at the trailing edge, the design pressures were discontinued fore of the trailing edge.

The error between the design pressure distribution and the computed pressures is defined as the summation of the squares of the difference between the computed and design pressures at chord stations for which the Bailey-Ballhaus pressures were computed. While minimizing this error, the Vanderplaats routine will cause changes in the airfoil section in such a way as to modify those pressures the furthest from the design objective. As the design is refined, an error value calculated from the absolute value of the pressure difference (as opposed to the square) is utilized.

An example of various methods employed in numerically modeling constraints is also depicted in Fig. 10. In addition to point constraints like Mach number and lift coefficient (constraints which can be handled directly by the Vanderplaats routine), performance constraints can be imposed by utilizing an appropriate model. High lift coefficients at subsonic Mach numbers (crucial to most trainers) can be sustained by preserving the initial leading edge radius and the airfoil thickness. This is numerically achieved by applying the upper surface thickness perturbations to the lower surfaces. It was assumed (correctly) that the airfoil modifications effects on the lower surface pressures, while favorably altering the upper surface pressures, would not significantly harm the lower surface pressures.

A first pass through the optimization cycle was undertaken to derive pressure distributions that closely resembled the design pressure distribution. Optimization was initiated at the root and progressed outboard in order that the supersonic flow zone of influence effect (on swept wings) would not alter the optimized pressures significantly.

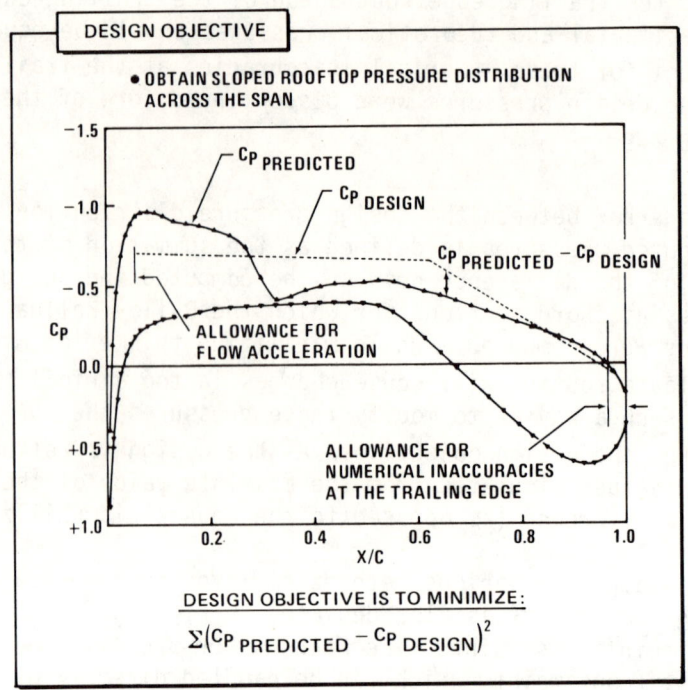

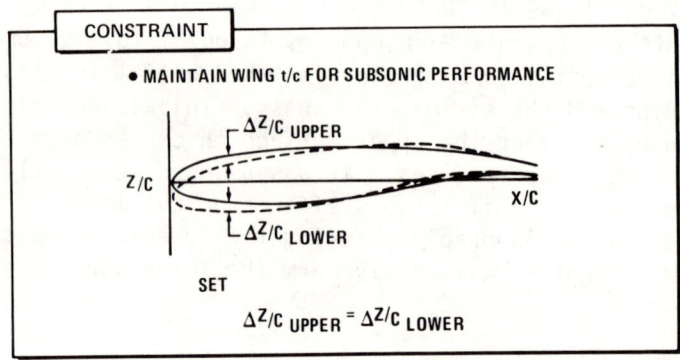

Fig. 10 Numerical modelling of problem.

Optimization modifications were made at seven span stations, corresponding to 7 of the 21 Bailey-Ballhaus generated wing grid spacings.

The optimization routine was first run at the root station (nondimensionalized span station = 0.2909). Perturbations (either positive or negative) added to the upper surface were also added to the airfoil lower surface to preserve thickness. The resulting airfoil section was incorporated into the design at that station and an identical section (thickness to chord ratio increased to the $\eta = 0.0$ value of 14%) was input at the configuration centerline.

The optimization was next executed at $\eta = 0.4000$. In order to quicken the first pass design, an identical section was input at $\eta = 0.5455$. In a similar fashion, optimizations were made separately at $\eta = 0.6909$ and $\eta = 0.9455$ with identical sections input at $\eta = 0.8364$ and $\eta = 1.00$, respectively.

Typical initial, final, and design pressure distributions for the first pass optimization are shown in Fig. 11 for span stations $\eta = 0.691$ and $\eta = 0.945$ respectively. The airfoil geometry modifications consisted primarily of a leading edge droop and a rearward displacement of camber.

Case II - Advanced Fighter Flap Design

The ability to predict the spanwise pressures on a three-dimensional wing with the Bailey-Ballhaus code is verified in detail by Bhateley et al.[19] Haney et al.[22,23] utilized the optimization capabilities of CONMIN (developed by Vanderplaats) in conjunction with the Bailey-Ballhaus code to optimize a variable camber wing consisting of a segmented leading and trailing edge.

This idea was taken one step further by incorporating a finite flap deflection in the Bailey-Ballhaus code instead of the relatively smooth camber employed by Haney.

Airfoil coordinates with leading and trailing edge flaps deflected were generated and input into the Bailey-Ballhaus code in an attempt to model an advanced fighter wing with flaps. Since the Bailey-Ballhaus code is based on small disturbance theory it is unrealistic to expect good results for large flap deflections. Determination of what constitutes a "large" flap deflection would be configura-

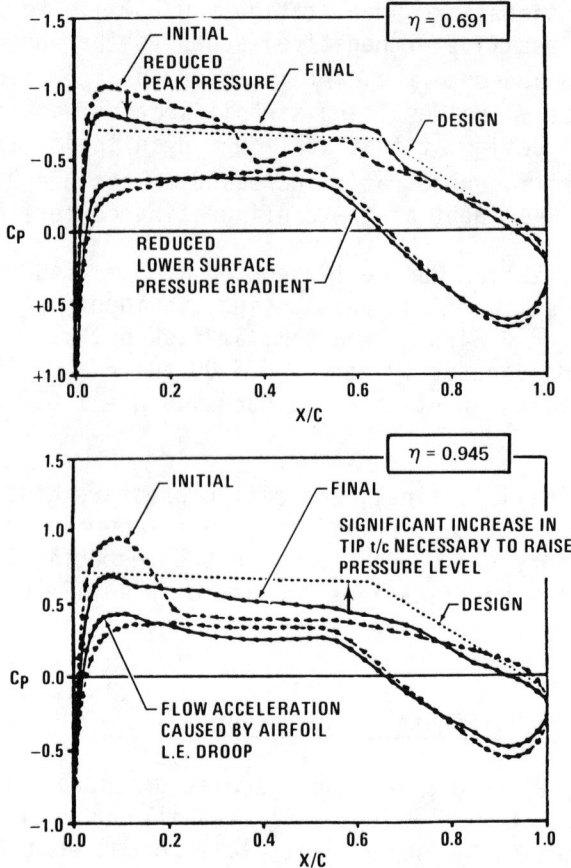

Fig. 11 Inviscid optimizations.

tion dependent and the user must be cautious and rely on test-theory experience to determine the validity of the results.

Various Bailey-Ballhaus operation modes were investigated for both the clean and the flapped wing in an effort to most accurately model the real flow. The planform, flap locations, and a representative midspan airfoil section are shown in Fig. 12.

At present, only the baseline analysis of the transonic wing with deflected flaps is complete. Future work envisions the capability of designing flap area and deflection in order to decrease maneuver drag while maintaining current levels of trim drag.

Typical three-dimensional pressures for a clean wing and the same wing with the leading and trailing edge flaps deflected to 7.5 deg are shown in Fig. 13. Flap deflection angle is denoted as δ. The clean wing exhibits a leading edge pressure peak which increases in magnitude toward the tip. Also visible is the unsweeping of the shock towards the tip.

Deflecting the flaps suppresses the leading edge pressures and allows an expansion of the flow on the trailing edge. Convergence instabilities are realized at the tip and were observed for both viscous and inviscid runs. Time damping coefficients and relaxation factors were altered to improve convergence and higher angles of attack were attained (from converged solutions) by increasing the angle of attack in discrete steps, yet the inviscid/7.5 deg flap configuration still diverged at α = 7.5 deg. It was possible, however, to get the solution to converge by running the boundary layer. Evidently the decrease in the airfoil slopes due to the boundary-layer displacement thickness addition was significant enough to allow convergence, but oscillations in the solution were still present.

TRANSONIC DESIGN GOALS
- DECREASE TRANSONIC MANEUVER DRAG
- DETERMINE "OPTIMUM" FLAP DEFLECTION AND SIZE
- KEEP FLAP SEPARATION TO A MIMIMUM
- MAINTAIN CURRENT TRIM DRAG LEVEL

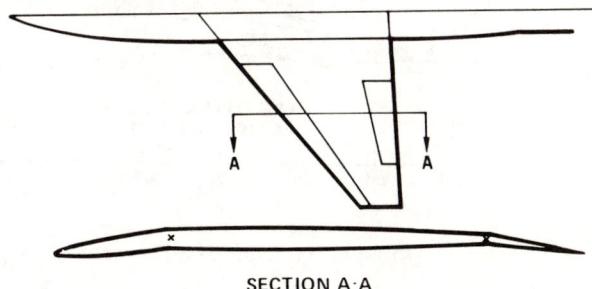

SECTION A-A

Fig. 12 Application to advanced fighter flap refinement.

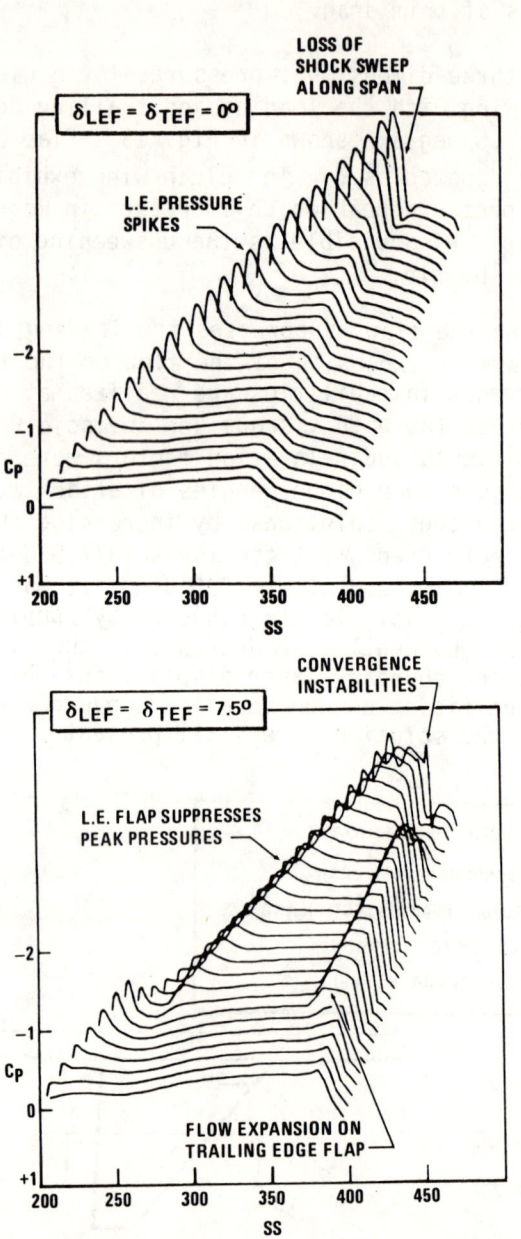

Fig. 13 Fighter baseline analysis.

COMPUTATIONAL TRANSONIC WING DESIGN

The Bailey-Ballhaus calculated lifts, drags, and moments are computed by an integration of the wing surface pressures. The total drag value is obtained by summing the Bailey-Ballhaus computed wing (pressure) drag with a test derived minimum drag, strake-body drag, and tip missile drag.

Figure 14 displays Bailey-Ballhaus computed lifts and drags for a variety of flap configurations at several an-

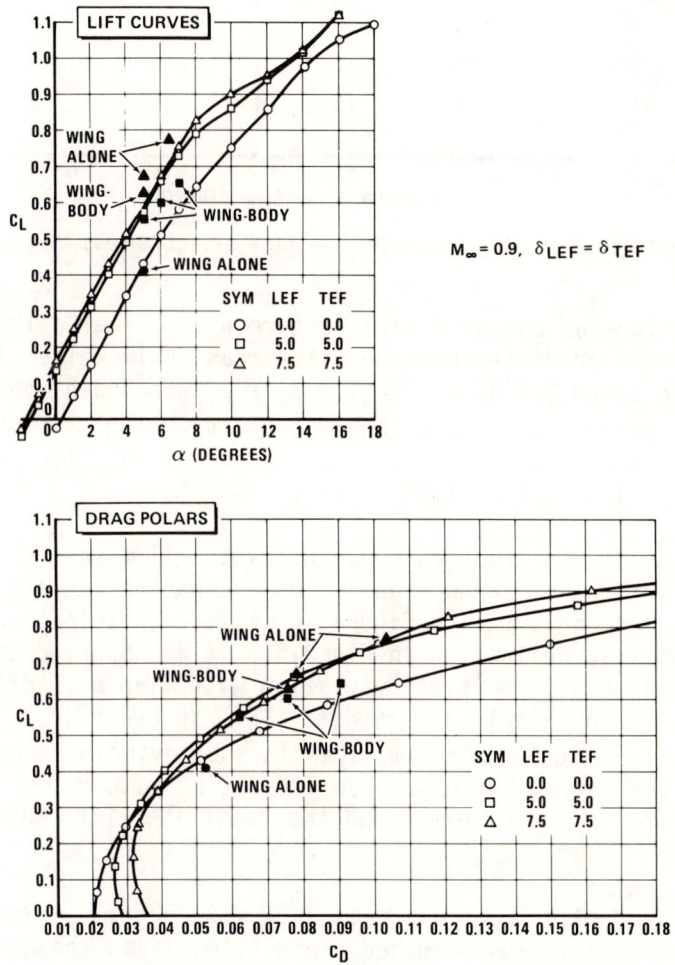

Fig. 14 Baseline analysis — forces.

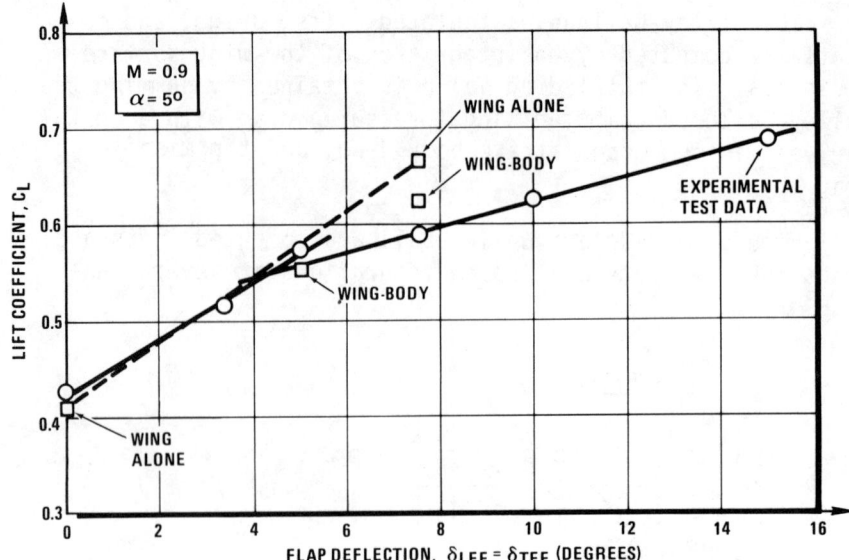

Fig. 15 Baseline analysis — flap effectiveness.

gles of attack as compared with test results. The Bailey-Ballhaus code predicts the wing lifts reasonably well. The lift curve slope for δ_F = 5 deg (δ_{LEF} = δ_{TEF} = 5 deg) does not agree well. It is felt that the Bailey-Ballhaus body (which was employed for the δ_F = 5 deg condition) affected the wing pressures too significantly.

The Bailey-Ballhaus code overpredicted lift for a 7.5 deg flap deflection (wing alone). At a constant angle of attack (α = 5 deg), the configuration lift coefficient is plotted vs flap deflection in Fig. 15. It is observed that the (experimental data) lift variation with flap deflection breaks at approximately δ_F = 5 deg. This shock-induced separation "break" is not simulated in the inviscid computations of the Bailey-Ballhaus code, resulting in the continued linear trend observed and the resulting lift overprediction.

A test derived $C_{D_{min}}$ increment of 202 counts was added to all Bailey-Ballhaus computed drags (Fig. 14). The clean wing and 7.5 deg flap drags (calculated inviscidly) compare well with experiment and the trends with flap deflection are encouraging, however, difficulties exist for the 5 deg flap deflection. Employment of viscous effects in the code had little effect on the drag polar.

The feasibility of design by flap-hingeline perturbation was investigated and initial findings are encouraging. For the type of improvements sought, it would be necessary to incorporate viscous effects and additionally, the solutions would have to be converged to an extent such that iteration errors would be an order of magnitude less than the hingeline perturbation effects on the lift, drag, and moment.

Conclusions and Recommendations

The use of computational methods by an experienced engineer can provide vital design guidance in the transonic flow regime. For moderate to low lift coefficients, the computed flowfield for typical tactical aircraft configurations have shown good correlation with experimental data. Most methods run into convergence problems at high lift coefficients, Mach numbers near sonic, and for severe configurations (wings with high sweep or with highly swept strakes, etc.).

Two other drawbacks of the computational methods need to be emphasized. One is their inability to treat complex configurations, especially arbitrary fuselages, and the other is the very large computational times required for problem solution.

The wing design codes currently being developed utilize a generalized optimizer code coupled with an analysis technique. There is no satisfactory way of selecting the design variables that will allow an arbitrary variation in the wing section geometry, especially near the leading edge. Furthermore, an adequate objective function based on satisfying multiple design criteria has to be found by trial and error. In general, the design changes made by the optimizer code are so small that the change in the flowfield is either insignificant, or so large that it leads to a configuration for which a converged solution cannot easily be obtained.

To enhance the use of computational methods by the airframe designers, we need to develop faster executing, user oriented, analysis codes with the capability to handle realistic configuration geometries. We also need to develop optimizer codes specifically tailored to the aerodynamic analysis methods.

References

[1] Ballhaus, W.F., Bailey, F.R. and Frick, J., "Improved Computational Treatment of Transonic Flow About Swept Wings," *Advances in Engineering Science*, Vol. 4, NASA CP-2001, 1976, pp. 1311-1320.

[2] Ballhaus, W.F. and Bailey, F.R., "Numerical Calculation of Transonic Flow About Swept Wings," AIAA Paper No. 72-677, June 1972.

[3] Bailey, F.R. and Ballhaus, W.F., "Relaxation Methods for Transonic Flow About Wing-Cylinder Combinations and Lifting Swept Wings," *Lecture Notes in Physics*, Vol. 19, Springer, New York, 1972, pp. 2-9.

[4] Bailey, F.R. and Ballhaus, W.F., "Comparison of Computed and Experimental Pressures for Transonic Flows about Isolated Wings and Wing-Fuselage Configurations," NASA SP-347, 1975, pp. 1213-1231.

[5] Mason, William H., Mackenzie, Donald, Stern, Mark, Ballhaus, William F., and Frick, Juanita, "An Automated Procedure for Computing the Three-Dimensional Transonic Flow Over Wing-Body Combinations, Including Viscous Effects," Vol. I - Description of Analysis Methods and Applications (Available from DDC as AD A055 899) AFFDL-TR-77-122, Vol. I, II and III, Feb. 1978.

[6] Murman, E.M. and Cole, J.D., "Calculation of Plane Steady Transonic Flow," *AIAA Journal*, Vol. 9, No. 1, Jan. 1971, pp. 114-121.

[7] Boppe, C.W., "Calculation of Transonic Wing Flows by Grid Embedding," AIAA Paper No. 77-207, Los Angeles, Calif., Jan. 1977.

[8] Boppe, Charles W., "Towards Complete Configuration Using an Embedded Grid Approach," NASA CR-3030, 1978.

[9] Boppe, C.W., "Computational Transonic Flow About Realistic Aircraft Configurations," AIAA Paper No. 78-104, Huntsville, Ala., Jan. 1978.

[10] Boppe, C.W. and Stern, M.A., "Simulated Transonic Flows for Aircraft With Nacelles, Pylons, and Winglets," AIAA Paper No. 80-0130, Pasadena, Calif., Jan. 1980.

[11] Jameson, Antony, Transonic Flow Calculations - Numerical Methods in Fluid Dynamics, edited by H.J. Wirz and J.J. Smolderen, Hemisphere Pub. Corp., New York University, 1978, pp. 1-87.

[12] Jameson, Antony and Caughey, D.A., "Numerical Calculation of the Transonic Flow Past a Swept Wing," NYU Report No. COO-3077-140, Courant Inst. Math. Sci., New York University, June 1977 (Available as NASA CR-153297).

[13] Jameson, A., Caughey, D.A., Newman, P.A., and Davis, R.M., "A Brief Description of the Jameson-Caughey NYU Transonic Swept Wing Computer Program - FLO 22," NASA TMX-73996, 1976.

[14] Hicks, Raymond M. and Henne, Preston A., "Wing Design by Numerical Optimization," AIAA Paper No. 77-1247, Seattle, Wash., Aug. 1977.

[15] Smith, Robert E., Pitts, Joan I., and Lambiotte, Jules J., "A Vectorization of the Jameson-Caughey NYU Transonic Swept-Wing Computer Program FLO-22-VI for the STAR-100 Computer," NASA TM-78665, 1978.

[16] Caughey, D.A. and Jameson, Antony, "Numerical Calculation of Transonic Potential Flow About Wing-Fuselage Combinations," AIAA Paper No. 77-677, Albuquerque, New Mex., June 1977.

[17] Jameson, Antony and Caughey, D.A., "A Finite Volume Method for Transonic Potential Flow Calculations," AIAA Paper No. 77-635, Albuquerque, New Mex., June 1977.

[18] Jameson, Antony, "Numerical Calculation of Transonic Flow Past a Swept Wing by a Finite Volume Method," Third IFIP Conference on Computing Methods in Applied Science and Engineering, Versailles, France, Dec. 1977.

[19] Bhateley, I.C., Mann, M.J. and Ballhaus, W.F., "Evaluation of Three Dimensional Transonic Methods for the Analysis of Fighter Configurations," AIAA Paper No. 79-1528, Williamsburg, Va., July 1979.

[20] Lock, R.C. and Bridgewater, J., "Theory of Aerodynamic Design for Swept-Winged Aircraft at Transonic and Supersonic Speeds," from Progress in Aeronautical Sciences, Vol. 8, edited by D. Kuchemann, Pergamon Press, New York, 1967, pp. 144-149.

[21] Anon, "Drag-Rise Mach Number of Aerofoils Having a Specified Form of Upper-Surface Pressure Distribution: Charts and Comments on Design," ESD Item No. 71019, London, 1971, pp. 38-43.

[22] Haney, H.P., Waggoner, E.G., and Ballhaus, W.F., "Computational Transonic Wing Optimization and Wind Tunnel Test of a Semi-Span Wing Model," AIAA Paper No. 78-102, Huntsville, Ala., Jan. 1978.

[23] Waggoner, E.G., Haney, H.P., and Ballhaus, W.F., "Computational Wing Optimization and Comparisons with Experiment for a Semi-Span Wing Model," NASA TM 78480, June 1978.

Chapter IX.

A-7 Transonic Wing Designs

H.P. Haney[*]
Lockheed-Georgia Company, Marietta, Ga.

Introduction

The intent of this paper is to open a dialogue between code developers and code users. The hope is that more practical codes will result if the developers have a better understanding of the environment in which their codes are used and the problems encountered with existing codes. Computational fluid dynamics has come a long way in the last five years, and these methods are used extensively throughout the aircraft industry. Yet code developers, sponsors, and users are frustrated by the limited impact that these codes have had upon aerodynamic design. Much of our frustration and some of our failures stem from a misunderstanding and oversimplification of aerodynamic design and its relationship to the overall design process.

This paper will review a research study which used computational method to design two new wings for the Vought A-7 aircraft. This study comprised six separate tasks which were performed under a series of Navy and NASA contracts from 1976 through 1980:

1) Correlation of FLO-22 wing analysis with existing A-7 wind tunnel pressure data.

2) Transonic airfoil optimization to provide starting sections for later wing optimization.

Presented at the Transonic Perspective Symposium, NASA/Ames Research Center, Moffett Field, Calif., Feb. 18-20, 1981. Copyright © 1982 by H. P. Haney. Published by the American Institute of Aeronautics and Astronautics with permission.
[*]Staff Engineer, Aerodynamics Department.

3) Formulation of a wing design procedure combining transonic wing analysis with numerical optimization.

4) Wing design exercises in which two wings were optimized for the A-7 for different design criteria.

5) Wind-tunnel testing of pressure models of the new designs in the NASA Ames 11 Foot Wind Tunnel.

6) Comparison of experimental data with theoretical predictions. This correlation study used the Flo-22 code initially and then transonic wing-body analyses as they became available.

Early study results have been published previously,[1] but this paper will emphasize problems encountered and even admit a mistake or two. Raymond Johnson of Vought Corporation and Raymond Hicks of NASA/Ames Research Center were largely responsible for the ultimate success of this venture.

Correlation of FLO-22 Analysis

A comparison of A-7 experimental wing pressures with FLO-22 predictions was made before any design work was initiated. Jameson's FLO-22 is a full potential, nonconservative, finite-difference transonic wing analysis.[2] This code was selected over a small perturbation analysis because of its capability to model leading edge geometry and thick wings.

At the time of this study, FLO-22 was a new code that had only been correlated with limited transport planforms. The Navy prudently demanded proof that this code could model the complex A-7 geometry before funding design studies on this aircraft. The A-7 configuration is particularly difficult to model with its high-mounted moderate aspect ratio wing having a 35 deg quarter-chord sweep. A 12% leading edge chord extension at 60% semispan produces a streamwise vortex which prevents pitch-up by controlling stall propagation. This discontinuous leading edge was modeled by placing defining stations on either side of the discontinuity which was then approximated by a smooth curve. A fixed leading edge droop produces a discontinuity in the upper and lower surface geometry at 28% chord. All of these features of the A-7 wing presented a difficult challenge for the theory.

Surprisingly FLO-22 analysis agreed quite well with test data,[3] even near geometric discontinuities. No boundary-layer corrections were made. The good agreement between theory and experiment provided confidence to proceed with the design study. In light of latter difficulties, these results were probably misleading.

Airfoil Optimization

Starting airfoils for each wing design were optimized using the procedure of Hicks.[4] This work will be described in order to set the stage for subsequent three-dimensional method development.

The airfoil optimization procedure combines Jameson's potential flow analysis, FLO-6, with Vanderplaats' constrained minimization routine, CONMIN.[5] The starting geometry is systematically perturbed using shape functions, Fig. 1, which insure surface and slope continuity and smoothness. The optimized airfoil is defined by the starting geometry and coefficients of each shape function which give minimum airfoil drag at the design condition. Design constraints may be imposed upon a problem to obtain a physically practical solution. Geometric variables such as thickness and volume may be limited as well as computed characteristics such as lift and pitching moment. A displacement thickness is added to the initial geometry and subtracted from the optimized geometry to determine the final airfoil.

The 12% thick airfoil optimization illustrates the importance of imposing problem constraints, Fig. 2. Design conditions for this airfoil are 0.4 C_L and 0.78 Mach number. A large drag reduction was achieved by the optimized section, 412M1, at the design Mach number. However, when this airfoil was analyzed at off-design conditions, it was found to have significant drag creep.

The optimized airfoil is inferior to the starting airfoil because of its off-design performance. A drag constraint was then imposed upon the problem at 0.77 Mach number, and drag was again minimized at 0.78 Mach number. The reoptimized section, 412M2, retains most of the drag divergence improvement of the first design while eliminating drag creep. Models of the starting airfoil and 412M2 were built and tested in the NASA Ames 2 x 2 wind tunnel. Test results verified the drag improvements predicted by theory. Only the upper surface geometry ahead

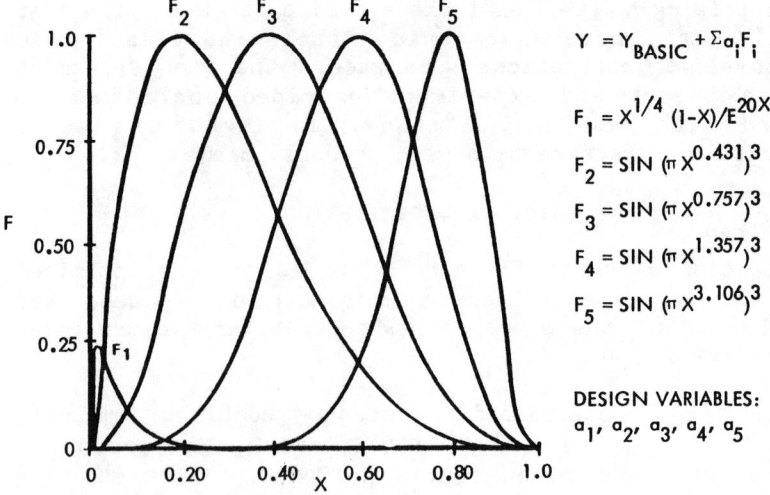

Fig. 1 Airfoil shape functions.

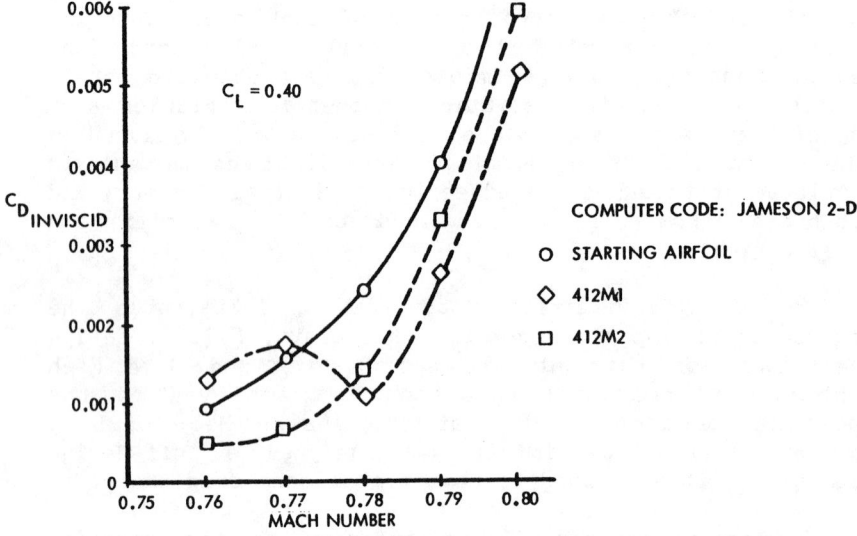

Fig. 2 Airfoil section optimization.

of the maximum thickness location was modified. Figure 3 compares the starting and final airfoils; drag divergence Mach number can change significantly with very small changes in surface contour.

Off-design performance is important for all practical airplane design problems. Cruise efficiency must be

Fig. 3 Airfoil geometry comparison.

balanced against buffet-free maneuver for a fighter aircraft for instance. Unfortunately this type of constraint doubles the already high computer cost. Faster computers and more efficient codes will be needed before three-dimensional multiple transonic design point problems can be solved. One intriguing possibility would be to use relatively inexpensive linear theory to compute low-speed or supersonic off-design performance while optimizing transonically.

Wing Design Procedure

The wing design procedure assembled for this study was an extension of the airfoil optimization procedure. Jameson's FLO-22 transonic wing analysis was substituted for the two-dimensional FLO-6 code. Strip boundary-layer displacement thickness corrections were added to the initial defining sections and removed after optimization. It was recognized that an active boundary-layer solution which accounted for changes to geometry and pressures was preferable, but time and money did not allow this sophistication. A total of 25 shape functions was allowed at each of six spanwise defining stations.

The upper surface was optimized first beginning at the root and progressing to the tip. Then the lower surface was designed in a similar manner. Initial optimizations were successful in significantly reducing computed drag. However the resulting pressure distributions had strong shocks which would obviously separate the flow. Attempts to constrain minimum pressure while minimizing drag did not produce an acceptable pressure distribution. Another problem became apparent at this time. Neither the magnitude nor even the sign of incremental drag due to small geometry changes could be trusted. Earlier drag minimization studies,[6,7] had limited success but encountered problems during design refinement. Thus the uncertainty in predicted drag precludes its use as the object function in a minimization procedure. An entirely new design approach was required.

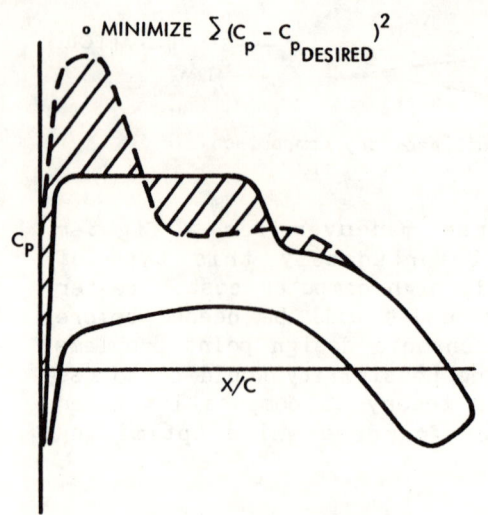

Fig. 4   Use of optimization code for wing design.

The design technique which finally worked was specifying the pressure distribution over the entire wing. The function to be minimized was the square of the difference between the actual and desired pressures, Fig. 4. It is uncertain whether to call this breakthrough a success or failure. This procedure does permit successful design improvements and the solution appears to be numerically stable. On the other hand, it fails to address directly the key parameter drag. Few people care what the pressure distribution looks like, but everyone is very interested in drag levels. There is no one who is qualified to define the pressure distribution which is associated with minimum drag. Despite these drawbacks, optimizing on pressures works and will have to suffice until better analysis codes become available.

The specified pressure distributions used in this study were obtained from the airfoil optimization work described earlier. Two-dimensional design conditions were derived from three-dimensional design conditions using simple sweep theory. The minimum drag airfoil optimized at that condition was then used as the starting airfoil, and its upper surface pressure became the specified wing pressure distribution. The airfoil lower surface pressure distribution was adjusted in magnitude to achieve the desired spanwise load distribution and total wing lift. This method of selecting design pressures was an expedient and not the result of an exhaustive study of alternatives.

Wing Design Studies

The wing design contract called for two wings to be optimized for the A-7 aircraft at different design conditions. The goal of the first design exercise was to maximize wing thickness without degrading drag divergence Mach number. Increasing the thickness of thin wings reduces structural weight, increases wing fuel volume, and improves high lift characteristics. The planform was the same as the existing A-7 except that the chord extension was eliminated, Fig. 5. The goal was to increase wing maximum thickness from 0.07 to 0.12 t/c, a 71% increase.

The goal of the second study was to reduce induced drag by 25% and increase maximum thickness by 28% while maintaining the same drag divergence Mach number. Quarter chord sweep was reduced from 35 deg to 20 deg and aspect ratio increased from 4 to 5, Fig. 5, in order to achieve the induced drag goal. The sweep and thickness changes would significantly reduce drag divergence Mach number if the A-7 airfoil were retained. The starting and final geometry and pressure distribution are shown in Fig. 6. The largest changes are at the root where 6 deg of wash-in have been added, the trailing edge decambered, and location of maximum thickness moved from 0.38 X/C to 0.20 X/C. In contrast, Wing No. 1 has no twist at the root because a different mechanization of the decision variables was used. The design program used camber line changes instead of twist to increase the leading edge loading. These results illustrate the nonuniqueness of an inverse-type solution.

Wing sections were optimized sequentially, root-to-tip on the upper surface followed by root-to-tip on the lower surface. A single pass was sufficient for Wing No. 1 which has 35 deg of sweep. But Wing No. 2 with 20 deg of sweep required two passes because outboard modifications altered the pressure distribution on the inboard wing which had already been optimized.

Numerous problems were encountered during the optimization process. Although 25 decision variables were available for each section, only a few could be used at one time because of cost. A separate FLO-22 analysis is required for each active decision variable whenever gradients are needed. New gradients are required at the beginning of a run, whenever a local minimum is found in a particular search direction, and when a constraint is encountered. In practice the design study consisted of many optimizations

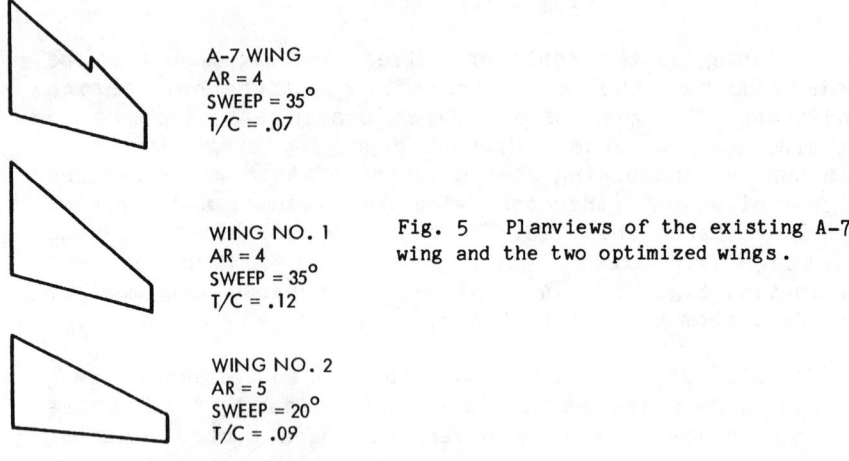

Fig. 5 Planviews of the existing A-7 wing and the two optimized wings.

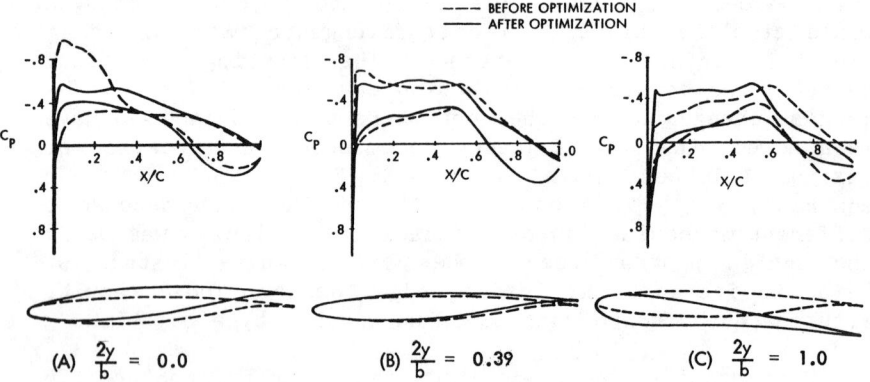

Fig. 6 Wing No. 2 starting and optimized geometry and pressures.

using four to six decision variables in order to minimize cost. The particular decision variables turned on for any particular run were selected manually by studying the desired and actual pressure distributions.

Initial improvements in the pressure distribution were relatively rapid, but each refinement was slower and more costly than the preceding one. If, on the other hand, the step size of each geometry change became too small, the solution converged rapidly to a meaningless answer. This problem results because each design step in a run is a FLO-22 restart from the previous step. If the step size was too small, the starting residual was too small to give a meaningful solution. When this happened, the solution could be restarted or the decision variables changed slightly to allow the design process to proceed.

Clearly there is much room for improvement in the design procedure used for this study. Refinements are needed to the minimization routine, and more efficient decision variables may be possible. The amount of manual intervention required for a solution should be reduced, but some designer interaction will always be necessary and desirable. Direct optimizing appears to be well suited for early design iterations involving large geometry changes subject to design constraints. Later refinements can probably be accomplished more efficiently with an inverse solution.

Wind-Tunnel Test Results

Pressure models of the two new wings were built and tested on a 0.1 scale Navy A-7 model along with a force model of the existing wing. Drag as a function of Mach number is presented in Fig. 7 for all three wings for 0.4 and 0.6 C_L. The drag divergence Mach number of all three wings is identical, not only at the cruise condition, but also at the off-design maneuver condition. All design goals were met for each new wing.

Lift-to-drag ratio is compared in Fig. 8 at 0.6 and 0.85 Mach number. No horizontal tails were tested because of model structural limitations, and all data presented are

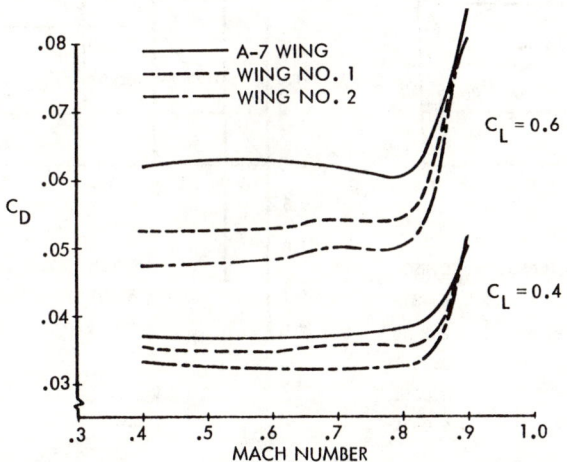

Fig. 7 Experimental drag divergence at constant lift coefficients.

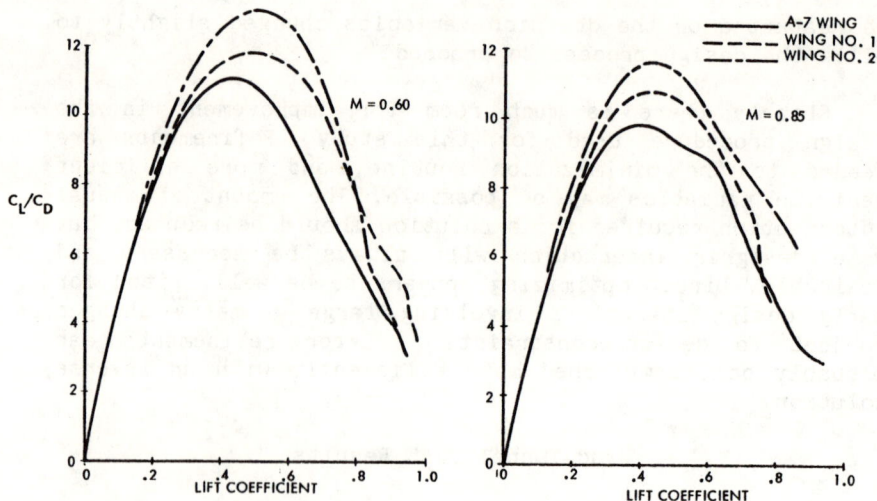

Fig. 8   Experimental lift-to-drag ratio.

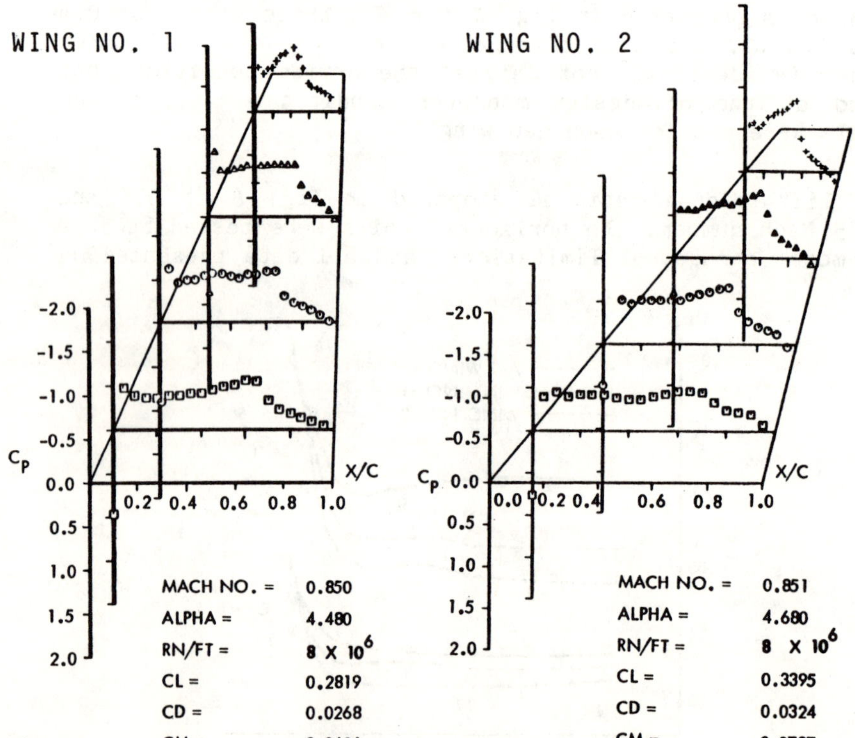

Fig. 9   Upper surface pressure distribution.

untrimmed. Trim drag for the new wings which have applicable trailing edge camber would negate most of the drag improvements of these wings. However, recall that drag reduction was not a design goal. Reduced drag could be achieved by trading off part of the thickness or sweep change for reduced camber and trim drag. The lower drag of the 12% thick Wing No. 1 relative to the 7% thick A-7 wing, both of which have the same planform, is due largely to increased leading edge suction and removal of the chord extension.

Elation at the apparently unqualified success of the test was tempered when the pressure data were plotted. Figure 9 presents upper surface pressure distributions for both new wings near the cruise design conditions. While test pressures are similar to the specified pressures, the shocks are considerably stronger near the tip. This error is very disconcerting because the entire design procedure depends upon the accurate prediction of wing pressures.

Comparison of Theory with Experiment

Transonic analyses were run at wind-tunnel test conditions to permit comparison of theoretical and experimental pressures. Predictions from FLO-22, the analysis used in the wing design procedures, were evaluated first. Later when wing-body analyses became available FLO-28, FLO-30, and WIBCO were compared.[8] These results will be discussed briefly in the following paragraphs.

FLO-22

Pressures predicted for the 12% thick wing by FLO-22, an isolated wing analysis, are compared with wind-tunnel test data at the design condition in Fig. 10. Not surprisingly, significant differences occur near the model wing-body intersection at 0.12 semispan. The FLO-22 wing has been extended to the centerline. The same boundary-layer correction used during design was used for this analysis. Both theory and experiment are compared at the same angle of attack in Fig. 10. Agreement away from the body is very good except in the region of the shock. Theory predicts a weak compression whereas experiment shows a strong shock on the outboard portion of the wing. Matching lift for this case does not appreciably increase the predicted shock strength while degrading the agreement elsewhere on the chord.

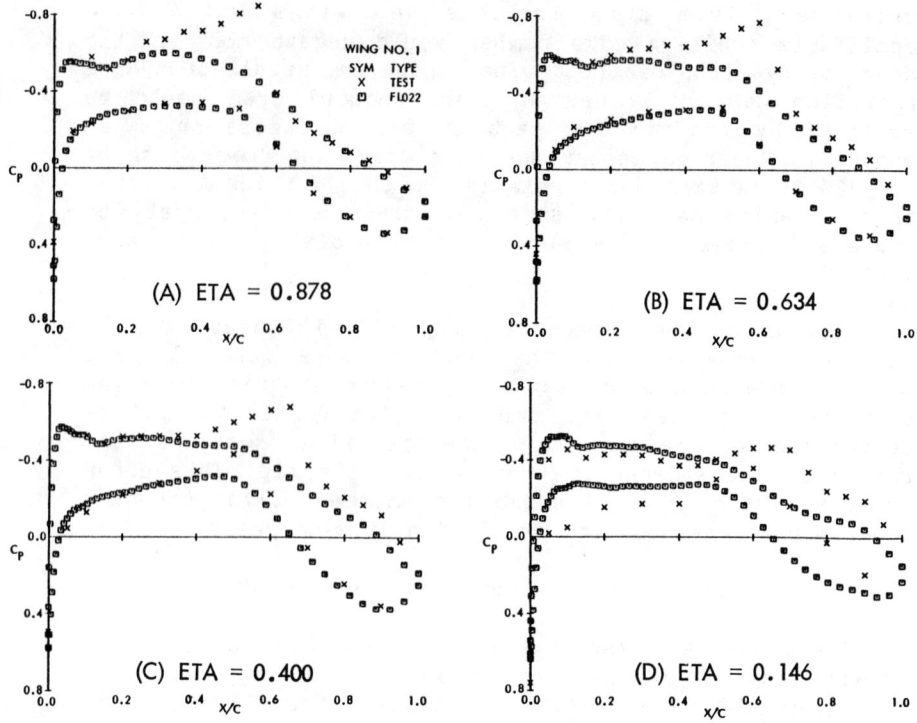

Fig. 10 Comparison of FLO-22 wing pressures with experiment.

One possible source of error is the crude boundary-layer representation used with the potential flow solution. The displacement thickness was computed based upon the specified pressure distribution instead of actual pressures and added as a first-order correction. This case was reanalyzed using FLO-22NM which computes a displacement thickness using predicted pressures and updates the boundary layer during the solution. FLO-22NM predictions were in close agreement with those of FLO-22. Thus the error probably lies with the potential flow solution, assuming that strip boundary-layer corrections are adequate.

FLO—28

Jameson's FLO-28 code is a full potential,[9] conservative, finite-volume transonic wing-body analysis. Wing No. 1 mounted on the A-7 fuselage was run on FLO-28. Erratic oscillations were predicted in the chordwise pressure distribution across the entire span. The program printed

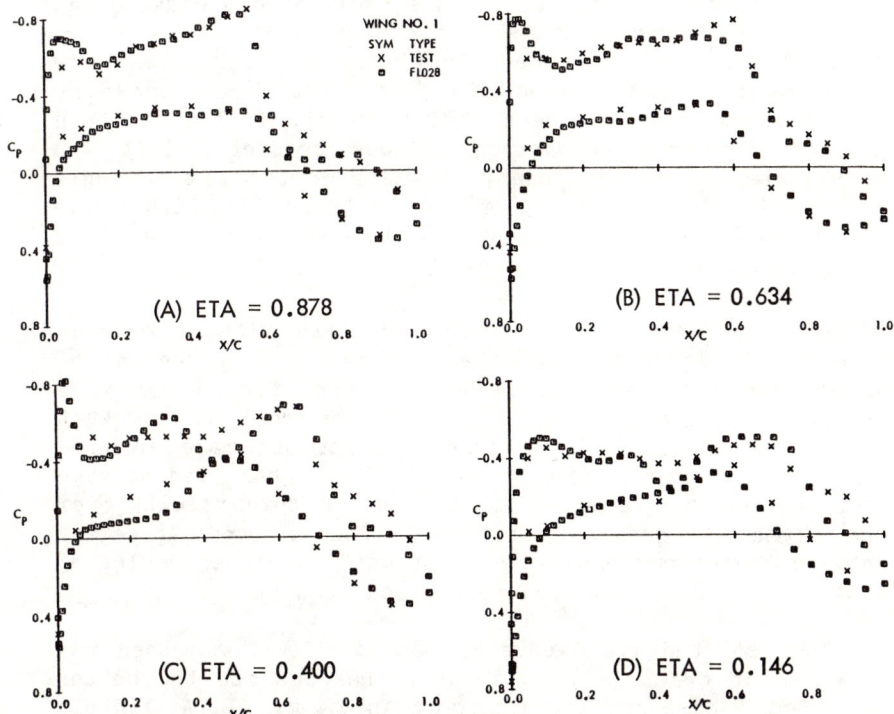

Fig. 11 Comparison of FLO-28 wing-body pressures with experiment.

out transformed wing coordinates at the root and tip; these coordinates were also very erratic. Since FLO-28 uses a Joukowski body transformation which is best suited for a circular cross section, the nearly rectangular A-7 fuselage was a possible cause of this problem.

Progressive rounding of the body corners had a smoothing effect on both the transformed coordinates and pressures. These fuselage modifications necessitated a lowering of the high mounted wing to maintain a reasonable wing-body intersection. When this process had smoothed the root and tip transformed coordinates, large pressure oscillations remained. The worst pressures were in the vacinity of midspan. The program was modified to print out transformed coordinates at every computational station, and coordinate oscillations were present with the largest being near 40% semispan.

Out of desperation, a run was made with the wing mounted in the midposition on the body. For some reason, most of the transformed coordinate oscillations disappeared and the pressures smoothed out dramatically. These FLO-28 predictions are compared in Fig. 11 with the same experimental data shown in Fig. 10. Some pressure oscillations are still seen at 0.4 semispan. Lift has been matched for this comparison because predicted lift was substantially low when analysis was run at the test angle of attack. Based on only a few analyses the lift curve slope was close to experiment and there was approximately a 1 deg shift in angle of attack.

FLO-28 predictions agree more closely with experiment than FLO-22, especially at the root and tip. Agreement of shock strength and position at the tip challenges credulity. The close comparison at the root is somewhat surprising in view of the large compromises made in wing position and intersection geometry. The higher leading edge acceleration predicted across the entire span results from the higher theoretical angle of attack. Good agreement between theory and experiment continued to higher angles of attack until shock-induced separation occurred.

The isolated wing geometry used in FLO-22 was then run in FLO-28 in order to help identify reasons for the better agreement. These results are shown in Figure 12. A slightly stronger compression is predicted by FLO-28 than FLO-22. This affect may be caused by the conservative solution of FLO-28 or may simply result from the higher lift coefficient. Basically, both isolated wing predictions are very similar. This fact would seem to identify the error then as a body effect. It is not obvious whether this body effect is a physical phenomenon or a numerical characteristic of the theoretical solution.

FLO—30

FLO-30 is a finite-volume wing-body transonic analysis developed by Caughey.[9] This code uses body-oriented coordinates instead of the Joukowski transformation of FLO-28. Input geometry is shown in Fig. 13. FLO-30 predictions are compared with experiment in Fig. 14 for the same test case shown previously. The code was run at the experimental angle of attack, and no modifications to the geometry were necessary. Agreement of theory and experiment is much better than with FLO-22. Lift is slightly low, and the shock is predicted too far aft near the tip. FLO-30 was run using a medium grid resulting from only one grid halving because of cost. FLO-22 and FLO-28 were run on a fine

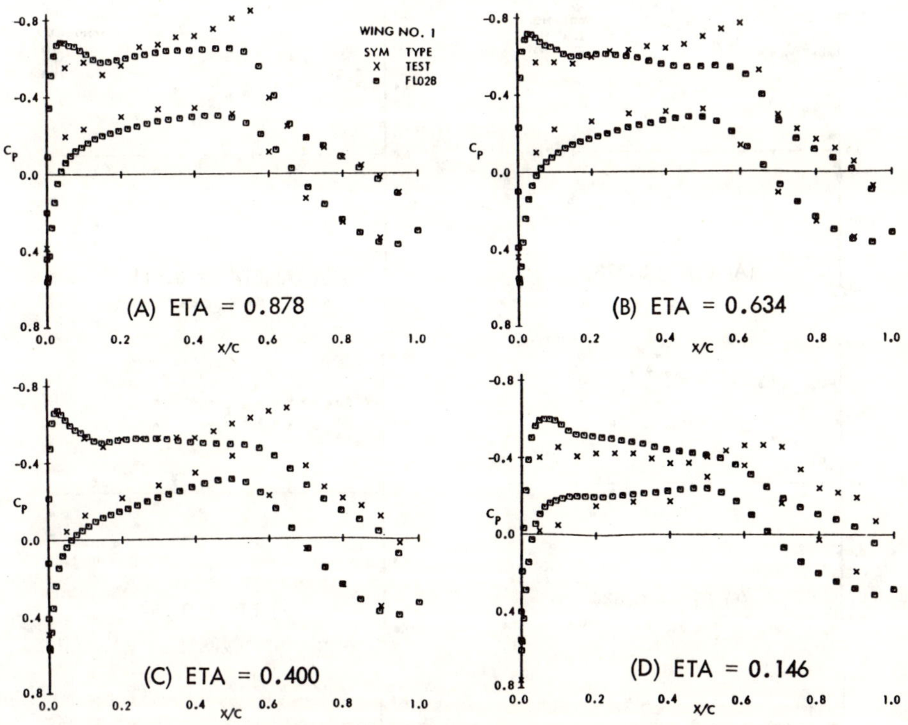

Fig. 12 Comparison of FLO-28 wing pressures with experiment.

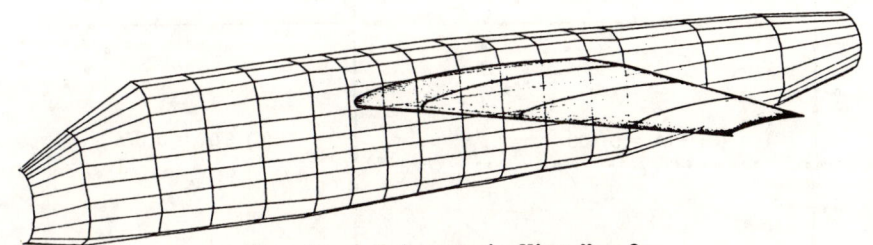

Fig. 13 A-7 transonic Wing No. 2.

grid with two grid halvings. A comparison of FLO-30 pressure predictions for a medium and fine grid is presented in Fig. 15 for the Wing No. 2 design case. Pressures are nearly converged after 200 iterations on the medium grid, and the cost is one-fourth that of the fine grid solution.

Both FLO-30 and FLO-28 results agree well with experiment. However, FLO-28 required adjustment to the angle of attack and extensive geometry manipulation to obtain results which agreed with test data. It certainly seems

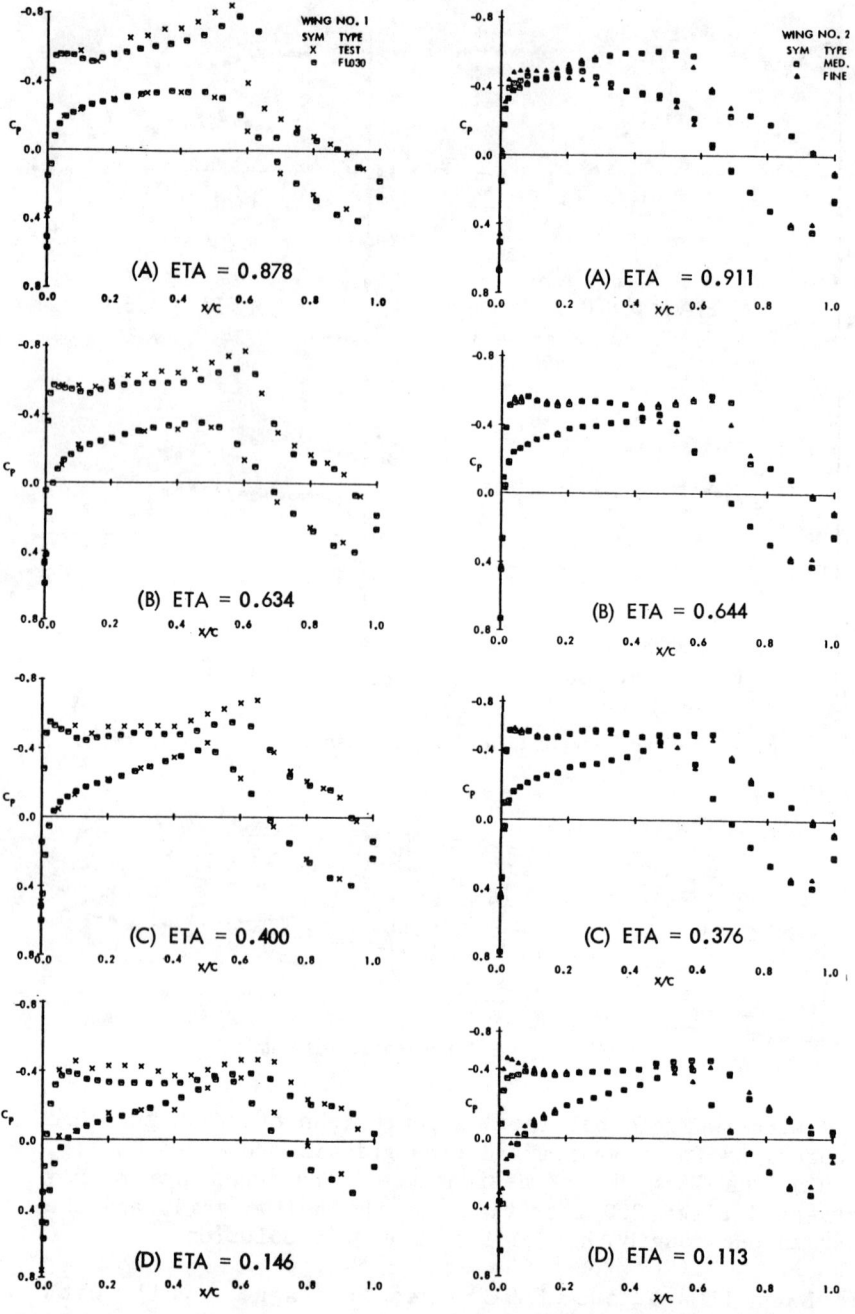

Fig. 14 Comparison of Flo-30 wing-body pressures with experiment.

Fig. 15 Comparison of FLO-30 medium and fine grid solutions.

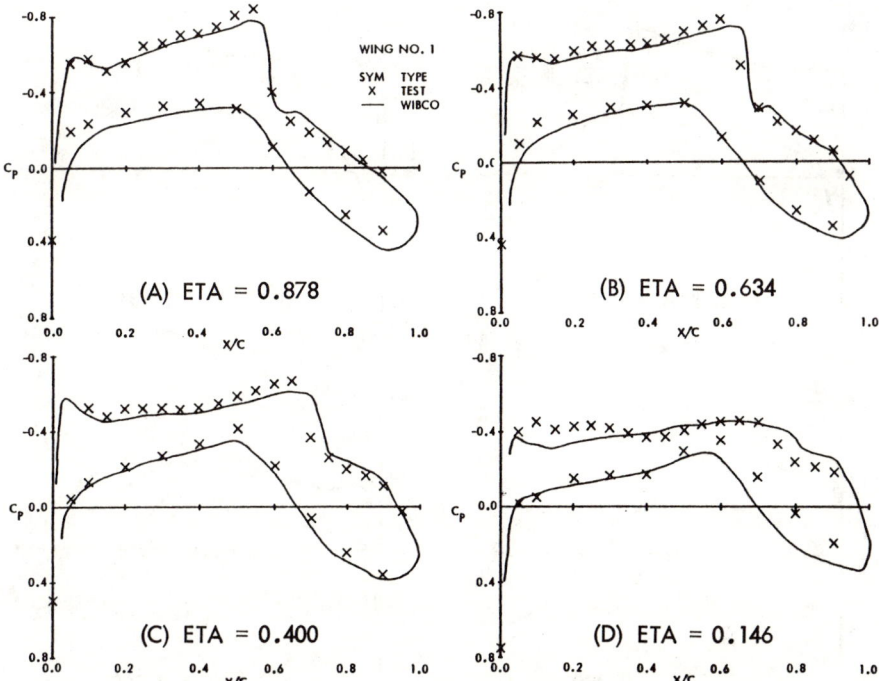

Fig. 16 Comparison of WIBCO wing-body pressures with experiment.

doubtful that good agreement would have been possible for this case without prior knowledge of the experimental data to be matched. FLO-30 on the other hand gave good results without manipulation.

WIBCO

WIBCO is a small perturbation analysis[10] which incorporates an embedded body grid system. A two-dimensional boundary-layer analysis is linked to this code, and the displacement thickness is updated during the solution. Wing No. 1 was analyzed as a midwing mounted on a body of revolution. This simplified body approximation was chosen because it could be quickly input without using the specialized WIBCO arbitrary geometry option.

Predicted pressures are compared with experiment in Fig. 16 for the cruise design condition. Agreement is surprisingly good considering the crude body representation used. The code was run at the test angle of attack, and wing lift is matched quite well. An isolated wing

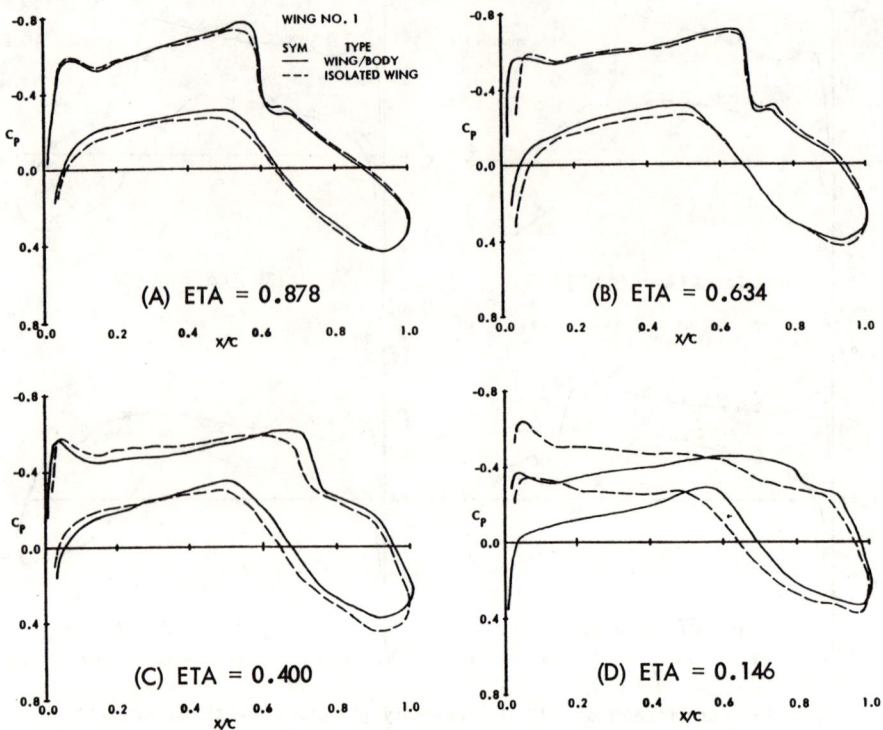

Fig. 17 Comparison of WIBCO wing-body and isolated wing pressures.

representation of this configuration was then run, and the results were compared to the wing-body prediction (Fig. 17). Relatively little difference is seen between these two geometric models except near the root. Practically no difference occurs near the tip compared with FLO-28 results (Figs. 11 and 12).

The pressure predictions of FLO-28, FLO-30, and WIBCO wing-body analyses have been shown to be in good agreement with experiment for the configurations tested in this study. These comparisons raise more questions than they answer, particularly the WIBCO isolated wing results. Since it is very dangerous to draw conclusions based upon such a limited correlation study, these comparisons are presented as a few pieces to a complex puzzle. Wing-body transonic analyses offer an important aid to the aircraft design problem. Unfortunately all of these codes are too expensive to be used with direct optimization. But transonic wing-body analyses can be used to evaluate isolated wing optimization results and to help define specified pressure distributions.

Concluding Remarks

The preceding sections described a transonic wing research study which spanned five years from design procedure formulation through verification studies. Certain difficulties in using computational fluid dynamics codes for design were highlighted. This research effort will now be contrasted with more practical aerodynamic design problems in order to explain the environment in which these codes must function.

The design criteria for both A-7 wings were difficult to meet but easy to define. Practical problems involve complex tradeoffs within aerodynamics and between many other disciplines. Conflicting aerodynamic requirements could involve efficient cruise, high transonic buffet-free maneuver, acceleration, good stall characteristics, high-speed performance, and good high-lift characteristics. The entire flight envelope must be considered. However, aerodynamic considerations alone do not dictate the optimum configuration. For instance an aft-loaded supercritical pressure distribution can result in penalties due to structural weight, actuator sizing, trim drag, and control effectiveness. The optimum configuration must be sought through repetitive analyses of all these factors. But CFD codes are uniquely ill-suited for this work because of their long run times and limitations on the geometry they can model.

The largest constraints to using CFD codes for aircraft design can be summed up by one word, risk. Risk that the design cannot be completed on time. Risk that design changes based on analysis predictions will result in a drag increase due to the limits of potential theory or interference effects of components which cannot be modeled. Risk is what separates research and development from practical design exercises, and risk can have an adverse effect on career longevity. Once CFD codes establish a track record of consistent reliability, they will have a larger impact upon aerodynamic design.

The A-7 wing study was built upon the experience of other users. Conversely, lessons learned during this study were readily available to other users and code developers. This interchange of information was made possible through the efforts of Raymond M. Hicks who has had a significant impact on the use of CFD codes by industry. If more of this work were done, it could have a very positive effect on both code development and implementation.

Acknowledgment

This work was performed while the author was employed by Vought Corporation, Dallas, Texas.

References

[1] Haney, H. P. and Johnson, R. R., "Application of Numerical Optimization to the Design of Wings with Specified Pressure Distributions," NASA CR 3238, Feb. 1980.

[2] Jameson, A., Caughey, D. A., Newman, P. A., and Davis, R. M., "A Brief Description of the Jameson-Caughey NYU Transonic Swept-Wing Computer Program - FLO22," NASA TM X-73, 996, Dec. 1976.

[3] Haney, H. P., Johnson, R. R., and Hicks, R. M., "Computational Optimization and Wind-Tunnel Test of Transonic Wing Designs," Journal of Aircraft, Vol. 17, July 1980, pp. 606-610.

[4] Hicks, R. M. and Vanderplaats, G. N., "Application of Numerical Optimization to the Design of Supercritical Airfoils without Drag Creep," SAE Paper 770440, Wichita, Kans., 1977.

[5] Vanderplaats, G. N., "CONMIN - A Fortran Program for Constrained Function Minimization, User's Manual," NASA TM X-62,282, Aug. 1973.

[6] Haney, H.P., Waggoner, E.G., and Ballhaus, W. F., "Computational Transonic Optimization and Wind-Tunnel Test of a Semi-Span Wing," Journal of Aircraft, Vol. 15, Sept. 1978, 457-463.

[7] Haney, H. P., Waggonner, E. G., and Ballhaus, W. F., "Wind-Tunnel Investigation of Computationally Optimized Variable Camber Wing Configurations," NASA TM 78478, June 1978.

[8] Haney, H. P. and Hicks, R. M., "A Comparison of Theoretical and Experimental Pressure Distributions for Two Advanced Fighter Wings," NASA TM 81331, Oct. 1981.

[9] Caughey, D. A. and Jameson, A., "Recent Progress in Finite-Volume Calculations for Wing-Fuselage Combinations," Paper No. 79-2523, AIAA 23rd Fluid and Plasma Dynamics Conference, Williamsburg, Va., July 1979.

[10] Boppe, C. W., "Computational Transonic Flow About Realistic Aircraft Configurations," Paper No. 78-104, AIAA 16th Aerospace Sciences Meeting, Huntsville, Ala., Jan. 1978.

Chapter X.

Transonic Computational Experience for Advanced Tactical Aircraft

E. Bonner* and P. B. Gingrich*
North American Aircraft Division, Rockwell International, Los Angeles, Calif.

Nomenclature

AR	=	aspect ratio
b	=	surface span
c	=	local chord
C_D	=	drag coefficient
C_ℓ	=	section lift coefficient
C_L	=	lift coefficient
C_P	=	surface pressure coefficient
CSD	=	classical small disturbance
FPE	=	full potential equation
ITER	=	relaxation iterations
M	=	freestream Mach number
MSD	=	modified small disturbance
P	=	static pressure
q	=	freestream dynamic pressure
U	=	freestream velocity
w	=	vertical perturbation velocity
X,Y,Z	=	axial, lateral, vertical body axis coordinates
α	=	angle of attack
Γ	=	dihedral angle
η	=	$Y/_{b/2}$
λ	=	taper ratio

Presented at the Transonic Perspective Symposium, NASA/Ames Research Center, Moffett Field, Calif., Feb. 18-20, 1981. Copyright © American Institute of Aeronautics and Astronautics, Inc., 1981. All rights reserved.
  *Member Technical Staff, Aerodynamics.

Subscripts

u   = upper surface

Superscripts

*   = critical (sonic) condition

Introduction

The aerodynamic design of wings for transonic flight has, until recently, emphasized supercritical airfoils in conjunction with extensive experimental tests. The development of theoretical transonic methods to aid the design process was hampered by the nonlinear mathematical character of such flows. Recently, several computational methods have been developed for three-dimensional transonic flows about wings and simple body-wing combinations.

The applications of these techniques to a wide variety of planforms and conditions is required in order to define their reliability and limitations. The HiMAT development[1] and the Forward Swept Wing Program provide this opportunity for a highly maneuverable fighter. A similar study for transport-type configurations has recently been reported.[2] Since three-dimensional transonic codes are undergoing continual evolution, the validation effort will be an essential requirement for the forseeable future.

The usefulness of transonic computational methods should be considered in terms of cost and time to derive an efficient transonic flow by alternative means. The cost of high Reynolds number wind-tunnel testing is high and continuously increasing. Parametric model modifications to improve flow quality are expensive with turnaround time much slower than equivalent numerical simulation changes. Since these trends can be expected to continue, computational methods will receive increasing emphasis in the future. Theoretical analysis has the further benefit of permitting experimental refinement and verification to be performed for a much smaller design space.

Discussion

The effort to be described is concerned with the three-dimensional transonic numerical analysis performed in support of the HiMAT and FSW maneuver wing definition (Table I). In order to place these results in their proper context, it is emphasized that computational phases were iteratively coupled with experimental tests to assess the wing performance and verify (or in certain cases dispute as the case turns out) the numerical analysis. To further understand the approach used, the status of the available transonic codes at the time must be kept in mind. Early refinement was accomplished by parametric variation of the wing geometry, i.e., numerical filing. Currently inverse solvers are used to systematize and accelerate the process.

The transonic wing design effort conclusively established the importance of a balanced approach between theory and experiment. Measurements directly led to theoretical code extensions to reconcile test results and refine imposed pressure distributions to minimize or eliminate separation.

HiMAT Wing Analysis

The computational model and associated geometric parameters under consideration are presented on Fig. 1. Typical maneuver airfoil sections for the root, midspan, and tip region and the twist distribution are indicated to further define the problem.

Numerical results will first be presented for the isolated wing. Initial analysis utilized the Bailey-Ballhaus classical small disturbance (CSD) theory formulations.[3] At the time, it was suggested that the CSD formulation would not adequately capture highly swept shocks. Some calibration would be necessary to determine those flow conditions or regions of the planform where the predictions could be used with confidence. Wind-tunnel tests indicated that the computations adequately predicted the location and strength of the inboard shock. The existence of the outboard shock however was not produced numerically. These results are summarized on Fig. 2. This finding led to several approximate theoretical modifications[4] to "enrich" the Karman-Guderly transonic small disturbance equation of motion by the addition of certain nonlinear cross terms. The extended analysis was successful in predicting the entire upper surface wing flow. The before and after comparison is presented on Fig. 3. The modified small disturbance (MSD) theory derived in response to the HiMAT tests has been adopted in more recent code developments[5,6] concerned with improving computational efficiency, including fuselage effects, etc.

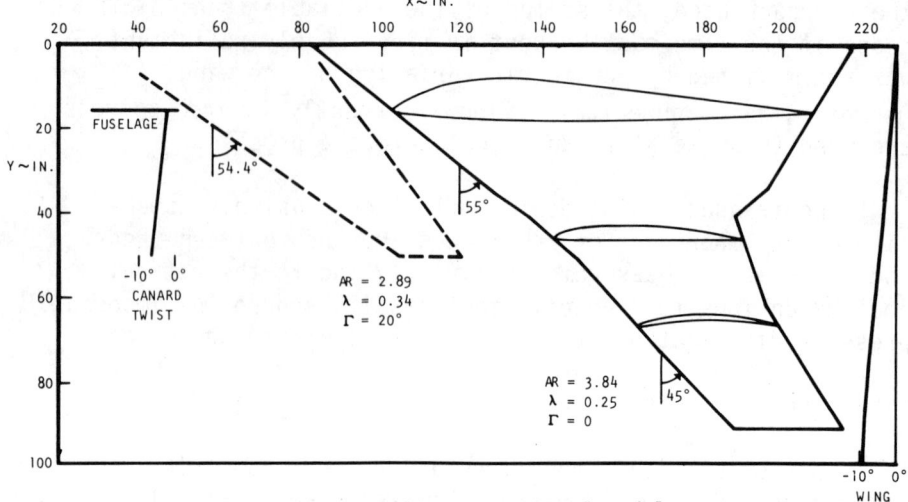

Fig. 1 HiMAT computational model.

Table 1. Numerical details

Case	Computation	Fine mesh, X, Y, Z	Fine relaxation cycles	Boundary-layer iterations
HiMAT	CSD	80 x 40 x 50	400	--
	MSD	80 x 40 x 50	400	--
	FLO-22	160 x 32 x 16	250 (α = 10 deg)	--
	FLO-27	160 x 32 x 16	100 (α = 6 deg)	--
FSW	MSD	60 x 30 x 20[a]	300	--
		30 x 20 x 20[b]		
	FLO-27	128 x 24 x 16	200	--
	FLO-28	128 x 24 x 16	200	--
	Boundary layer	23 x 10 x 20	40	3

[a] Inner mesh.   [b] Outer mesh.

TRANSONIC EXPERIENCE FOR TACTICAL AIRCRAFT

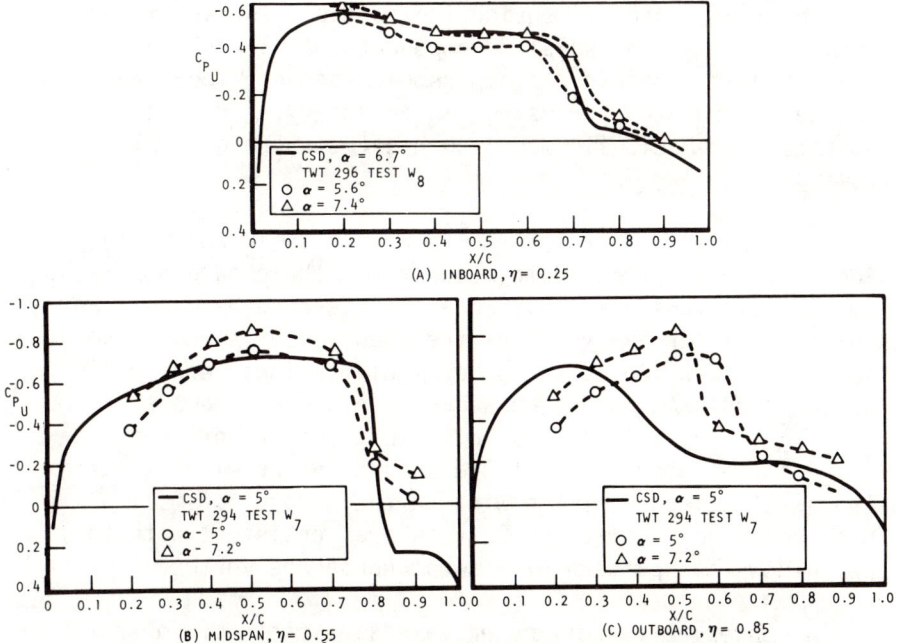

Fig. 2 Calculated and measured canard-off wing pressures, M = 0.9: a) inboard, η = 0.25; b) midspan, η = 0.55; c) outboard, η = 0.85.

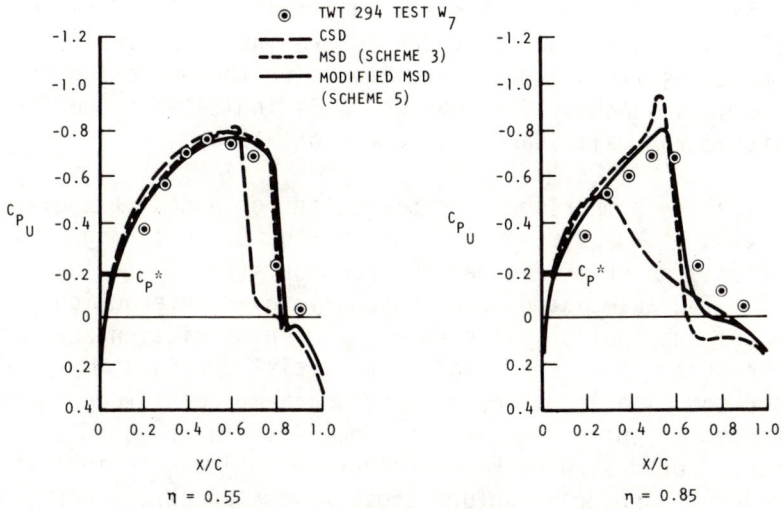

Fig. 3 Comparison of small disturbance theory and measured HiMAT wing upper surface pressure distribution at M = 0.9, α = 5 deg.

The importance of considering multiple surface interactions rather than isolated components has long been recognized at subsonic and supersonic speeds. They are routinely accounted for using linear theory. Transonic codes to systematically model such effects are generally lacking.

An approximate procedure was developed in which the wing twist was corrected to account for the canard downwash. Computations indicated only modest difference between transonic and linear analysis for these characteristics. The results further indicated camber modifications were small for the conditions under consideration and consequently neglected. Typical downwash distributions are presented on Fig. 4. The interaction of the canard on the wing is quite strong. The impact of the wing on the canard is small. These relative interaction effects are consistent with the upwash-downwash pattern of a swept horseshoe vortex.

A conservative full potential assessment[7] of correcting the wing for canard downwash is presented in Figs. 5 and 6. The chordwise pressure distributions are modified by changes in the local and gross wing circulation.

Comparison of measured wing upper surfaces pressures and conservative full potential analysis[7] at an off design condition of $M = 0.9$, $\alpha = 6$ deg is presented in Fig. 7. The test results are well predicted with the exception of the forward shock at 70% span which is indicated to be sensitive to small changes in angle of attack.

Similar comparisons are presented for a near maneuver condition of $M = 0.9$, $\alpha = 10$ deg in Fig. 8. A strong computational shock of increasing spanwise strength is indicated. The associated shock/boundary-layer interaction produces substantial differences between prediction and measurement. A nonconservative analysis[8] provides improved comparison as a result of forward shock movement (due to the mass source) which simulates the neglected viscous interaction. For design work, conservative analysis is indicated in order to properly capture shock waves and pursue corrective action to weaken or eliminate them. For flows with existing strong shocks, correlation success will require

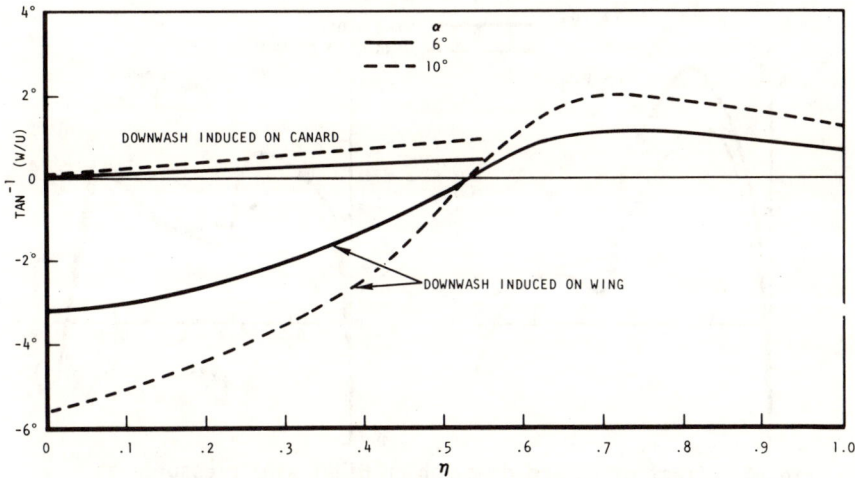

Fig. 4 Average downwash induced on wing and canard at M = 0.9.

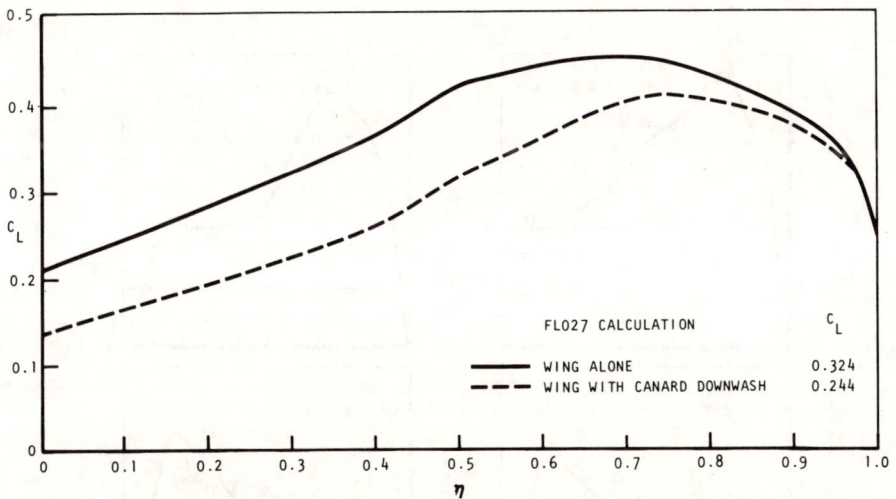

Fig. 5 Effect of canard on HiMAT wing span load at M = 0.9, α = 6 deg.

explicit accounting for viscous effects or the interim use of nonconservative shock point operator approximations.

Measured transonic lifting efficiency is presented in Fig. 9 to globally judge the supercritical flow quality. Vortex drag levels corresponding to wing elliptic span loading are shown for comparison purposes. The impact of wing-

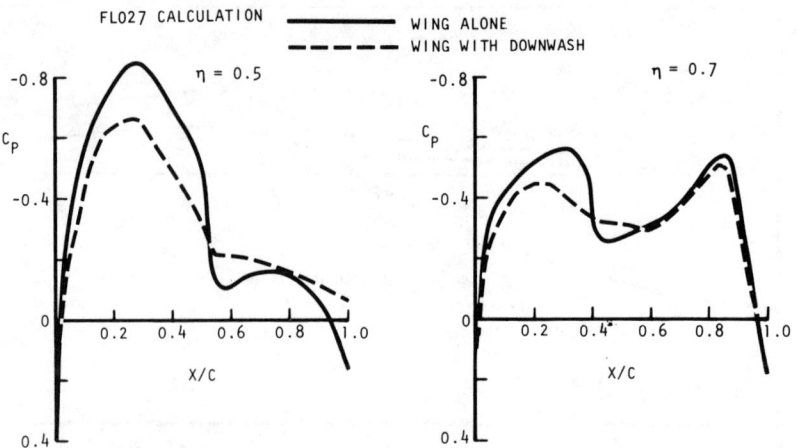

Fig. 6 Effect of canard downwash on HiMAT wing pressures at M = 0.9, α = 6 deg.

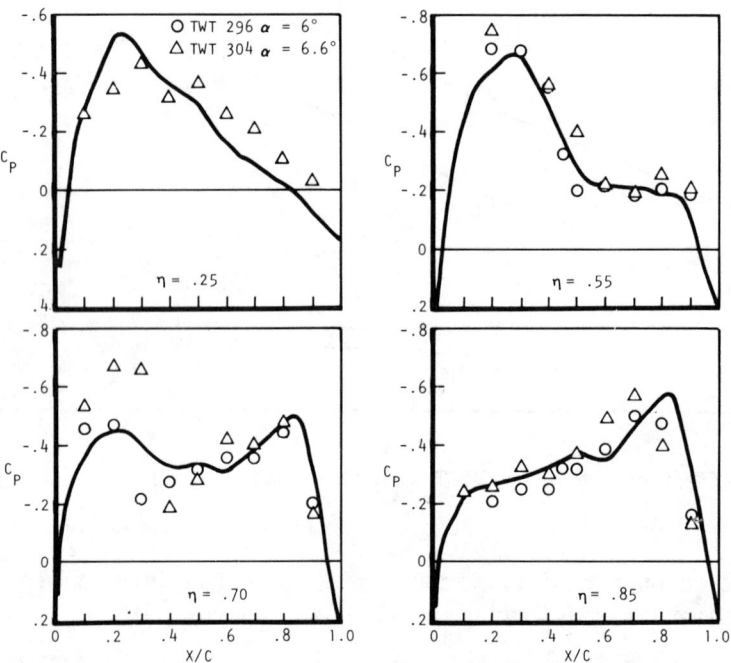

Fig. 7 Comparison of full potential theory and HiMAT wing upper surface pressure distributions at M = 0.9, α = 6 deg.

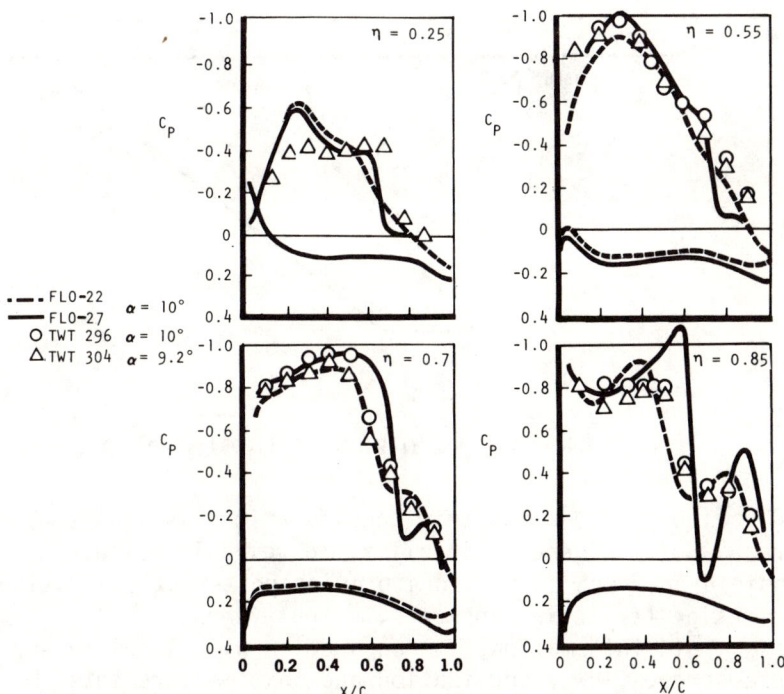

Fig. 8 HiMAT wing pressures near maneuver; $M = 0.9$, $\alpha = 10$ deg.

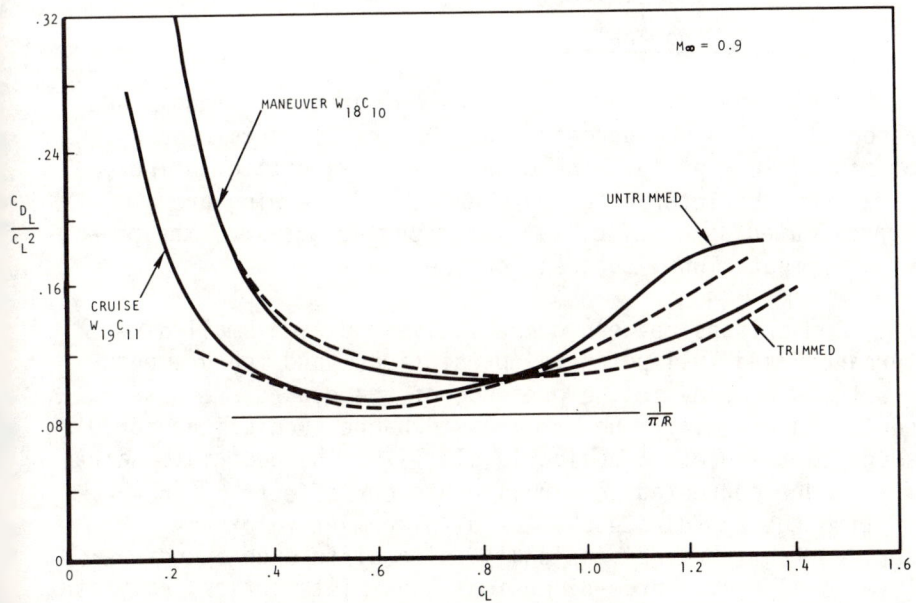

Fig. 9 HiMAT transonic drag experience.

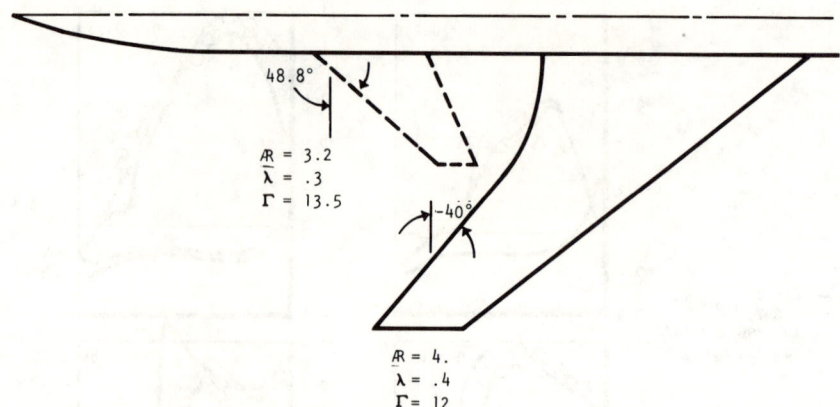

Fig. 10 Forward swept wing computational model.

canard variable camber and trim considerations is indicated. As a result of the general inability to reliably calculate transonic drag, such comparisons provide post-test evaluation of the design (criteria, approach, computations, etc.) success. This information, in conjunction with local surface pressures, flow visualization and rake measurements, is used to initiate further computational design as deemed necessary.

Forward Swept Wing Analysis

The wing-body and canard-body computational models and associated planform geometric parameters for this case are presented in Fig. 10. The canard is shown dotted to indicate that the interaction between it and the wing are approximated as an effective twist on the basis of the previous computational success for the HiMAT.

Comparisons between conservative modified small disturbance[6] and full potential predictions[7] and measurements for the wing-body in the presence of the canard are presented in Fig. 11. The small disturbance predictions correspond to the pretest design levels.[9,10] The associated airfoils were corrected to minimize viscous effects by undercutting the coordinates by the displacement thickness derived from three dimensional finite difference boundary-layer analysis. Three-dimensional weak interaction analysis for the test coordinates indicates (Fig. 12) that viscous

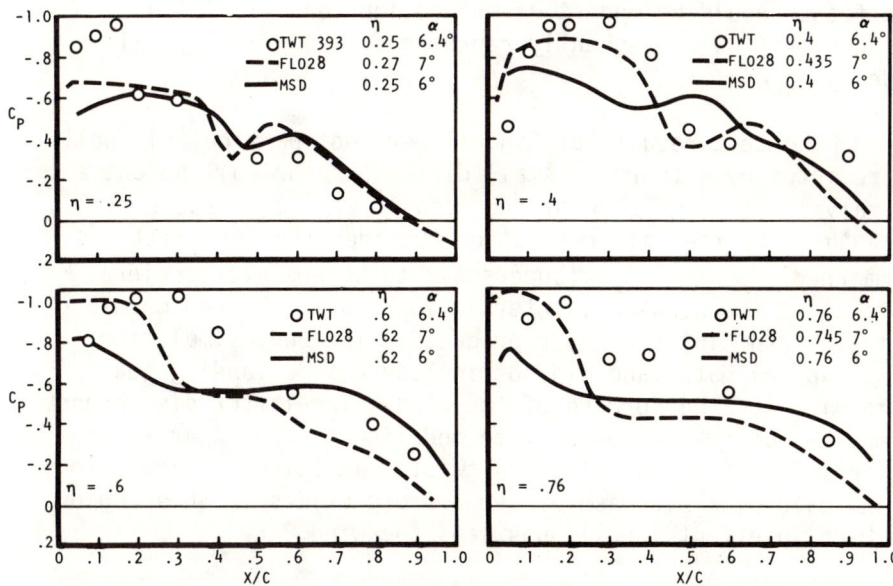

Fig. 11 Comparison of calculated and measured wing upper surface pressure distributions at M = 0.9, α = 6 deg.

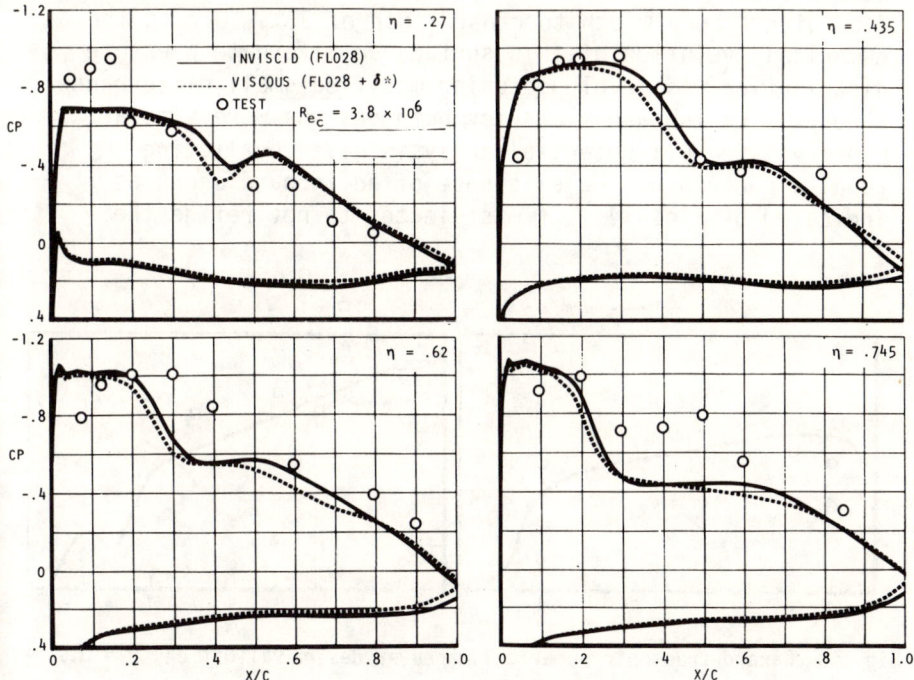

Fig. 12 Impact of viscous effects at M = 0.9, α = 7 deg.

effects should be modest at the design condition. The maneuver flow consequently represents a numerically well-designed one.

The test results of Fig. 11 were not particularly well predicted by either the small disturbance or full potential analysis although some of the details were more successfully captured by the latter solution provided the test lift was matched. Weak shocks (supersonic to supersonic deceleration) at the wing-body intersection and in the region of the canard tip were not reproduced. The body simulation was approximate, and did not include a side landing gear fairing located forward of the wing. The latter discrepancy may be due to the approximate modeling of the canard induction effects as an effective twist. Additional calculations indicate peak sidewash of about 8 deg exists in this region which in effect locally unsweeps the wing.

Similar comparisons for the canard-body in the presence of the wing are presented in Fig. 13. The modified small disturbance pretest design levels[9,10] are reasonably well produced for this heavily loaded, highly supercritical flow with a body width to canard span of 30%. During the numerical evolution of this surface an alternate pressure distribution was developed using modified small disturbance theory. The inboard flows for both designs were similar. Conservative full potential analysis gave conflicting results concerning the existence of an inboard shock as indicated by Fig. 14. It was elected to not refine the

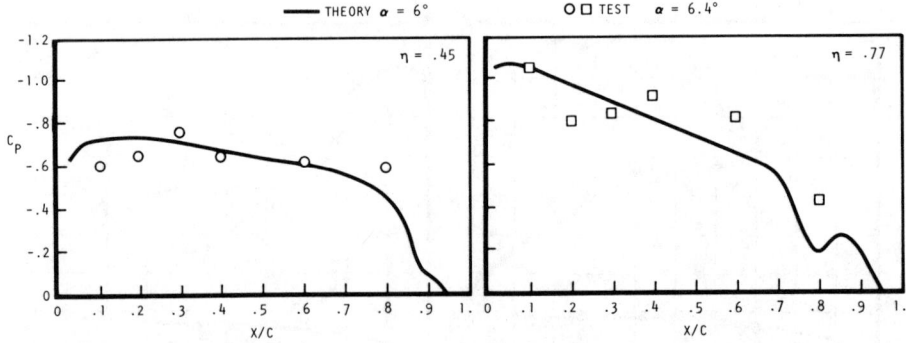

Fig. 13 Canard transonic numerical maneuver design validation, M = 0.9.

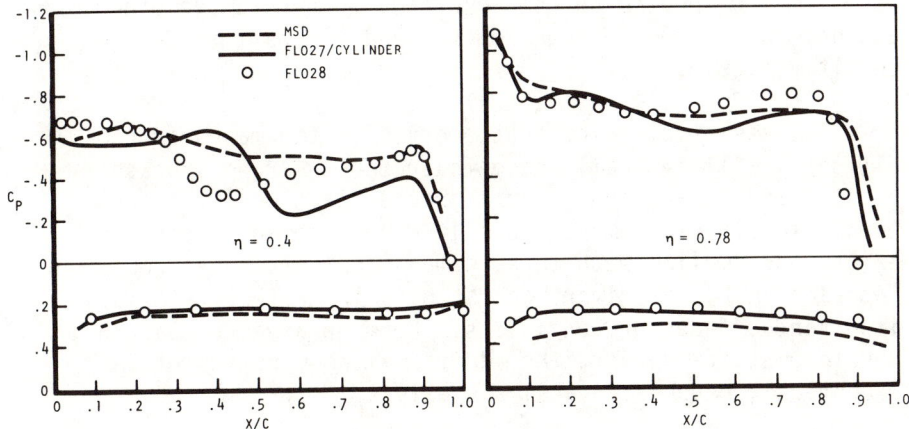

Fig. 14 Comparison of transonic canard predictions at M = 0.9, α = 6 deg.

modified small disturbance design. Although the latter design was not tested, experimental results for the initial design did not indicate a midchord root shock.

The previously cited differences in prediction success between the wing and canard and small disturbance and full potential analysis tend to inhibit decisions concerned with design refinements to weaken shock waves, develop prescribed supercritical upper surface pressure distributions, control and minimize separation, etc. The importance of iteratively coupled tests to sort out these anomalies is clearly indicated.

Design Considerations

The use of controlled supercritical flow technology for maneuver results in performance penalties for subsonic penetration and supersonic acceleration/cruise conditions. Variable camber (e.g., deflectable leading and trailing-edge devices) and aeroelastic twist is commonly used to improve the compatibility between widely different operating points.

Such approaches however do not manage surface spar box camber. Current design trends are to adopt a compromise between the low and high design lift conditions. Emphasis (bias) may be placed on one condition more than the other and depends on design requirements and aircraft trade sensitivities.

The essential point that emerges is the maneuver condition(s) will typically have moderate-to-strong shocks present and separation extents greater than well-designed transonic flows. Computationally a requirement therefore exists for dealing with three-dimensional strong viscous interactions at a primary operating point rather than less important off-design conditions. From an advanced aircraft development standpoint, the additional requirement of an economical, responsive analysis exists.

Conclusions

1) Addition of nonlinear cross terms to classical small disturbance theory was required to capture the HiMAT outboard wing shock.

2) Full potential analysis successfully predicted HiMAT wing pressure at intermediate and near maneuver conditions. Successful approximation of canard-wing transonic interactions using an immersion (effective twist) approach is a corollary result.

3) Strong shock measurements were more accurately approximated by nonconservative analysis. Use of conservative analysis is indicated for design since geometric modifications to weaken or eliminate shocks requires accurate capture.

4) Neither modified small disturbance or full potential analysis was particularly successful in predicting forward swept wing maneuver design pressures.

5) A thin heavily loaded surface with a relatively large body ($W/b \simeq 0.30$) was successfully designed using inverse modified small disturbance theory.

6) An improved accounting of the transonic interaction between multiple surfaces is required.

TRANSONIC EXPERIENCE FOR TACTICAL AIRCRAFT

7) A need to computationally treat strong viscous interactions at a maneuver operation point is emerging.

References

[1] Gingrich, P.B., Child, R.D., and Panageas, G.N., "Aerodynamic Development of the Highly Maneuverable Aircraft Technology/ Remotely Piloted Research Vehicle," NASA CR-143841, June 1977.

[2] Henne, P.A. and Hicks, R.M., "Transonic Wing Analysis Using Advanced Computational Methods," AIAA Paper No. 78-105, Jan. 1978.

[3] Ballhaus, W.F. and Bailey, F.R., "Numerical Calculation of Transonic Flow About Swept Wings," AIAA Paper No. 72-667, Boston, Mass. June 1972.

[4] Ballhaus, W.F., Bailey, F.R. and Fricks, J., "Improve Computational Treatment of Transonic Flow About Swept Wings," Advances in Engineering Sciences, NASA CP-2001, Nov. 1976.

[5] Boppe, C.W., "Calculation of Transonic Wing Flows by Grid Embedding," AIAA Paper No. 77-207, Los Angeles, Calif., Jan. 1977.

[6] Mason, W., MacKenzie, C., Stern, M., Ballhaus, W.F., and Frick, J., "An Automated Procedure for Computing the Three-Dimensional Transonic Flow Over Wing-Body Combinations, Including Viscous Effects," AFFDL-TR-77-122, Feb. 1977.

[7] Jameson, A. and Caughey, D.A., "A Finite Volume Method for Transonic Potential Flow Calculations," AIAA Paper No. 77-635, Albuquerque, N. Mex., June 1977.

[8] Jameson, A., "Iterative Solution of Transonic Flows Over Airfoils and Wings, Including Flows at Mach 1," Communications in Pure and Applied Mathematics, Vol. 27, 1974, pp. 283-309.

[9] Shankar, V. and Malmuth, N.D., "Computational Transonic Design Procedure for Three-Dimensional Wings and Wing-Body Combinations.," AIAA Paper No. 79-0344, New Orleans, La., Jan. 1979.

[10] Shankar, V., "Computational Transonic Inverse Procedure for Wing Design with Automatic Trailing Edge Closure," AIAA Paper No. 80-1390, Snowmass, Colo., July 1980.

Chapter XI.

Extension of FLO Codes to Transonic Flow Prediction for Fighter Configurations

A. Verhoff* and P. J. O'Neil†
McDonnell Aircraft Company, McDonnell Douglas Corporation, St. Louis, Mo.

Nomenclature

AR	=	aspect ratio
C_p	=	pressure coefficient
i	=	additional fuselage incidence
M_∞	=	freestream Mach number
q_∞	=	freestream velocity
u	=	streamwise velocity perturbation
v	=	upwash velocity perturbation
w	=	spanwise velocity perturbation
x	=	streamwise coordinate
y	=	normal coordinate
z	=	spanwise coordinate
α	=	aircraft angle of attack

Introduction

The development of transonic prediction methods for wing/fuselage combinations having low aspect ratio and wide, irregular-shaped fuselage cross sections is of great importance because of potential application in the fighter aircraft design process. The most suitable computer programs at the present time for calculating three-dimensional, transonic, potential flow are the series of finite-volume FLO codes (i.e., FLO-27,28,30) developed by Jameson and Caughey.[1-3] These programs can predict transonic flow, including the approximation of embedded shock wave location and strength, about a swept wing attached to either 1) a

---

Presented at the Transonic Perspective Symposium, NASA/Ames Research Center, Moffett Field, Calif., Feb. 18-20, 1981. Copyright © 1981 by the American Institute of Aeronautics and Astronautics. All rights reserved.
*Technical Specialist - Technology, Aerodynamics.
†Engineer - Technology, Aerodynamics.

wall (plane of symmetry) or 2) a simplified fuselage geometry. The simplified fuselage geometry option is an infinite circular cylinder in the FLO-27 code, a finite-length body with smoothly varying arbitrary cross sections in the FLO-28 code, and a finite-length body with moderate cross-sectional complexity in the FLO-30 code.

An extensive study was conducted to validate the applicability of the FLO codes to simple low aspect ratio wing/fuselage configurations. However, when applied to a high-performance fighter configuration having a wide, complex fuselage (such as the F-15), none of the programs is capable of straightforward prediction of wing surface pressures, even at outboard span stations. On the other hand, accurate predictions can be obtained at subcritical Mach numbers using panel methods which have the capability for detailed fuselage modelling. These methods allow detailed investigation of the complicated wing/fuselage interactions which the current transonic FLO codes fail to predict accurately.

An approach is being developed by the McDonnell Aircraft Company (MCAIR) which greatly extends the applicability of these programs by using an equivalent simple body (ESB) in the FLO codes to duplicate approximately the perturbation velocity distributions of a complex fuselage. The streamwise, spanwise, and normal (upwash) velocity perturbation distributions are determined at subcritical Mach numbers using a panel program with accurate fuselage modelling. For small aircraft angles of attack, the perturbations are small relative to freestream velocity so that their effects can be assumed linear and Mach number independent. Correlations can therefore be developed at subcritical Mach numbers by matching FLO code results with those of a panel program. These correlations can then be used within the FLO code for transonic flow calculations.

A very simple ESB for the F-15 aircraft has been used to predict transonic wing pressure distributions over a wide range of Mach numbers and angles of attack. These results compare favorably with existing wind-tunnel test data. The ESB method has also been applied to the study of aileron effectiveness and wing/turret interaction.

FLO Code Validation for Low Aspect Ratio Wings

The accuracy of the FLO codes in predicting low aspect ratio wing pressure distributions was verified by an extensive comparison with test data. Flowfields at various Mach numbers and angles of attack were predicted for the aspect ratio 3.0 wing-alone configuration of Ref. 4 using the FLO-27 code. A sample comparison of several predicted chordwise pressure distributions with test data is shown in Fig. 1. Flowfields at various Mach numbers and angles of attack were predicted for the aspect ratio 4.0 wing/cylinder configuration of Ref. 5 using the FLO-28 code. A sample comparison of predicted results with test data is shown in Fig. 2.

FLO Code Application to F-15 Geometry

The various FLO codes were used to predict transonic wing pressure distributions for the F-15 wing/fuselage geometry shown in Fig. 3 in order to evaluate their accuracy for such configurations. No provision was made to model the horizontal and vertical tail surfaces, as existing test data[6] are for a tail-off arrangement. The wing geometry is shown in greater detail in Fig. 4. The root section is 6% thick while the tip section is 3% thick. The leading edge sweep angle is 45 deg and the aspect ratio is 3.0 for the wing/fuselage combination.

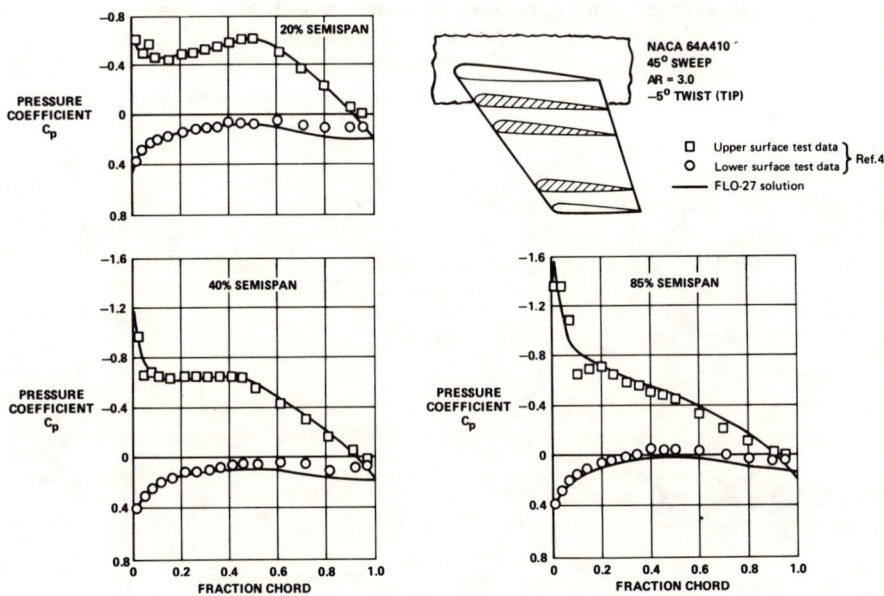

Fig. 1 Comparison of FLO-27 results with test data for wing-alone configuration; M_∞=0.80, α=8.0 deg.

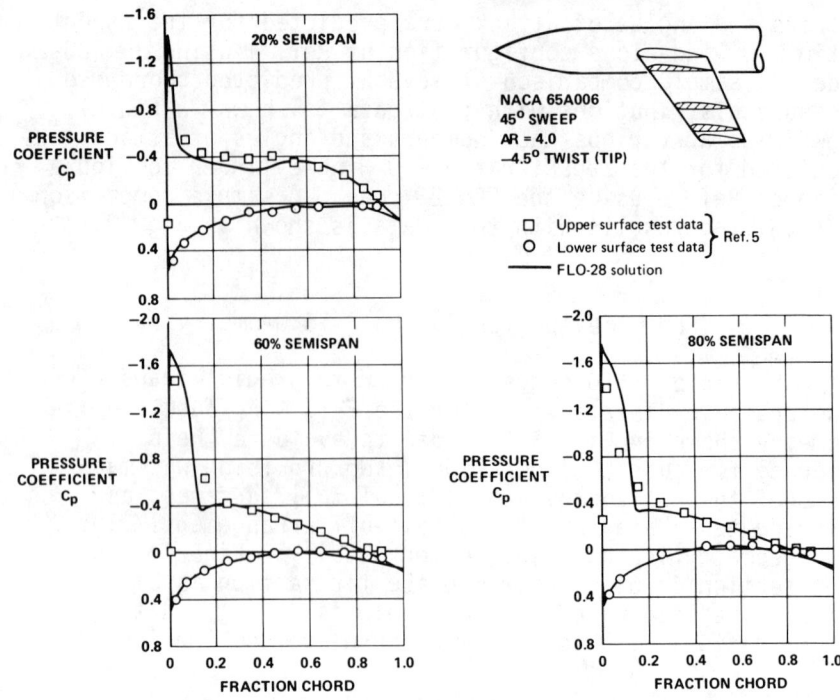

Fig. 2 Comparison of FLO-28 results with test data for wing/cylinder configuration; $M_\infty=0.85$, $\alpha=3.9$ deg.

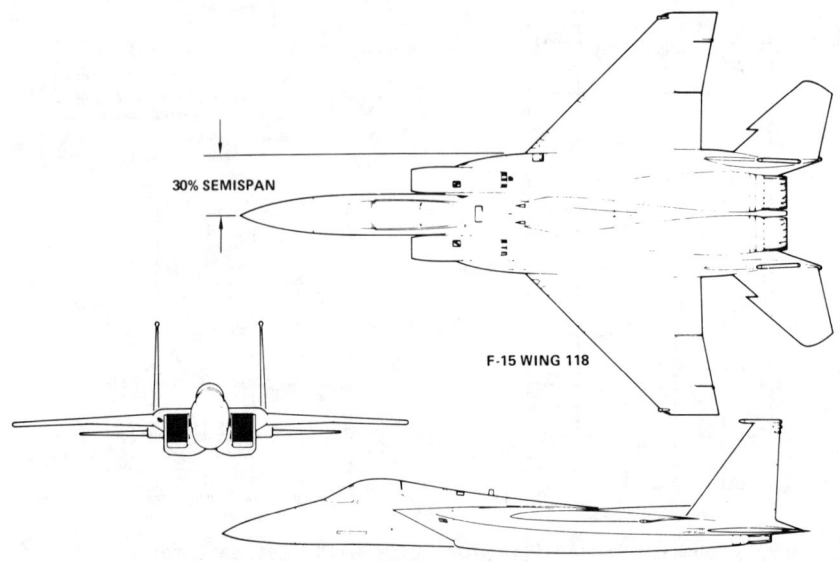

Fig. 3 F-15 fighter configuration.

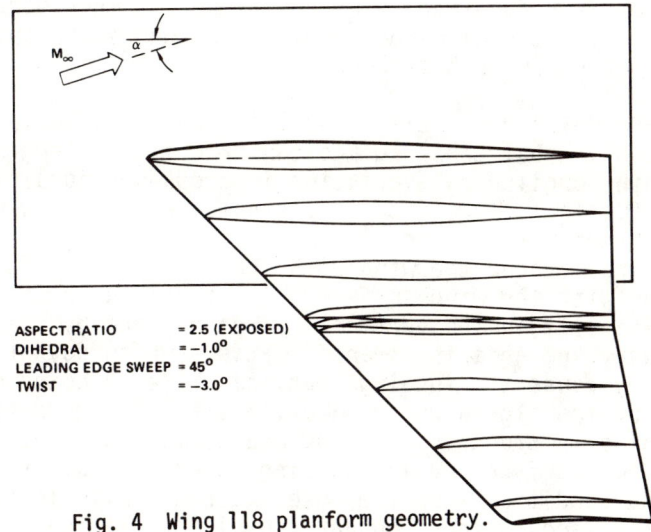

Fig. 4 Wing 118 planform geometry.

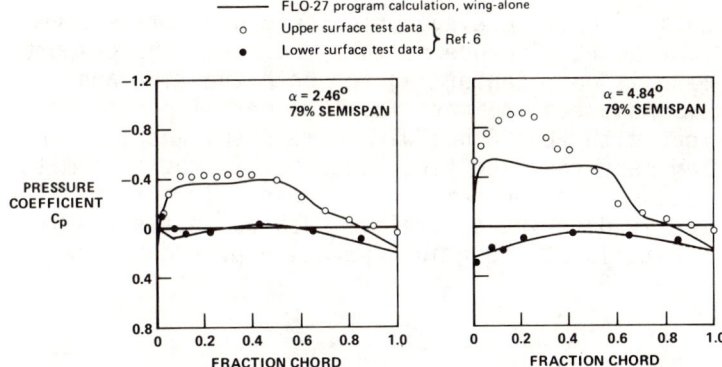

Fig. 5 Comparison of FLO-27 results with test data for F-15 Wing 118; $M_\infty = 0.90$.

Figure 5 shows FLO-27 wing-alone results compared with test data at the 79% semispan location. A similar comparison of FLO-27 wing/infinite-cylinder results is shown in Fig. 6. The cylinder diameter was chosen so that the wing root was at the correct 30% semispan location (see Fig. 3) with midwing attachment. Some improvement is achieved with the wing/cylinder modelling, but at the higher angle of attack (α = 4.84 deg), neither model is capable of predicting the increase in wing leading edge pressure peak caused by the predominantly inviscid interaction due to the complex F-15 fuselage. The deficiency is much more pronounced near the wing root (not shown) where the interaction effects are much stronger.

A more detailed fuselage modelling can be used in the FLO-28 code in that the fuselage can have finite length with smooth longitudinal variation in cross-sectional shape. However, fuselage representation is hampered by the wing-oriented C-grid system used in the program. The C-grid surfaces wrap around the wing with the innermost surface coinciding with the wing surface. The fuselage is collapsed to a slit in the vertical symmetry plane by means of Joukowsky and shearing transformations which do not necessarily cause the fuselage outline shape to coincide with one of the C-grid coordinate lines. This leads to difficulties in properly applying boundary conditions on the fuselage surface and in applying an appropriate fuselage Kutta condition when the fuselage has finite length. Similar fuselage-surface boundary condition difficulties exist in the FLO-27 code.

A cylindrical, fuselage-oriented coordinate system is used in the FLO-30 code so that the resulting computational mesh is body-conforming for both the wing and fuselage. As a result, more fuselage detail can be included than with FLO-28 but with some resultant loss in resolution near the wing tip. Because of program limitations, however, the number of grid points is not sufficient to completely describe a fuselage such as that shown in Fig. 3. Details of the grid generation procedure are given in Ref. 3.

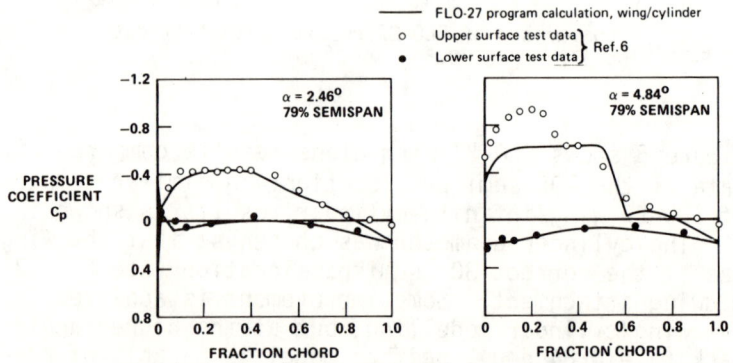

Fig. 6 Comparison of FLO-27 results with test data for F-15 Wing 118; M_∞=0.90.

Both FLO-28 and FLO-30 were used to predict F-15 wing surface pressures at various Mach numbers and angles of attack using several approximate fuselage modellings. The same wing geometry (Fig. 4) was used in each instance. Sample results compared with test data are shown in Fig. 7. The FLO-27 results of Fig. 6 are also shown for reference. The fuselage modelling used to generate the FLO-28 results in Fig. 7 was axisymmetric with a longitudinal area variation approximating that of the actual F-15 fuselage. The modelling used for the FLO-30 results is sketched in Fig. 7, and its greater detail produces some improvement in the predicted leading edge pressure peak. However, none of the predicted results shown in Figs. 5 through 7 are adequate in the wing leading edge region; the trend does indicate that accurate modelling of fuselage interaction effects is required for correct prediction of wing pressure distributions.

Panel Method Application to F-15 Geometry

Accurate, detailed fuselage modelling can be achieved by panel methods; however, linear methods are not valid for freestream Mach numbers for which the flowfield is mixed subsonic-supersonic. To test the ability of panel methods to calculate F-15 subcritical potential flowfields, an accurate representation of the F-15 wing/fuselage geometry was constructed for input to the PAN AIR program.[7,8] This program has been found by MCAIR to be very versatile and accurate. The half-space wing/

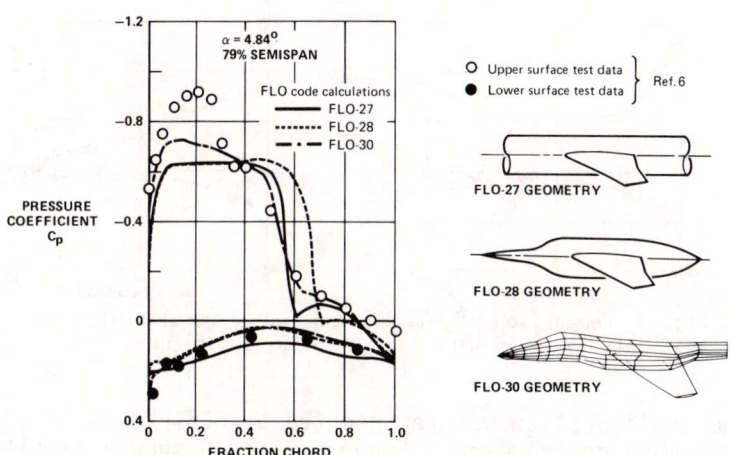

Fig. 7 Comparison of FLO code results with test data for F-15 Wing 118; $M_\infty = 0.90$.

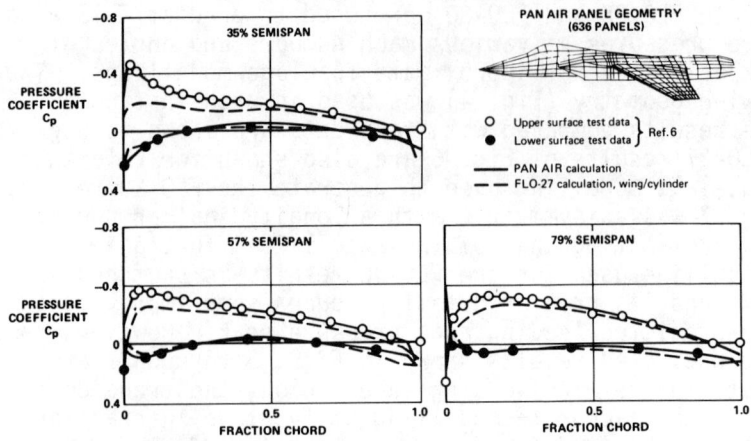

Fig. 8 Comparison of PAN AIR and FLO-27 results with test data for F-15 Wing 118; $M_\infty=0.60$, $\alpha=2.21$ deg.

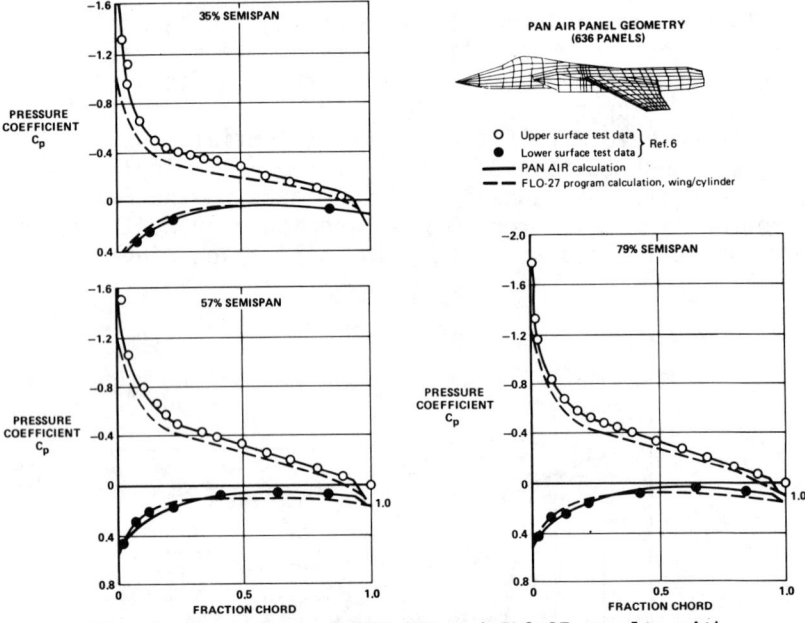

Fig. 9 Comparison of PAN AIR and FLO-27 results with test data for F-15 Wing 118; $M_\infty=0.60$, $\alpha=6.51$ deg.

fuselage configuration was represented by 636 panels of which 208 were on the wing. Figures 8 and 9 show a sample comparison of these results with the test data of Ref. 6. At these low angles of attack the agreement is very good.

FLOW PREDICTION FOR FIGHTER CONFIGURATIONS

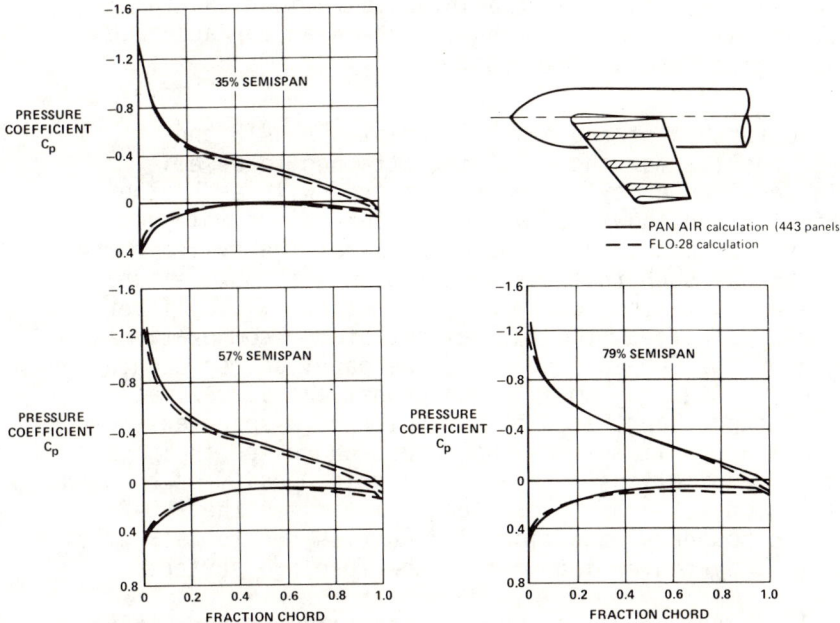

Fig. 10 Comparison of PAN AIR and FLO-28 results for semi-infinite cylinder with F-15 Wing 118; $M_\infty=0.60$, $\alpha=6.51$ deg.

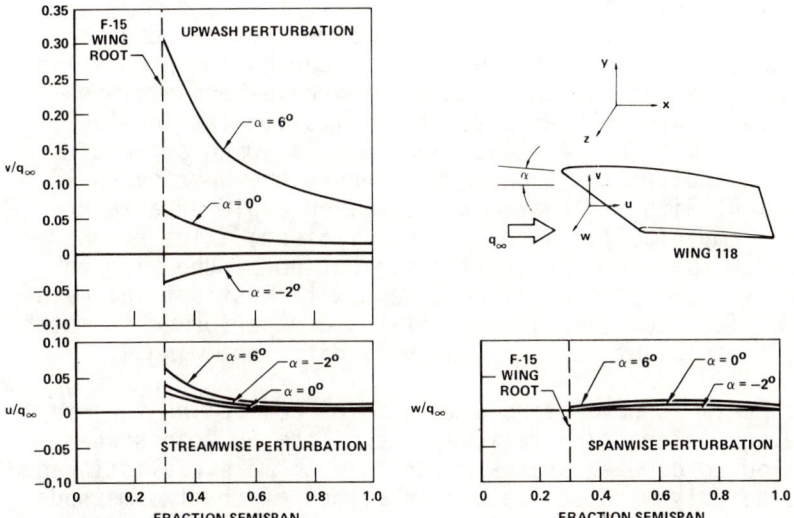

Fig. 11 Velocity perturbations in neighborhood of F-15 Wing leading edge; $M_\infty=0.60$.

The FLO-27 results using the infinite cylindrical fuselage model are also shown to emphasize the need for accurate modelling of fuselage effects.

To further strengthen this point, PAN AIR and FLO-28 results for the F-15 wing attached to a semi-infinite circular cylinder are compared with each other in Fig. 10 and show good agreement for this simple geometry. The wing was attached at the midpoint of the cylinder whose diameter was such that the wing root was at 30% semispan consistent with the actual F-15 geometry. The cylinder extended downstream the full extent of the computational domain in the FLO-28 code while the panelled representation in the PAN AIR program extended downstream approximately 13 root-chord lengths. The parabolic nose shape extended upstream approximately one half root-chord length in both program descriptions. Because the grid density is high in this region, the short nose length minimizes the FLO-28 fuselage boundary condition difficulties due to lack of a totally body-conforming grid at the fuselage surface.

Fuselage Representation by Equivalent Simple Body (ESB)

The inability of the FLO codes to represent complex fuselage geometries properly and to model the inviscid wing/fuselage interaction correctly causes inaccurate prediction of the F-15 wing leading edge pressure peak as seen in Figs. 5 through 9. The simple fuselage models used in the FLO code calculations simply do not create the proper flow environment experienced by the wing. Two common remedies for this shortcoming have been employed by various users in an attempt to simulate the fuselage interaction. The first is to adjust the wing angle of attack input to the program to produce the best overall agreement with test data at each given angle of attack (i.e., empirically match lift coefficient). The second is to alter the spanwise twist distribution of the wing to produce the best overall agreement with test data at each angle of attack (i.e., calculations are performed to match lift coefficient using an entirely different wing.)

MCAIR is developing a method for correcting the wing/fuselage interaction deficiencies of the FLO codes when applied to complex shapes which is much more realistic and considerably more accurate than either of the two methods described above. More important is the fact that it is independent of test data. The method uses an accurate panel method, such as PAN AIR or Douglas-Neumann, to determine the perturbation velocity distributions induced by the interaction of the wing with a complex fuselage shape at subcritical Mach number. An ESB which closely duplicates these perturbation distributions in the streamwise, spanwise, and upwash directions at each aircraft angle of attack is constructed. For small angles of attack, the perturbations are small relative to freestream velocity so that linearity and Mach number independence of their effects are reasonable assumptions. Correlations can therefore be developed at subcritical Mach number by matching FLO code results using the ESB with those of the panel method. The correlations can then be applied in the transonic flow regime. Since the FLO codes can accurately predict flowfields for simplfied wing/fuselage geometries (i.e., wing plus ESB), this method greatly extends their geometric range of applicability.

Application of this concept can be demonstrated quite simply for the F-15 wing/fuselage configuration of Fig. 3. The perturbation velocity distributions relative to freestream in the streamwise, spanwise, and upwash directions immediately upstream of the wing leading edge as determined from the PAN AIR program are shown in Fig. 11 for freestream Mach number of 0.60 and various angles of attack. These results were generated with the engine inlet completely blocked; wing pressure distributions were found to be relatively insensitive to inlet flow conditions for the F-15. The perturbation velocities were calculated at each span location by subtracting the three wing-alone solution components of velocity immediately upstream of the leading edge from the corresponding components of the wing/fuselage solution. The wing-alone results first were corrected for aspect ratio since the exposed wing has an aspect ratio of only 2.5 while the wing/fuselage has an aspect ratio of 3.0. The corrections were determined from classical wing theory and were found to be small for the angle of attack range shown in Fig. 11.

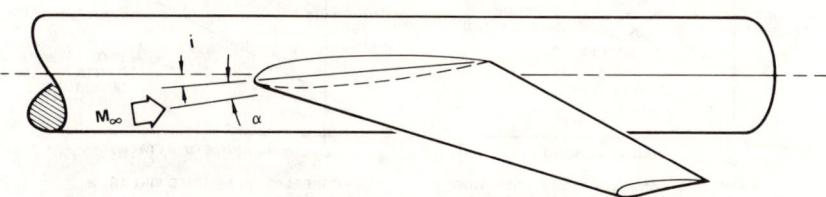

Fig. 12 Additional fuselage incidence using FLO-27 wing/cylinder; mid-wing attachment.

The predominant effect of the F-15 fuselage is to induce strong upwash along the wing leading edge with much smaller perturbations in the other two directions. This suggests that the F-15 fuselage can be modelled by an infinite circular cylinder (ESB) with its own angle of incidence varied with wing angle of attack to approximate the actual fuselage upwash distribution along the wing leading edge. Such an arrangement is shown schematically in Fig. 12. The additional fuselage incidence angle i required in the FLO code calculation (e.g., FLO-27) can be determined by trial-and-error comparison of FLO code results with those of a panel program such as PAN AIR at a subcritical Mach number. This is illustrated schematically in Fig. 13. Since these interactions can be assumed linear for small wing angles of attack, correlation between the additional fuselage incidence i and wing angle of attack α can be determined by carrying out the trial-and-error process at two different values of α. This is indicated in Fig. 13. It should be emphasized that the wing remains at the correct geometric angle of attack. Since the (i, α) correlation can be assumed Mach number independent, the F-15 ESB can be used with the FLO codes to predict flowfields at transonic Mach numbers.

The (i, α) correlation was developed as outlined above for the F-15 fuselage with Wing 118 shown in Fig. 4 This result is presented in Fig. 14. The cylinder diameter was chosen so that the overall span was maintained and the wing root was at the correct 30% semispan location. The wing was attached at the midpoint of the cylinder. Also shown is a similar correlation developed for the research wing (Wing 153) described in Ref. 9. This wing has a sweep angle of 35 deg, has a larger leading edge radius of curvature, and is considerably thicker than Wing 118. It is

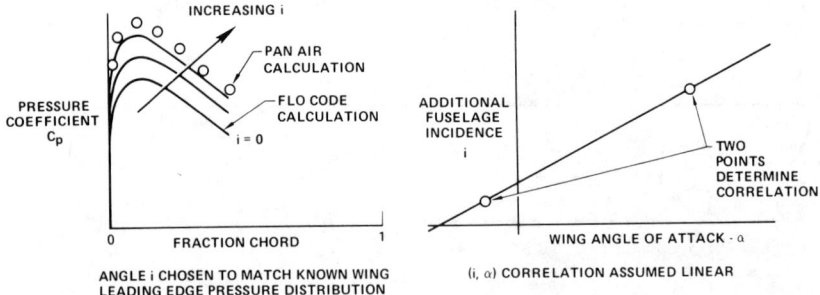

Fig. 13 Development of additional fuselage incidence correlation.

FLOW PREDICTION FOR FIGHTER CONFIGURATIONS

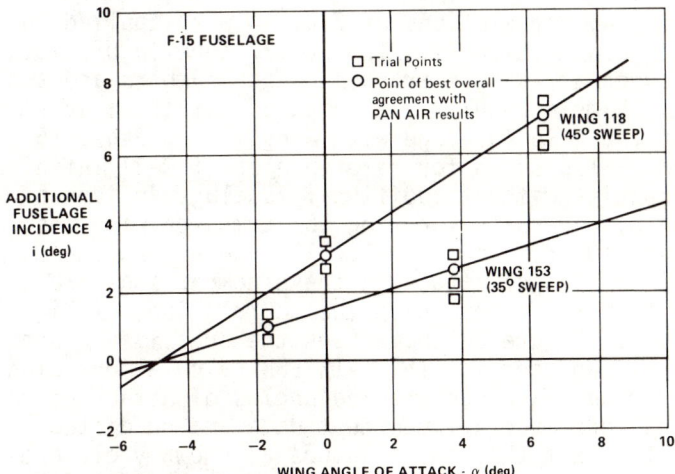

Fig. 14 Additional fuselage incidence correlation using FLO-27 and PAN AIR; $M_\infty=0.60$.

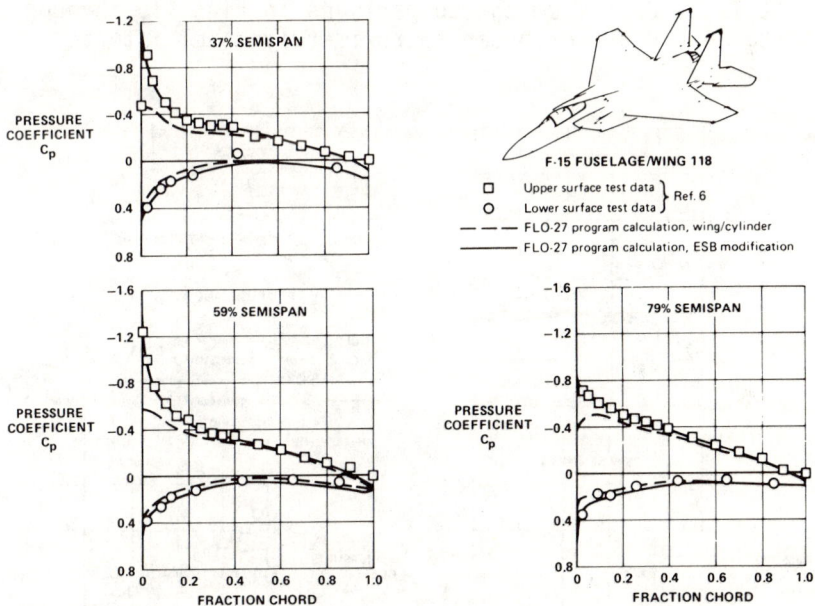

Fig. 15 Comparison of FLO-27 results with test data for F-15 Wing 118; $M_\infty=0.60$, $\alpha=4.36$ deg.

interesting that both correlations show zero additional fuselage incidence required at the same negative angle of attack. Possible explanations for this are being studied, but it may be due merely to the fact that the fuselage geometry is common to both configurations.

The (i, α) correlations of Fig. 14 were applied to the F-15 ESB in conjunction with the FLO-27 code to predict wing pressures at various transonic Mach numbers and angles of attack. Sample results are compared with test data of Ref. 6 for the F-15 fuselage with Wing 118 in Figs. 15 through 18. Also shown for reference are FLO-27 infinite-cylinder results without additional fuselage incidence (i.e., i = 0). Results for Wing 153 compared with test data of Ref. 10 are shown in Fig. 19. Note that the Mach numbers and angles of attack in these comparisons are different from those for which the correlations of Fig. 14 were derived. Figure 20 summarizes the additional fuselage incidence values used for the F-15 ESB calculations along with the various Mach numbers and angles of attack at which comparisons were made between test data and predicted results. The fact that the i values as shown yield predictions which compare well with test data verifies the linearity and Mach number independence of the interaction effects for small aircraft angles of attack.

It is evident from the comparisons in Figs. 15 through 19 that the ESB approach can approximately model interac-

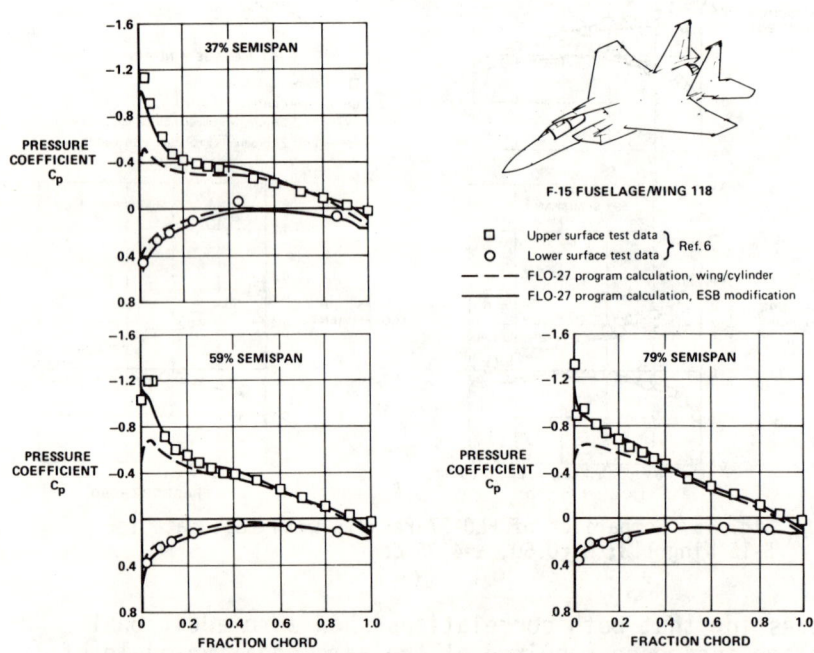

Fig. 16 Comparison of FLO-27 results with test data for F-15 Wing 118; M_∞=0.81, α=4.84 deg.

tion effects of more complex fuselage shapes. Even though a very simple ESB was used to model the F-15 fuselage in these comparisons, the overall results are good, especially in the vicinity of the leading edge pressure peak. Some degradation in prediction of embedded shock strength and location is seen in the higher Mach number results in Figs. 17 and 18. This might be corrected by refinement of the simple ESB representation used to generate these results. Preliminary results indicate that the shock position and strength as calculated by FLO-27 are altered significantly by slight change in the vertical attachment point of the wing on the cylinder.

Obviously, not all fighter fuselage shapes produce perturbation velocity distributions such as those shown in Fig. 11 for the F-15. The perturbation velocity distributions of another configuration, the F/A-18 (shown in Fig. 21), were studied using the PAN AIR program. The half-space wing/fuselage configuration was represented by 570 panels of which 156 were on the wing. The streamwise, spanwise, and upwash distributions immediately upstream of

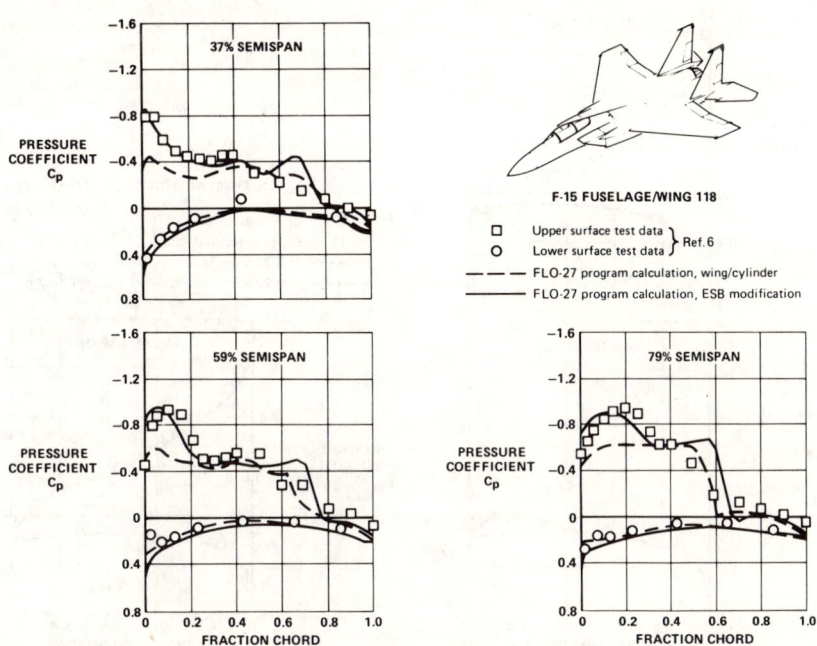

Fig. 17 Comparison of FLO-27 results with test data for F-15 Wing 118; $M_\infty=0.90$, $\alpha=4.84$ deg.

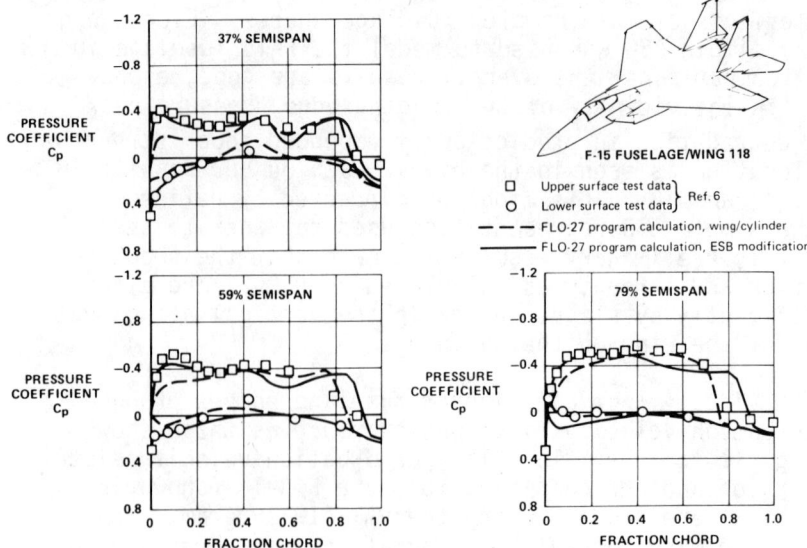

Fig. 18 Comparison of FLO-27 results with test data for F-15 Wing 118; M_∞=0.95, α=2.58 deg.

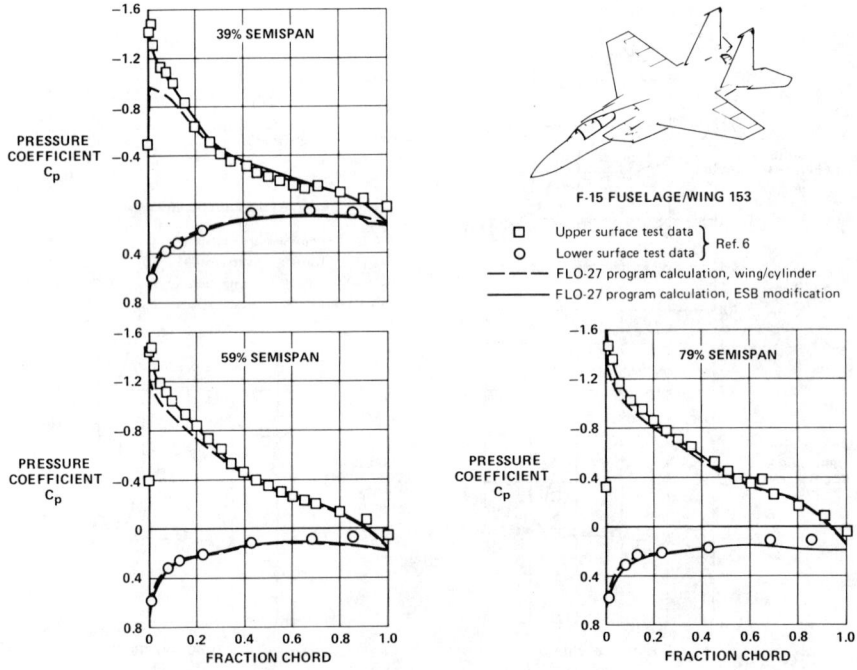

Fig. 19 Comparison of FLO-27 results with test data for F-15 Wing 153; M_∞=0.60, α=6.84 deg.

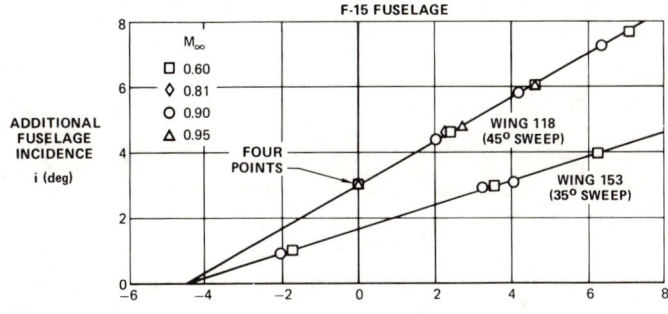

Fig. 20 Additional fuselage incidence values used for F-15 ESB calculations.

the wing leading edge with engine inlet blocked are shown in Fig. 22 along with the previously shown F-15 results for comparison. The distributions for flow-through inlet conditions (not shown) differed very little for the F-15 but were significantly different for the F/A-18. The F/A-18 results of Fig. 22 show a sizeable spanwise perturbation in addition to the upwash perturbation, while the streamwise perturbation remains relatively small as in the case of the F-15. The ESB model used for the F-15 (i.e., infinite circular cylinder with additional fuselage incidence) is not adequate for the F/A-18 since it primarily provides an upwash correction, and a more refined ESB model must be used to include the other perturbation effects.

In developing a generally valid ESB method, the following geometric properties and their primary effects are being analyzed:

1) Additional fuselage incidence - upwash perturbation.
2) Nose shape and length - streamwise perturbation.
3) Wing vertical position - spanwise perturbation.
4) Elliptic cross section - local upwash perturbation.

Parabolic or higher-order nose shapes of finite length are being investigated to determine their relationship to streamwise velocity perturbations. Results such as those of Fig. 23 show a strong inboard streamwise acceleration effect when compared with infinite-cylinder results, which may affect embedded shock location and strength in this region. A positive streamwise velocity

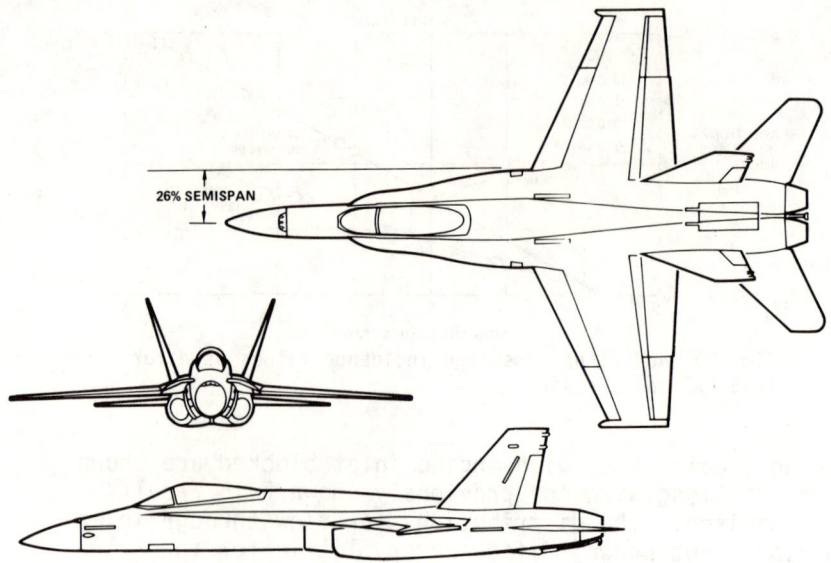

Fig. 21 F/A-18 Fighter configuration.

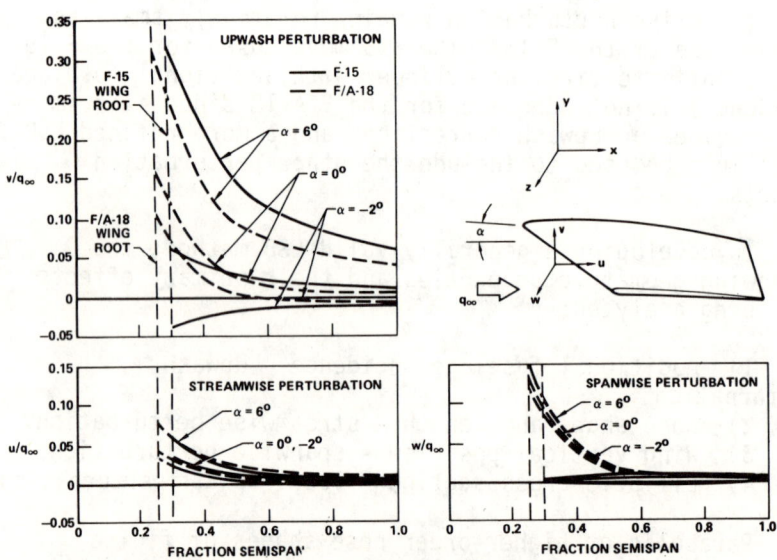

Fig. 22 Velocity perturbations in neighborhood of F-15 and F/A-18 wing leading edges; $M_\infty=0.60$.

increment causes a negative shift in the upper and lower surface C_p distributions as shown in Fig. 23. Use of an elliptical cross-section ESB model allows the longitudinal area variation of the actual fuselage configuration to be

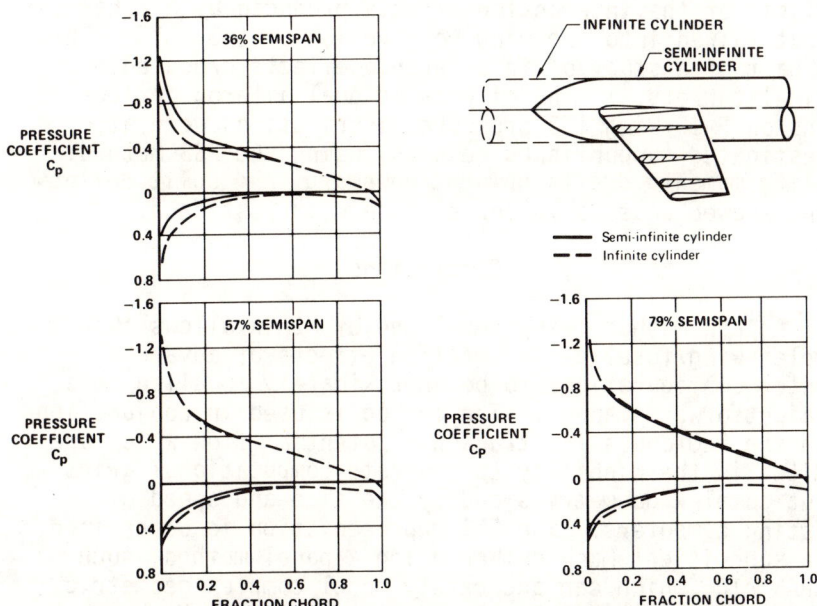

Fig. 23 Comparison of PAN AIR results for infinite and semi-infinite cylinder with F-15 Wing 118; $M_\infty=0.60$, $\alpha=6.51$ deg.

approximately modelled which is important transonically. It is believed that the overall width of the fuselage should be approximated as close as practical in order to maintain the correct aspect ratio, which means that the elliptical eccentricity becomes the longitudinally varying parameter.

As mentioned earlier, the vertical position of wing attachment on the fuselage (ESB) has a strong influence on embedded shock position and strength, at least for the F-15. However, any correlation between wing position and shock location would have to be developed in conjunction with test data. The effects of a small amount of camber in the ESB fuselage representation are also being investigated. The effect of inlet mass flow on wing pressures has been found to be very important for some fighter configurations. Possible ways of modelling these inlet effects using the ESB method are currently being studied.

Applications

The ESB method outlined above has been applied to a number of practical problems. One of these deals with the analysis of the interaction effects produced by a spherical turret faired into the wing root region of the F-15. This was part of a study of interference effects produced by optical turrets.[11] The effects of dual aileron deflections on F-15 Wing 153 pressure distributions were also investigated (unpublished results) using the ESB method. Rolling moments due to upward, downward, and split deflections showed reasonable correlation with test data.

Conclusions

An ESB method being developed by MCAIR allows the complex wing/fuselage interaction effects of advanced fighter configurations to be approximately modelled in a straightforward manner. The method is used in conjunction with the FLO codes for transonic potential flow which are limited in their ability to generate computational grids about complex geometries and by the size and speed of existing computers. The ESB representation is determined at a subcritical Mach number using a panel method, such as PAN AIR, which can accurately model complex geometric shapes. Correlations developed at subcritical Mach number were shown to be linear and Mach number independent for small aircraft angles of attack and are therefore valid in the transonic flow regime. The ESB approach is not restricted to use with the FLO codes and is equally applicable to other currently available transonic programs which also lack the capability to resolve complex fuselage shapes adequately.

An ESB representation of the F-15 fighter aircraft fuselage consisting of an infinite circular cylinder with additional incidence has been used to generate transonic wing pressure distributions which agree well with experimental data. Extension of this F-15 ESB model to include additional body parameters is being studied so that the prediction of embedded shock strength and location can be refined. The ESB approach in general is being applied to other fighter configurations, namely the F/A-18, AV-8B, and F-4, to assess its range of applicability for transonic wing/fuselage interactions. The possibility of applying viscous boundary layer and wake corrections to accurate inviscid wing flowfield predictions obtained using the ESB method is also being evaluated.

Acknowledgments

This work is being supported by the McDonnell Douglas Independent Research and Development program under the sponsorship of the Aerodynamics Department of McDonnell Aircraft Company. A portion of the computer resources used in this study is being supplied by NASA/Ames Research Center. The FLO-30 calculations shown herein were provided by Prof. D. A. Caughey of Cornell University who also is a McDonnell Douglas Corporation consultant.

References

[1] Jameson, A. and Caughey, D.A., "Numerical Calculation of the Transonic Flow Past a Swept Wing," New York University, ERDA Report No. COO 3077-140, June 1977.

[2] Jameson, A. and Caughey, D.A., "A Finite-Volume Method for Transonic Potential Flow Calculations," *Proceedings of AIAA 3rd Computational Fluid Dynamics Conference*, Albuquerque, N. Mex., June 1977, pp. 35-54.

[3] Caughey, D.A. and Jameson, A., "Numerical Calculation of Transonic Potential Flow About Wing-Body Combinations," *AIAA Journal*, Vol. 17, No. 2, Feb. 1979, pp. 175-181.

[4] Boltz, F.W. and Kolbe, C.D., "The Forces and Pressure Distribution at Subsonic Speeds on a Cambered and Twisted Wing Having 45° of Sweepback, an Aspect Ratio of 3, and a Taper Ratio of 0.5," NACA RM-A52D22, July 1952.

[5] Robinson, H.L., "The Effects of Wing Incidence on the Aerodynamic Loading Characteristics of a Sweptback Wing-Body Combination at Transonic Speeds," NACA RM-L52G23b, Oct. 1954.

[6] Anderson, R.M., "Wind Tunnel Test on the 4.7 Percent Scale F-15 Model in the McDonnell Douglas Polysonic Wind Tunnel," McDonnell Aircraft Co., Report No. MDC A0974, Aug. 1971.

[7] Ehlers, F.E., et al, "A Higher Order Panel Method for Linearized Supersonic Flow," NASA CR-3062, Feb. 1978.

[8] Moran, J., et al, "User's Manual - Subsonic/Supersonic Advanced Panel Pilot Code," NASA CR-152047, Feb. 1978.

[9] Pavelka, J. and Tatum, K.E., "Wind Tunnel Force and Moment Test Results of Short Laminar Separation Bubble Wing Design," McDonnell Aircraft Co., Report No. MDC A5730, Vol. I, Aug. 1979.

[10] Tatum, K.E. and Pavelka, J., "Wind Tunnel Pressure Distribution Test Results of Short Laminar Separation Bubble Wing Design," McDonnell Aircraft Co., Report No. MDC A5983, Sept. 1979.

[11] Lemley, C.E., Triplett, W.E. and Verhoff, A., "Aerodynamic Interference Due to Optical Turrets," AFWAL TR-80-3058, Sept. 1980.

Chapter XII.

A Series of Airfoils Designed by Transonic Drag Minimization for Gates Learjet Aircraft

M. L. Hinson*
Gates Learjet Corporation, Wichita, Kans.

Introduction

Since the introduction of numerical optimization for the design of airfoil sections,[1] and three-dimensional wings,[2] the ability to tailor wing shapes to meet highly specific design goals has greatly increased. This is a consequence of the generality of the optimization technique, CONMIN,[3] and of the capability of the potential flow methods[4] to predict the flow under many conditions of interest to the designer. The use of numerical optimization to minimize drag at transonic cruise flight conditions[5] is particularly interesting to a manufacturer of high-speed business aircraft, such as Gates Learjet. This paper describes a series of airfoils designed for reduced transonic drag using various formulations of the optimization problem. Experimental results are included and compared with theoretical calculations.

Description of Production Aircraft Wings

Figure 1 is a recent family portrait of several Gates Learjet aircraft, showing the resemblance among all of the models. Development of the new Learjets has been evolutionary, using common components wherever feasible, in the interest of minimizing the cost of design, engineering, certification, tooling, and manufacture. In aerodynamic terms, the largest change to date was the introduction of the Longhorn wing for Models 28, 29, 55,

---

Presented at the Transonic Perspective Symposium, NASA/Ames Research Center, Moffett Field, Calif., Feb. 18-20, 1981. Copyright © 1981 by the American Institute of Aeronautics and Astronautics. All rights reserved.
*Senior Engineer, Basic Aerodynamics.

and 56. As shown in Fig. 2, the Longhorn wing is a high aspect ratio wing of low sweep (13 deg at the quarter chord). The aspect ratio, 6.7, is increased by the winglet effect to 7.9. For such a planform, three-dimensional effects are minimal and two-dimensional airfoil section characteristics are applicable to a large extent. The external configuration of the wing, from the aileron inboard, dates back to the original Learjet Model 23 except for leading edge details. Consequently, a new airfoil section designed for the Longhorn wing could conceivably be applied to any Learjet model.

The wing sections currently in production (Fig. 2) are derived from the NACA 64A109 (a=0.8, modified) airfoil section. The relatively sharp NACA leading edge has been replaced by a large radius of constant dimension (1.35 in.) between W.S. 0 and W.S. 181 and constant percent chord (2.21%) on the outboard panel of the Longhorn wing. The radius is faired tangent to the upper surface resulting in a progressive leading edge droop as the chord tapers outboard. This modification is called the "Century III" leading edge and results in a significant reduction in stall speeds compared to what would be possible with the basic 64A109. High-speed characteristics of the Century III sections are similar to the unmodified 6-series section.

Wing Design Considerations

If the Century III wing sections are to be replaced by an improved airfoil, several factors must be considered. Many Learjet operators place a premium on the time saved

Fig. 1   Representative Gates Learjet aircraft; from top: Models 23, 25, 35, 55.

by flying their aircraft at maximum operating Mach number (M_{MO} = 0.81). Since this exceeds the drag divergence Mach number of the wing, considerable potential exists for drag reduction at Mach 0.81. However, for longer flights, the maximum range cruise Mach number, 0.70, is used. Any new design would be unacceptable if it had a significant adverse effect on the drag at Mach 0.70. Likewise, the low stall speeds achieved with the Century III leading edge should at least be maintained. At the other extreme of the speed range, shock-induced separation effects, including aileron "buzz," exist which might be alleviated by reducing the shock strength at the design dive Mach number, 0.86. Elimination of aileron "buzz" would have a synergistic effect by allowing deletion of vortex generators and other devices now used to correct the problem.

The nonaerodynamic factors also figure prominently in the definition of an acceptable alternative to the current wings. In particular, the aft-loading typical of Whitcomb supercritical sections would significantly increase the scope of engineering and certification tasks, with attendant cost increases. The technical problems and costs of aft-loading are more appropriate to a complete airplane redesign effort than to the limited wing modification program envisioned here. Thus one ground rule for the present study is to make no change to the trailing edge aft of 70% chord. Costs can further be reduced by minimizing the physical extent of the modification, so that fewer structural components need be changed. Fin-

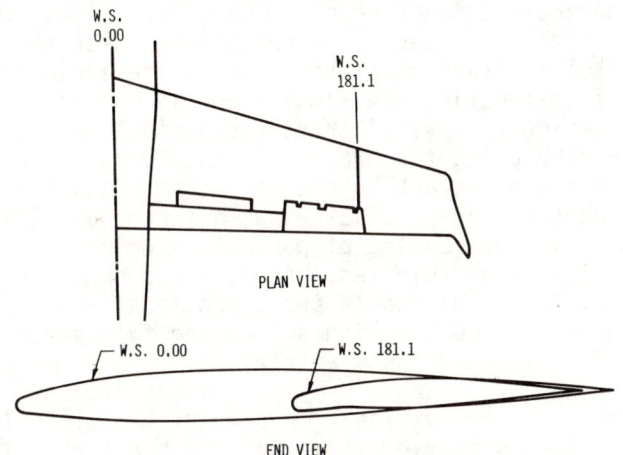

Fig. 2 Longhorn wing planform and Century III airfoil sections.

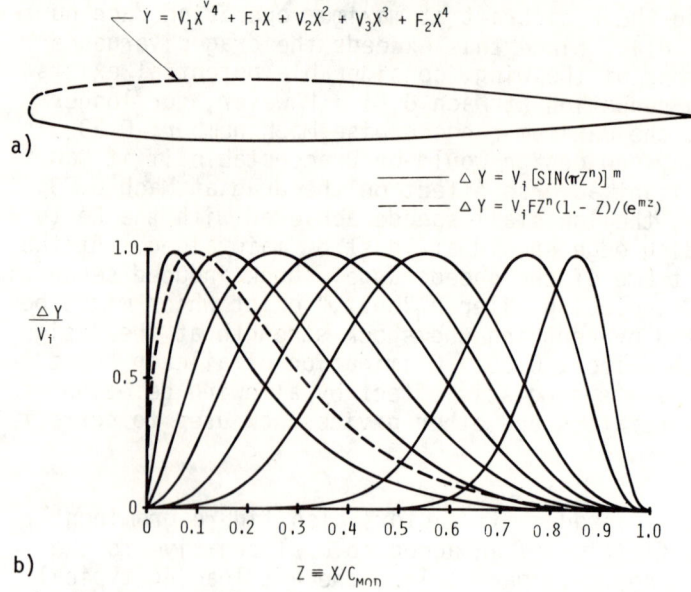

Fig. 3 Design variables for numerical optimization of airfoil shape. a) Polynomial spline; b) Sine and exponential shape functions.

ally, to facilitate development flight testing and possible retrofit, the existing wing structure should fit inside the new contours.

Optimization Technique

The constrained function minimization program, CONMIN,[3] employs the method of feasible directions to seek the combination of design variables which yields the minimum value of an objective function and simultaneously satisfies a set of constraints. Several transonic airfoil designs have been obtained by using the drag coefficient at one Mach number and angle of attack as the objective, usually with a constraint on the drag coefficient at a second condition. The design variables consist of parameters which define or modify the airfoil coordinates. Aerodynamic parameters are calculated by FLO-6, Jameson's two-dimensional inviscid transonic analysis code,[4] which solves the full potential equations of fluid flow about an airfoil.

Test results are available on some of the airfoils designed by transonic drag minimization. Lores et al.[6] reported success with a minimal leading edge modification involving only the forward 12% chord of the upper surface. The design variables consisted of three coefficients and one exponent of a polynomial spline similar to the one illustrated in Fig. 3a. A more general design was also obtained using incremental shape functions (like those of Fig. 3b) to modify the entire upper surface. The second airfoil did not perform as predicted, due to its large trailing edge included angle which probably exceeded the limitations of boundary-layer theory. Johnson and Hicks used similar incremental shape functions to modify a 12% thick supercritical section, which performed well in the wind tunnel. Although some of the variables used were capable of affecting the trailing edge shape, the final optimized airfoil did not differ significantly from the starting airfoil aft of about 60% chord. Evidently, restriction of the modified region to the forward part of the airfoil is not only advantageous for cost reasons, but may sometimes be necessary to avoid undesirable viscous effects since those effects are not considered during drag minimization with an inviscid code.

Inviscid drag minimization has been found inadequate for the design of three-dimensional swept wings. This is because the inviscid drag calculation is insensitive to steep pressure gradients which would cause separation if not improved. Haney et al.[8] overcame this problem by defining objective functions which reached minimum values when the computed pressures matched arbitrarily specified chordwise pressure distributions. The technique was expensive and did not allow simultaneous constraints at an off-design Mach number, because of the large computer requirements of the three-dimensional flow calculations. For the present study, three-dimensional effects are not considered during the optimization, but are included in the evaluation of the airfoils by testing them as wings on complete aircraft configurations. This approach is feasible because of the high aspect ratio and low sweep of the Learjet wings. It is desirable because it combines the lower cost of two-dimensional computations with the increased utility of three-dimensional test results. It also allows simultaneous consideration of off-design conditions during the optimization process, and avoids the difficulty of specifying pressure distributions by using drag as the objective. The major disadvantage is the lack of direct comparison between the theory and experiment.

As Hicks and Vanderplaats[5] pointed out, the number of design points that can be handled by numerical optimization is theoretically unlimited, but practical considerations of computer core and CPU time dictate that the optimization scheme be kept simple. Simultaneous consideration of the four flight conditions of interest to the present problem would be prohibitively complex. Instead it is necessary to consider only one or two conditions at a time. Allowance for the other conditions can be made by selecting design variables that have minimal impact on the off-design characteristics or by appropriate geometric constraints. For example, empirical methods can be used to convert a stall speed contraint into a simple geometric constraint by use of the Δy parameter.[9] Alternatively, some of the design conditions may be omitted because they happen to be compatible with one or more of the others. For the present study, a variety of design points have been considered.

Single-Point Optimization Results

The first airfoil optimization for application to Gates Learjet aircraft was done by Hicks of NASA/Ames Research Center. The design point was Mach 0.85, zero angle of attack, which is near the aileron "buzz" boundary. The initial airfoil was the NACA 64A109. A polynomial spline was used to replace the upper surface forward of 37.5% chord (Fig. 3a). The inviscid section drag coefficient calculated by FLO-6 was the objective and there were no constraints.

The resulting airfoil, the GLC 301, is compared with the 64A109 in Fig. 4 at the design condition. The increased bluntness of the leading edge causes increased suction on the nose, increased lift, and reduced drag. Though the coordinates aft of the maximum ordinate remained the same, the pressure distribution changed as the shock was weakened slightly and moved forward by the increased expansion on the nose.

The GLC 301 was never tested. Instead, its nose shape was stretched to create an 8% thick section with a 12.5% chord leading edge extension, the GLC 302. Calculations by FLO-6 at $M = 0.85$, $\alpha = 1$ deg, indicate that the basic shape of the upper surface pressure distribution survived the stretch (Fig. 5). On the lower surface, a small supercritical zone near midchord is eliminated. The force coefficients, based on the original unstretched chord, show increased lift and reduced drag for the GLC 302 relative to the GLC 301.

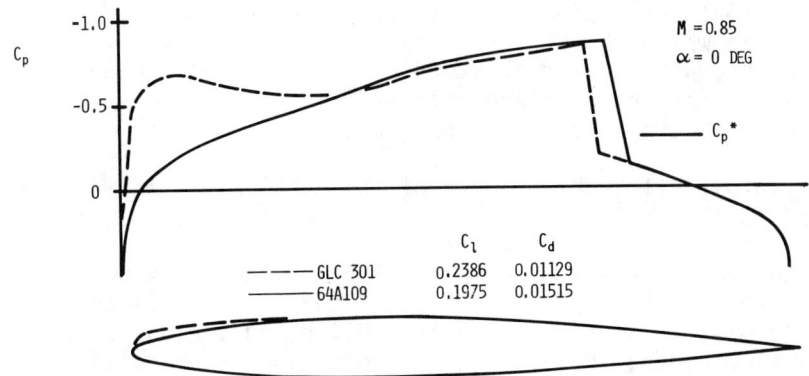

Fig. 4   The GLC 301 airfoil section compared to the NACA 64A109, with upper surface pressure distributions calculated by FLO-6.

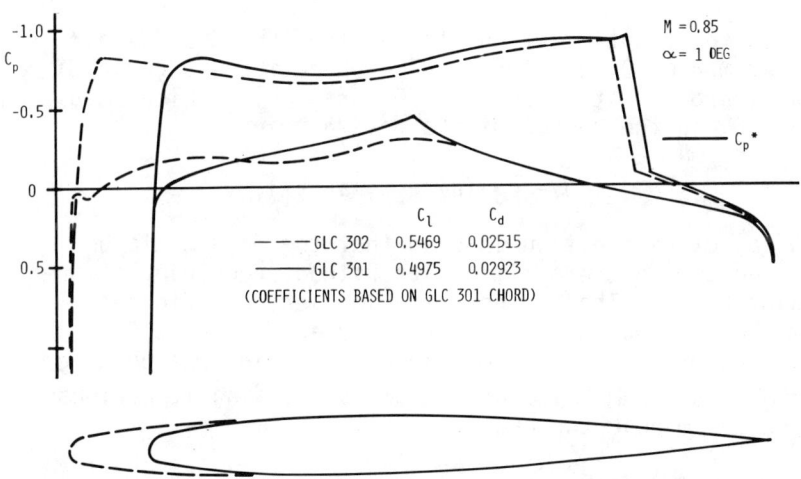

Fig. 5   The GLC 302 airfoil section compared to the GLC 301, with FLO-6 pressure distributions.

A wing having the GLC 302 section was tested in the Boeing Transonic Wind Tunnel. The wing was mounted on the Gates Learjet 0.09 scale high-speed wind-tunnel model of the Model 35 airplane. The model leading edge was removable, allowing direct comparison between the wing with the GLC 302 section and the production Century III wing. The drag increments obtained from the test are shown in Fig. 6. There was a significant drag reduction in the vicinity of the airfoil design point. The drag was increased at lower Mach numbers, especially at lift coefficients exceeding 0.2. In fact, the region of M and C_L in which the GLC 302 showed a drag penalty relative to the production wing included every practical cruise condition for the airplane. It was a simple matter to deduce that the design was unacceptable.

Intuitively, the failure of the GLC 302 was attributed to the choice of a design point which was too far beyond the drag divergence Mach number. Stretching the GLC 301 nose shape to create the GLC 302 increased wetted area and may also have deteriorated the flow at off-design conditions. Analysis of the GLC 301, using various potential flow codes, showed that it, too, had drag penalties at cruise Mach numbers. Based on these results, both the GLC 301 and 302 were rejected, and a new design was initiated with the design point moved down to M = 0.80, near the high-speed cruise condition.

The next optimization started from the GLC 301 airfoil at M = 0.80, α = 1 deg. The objective was defined by a parabolic fit of the GLC 301 drag polar calculated by FLO-6, in the region of the design point.

$$OBJ = c_d - (a_1 + a_2 c_\ell + a_3 c_\ell^2)$$

This objective function insured that the drag would be improved during the optimization without requiring a constraint on c_ℓ. The design variables were coefficients from the same polynomial spline as used for the GLC 301 optimization. The optimum polynomial spline was obtained in the first CONMIN iteration; two succeeding iterations failed to improve the drag.

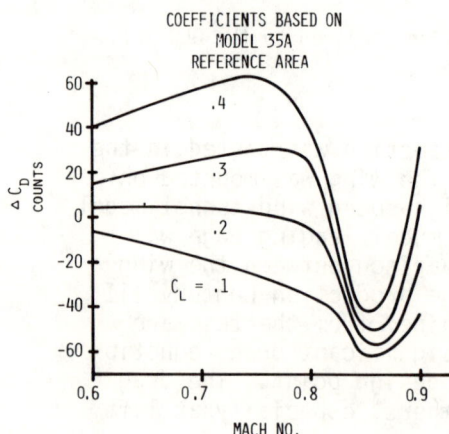

Fig. 6 Wind-tunnel drag increments due to GLC 302 section on a Gates Learjet Model 35A.

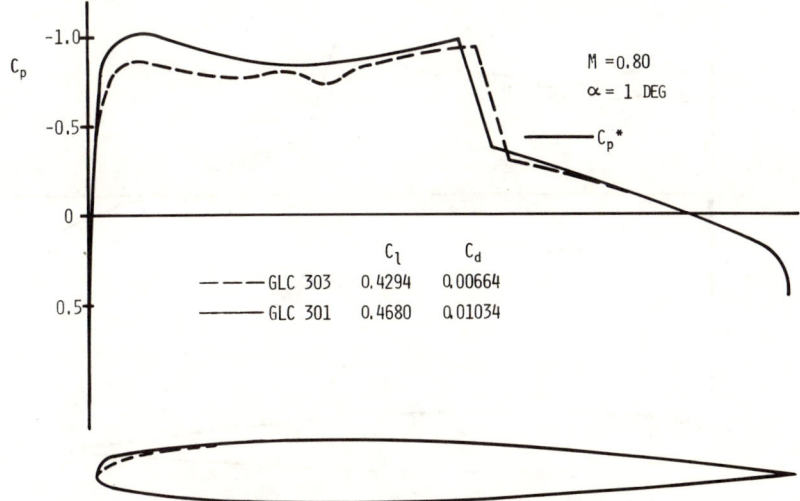

Fig. 7 The GLC 303 airfoil section compared to the GLC 301, with FLO-6 upper surface pressure distributions.

A further refinement of the airfoil was obtained by using sine and exponential decay functions as increments to the shape (Fig. 3b). Note that the X coordinates of these functions are normalized using an arbitrary chord, C_{MOD}. In this case the functions affected only the upper surface forward of the maximum ordinate by selected C_{MOD} = 40% of the total chord. The remainder of the airfoil was not allowed to vary. The design point remained at M = 0.80, but the drag at M = 0.775 was also monitored. In this case, it turned out that the drag at M = 0.775 decreased along with the drag at the design point so it was not necessary to add a constraint at M = 0.775. The optimization was terminated rather arbitrarily after five days of running, resulting in the GLC 303 design.

The GLC 303 is compared with the GLC 301 in Fig. 7. The lower design Mach number has resulted in less leading edge bluntness, less lift at constant angle of attack, and less drag. The GLC 303 has nine counts (0.0009) less drag at C_ℓ = 0.43 than the GLC 301. The pressure distribution is less smooth probably due to the intrinsic waviness of the sine functions and the insensitivity of the drag calculation to minor pressure variations. It appears, however, that CONMIN may have achieved reduced shock strength and a longer supercritical zone by reducing the local Mach number at the crest.

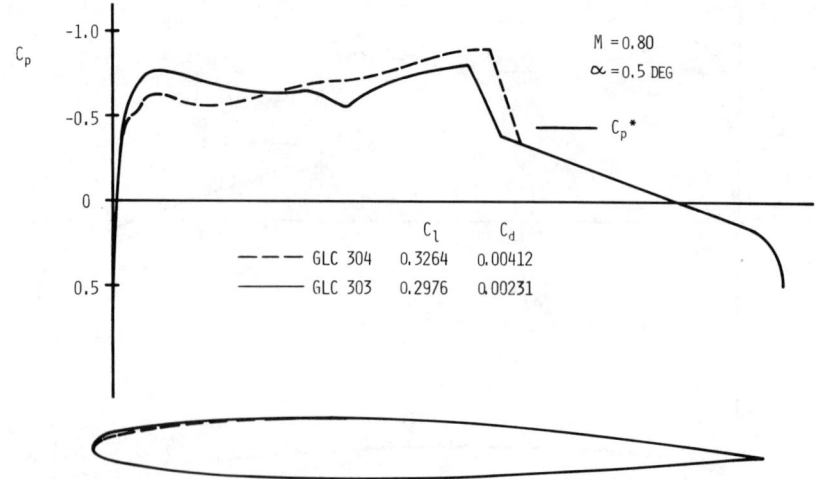

Fig. 8 The GLC 304 airfoil section compared to the GLC 303, with upper surface pressures from FLO-6.

Calculations of the GLC 303 lift and drag were made using FLO-6 over a range of Mach numbers from 0.70 to 0.85 and angles of attack from 0.5 deg to 1.5 deg. By comparing these results with similar calculations for the average section from the Longhorn wing (Wing Station 111) it was concluded that off-design penalties still existed, for Mach numbers less than 0.765 when the lift coefficient exceeded 0.3. This indicated a need to constrain the drag at a lower Mach number during the next optimization.

Two-Point Optimization Results

A constraint was introduced consisting of the local parabolic drag polar of the average Longhorn wing section at M = 0.75, α = 1.5 deg. Four geometric constraints were also added to prevent undercutting of the production structural box which had occurred with the GLC 303 at 35% chord. The objective was the drag polar at M = 0.80 of the GLC 303 airfoil, with the design angle of attack at 0.5 deg. Fifteen design variables were used, eleven sine functions and four exponential decay functions, all limited to affect only the upper surface forward of 40% chord. It took thirteen iterations for CONMIN to satisfy the constraint. An additional seven iterations were completed before work was stopped to satisfy schedule requirements.

The airfoil resulting, the GLC 304, is compared with the GLC 303 in Fig. 8. The leading edge is slightly less blunt, with less suction on the forward upper surface. In contrast to earlier results, the expansion in the middle of the airfoil has increased, causing a stronger shock, more lift, and more drag. The increased expansion was most likely a consequence of satisfying the geometric constraint at 35% chord. The drag at constant $c_\ell = 0.33$ was 11 counts higher for the GLC 304 than for the GLC 303.

The GLC 304 was tested in flight, using aircraft 35-001. The wing had been modified by removing the production Century III leading edge and adding a thinned microballoon epoxy and body putty buildup to the leading edge and upper surface forward of 40% chord. The wing had no twist and the GLC 304 section shape was constant full span. Stall, buffet, and speed-power tests were flown for comparison with similar tests performed earlier on the same airplane with the standard Century III wing.

The flight test drag increments are compared with calculated drag increments in Fig. 9. The flight tests showed drag penalties up to 30 counts for the GLC 304. The penalty is greatest at the lowest cruise Mach number (M = 0.7) and increases with increasing lift coefficient. At high cruise Mach and low C_L there is some benefit. These results contrast with the calculated drag increments from FLO-6 which show no appreciable effect of c_ℓ for $0.3 \le c_\ell \le 0.45$, no large penalty at M = 0.7, and a large drag reduction at M = 0.80 (40 counts). The wave drag increments from the Bauer code[10] show trends similar to the flight-test data, although the magnitudes of the calculated increments are smaller and they are biased slightly in favor of the GLC 304. The fact that the Bauer code predicts the Mach number trends for the GLC 304 drag increments better than FLO-6 can be understood from the comparison in Fig. 10. The thickening of the boundary layer on the rear of the airfoil, included in the Bauer calculation, causes a lift loss which has to be made up by increased angle of attack. Consequently, the peak velocity and shock strength are greater than predicted by the inviscid code, FLO-6.

To improve the FLO-6 calculations for the next optimization, the boundary-layer displacement thickness from the Bauer code was added to the GLC 304 airfoil coordinates. The FLO-6 drag polar of the GLC 304 with this passive vis-

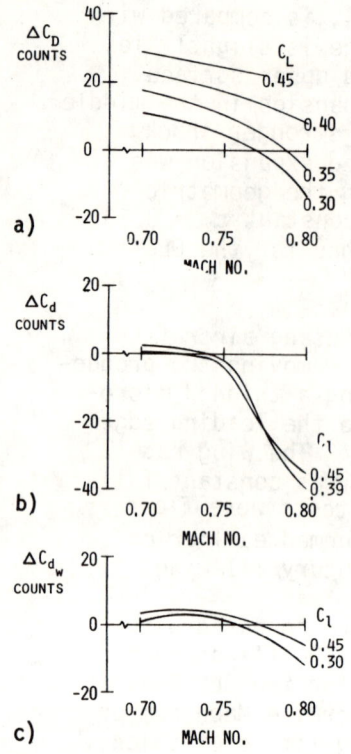

Fig. 9 Drag increments due to GLC 304 replacing current production sections. a) Flight test; b) FLO-6; c) Bauer code wave drag.

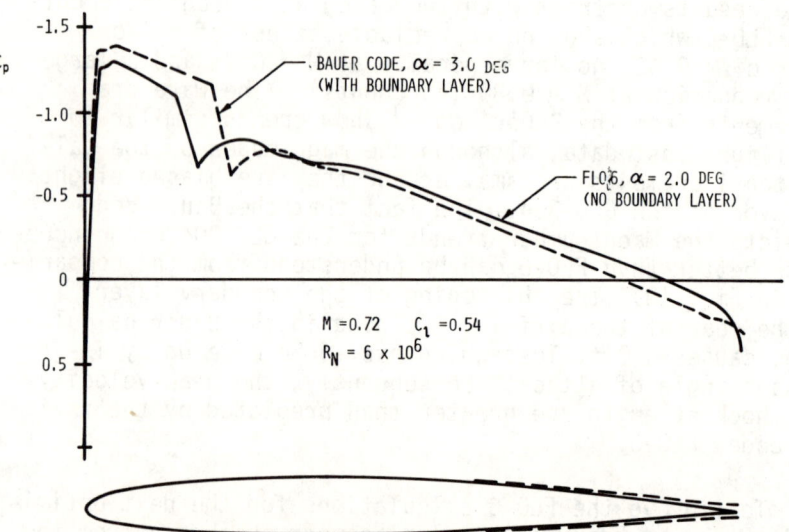

Fig. 10 Comparison of upper surface pressures showing the effect of viscosity on the GLC 304 airfoil at Mach 0.72.

cous correction was calculated at M = 0.8. A parabolic fit of this polar defined the objective function. The off-design constraint was defined by a parabolic fit of the Mach 0.72 drag polar computed by FLO-6 for the average Longhorn wing section including the same boundary-layer displacement thickness.

Optimization with the boundary-layer corrections, the new objectives, but the same sine and exponential design variables as had been used for the GLC 304, led to an airfoil which satisfied the contraint at M = 0.72. The drag at M = 0.80 was significantly higher than the GLC 304, and the airfoil bore an uncanny resemblance to the 64A109. Thus, after more than two years of intermittent effort, five "optimized" airfoil designs, a transonic wind-tunnel test, and a flight test program with a modified aircraft, the loop had closed back to the starting point. Clearly, the design problem to this point had been over constrained.

The most favorable results of the GLC 304 flight tests had been the stall speeds and stall characteristics. The low-speed performance benefited from the increased leading edge bluntness which was sufficient to cause a combined leading edge/trailing edge type of stall, instead of the abrupt short bubble leading edge stall of the Century III wing. To preserve the desirable airfoil thickness distribution for stall performance, while allowing new shape possibilities for cruise performance, leading edge cambering functions were introduced as design variables in the optimization. These were of two types (Fig. 11): a simple polynomial expression, and a cosine function of a format similar to the sine functions. These functions allowed the leading edge to be drooped or raised, and increased the generality of the airfoil shapes that could be produced.

Further generality was achieved by using functions that modified the X coordinates rather than the Y coordinates. One type was a sine function:

$$\Delta X = V_i \{\sin[\pi(\frac{Y-Y_{min}}{Y_{max}-Y_{min}})^n]\}^m$$

With these sine functions it was necessary to imposed side constraints on the design variables, V_i. Otherwise, too large a modification of this type would result in leading edge shapes which could not be mapped. The other type

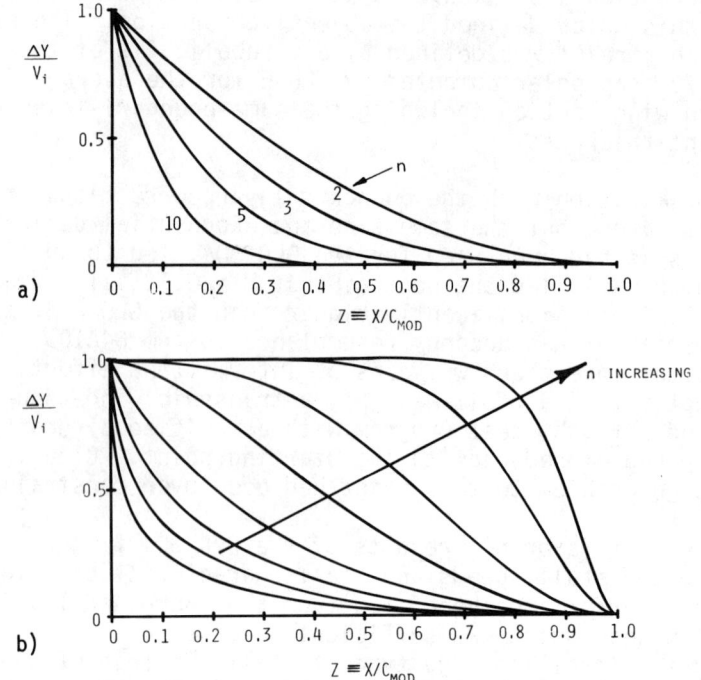

Fig. 11 Leading edge cambering functions. a) Polynomial cambering functions $\Delta Y = V_i (1.0 - z)^n$; b) Cosine Cambering functions, $\Delta Y = V_i \{z - 1.0 + [0.5 \cos(\pi z^n) + 0.5]^m\}$

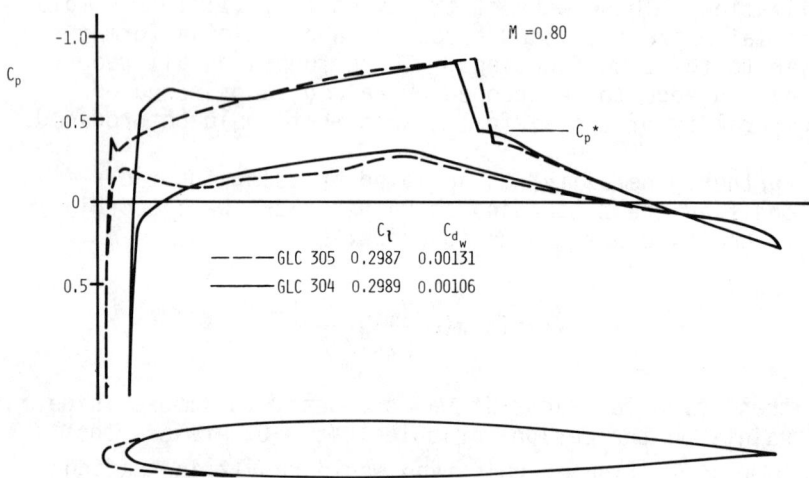

Fig. 12 Bauer code comparison between the GLC 305 and the GLC 304.

of X modifier was a simple linear shearing function:

$$\Delta X = V_i \left[\frac{X_{max} - X}{X_{max} - X_{min}}\right]$$

The shearing function allowed forward extension of the leading edge, which resulted in a reduction in overall thickness ratio.

Many combinations of these variables were tried over a four-week period until a satisfactory airfoil was obtained, the GLC 305. This final airfoil (Fig. 12) has a 3.7% chord leading edge extension, which results in an 8.68% thickness ratio. The Bauer code pressure distributions at M = 0.8, c_ℓ = 0.3, show reduced upper surface leading edge suction for the GLC 305 compared to the GLC 304. The shock is slightly stronger and moved aft. Lower surface pressures are more positive over most of the chord, but less positive near the leading edge. The calculated wave drag is 2.5 counts higher, continuing the trend of reduced benefit at M = 0.8 as airfoils have improved at the lower cruise Mach numbers.

The GLC 305 was evaluated in the Calspan 8 ft Transonic Wind Tunnel by testing a modified Longhorn wing on a 0.09 scale Model 55. The modified wing had the GLC 305 contour full span with no twist. The production Longhorn wing was tested on the same model during the same test for comparison. The drag increments from the test followed trends similar to the Bauer code predictions (Fig. 13). The results were encouraging, as performance calculations based on the wind-tunnel drag increments showed a 21% increase in fuel efficiency at M = 0.81, and the range penalty at M = 0.70 was less than 1%. To further evaluate and verify the GLC 305 airfoil, flight tests are planned with a modified airplane.

Theory-Experiment Correlation

One correlation observed in the course of this study is illustrated in Fig. 14. Contours of constant shock Mach number, calculated by the Bauer code, are superimposed with experimental data which define the separation boundary in terms of lift coefficient and Mach number. The GLC 304 separation boundary is based on pilot-reported buffet onset. The scatter is due to the vagaries of defining "perceptible" buffet and recording load factor and Mach number during a pullup maneuver which is necessarily im-

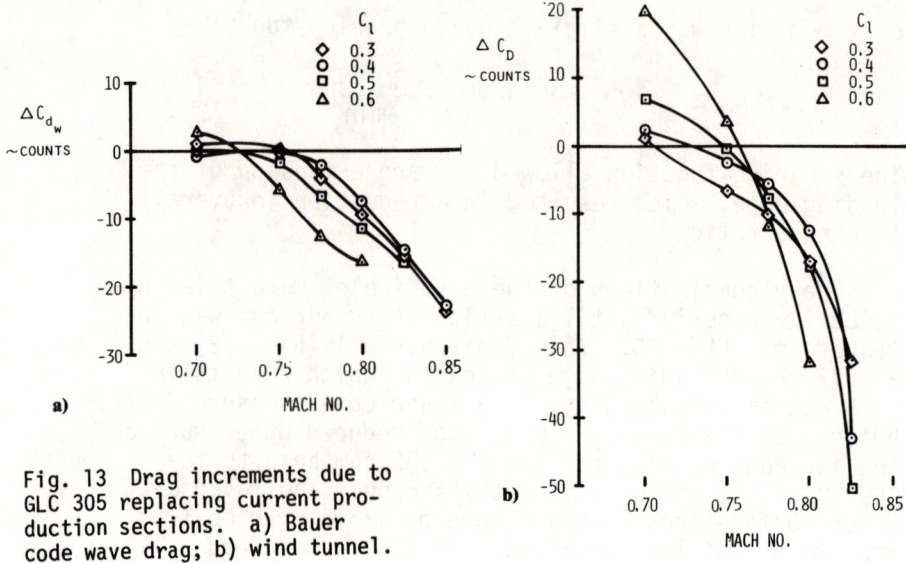

Fig. 13 Drag increments due to GLC 305 replacing current production sections. a) Bauer code wave drag; b) wind tunnel.

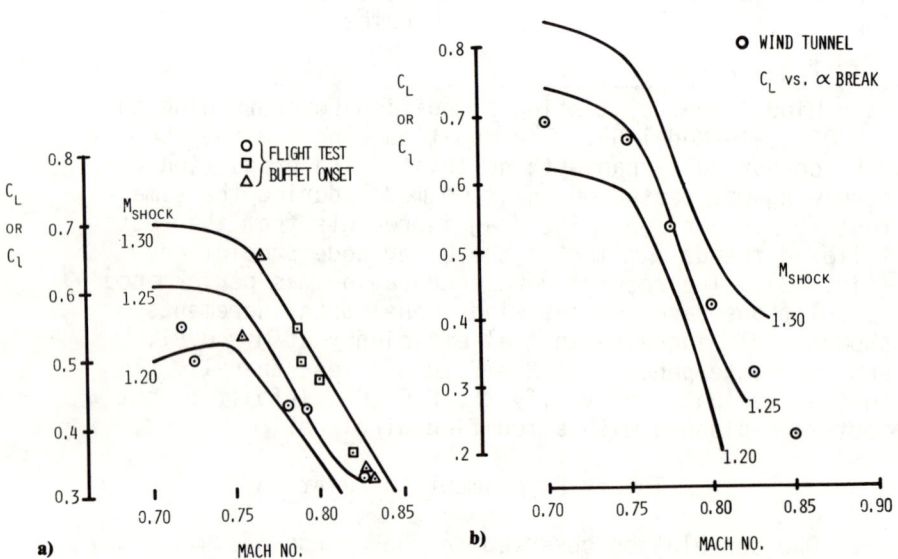

Fig. 14 Comparison of shock-induced separation boundaries with Bauer code shock Mach number contours. a) GLC 304; b) GLC 305.

precise. The GLC 305 separation boundary corresponds to the points at which the wind-tunnel lift curves change slope, or become nonlinear. Correlations of this type are useful for deriving design criteria for future airfoil sections.

As noted previously, for the GLC 304 and GLC 305 airfoils, the wave drag increments calculated by the Bauer code were found to imitate trends seen in the experimental drag increments for aircraft configurations tested with these sections. The test increments compared modified wings with standard production wings. The Bauer code increments compared the new sections with the average section from the production Longhorn wing. Correlation between the Bauer code and the experiments (Fig. 15) shows a scatter band up to ±10 counts. The slopes of the lines of correlation are consistent at 2.5 counts tested drag to 1 count computed drag. The two data sets both show a zero shift, but the GLC 304 test results are shifted 17 counts while the GLC 305 shift is 8 counts.

Since the GLC 304 was flight tested on a Model 35 configuration while the GLC 305 was tested on a Model 55 in the wind tunnel, the wing planform could have had an effect. This possibility was investigated using an isolated wing analysis code, FLO22NM.[11,12] This code, the three-dimensional counterpart to the Bauer code, is known to predict surface pressures accurately for isolated wings.[13] The drag correlations obtained with FLO22NM at Mach 0.70 and Mach 0.80 are shown in Fig. 16. The shock drag increments are used since they correlate with the test data slightly better than the total drag increments. The GLC 305/Model 55 wind-tunnel drag increments line up well with the FLO22NM predictions. In contrast, the GLC 304/Model 35 flight test increments again show a 17 count zero shift. This indicates either a problem with the flight test drag data or, possibly, that the nacelle and/or tip tank inter-

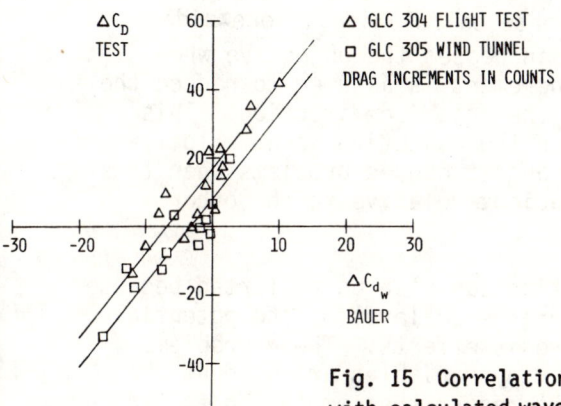

Fig. 15 Correlation of tested drag increments with calculated wave drag increments from the Bauer code.

ference effects associated with the Model 35 configuration were aggravated by the GLC 304 airfoil. These interference effects do not exist on the Model 55 configuration. Additional interpretation of these comparisons may be possible after flight test data on the GLC 305 are obtained.

Assessment of the Optimization Technique

The fact that the present study required five attempts to arrive at an acceptable airfoil design should not be construed as an indictment of the optimization technique. The four unsuccessful airfoils represent steps in the process of learning the limitations of the design method and of the flow analysis codes. The GLC 301 and 302 predated the development of the multipoint design technique which turns out to be essential for transonic drag minimization problems. The GLC 303 was a substep whose defficiencies were discovered in time by analysis of its off-design performance. The GLC 304 suffered from reliance on inviscid analysis. These mistakes would be avoided in future design efforts.

If the constrained drag minimization technique can be faulted for anything, it is inefficiency. CONMIN, which was conceived for structural optimization, does not work well in the nonlinear environment of transonic flow, especially when drag is the objective. Sometimes, design variables which would each reduce the objective if taken separately, actually increase the objective when combined. This can occur because the gradient calculation is necessarily a finite-difference ratio rather than a true partial derivative.

Further, the difference calculation is one-sided, so that any variable which increases the objective when stepped by a positive increment is assumed to reduce the objective when moved in the opposite direction. This assumption is untrue when the objective is at a local minimum of one variable and it causes problems when this variable has a large gradient relative to the other variables.

The objective function can also be distorted because of variations in the convergence level of the potential flow solution and hysteresis effects. These problems are compounded when too many variables are used at the same time. The user must resist the temptation to try to complete the design in a single run, and instead plan on a

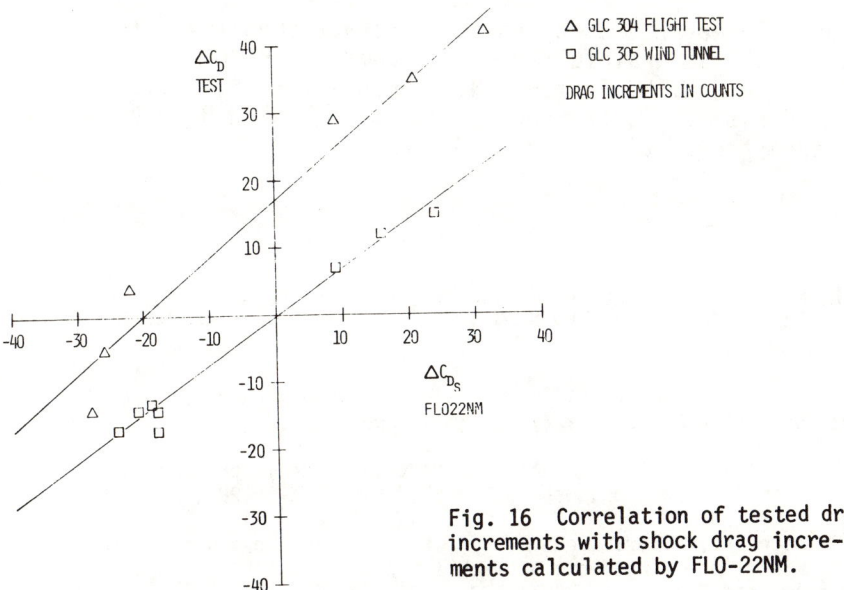

Fig. 16 Correlation of tested drag increments with shock drag increments calculated by FLO-22NM.

series of runs with a few variables at a time. There is a need for an optimization technique specially adapted to the peculiarities of transonic airfoil and wing design.

Conclusion

Five airfoils featuring limited contour changes were developed using numerical minimization of transonic drag. In all five cases, significant drag reductions were obtained at the design points. However, off-design performance was unacceptable for the first four airfoils, and this consideration dominated the optimization process. Satisfaction of off-design constraints reduced the benefit at the design point, according to calculated drag comparisons. It was essential that weak viscous interactions be included for successful design and evaluation of the final airfoil. The Bauer code proved most useful in this respect, and two-dimensional airfoil analyses with the Bauer code represented trends observed in flight and wind-tunnel test results on complete aircraft configurations.

Acknowledgments

The work summarized in this paper would not have been possible without the generous contribution of R. M. Hicks through the cooperative agreements between NASA/Ames Aerodynamic Research Branch and the aircraft industry. The author is also grateful to D. J. Grommesh, R. E. Etherington, W. M. Gertsen, N. E. Conley, K. P. Dill, L. L. Christie, D. C. Howe, J. K. Worrell, and B. DiPietra for their cooperation and hard work.

References

[1] Hicks, R. M., Murman, E. M., and Vanderplaats, G. N., "An Assessment of Airfoil Design by Numerical OPtimization," NASA TMX-3092, 1974.

[2] Hicks, R. M. and Henne, P. A., "Wing Design by Numerical Optimization," AIAA Paper 77-1247, Seattle, Aug. 1977.

[3] Vanderplaats, G. N., "CONMIN - A Fortran Program for Constrained Function Minimization, User's Manual," NASA TMX-62,282, Aug. 1973.

[4] Jameson, A., "Iterative Solution of Transonic Flows over Airfoils and Wings Including Flows at Mach 1," Communications on Pure and and Applied Mathematics, Vol. 27, 1974, pp. 283-309.

[5] Hicks, R. M. and Vanderplaats, G. N., "Application of Numerical Optimization to the Design of Supercritical Airfoils without Drag-Creep," SAE Paper 770440, 1977.

[6] Lores, M. E., Burdges, K. P., Shrewsbury, G. D. and Hicks, R. M., "An Evaluation of Numerical Optimization in Supercritical Airfoil Modification and Design," NASA Advanced Technology Airfoil Research Conference, Hampton, Va., March 1978.

[7] Johnson, R. R. and Hicks, R. M., "Application of Numerical Optimization to the Design of Advanced Supercritical Airfoils," NASA Advanced Technology Airfoil Research Conference, Hampton, Va., March 1978.

[8] Haney, H. P., Johnson, R. R. and Hicks, R. M., "Computational Optimization and Wind Tunnel Test of Transonic Wing Designs," AIAA Paper 79-0080, Jan. 1979.

[9] Hoak, D. E. et al., "USAF Stability and Control DATCOM," Air Force Flight Dynamics Laboratory, Wright-Patterson Air Force Base, Ohio, 45433, Sept. 1970.

[10] Bauer, F., Garabedian, P., Korn, D. and Jameson, A., Supercritical Wing Sections II, Lecture Notes in Economics and Mathematical Systems, Vol. 108, Springer-Verlag, New York, 1975.

[11] Jameson, A. and Caughey, D. A., "Numerical Calculation of the Transonic Flow Past a Swept Wing," New York University ERDA Rept. COO 3077-140, 1977.

[12] Nash, J. F. and MacDonald, A. G. J., "The Calculation of Momentum Thickness in a Turbulent Boundary Layer at Mach Numbers up to Unity," ARC CP No. 963, 1967.

[13] Hinson, B. L. and Burdges, K. P., "An Evaluation of Three-Dimensional Transonic Codes Using New Correlation-Tailored Test Data," AIAA Paper 80-0003, Jan. 1980.

Chapter XIII.

Applied Computational Transonics— Capabilities and Limitations

P.A. Henne,[*] J.A. Dahlin,[†] and C.C. Peavey[‡]
*Douglas Aircraft Company, McDonnell Douglas Corporation,
Long Beach, Calif.*

Nomenclature

c	= chord
$\bar{c}$	= mean aerodynamic chord
c_d	= section drag coefficient
Δc_{d_c}	= section compressibility drag coefficient increment
c_{d_f}	= section profile drag at 0.5 Mach number
ΔC_{D_c}	= total compressibility drag coefficient increment
c_ℓ	= section lift coefficient
C_L	= total lift coefficient
$C_{L\,MAX}$	= maximum lift coefficient
C_p	= pressure coefficient
$C_{p\,min}$	= peak minimum pressure coefficient
dc_d/dM	= section drag divergence slope
dC_D/dM	= total drag divergence slope
M	= freestream Mach number
M_{dd}	= section drag divergence Mach number
M_{DD}	= total drag divergence Mach number
M_L	= local Mach number
M_1	= upstream Mach number normal to shock wave
$R_c, R_{\bar{c}}$	= Reynolds number based on chord

Presented at the Transonic Perspective Symposium, NASA/Ames Research Center, Moffett Field, Calif., February 18-20, 1981. Copyright © American Institute of Aeronautics and Astronautics, Inc., 1981. All rights reserved.
　[*]Unit Chief, Technology Programs, Aerodynamics Subdivision.
　[+]Senior Engineer, Technology Programs, Aerodynamics Subdivision.
　[†]Engineer-Scientist/Specialist, Technology Programs, Aerodynamics Subdivision.

x/c = chord fraction
α = angle of attack
δ* = boundary-layer displacement thickness

Introduction

Progress in the development of transonic flow computational methods has led to extensive applications of such procedures to real transonic design and analysis problems.[1-8] These applications have provided valuable contributions to the transonic configuration development process and are the basis for an increasing reliance on computational transonics for aerodynamic development studies.[9] Such applications also provide a means of evaluating the capabilities and limitations of computational schemes and establish a framework for further method improvement and enhancement. This paper is concerned with the latter context.

This paper summarizes recent applications of one class of computational transonic methods. This class is limited to methods employing the nonconservative solution of the full-potential equation for airfoils and wings. This particular class of computational transonics has reached a somewhat mature state and has been shown to be quite reliable.[2,10,11] In recent years, use of these methods has become routine. The results presented in this paper were selected to identify the level of utilization that this class has reached in a production environment. Both analysis and design applications are presented. Attention is restricted to subsonic freestream conditions and to airfoil and wing configurations representative of transport-type aircraft. The applications presented help to define the capabilities and limitations of this class of methods.

Summary of Computational Methods

Figure 1 illustrates the matrix of four computational methods that constitute the class of methods being examined. Both direct and inverse methods are included in the matrix. Direct solutions are used for the analysis-type problem; that is, given the aerodynamic surface, define the flow about it. The inverse solutions are used for the design-type problem; that is, given the desired surface flow characteristics, define the aerodynamic surface which produces such characteristics. The four

	TWO-DIMENSIONAL	THREE-DIMENSIONAL
DIRECT SOLUTION	BAUER, GARABEDIAN, AND KORN (PGM H)	JAMESON (FLO-22)
INVERSE SOLUTION	TRANEN	HENNE

Fig. 1 Set of four computational transonic flow methods.

methods identified are mathematically consistent; all are full-potential, nonconservative formulations. The direct and inverse methods utilize the same finite-difference scheme for the potential flow. Likewise, the two-dimensional and three-dimensional inverse methods are based on analogous schemes utilizing the concept of surface transpiration.[12,13]

The two-dimensional direct solution is the Douglas version of the Bauer, Garabedian, and Korn program known as Program H.[10] This program includes the simulation of viscous effects by boundary-layer displacement additions to the airfoil surface. The boundary layer is calculated using a Nash-Macdonald integral boundary-layer procedure.[14] The boundary-layer solution is repeatedly updated during the potential solution convergence. The final calculated flow solution includes both a converged boundary-layer simulation and a converged potential field. Viscous profile drag is computed using a compressible Squire-Young drag formula and the trailing edge boundary-layer quantities.

The two-dimensional inverse solution is the method of Tranen,[12] and is an extension of the two-dimensional direct solution. This procedure includes the capability to design airfoils for flows including shocks. The general nature of this procedure provides a useful method for a variety of design problems.

The three-dimensional solution is the Douglas version of the Jameson program known as FLO-22.[15] The original FLO-22 program was updated in 1978, in a cooperative program with R.M. Hicks at NASA/Ames. The enhancements included the addition of an accelerated iteration step,[16] iterated Nash-Macdonald two-dimensional strip boundary-layer simulation, and approximate fuselage volume and cross-flow effects. Similar to the two-dimensional direct solution, the viscous effects are simulated by adding the boundary-layer displacement thickness to the wing surface. The boundary-layer is also updated during the potential solution convergence. This particular implementation is an automated version of a process initially investigated at NASA/Langley.[17]

The three-dimensional inverse solution is the method of Henne.[13] Similar to the two-dimensional case, this method is an extension of the three-dimensional direct solution. Again, the general nature of the transpiration scheme allows the designer to address flows, including shocks, for a variety of problems.

The methods are implemented in a production status within the Douglas computing system. The computational procedures are supported by numerous auxiliary routines to supply input data or access output data. Auxiliary routines include geometry libraries, geometry manipulation routines, graphics routines, and related aerodynamic analysis methods such as a vortex lattice lifting surface procedure and a transonic buffet analysis procedure.

It is important to emphasize that this particular class of computational methods calculates transonic potential flow with viscous effects added by boundary-layer theory. The inherent assumptions of such simulations include weak shocks and no significant boundary-layer separations. Applications which violate such assumptions increase the risk of significant error.

Flow Analysis Applications

A number of two-dimensional and three-dimensional flow analysis applications have been compiled and are discussed in this section. This discussion includes the following major topics:

1) Airfoil performance assessment
2) Wing performance assessment
3) Configuration performance improvements
4) Lift effects
5) Reynolds number effects

Performance assessments and improvements are discussed in terms of both pressure distributions and drag

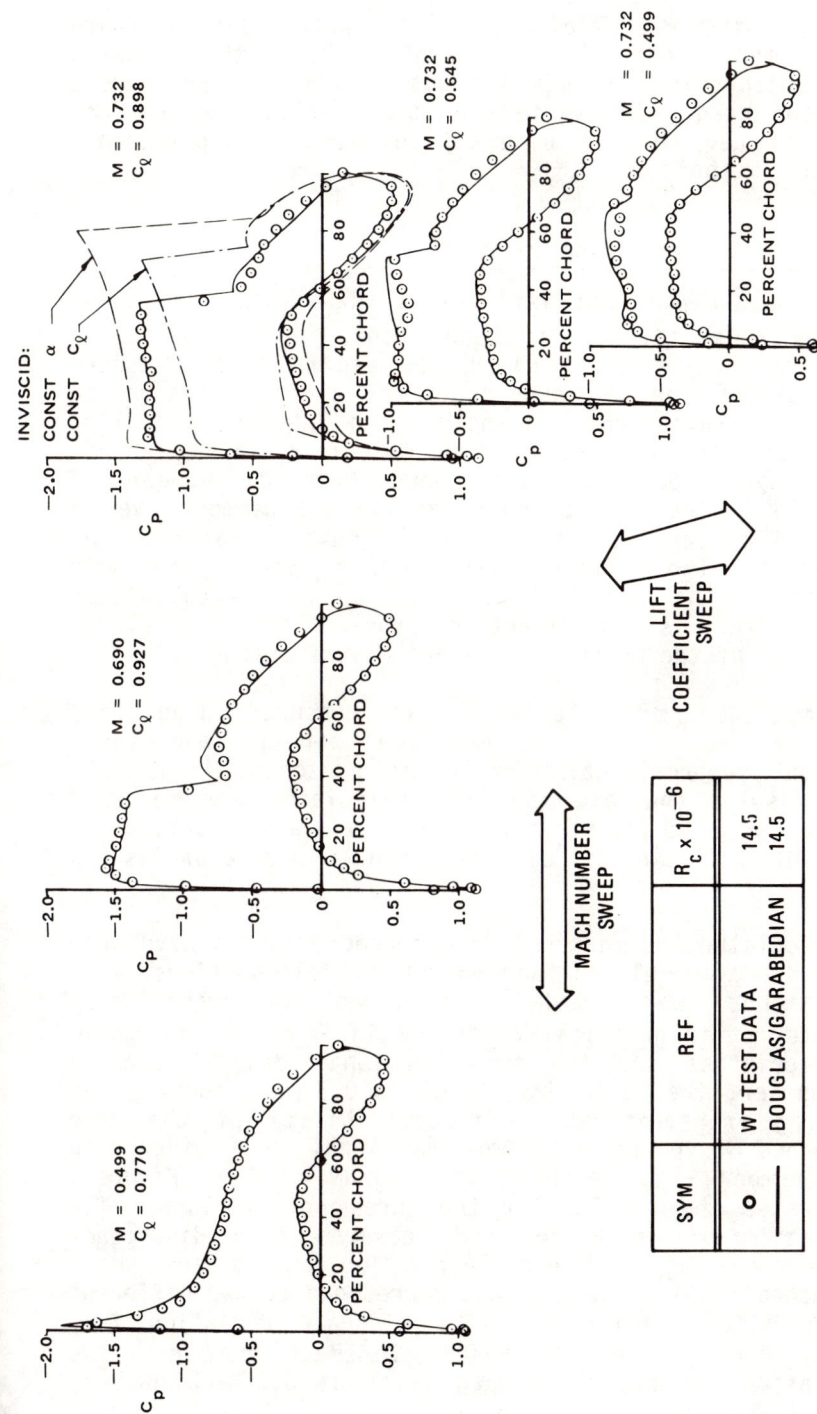

Fig. 2 Comparison of calculated and measured pressure distributions for supercritical airfoil DSMA 671.

characteristics. Lift effects include both buffet boundary and maximum lift analyses. Examples within the Reynolds number discussion include viscous effects on pressures, lift, and drag. The variety of applications reveals the extent of use that is becoming common for computational transonic methods.

Airfoil Performance Assessment

The two-dimensional solution has been used to analyze the numerous airfoil configurations. Figure 2 is a composite comparison of calculated and measured pressure distributions on a supercritical airfoil designated DSMA 671. Variations in both Mach number and lift coefficient are presented. The calculated and measured results are for the same chord Reynolds number of 14.5×10^6. The results show excellent agreement over a broad lift coefficient and Mach number range. For reference, the figure includes several inviscid calculations indicated by the dashed lines. The calculated viscous effect is significant and provides for an excellent simulation of the measured pressure distributions.

Comparison of calculated and measured drag rise characteristics for a conventional airfoil and three different supercritical airfoils is presented in Fig. 3. These results indicate that the calculated and measured section drag characteristics are in good qualitative agreement but exhibit somewhat random discrepancies in absolute drag level.

Calculated and measured drag characteristics have been analyzed and correlated for numerous airfoil sections. The correlations were made for three different drag characteristics graphically defined in Fig. 3. The three characteristics are the two-dimensional drag divergence Mach number, the basic drag level at 0.5 Mach number, and the drag increment due to compressibility at the drag divergence Mach number. Two-dimensional drag divergence Mach number is defined at $dc_d/dM = 0.1$. Figure 4 illustrates the results for the three correlations. The correlation of calculated and measured drag divergence Mach number, M_{dd}, is excellent. Two points are shown for each airfoil. These points correspond to two different section lift coefficients. The standard deviation from perfect correlation is 0.0066 in Mach number. The correlation for the basic drag level at 0.5 Mach number,

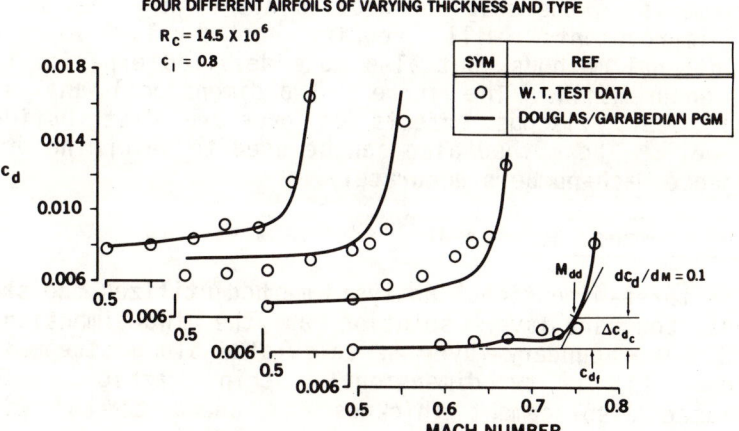

Fig. 3 Comparison of calculated and measured airfoil drag characteristics.

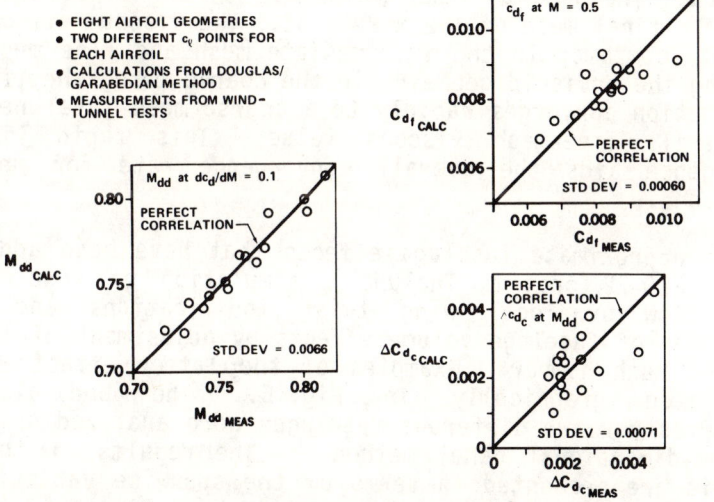

Fig. 4 Correlation of calculated and measured airfoil drag divergence Mach number, profile drag, and compressibility drag at drag divergence.

c_{d_f}, indicates a more significant scatter. The standard deviation from perfect correlation is 0.00060 in drag coefficient. The correlation of the remaining drag characteristic, the compressibility drag increment, Δc_{d_c}, also indicates significant scatter. The standard deviation is 0.00071 in drag coefficient.

The latter two figures quantify the earlier statement concerning somewhat random errors in absolute drag level.

Improvements in the drag correlation are highly desirable. Such improvements will require not only improved computational methods, but also more detailed experimental drag measurements. The current two-dimensional analysis method models viscous effects on pressure distributions quite well. The method also can be used to determine drag divergence Mach numbers accurately.

Wing Performance Assessment

The three-dimensional analysis method utilizes the same viscous boundary-layer solution as the two-dimensional method. The boundary-layer is calculated along streamwise sections in a two-dimensional strip fashion. The calculated displacement thickness is added to the wing surface and the surface is remapped. This procedure is repeated during the potential iteration convergence. Three computational meshes are used in the potential solution. The initial mesh (48 x 6 x 8) is refined twice to give a final mesh of 192 x 24 x 32. The viscous effects are only computed in the intermediate mesh and fine mesh. By using the inviscid geometry in the coarse mesh, the lift distribution converges rapidly to a coarse mesh level near the final, fine-mesh viscous value. This rapid lift convergence aids the overall convergence rate for most situations.

The approximate fuselage effects that have been added to the FLO-22 program include a simulation of fuselage cross flow by local wing twist modifications and a simulation of fuselage volume effects by adjustment of the solution Mach number. Examples of the latter effect are illustrated graphically in Fig. 5. The body-alone geometries for two different fuselages were analyzed using a three-dimensional panel method.[18] The results of this analysis are presented in terms of the spanwise variation of local Mach number along the wing midchord line. In the first case the body for the wing designated W_2 had a length-to-diameter ratio, L/D, of 11.5. The off-body flowfield in the wing region is perturbed approximately 0.005 in Mach number. The second case, designated W_8, is for a fuselage having an L/D of 7.1. This reduction of L/D has a large impact on the wing flowfield. This fuselage effect is approximated by calculating the transonic wing flow solution at an average local Mach number and correcting the results to the true freestream Mach number.

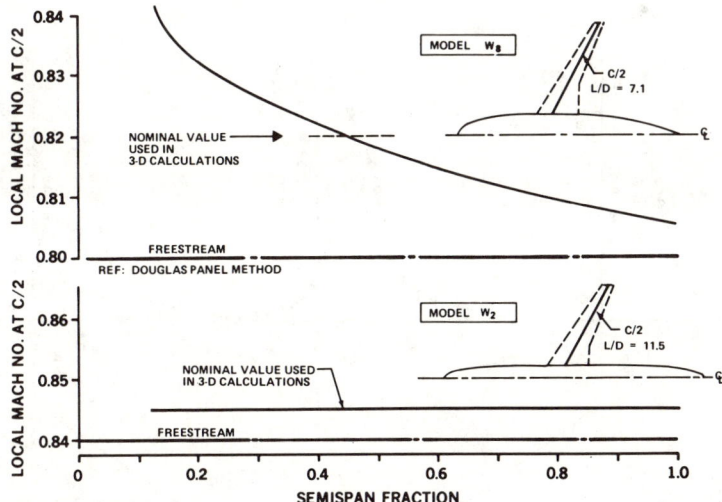

Fig. 5 Effect of body volume on wing flowfield for two different fuselage configurations.

Figures 6-9 present comparisons of calculated and measured pressure distributions for the W_2 wing-body geometry, the geometry with the smaller fuselage effect. This set of comparisons is a Mach number sweep. The strong double-shock, transonic character of this particular geometry is evident in both the calculated and measured pressure distributions at the higher Mach numbers. The evolution of the transonic character with Mach number is well defined. At the highest Mach number, the results include an inviscid calculation as well as a viscous calculation. Similar to the two-dimensional calculations, the viscous effects are substantial, and the introduction of the viscous modeling provides a significant improvement. This set of comparisons indicates that for the case of small fuselage effects, the current three-dimensional analysis is quite adequate for the calculation of pressure distributions.

A set of calculations has also been made for the W_8 wing-body geometry. This geometry has the reduced fineness ratio fuselage. The comparison of calculated and measured pressure distribution indicates a more significant fuselage effect. Figures 10 and 11 present a comparison of measured and calculated pressure distributions, using different fuselage effects, for this case at 0.8 Mach number. The calculations are ordered in the degree of fuselage effect simulation. Figure 10

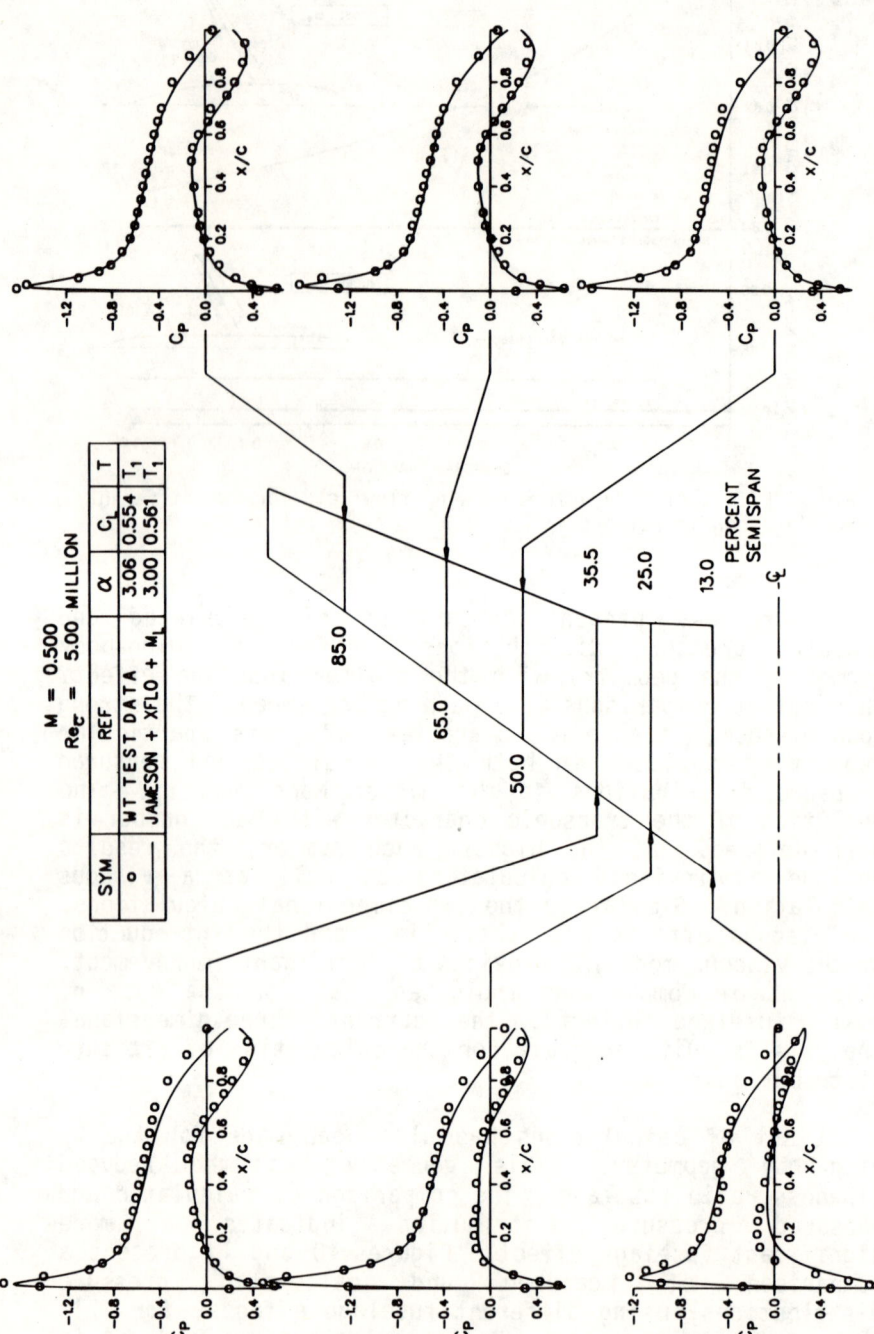

Fig. 6 Comparison of calculated and experimental chordwise pressure distributions for W_2.

APPLIED COMPUTATIONAL TRANSONICS

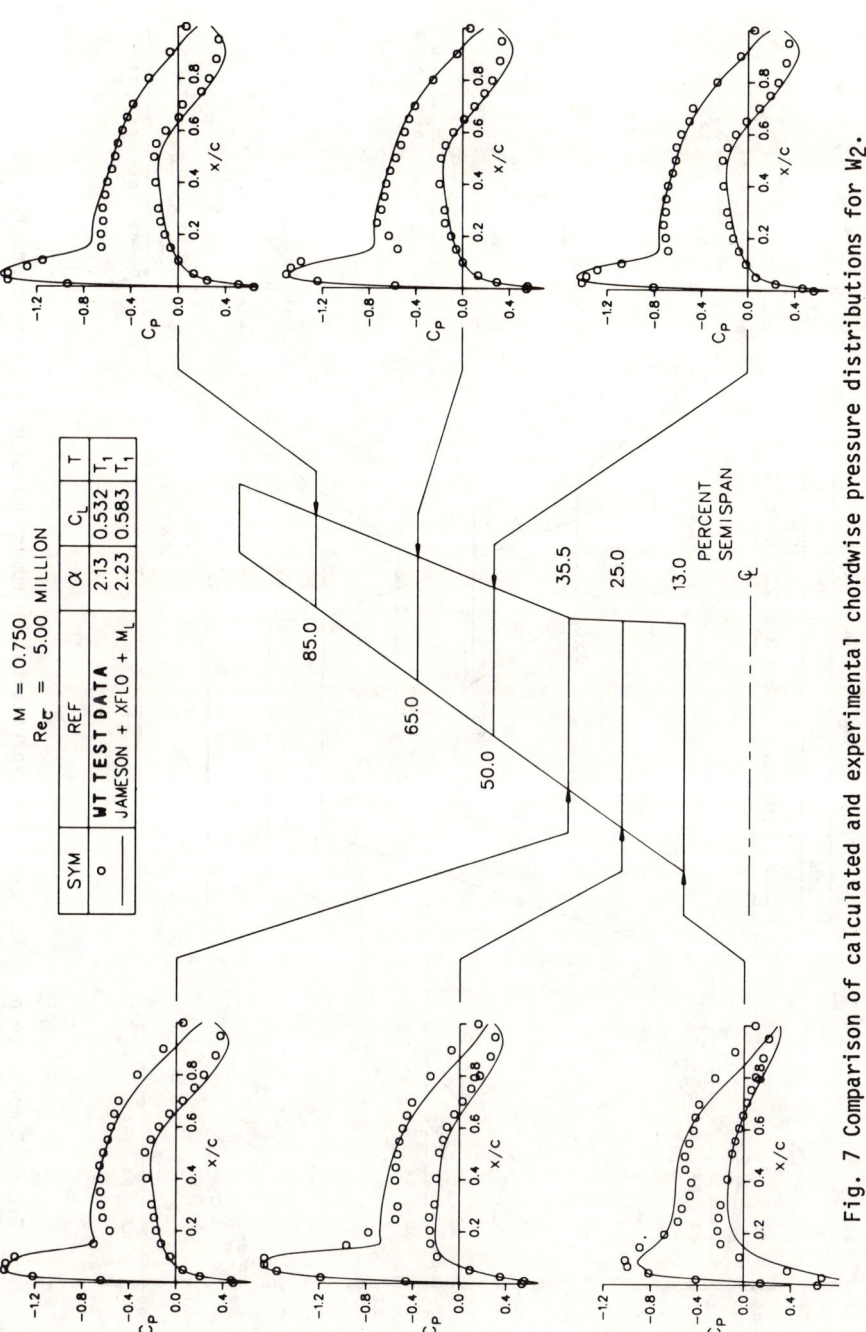

Fig. 7 Comparison of calculated and experimental chordwise pressure distributions for W_2.

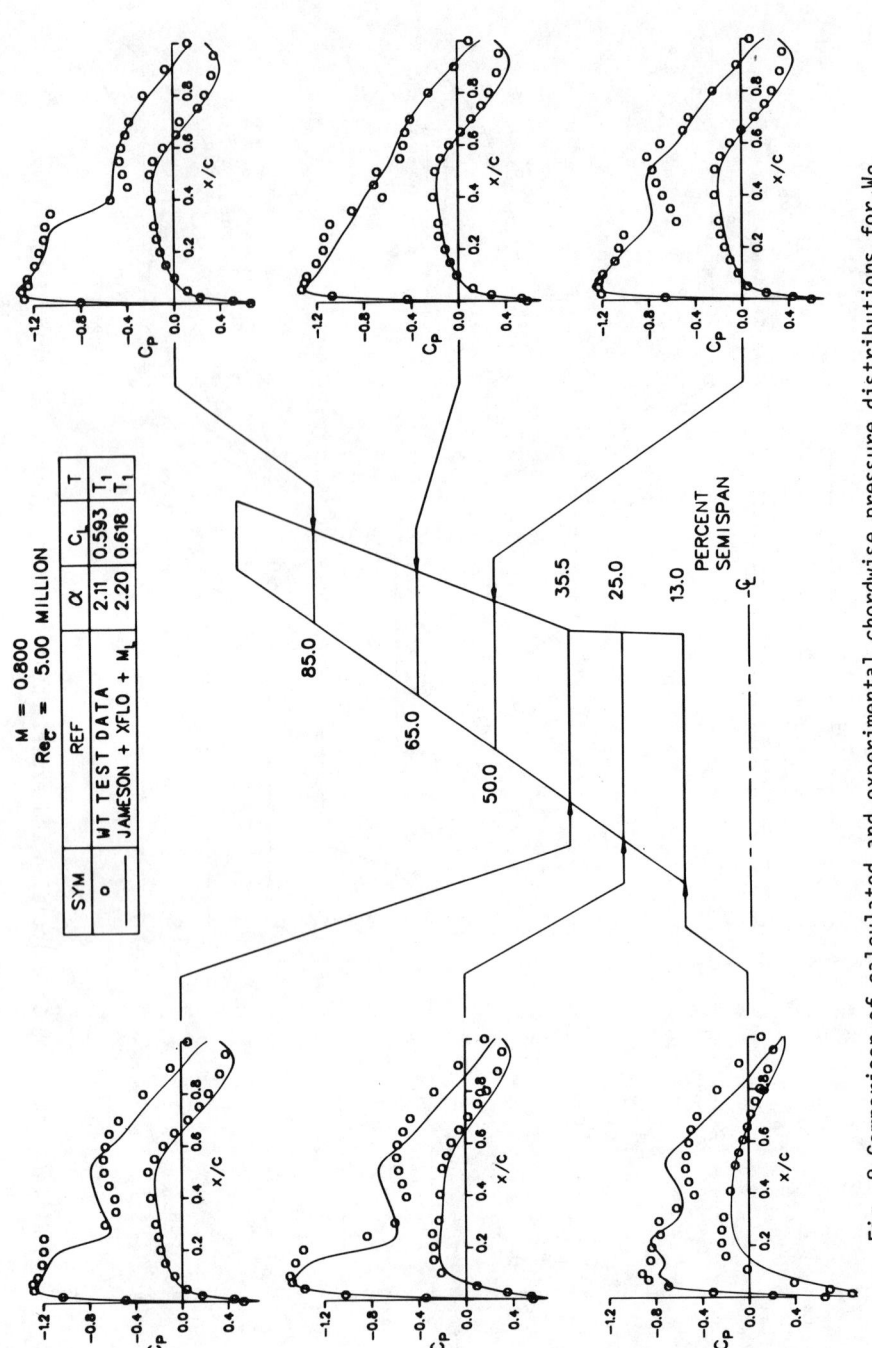

Fig. 8 Comparison of calculated and experimental chordwise pressure distributions for W_2.

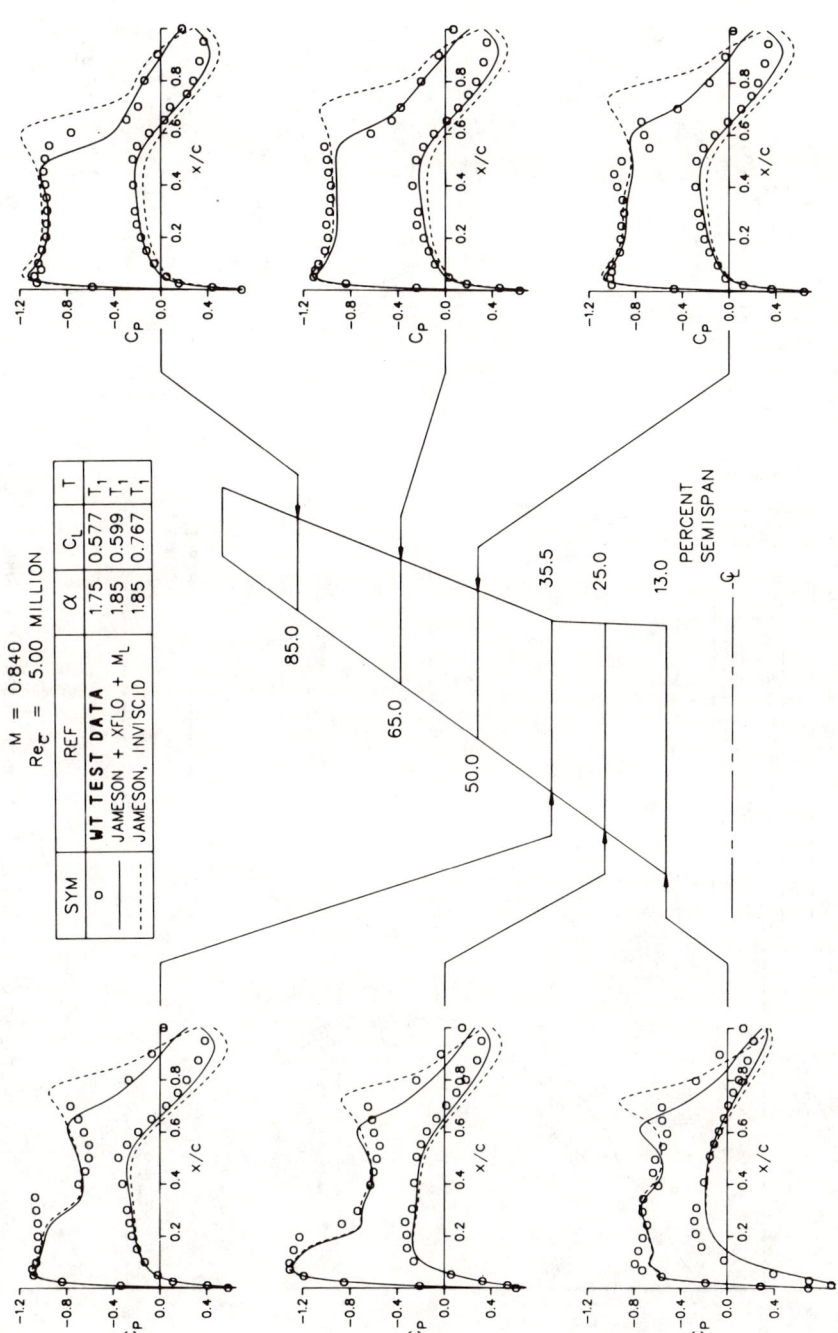

Fig. 9 Comparison of calculated and experimental chordwise pressure distributions for W_2.

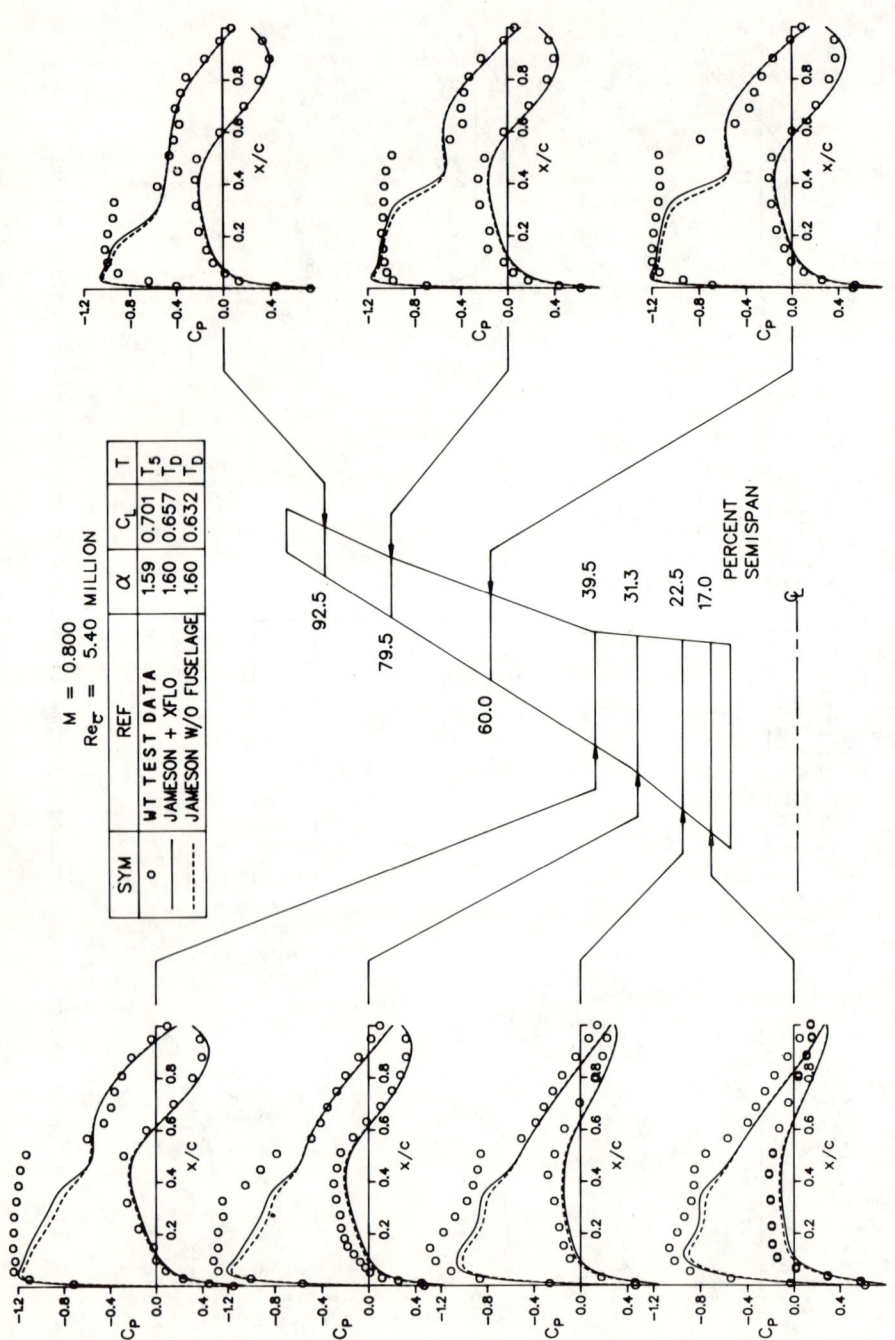

Fig. 10 Comparison of calculated and experimental chordwise pressure distributions for W_8.

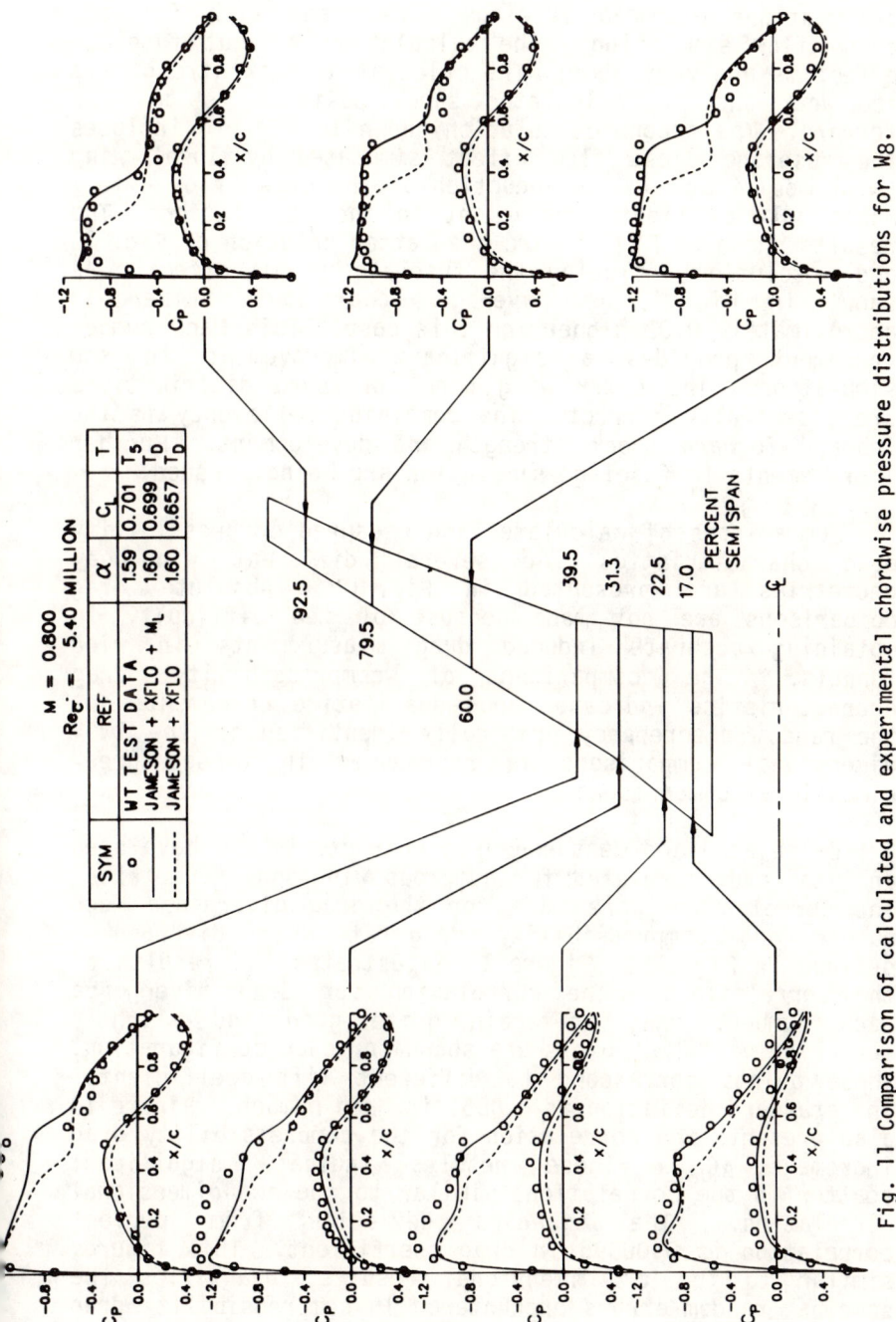

Fig. 11 Comparison of calculated and experimental chordwise pressure distributions for W8.

illustrates the comparison of measured results and calculations with no fuselage effect and with fuselage cross flow simulation. The calculation without fuselage effect is not very good. The calculated shock strength is too weak and the calculated shock position is too far forward. The second calculation shown in Fig. 10 includes the fuselage cross flow effect simulated by local wing twist modification. Introduction of the cross flow effect makes only a slight improvement to the calculation. The results shown in Fig. 11 are the latter solution of Fig. 10 and a solution including the fuselage volume effect. As shown in Fig. 5, the average local Mach number is approximately 0.02 higher for this case. This Mach number increment provides a significant improvement in the simulation. The outer wing panel pressure distributions are essentially correct. The remaining deficiency is the inboard forward shock strength and development. Further improvements in fuselage simulation are being pursued.

Comparisons of calculated and measured compressibility drag characteristics for several different wing-body geometries are presented in Fig. 12. Absolute drag comparisons are not made because of the difficulty in obtaining accurate induced drag measurements in wind tunnels. The comparison of compressibility drag characteristics indicates good qualitative agreement, but the random discrepancy originally identified in the two-dimensional comparisons is reproduced in these three-dimensional comparisons.

Calculated and measured drag characteristics have been analyzed and correlated for numerous wing-body geometries. The correlations were made for the drag divergence Mach number and compressibility drag at drag divergence, defined in Fig. 12. Figure 13 illustrates the results of the correlations. The correlation for drag divergence Mach number, M_{DD}, is again quite good and slightly conservative. Two points are shown for each configuration. These points correspond to different lift coefficients. The standard deviation is 0.0051 in Mach number. Figure 13 also presents the correlation for the compressibility drag increment, ΔC_{D_C}. These results indicate significant scatter in the correlation, similar to the two-dimensional correlation. The standard deviation from perfect correlation is 0.00095 in drag coefficient. This figure, similar to the two-dimensional results, quantifies the somewhat random errors encountered in compressibility drag increments. These inaccuracies are distributed in some

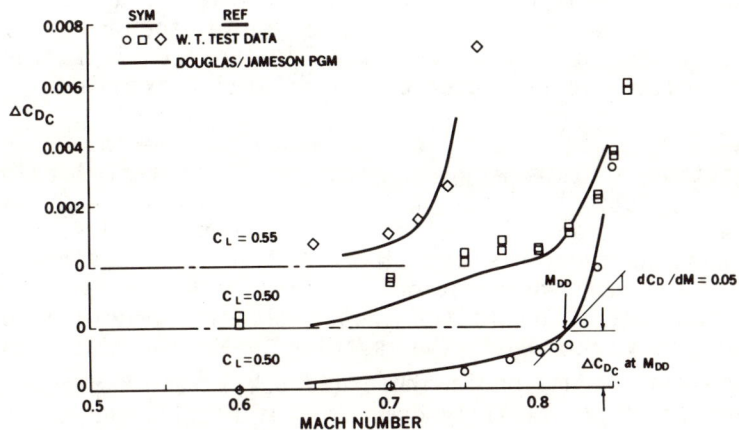

Fig. 12 Comparison of measured and calculated drag rise characteristics for three different supercritical wings.

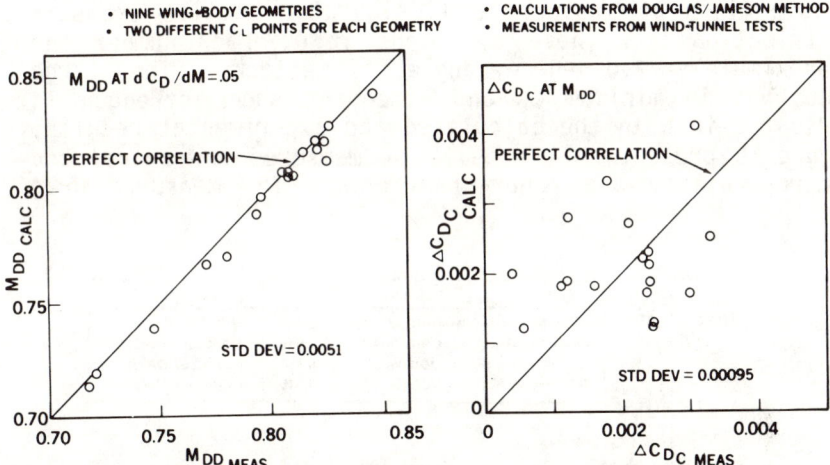

Fig. 13 Correlation of calculated and measured wing-body drag divergence Mach number and compressibility drag at drag divergence.

unknown fashion between the calculation method and the experimental measurements. The correlation helps to identify the level of accuracy that can be expected from drag calculations using the present method.

Configuration Performance Improvements

The drag correlations for airfoils and for wings indicate that the existing methods cannot be used to resolve the last drag count of absolute drag level. However, the methods do provide a rapid and effective means of reaching a relatively well-refined configuration. This capability is reflected by the following example.

The W_2 wing and a subsequent leading edge modification of this wing were designed and tested prior to the practical implementation of three-dimensional computational transonic methods. As shown in Figs. 6-9, the transonic development of W_2 was less than desirable. The leading edge modification, designated W_4, provided a significant reduction in the inboard leading edge suction peaks relative to W_2. The suction peak reduction and corresponding reduction in inboard shock strength produced an improved compressibility drag characteristic. Post-test analysis using the current three-dimensional direct solution verified both the improvement in surface pressure distributions, and the attendant improvement in compressibility drag characteristics. Figure 14 presents a comparison of measured and calculated section pressure distributions for these two wings at 0.7 Mach number and approximately 2.0 degree angle of attack. The large reduction in minimum C_p and attendant shock strength is indicated in both the calculated and experimental results. Figure 15 presents a composite of measured and calculated compressibility drag characteristics. The measured drag

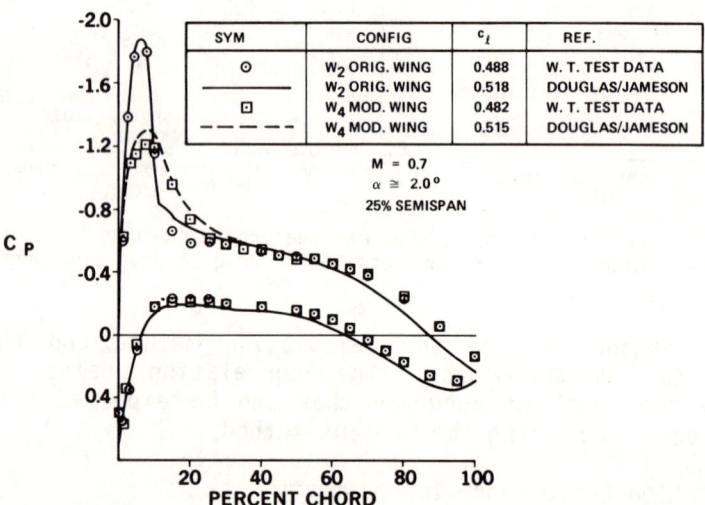

Fig. 14. Comparison of calculated and measured section pressure distributions for a wing leading edge modification.

improvement is simulated reasonably well by the calculations. Included in the figure is a summary of elapsed time and cost associated with the experimental analysis and the computational analysis. The comparison of the time and cost data underscores the significance and importance of applied computational transonics. The drag difference between W_2 and W_4 is presented in Fig. 16. While the calculated and measured results do not match perfectly, the W_4 configuration drag improvement is identified by the calculations. Had such analysis capability been available during the design process, the W_2 configuration would have been quickly rejected as an inferior aerodynamic surface.

Lift Effects

Two lift characteristics important to the design of transport type aircraft are the high speed buffet boundary and the low-speed, maximum lift coefficient, $C_{L\ MAX}$. The current analysis procedures have been applied to both areas.

A high-speed buffet analysis procedure has been developed for use with the two-dimensional and three-dimensional computational methods. The semiempirical procedure is based on an extensive correlation of high-

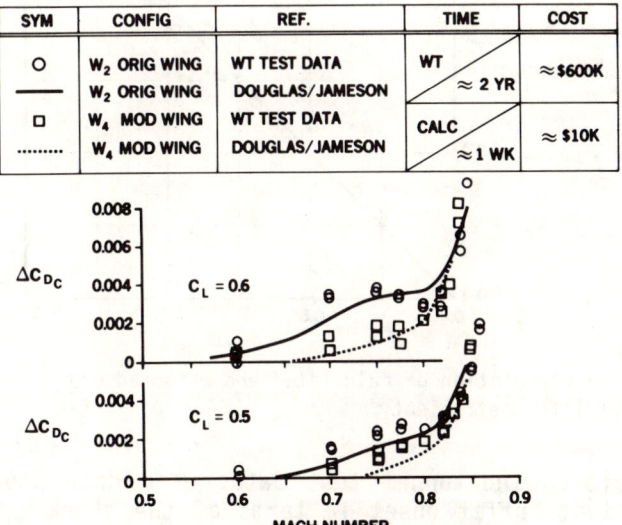

Fig. 15 Comparison of calculated and measured compressibility drag characteristics for wing leading edge modification.

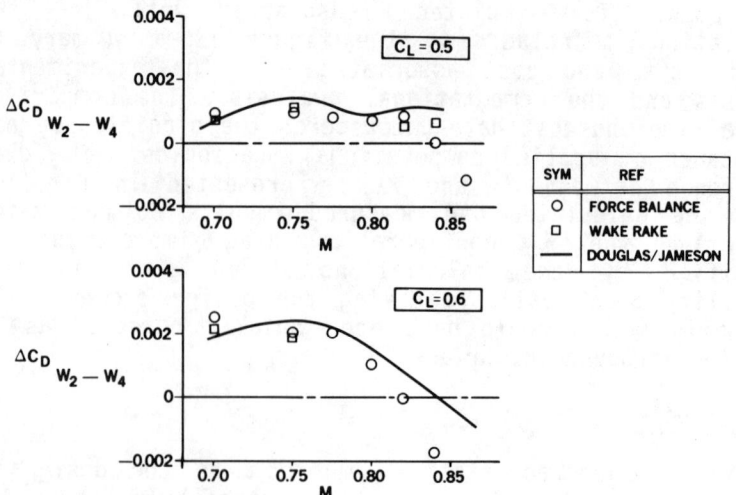

Fig. 16 Comparison of force balance, wake rake, and calculated drag increments between W_2 and W_4.

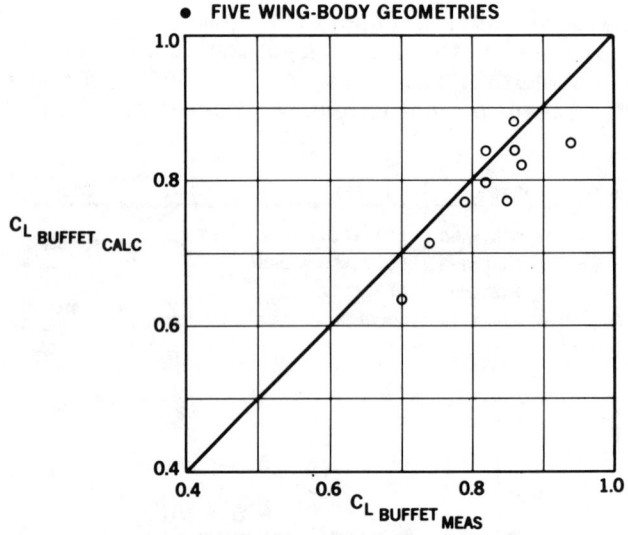

Fig. 17 Correlation of calculated and measured cruise regime buffet lift coefficient.

speed airfoil wind-tunnel test data. The data have been correlated at buffet onset in terms of the shock upstream Mach number, M_1, vs shock position as a fraction of airfoil chord. This correlation provides a simple criterion to make comparative studies of buffet boundary.

If calculated Mach numbers normal to the shock exceed the correlation line, the airfoil or wing is considered to be in buffet conditions. This procedure has been applied to numerous wing-body geometries using the three-dimensional direct solution to provide the calculated shock normal Mach number and shock position. An implicit assumption in this application is the existence of reasonably two-dimensional flow. Figure 17 presents a summary of calculated and measured wing-body buffet C_L characteristics. The measured buffet C_L characteristics were derived from lift break and moment break data. The comparison indicates that the buffet analysis is generally conservative. While the results display some scatter, the method has been successfully applied to several design studies to increase buffet C_L.

The buffet method is an attempt to use transonic potential flow with viscous effects to calculate flows at the onset of large shock-induced separation effects. Breakdown of the potential flow and boundary-layer theory assumptions can be expected in this flow regime. Consequently, the buffet analysis must be used with caution. The M_1 criterion, added to the basic computational procedures, is the simple model which provides the buffet analysis some physical realism without excessive complication.

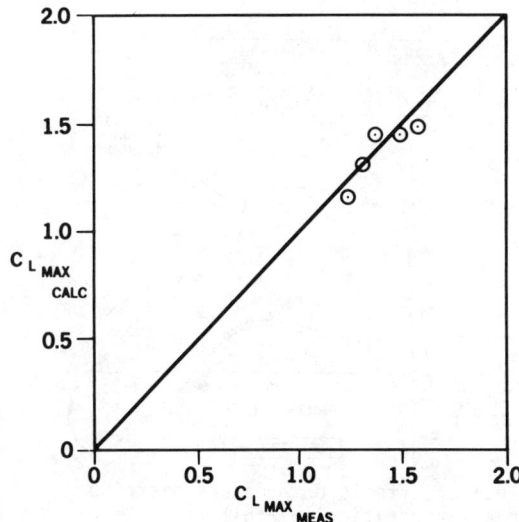

Fig. 18 Correlation of calculated and measured low speed C_L MAX for cruise wing configurations.

A low-speed $C_{L\ MAX}$ criterion based upon $C_{p\ min}$ has also been developed in conjunction with the current computational methods. The computational methods, due to the relatively good leading edge definition obtained in fine mesh solutions, are particularly amenable to calculation of this parameter. Several different low-speed, clean wing $C_{L\ MAX}$ estimates have been accomplished using this $C_{p\ min}$ criterion in conjunction with the current flow solutions. A correlation of calculated and measured clean wing $C_{L\ MAX}$ is presented in Fig. 18. While this correlation is limited, it does suggest that the procedure provides some degree of realism.

Reynolds Number Effects

One of the more interesting applications of current computational methods is the analysis of Reynolds number effects. Current wind-tunnel test capability typically is limited to subscale Reynolds number conditions. No such restriction applies to computational methods.

Figure 19 illustrates calculated pressure distributions for a conventional and a supercritical airfoil. Three pressure distributions are shown for each. These pressure distributions were computed at constant angle of attack and at different Reynolds numbers. The dashed line is an inviscid calculation. The solid line is a calculation at a chord Reynolds number of 30×10^6, a value representative of flight Reynolds

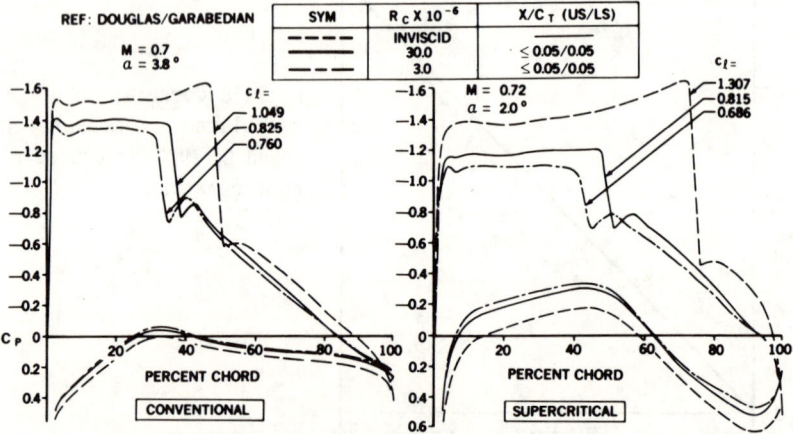

Fig. 19 Calculated Reynolds number effects on pressure distributions for a conventional and a supercritical airfoil.

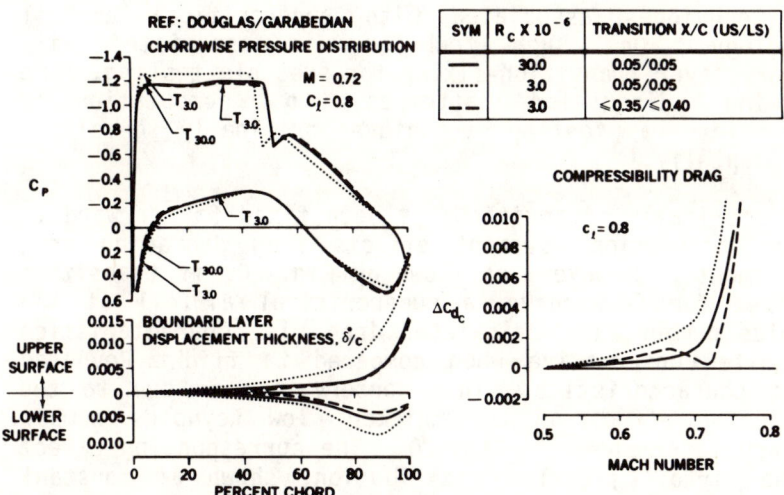

Fig. 20 Calculated Reynolds number and transition position effects for supercritical airfoil DSMA 671.

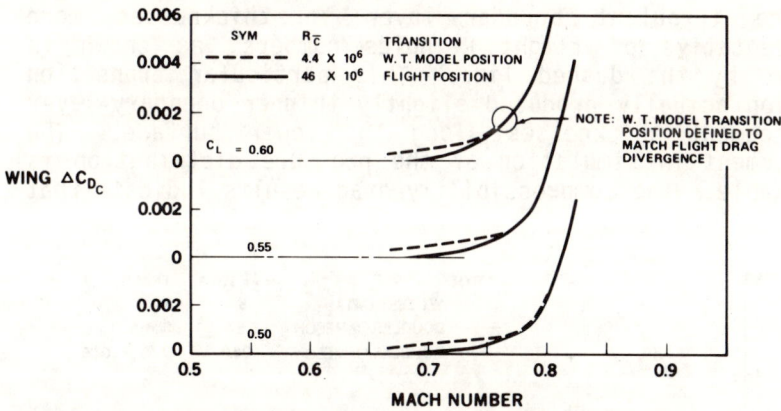

Fig. 21 Calculated wind-tunnel and flight Reynolds number supercritical wing compressibility drag characteristics.

number on a transport's outer wing panel. The dashed line is a calculation at a chord Reynolds number of 3×10^6, a value typical of a high-speed wind-tunnel model's outer wing panel. It can be noted that the supercritical airfoil section displays roughly twice the Reynolds number sensitivity of the conventional airfoil.

The increased Reynolds number sensitivity has been previously noted,[2] and has led to special wind-tunnel test techniques associated with supercritical airfoil technology. One such technique is associated with boundary-layer transition-fixing for subscale testing. The following discussion attempts to computationally substantiate a testing technique originally developed experimentally.[19]

Historically, transition has been fixed far forward to prevent transition movement effects from degrading drag measurements. However, as shown by Fig. 20, if transition is fixed far forward on a supercritical airfoil at low Reynolds number, the calculated drag rise characteristics are quite conservative when compared to flight Reynolds number characteristics. This conservatism is due to the decambering effect of the thicker, low Reynolds number boundary layer shown in Fig. 20. The corresponding effect on the airfoil pressure distribution, shown at constant section lift coefficient, is also presented in Fig. 20. The experimental technique used to improve the drag simulation is to allow transition to move aft on both airfoil surfaces. This aft movement suppresses the growth of the turbulent boundary layer to thicknesses more representative of flight Reynolds numbers, as shown in Fig. 20 by the dashed line. This particular transition position actually produced slightly thinner boundary-layer displacement thicknesses along the lower surface. The improvement in simulation of the pressure distribution is remarkable. The compressibility drag results indicate that

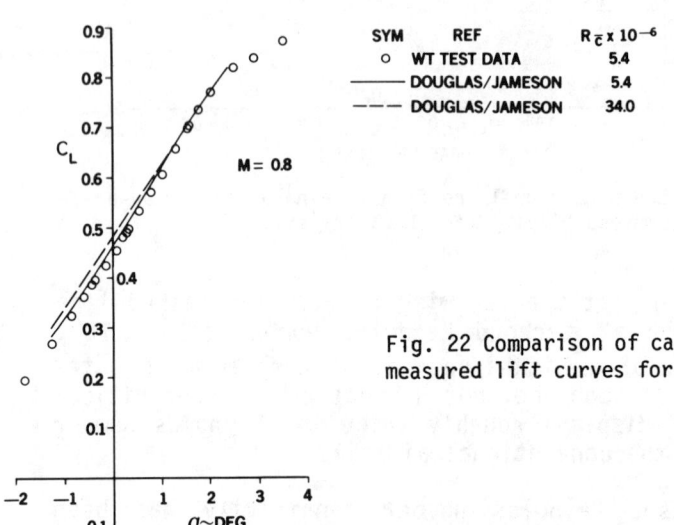

Fig. 22 Comparison of calculated and measured lift curves for W_8.

complete simulation of the drag characteristics cannot be achieved at all Mach numbers. However, a transition position can be found which closely reproduces the flight Reynolds number, drag divergence Mach number, and design point compressibility drag. This approach has been applied to three-dimensional calculations to define transition strip patterns for accurate wing-body drag performance testing at subscale Reynolds numbers. Figure 21 illustrates such a case. The solid line is a calculated compressibility drag at constant C_L and flight Reynolds number. The subscale Reynolds number calculation was obtained by adjusting the transition strip pattern until a drag characteristics match was obtained in the cruise Mach number regime. This pattern was subsequently used in a successful wind-tunnel test of this configuration.

A second area in which the Reynolds number capability of computational methods has been applied is in the analysis of the nonlinear lift curve associated with advanced airfoils. Regions of high local lift curve slope may result in flutter and gust load weight penalties for an aircraft wing design. Figure 22 presents a comparison of calculated and measured lift curves for the W_8 wing-body. The measured characteristics were obtained during wind-tunnel testing. The calculation, made for wind-tunnel Reynolds number, also indicated a simulation of transition position as a function of C_L. The flight Reynolds number calculation was made with transition far forward. The low Reynolds number calculation displays the same nonlinearity as the model test data. The high

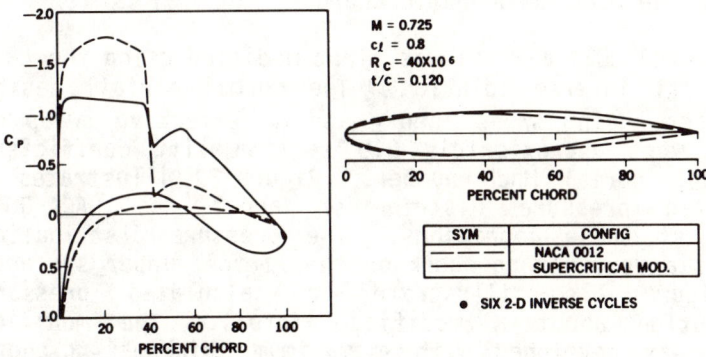

Fig. 23 Comparison of pressure distributions and airfoil geometries for NACA 0012 modification.

Reynolds number calculation does not indicate such nonlinearity. These calculations suggest that the nonlinearity is a viscous effect which should decrease with increasing Reynolds number.

Geometry Design Applications

Two geometry design applications of the inverse method capability are summarized in this section. The first case represents a sample design problem to illustrate the utility of the computational methods. The second case is an actual design problem in which an existing wing configuration was computationally modified and subsequently wind-tunnel tested.

Sample Design Problem

The first case includes application of both the two-dimensional and three-dimensional inverse methods. The design problem was to start with an initial symmetrical airfoil and use the two-dimensional and three-dimensional inverse methods to rapidly develop a supercritical wing design for a nominal cruise condition. The objective of this case was not to create a highly refined wing design. Rather, the case serves as an illustration of the inverse methods' capabilities to rapidly transform and develop aerodynamic contours.

The initial airfoil was taken to be the NACA 0012. The wing planform was defined as trapezoidal with 25 deg sweep. The aspect ratio and taper ratio were specified as 8.0 and 0.3, respectively. The design conditions were taken as 0.8 Mach number, 0.65 C_L, and 40 million Reynolds number based on the mean aerodynamic chord.

The NACA 0012 airfoil was first modified using the two-dimensional inverse solution. The normal airfoil design conditions, using sweep theory and an effective sweep of 25 deg, were approximately 0.8 section lift coefficient and 0.725 normal Mach number. Figure 23 illustrates a calculated pressure distribution about the NACA 0012 airfoil at these conditions. The pressure distribution exhibits a very strong shock on the airfoil upper surface. The figure also illustrates a calculated pressure distribution about a modified airfoil. The modified airfoil was developed with a maximum thickness-to-chord ratio constraint of 0.12. The modification includes the introduction of aft loading typical of supercritical

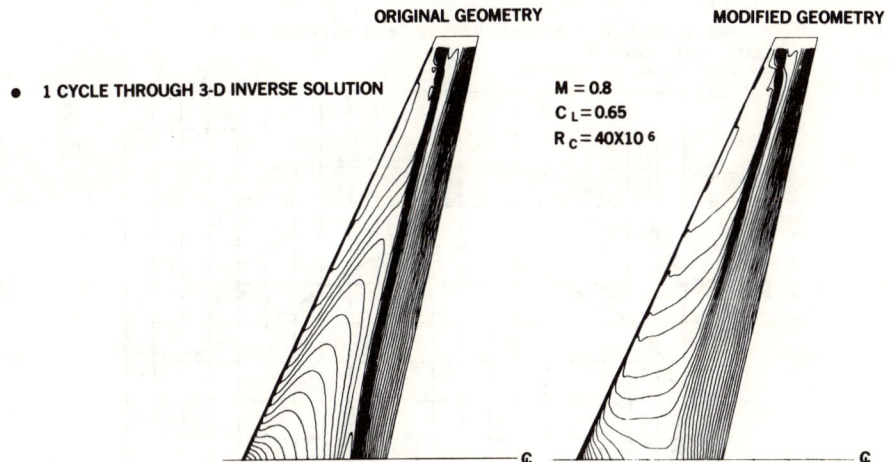

Fig. 24 Comparison of upper surface isobars for original and modified trapezoidal wing.

airfoils. The reduction in shock strength associated with the modified airfoil is substantial. The two-dimensional modification was accomplished in six cycles through the inverse solution and required less than two man-hours to complete.

The modified airfoil was used as a constant, normal airfoil section for the initial three-dimensional wing surface. The initial wing surface was defined by a streamwise root and tip airfoil including three degrees of geometric washout.

Figure 24 illustrates calculated upper-surface isobars for the initial wing surface at the design condition. Also shown are the corresponding isobars for a modified wing surface. The modified wing surface was developed with one cycle through the three-dimensional inverse solution. The original wing isobars indicated an inboard aft shock with reduced sweep. This inboard shock was repositioned further forward and weakened for the modified wing. Figures 25, 26, and 27 illustrate the original, specified, and new section pressure distributions and the original and new airfoil geometries for these inboard stations. The specified pressure distributions are significantly different from the original pressure distributions. The match obtained between the new and specified pressure distributions is remarkably good. The original and new airfoil section geometries are also illustrated. The

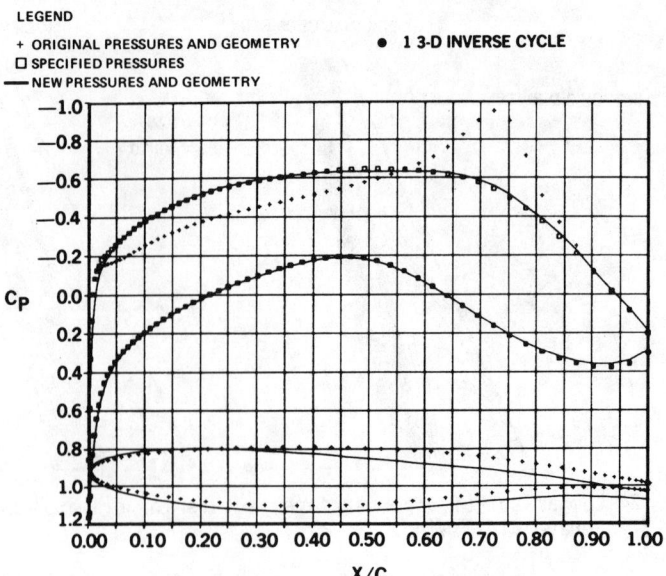

Fig. 25 Comparison of pressure distributions and airfoil geometries for trapezoidal wing modification at 0% semispan.

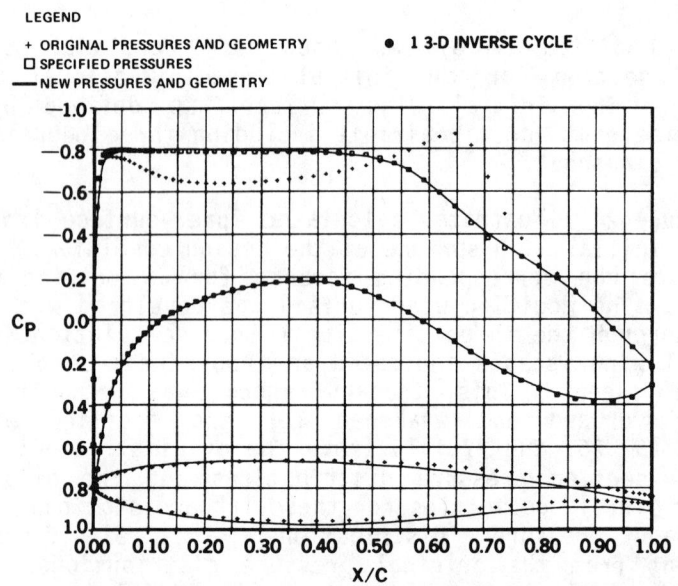

Fig. 26 Comparison of pressure distributions and airfoil geometries for trapezoidal wing modification at 20% semispan.

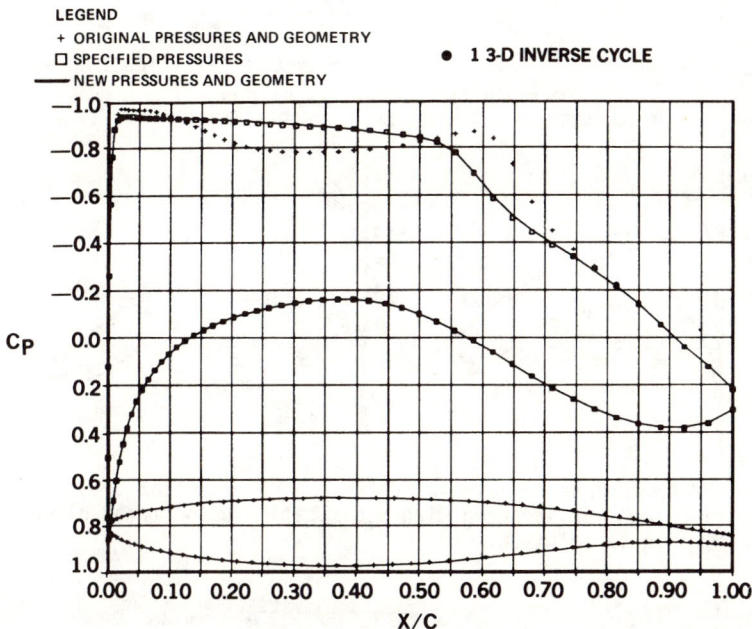

Fig. 27 Comparison of pressure distributions and airfoil geometries for trapezoidal wing modification at 39% semispan.

geometry modification required to achieve the specified pressure distribution is significant. The three-dimensional modification also required less than two man-hours to complete. Clearly, this wing surface is not a highly developed configuration. Additional direct and inverse solutions would be required to develop a well refined configuration. However, the rapid development of this configuration from the NACA 0012 airfoil and the planform specification illustrates the capabilities of the inverse solutions.

Actual Design Problem

The second geometry design application involves the use of the three-dimensional inverse solution in an actual design study. The inboard panel of a swept, supercritical wing was modified to reduce inboard suction peaks and shock strengths at the nominal cruise design point. A significant reduction was achieved computationally and is illustrated in Fig. 28. This reduction was experimentally verified as indicated by the measured pressure distribution comparison also presented in Fig. 28. A

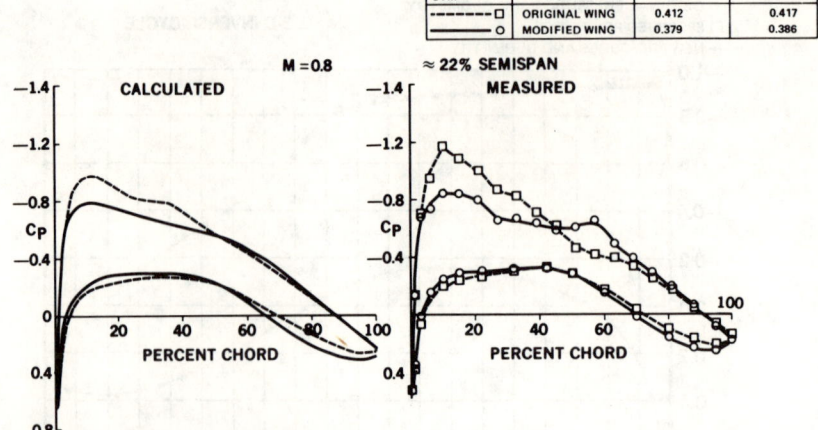

Fig. 28 Comparison of calculated and measured section pressure distributions for advanced wing modification using three-dimensional inverse solution.

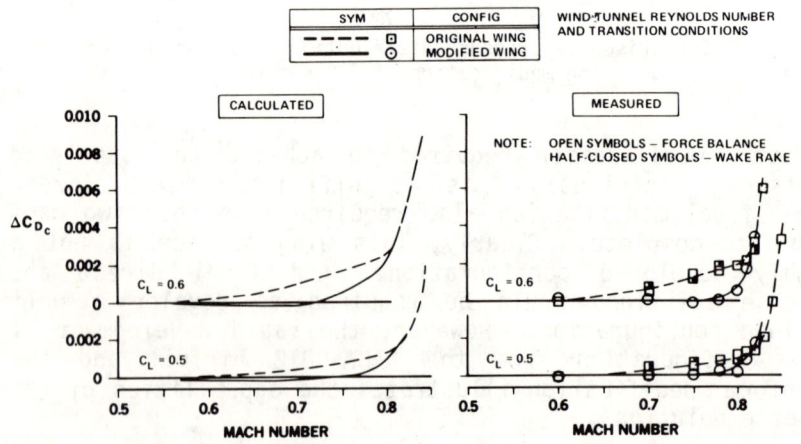

Fig. 29 Comparison of calculated and measured compressibility drag increments for advanced wing modification using three-dimensional inverse solution.

comparison of calculated and measured compressibility drag characteristics for these two wings is presented in Fig. 29. The calculated drag improvement due to the modified geometry is reasonably well substantiated by the measured characteristics. This application illustrates the real advantage of the inverse solutions in a production environment.

Summary

The current class of computational transonic flow methods is being applied to an ever-increasing range of aerodynamic problems. Within the context of subsonic-freestream conditions and transport type geometries, the following capabilities and limitations have been observed:

Substantial Capabilities

1) Cruise regime configuration design
2) Cruise regime configuration analysis
3) Determination of drag divergence Mach number
4) Determination of some Reynolds number effects

Marginal Capabilities

1) Determination of drag levels and increments
2) Determination of high speed buffet C_L (semiempirical)
3) Determination of low-speed $C_{L\ MAX}$ (semiempirical)

It should be further observed that situations such as substantial boundary-layer separation or significant transonic body effects must be carefully scrutinized. Such situations violate the basic assumptions inherent in the methods and the occurrence of significant errors is likely.

It must be recognized that the class of methods discussed in this paper represents a significant milestone, but only a milestone, in the application of computational transonics to real aerodynamic problems. Summaries such as the present example should serve to focus attention on areas requiring further improvements. Development efforts in such areas will provide further capabilities for the aerodynamic designer.

References

[1] Hicks, R.M. and Henne, P.A., "Wing Design by Numerical Optimization," AIAA Paper No. 77-1247, Seattle, Wash., August 1977.

[2] Henne, P.A. and Hicks, R.M., "Wing Analysis Using A Transonic Potential Flow Computational Method," NASA TM-78464, July 1978.

[3]DaCosta, A.L., "Application of Computational Aerodynamics Methods to the Design and Analysis of Transport Aircraft," ICAS Proceedings, Lisbon, Portugal, 1978.

[4]Lynch, F.T., "Recent Applications of Advanced Computational Methods in the Aerodynamic Design of Transport Aircraft Configurations," Douglas Paper No. 6639, 1978.

[5]Haney, H.P., Johnson, R.R., and Hicks, R.M., "Computational Optimization and Wind-Tunnel Test of Transonic Wing Designs," AIAA Paper No. 79-0080, New Orleans, La., January 1979.

[6]Lores, M.E., Smith, P.R., and Hicks, R.M., "Supercritical Wing Design Using Numerical Optimization and Comparisons with Experiment," AIAA Paper No. 79-0065, New Orleans, La., January 1979.

[7]Haines, A.B., "Computer-Aided Design: Aerodynamics," Aeronautical Journal, March 1979, pp. 81-91.

[8]Blackerby, W.T. and Johnson, J.K., "Application of Advanced Technologies to Improve C-141 Cruise Performance," AIAA Paper No. 79-0066, New Orleans, La., January 1979.

[9]Steckel, D.K., Dahlin, J.A., and Henne, P.A., "Results of Design Studies and Wind Tunnel Tests of High Aspect Ratio Supercritical Wings for an Energy Efficient Transport," NASA Contractor Report 159332, 1980.

[10]Bauer, F., Garabedian, P., Korn, D., and Jameson, A., "Supercritical Wing Sections II" Lecture Notes in Economics and Mathematical Systems, Vol. 108, Springer-Verlag, New York, 1975.

[11]Hinson, B.L. and Burdges, K.P., "An Evaluation of Three-Dimensional Transonic Codes Using New Correlation-Tailored Test Data," AIAA Paper No. 80-0003, Pasadena, Ca., January 1980.

[12]Tranen, T.L., "A Rapid Computer-Aided Transonic Airfoil Design Method," AIAA Paper No. 74-501, Palo Alto, Ca., June 1974.

[13]Henne, P.A., "An Inverse Transonic Wing Design Method," AIAA Paper No. 80-0330, Pasadena, Ca., January 1980.

[14]Nash, J.F. and Macdonald, A.G.J., "The Calculation of Momentum Thickness in a Turbulent Boundary Layer at Mach Numbers Up to Unity," ARC CP No. 963, 1967.

[15]Jameson, A. and Caughey, D.A., "Numerical Calculation of the Transonic Flow Past a Swept Wing," New York University ERDA Report COO 3077-140, 1977.

[16]Caughey, D.A. and Jameson, A., "Accelerated Iterative Calculation of Transonic Nacelle Flowfields," AIAA Paper No. 76-100, Washington, D.C., January 1976.

[17] Newman, P.A., Carter, J.E., and Davis, R.M., "Interaction of a Two-Dimensional Strip Boundary Layer with a Three-Dimensional Transonic Swept Wing Code," NASA TM-78640, 1978.

[18] Friedman, D.M., "A Three-Dimensional Lifting Potential Flow Program," McDonnell Douglas IRAD Report MDC J6182, 1974.

[19] Blackwell, J.A. Jr., "Preliminary Study of Effects of Reynolds Number and Boundary-Layer Transition Location on Shock-Induced Separation," NASA TN D-5003, 1969.

Chapter XIV.

Evaluation of Full Potential Flow Methods for the Design and Analysis of Transport Wings

Luis R. Miranda*
*Lockheed-California Company,
Burbank, California*

Introduction

The need to achieve highly efficient aerodynamic designs has become more crucial than ever due to the recent increases in the cost of fuel. This need is particularly accute for commercial aircraft; hence the attention that has been given to the design optimization of transonic transport wings. In the past, the wind tunnel was the main tool, or maybe the only tool, available to the aerodynamicist to perform this design optimization. Over the past decade, advances in computational aerodynamics have provided numerical methods that offer the potential of making the digital computer the main design optimization tool, relegating the wind tunnel to the role of verification and substantiation.

To realize this potential, it is necessary to know and understand the shortcomings and limitations of the present theoretical methodology so that proper emphasis can be placed on the future development of numerical methods. To contribute to this knowledge and understanding, the use of full potential flow computer codes for the analysis of transport-type wings in the transonic regime is discussed herein. Though the discussion centers on the use of these codes in an analysis or direct mode, they were used as part

---

Presented at the Transonic Perspective Symposium, NASA/Ames Research Center, Moffett Field, Calif., Feb. 18-20, 1981. Copyright © American Institute of Aeronautics and Astronautics, Inc., 1981. All rights reserved.
*Department Manager, Computational Aerodynamics.

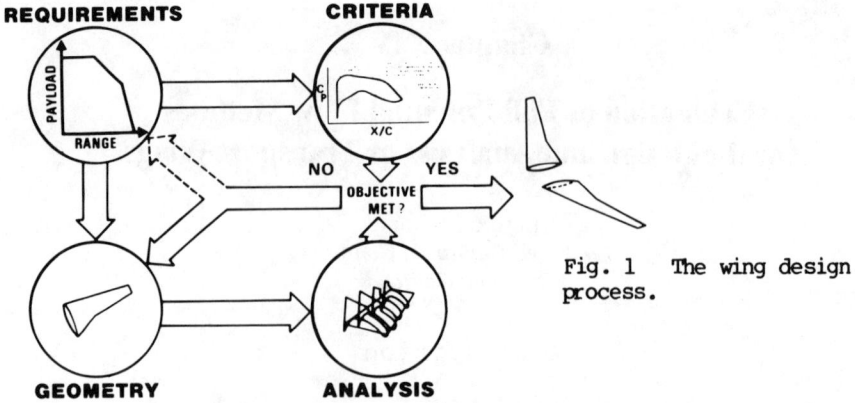

Fig. 1 The wing design process.

of the wing design process schematized in Fig. 1. In this process, the particular design requirements are converted into appropriate design criteria, e.g., a set of target pressure distributions.

A wing geometry definition system generates a tentative design that is analyzed by one of the subject computer codes. The resulting pressure distributions are then compared with the target ones; if not sufficiently close, changes are made to the corresponding geometry definition parameters and the cycle is repeated until convergence. The success of this design process rests on the predictive capability of the analysis code.

Computer Codes Evaluated

Three computer codes were evaluated; they are based on the finite-difference solution of the full potential three-dimensional flow equation. The first one of them, known as FLO-22, was originally developed by Jameson and Caughey[1] for the analysis of transonic flows about isolated wings, Fig. 2. It uses a nonconservative difference scheme and the computational grid is generated analytically by a series of sheared and parabolic coordinate transformations. The other two, designated FLO-28 and FLO-30, were also developed by Jameson and Caughey[2] and they are capable of calculating the transonic flow about wing-body combinations. They are both based on the finite-volume conservative difference scheme; this decouples the flow solution part of the codes from the grid generation process. The essential difference between FLO-28 and FLO-30 lies in the manner in which the finite-difference grid that surrounds the configuration is generated. FLO-28 uses a combination of sheared parabolic and Joukowsky transformations to map the physical space into the appropriate computational domain. Cylindrical and logarithmic-type transformations are used in FLO-30 to accomplish such mapping. All three codes calculate the surface pressure distribution in a potential (inviscid and isentropic) flow, and the aerodynamic force and moment coefficients are obtained by the corresponding surface integrations of the computed pressure distribution.

Prediction of Pressure Distributions

These codes were applied to the analysis of the two wing-body configurations shown schematically in Fig. 3. Fuselage, planform geometry, and airfoil thickness ratio are identical for both configurations, the only difference lying in the wing section geometry. The configuration denoted as W49 has a wing section geometry which is representative of the airfoil technology incorporated in present wide-body transports. The other configuration, designated W53, has airfoil sections representative of a more advanced supercritical technology with a larger chordwise extent of upper surface supercritical flow and higher aft loading.

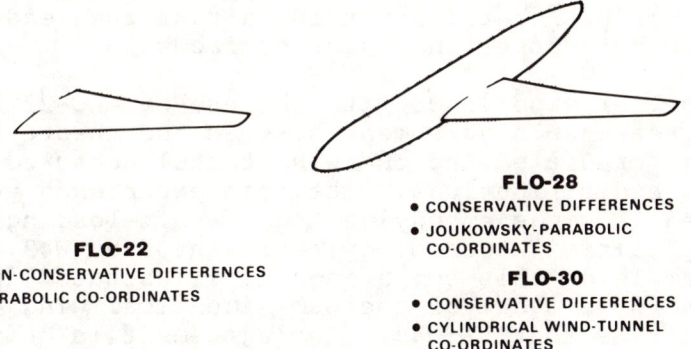

Fig. 2 Transonic codes evaluated.

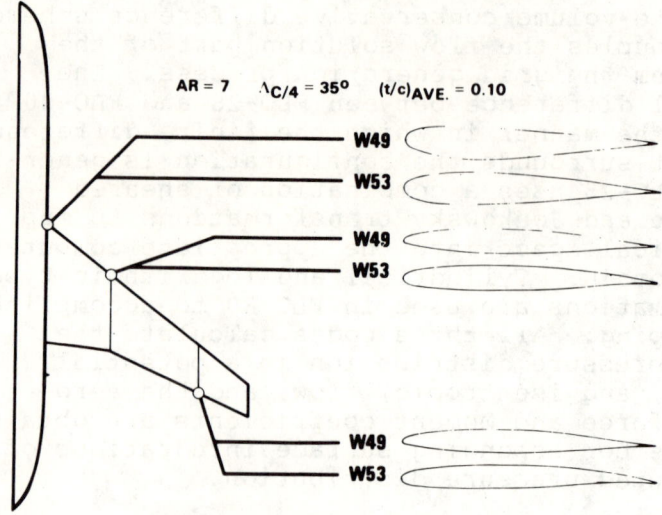

Fig. 3   Wing-body configuration.

Full-span, sting-mounted models of these configurations were built and tested in the NASA/Ames 14-Foot Transonic Tunnel as part of a not mentioned before collaborative program between the NASA/Ames Research Center and the Lockheed-California Company. The models were equipped with about 150 surface pressure taps distributed on both upper and lower surfaces of the wing at four span stations; a six-component strain gage balance was internally mounted in the body. The data were obtained at a Reynolds number of approximately 3.5 x 10 based on the model mean aerodynamic chord. Boundary-layer transition was artificially tripped by carborundum strips located about 10% aft of the leading edge on both upper and lower surfaces.

Earlier experience with the use of FLO-22 had shown reasonable agreement between the theoretically predicted and the wind tunnel measured pressure distributions. But this experience was limited to wings embodying the low aft-loading "peaky" airfoil technology represented by W49. An example of this comparison is illustrated in Fig. 4 for a model of the same identical wing geometry as that of W49. The agreement is quite satisfactory, particularly in view of the fact

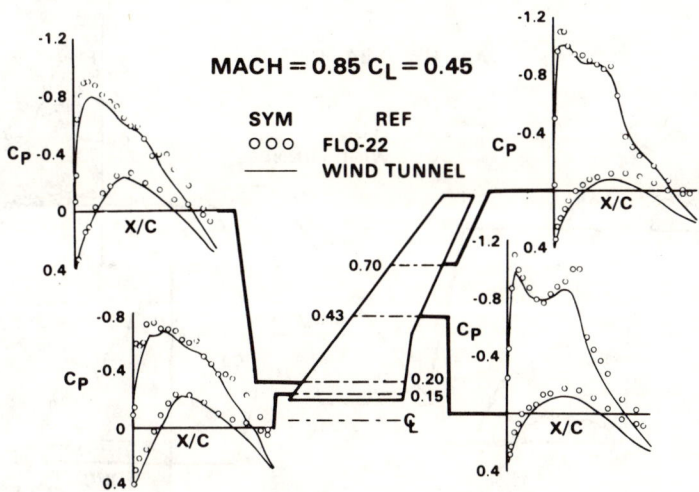

Fig. 4 Calculated vs experimental pressure distributions for wing 37B, FLO-22.

that no viscous correction was incorporated in the theoretical predictions; shock strength and location are captured fairly accurately. Similar success was encountered in the analysis of W49. Consequently, the following discussion concentrates on the more advanced wing design, W53.

FLO-22 was the principal analytical tool used in the design of wing W53, and it was used in an entirely inviscid mode, i.e., no attempt was made to correct for any boundary-layer displacement effect. Only the exposed wing was modelled in FLO-22: it was analyzed as a wall-mounted wing. The disparity between the predicted and measured pressure distributions at the wing design point ($C_{L_D} = 0.5$, $M_D = 0.84$) is shown in Fig. 5. In this comparison, as well as in all the others included in this paper, the value of the lift coefficient computed for the exposed wing has been matched with that measured in the wind tunnel for the wing-body combination. The following observations can be made about the comparison between FLO-22 predictions and experimental results:

1) Good agreement exists between experimental and the calculated values of sectional lift coefficient, given by the area enclosed by surface pressure distributions.

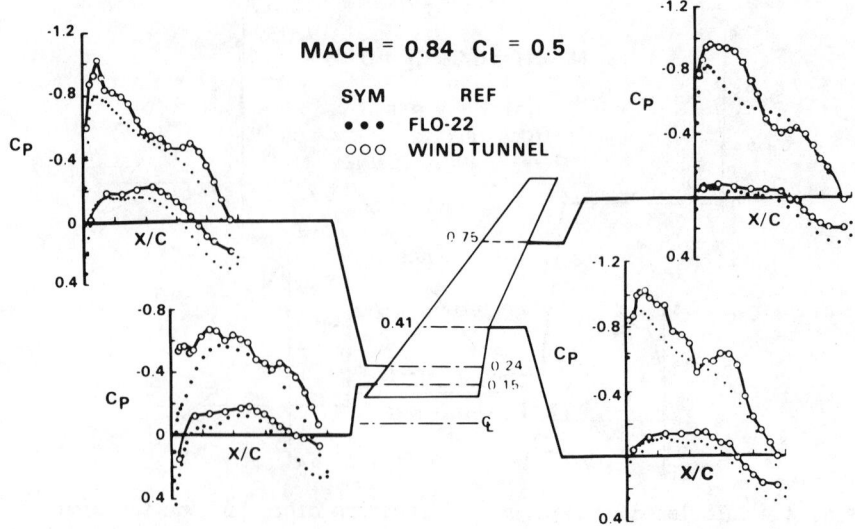

Fig. 5 Calculated vs experimental pressure distributions for wing 53, FLO-22.

2) The theoretically computed values of the upper surface pressure coefficients are less negative than the experimental values in the supercritical region of the wing.

3) No shock waves are clearly discernible in the theoretical pressure distributions, whereas the presence of a moderate shock running approximately along the 40% chord line is apparent in the experimental pressure distributions.

4) The experimental pressure distributions show less aft loading than the predicted distributions.

The analysis of the same wing-body configuration at the same test condition was then repeated using the more advanced codes, FLO-28 and FLO-30. This time the geometry of the finite fuselage was accurately included in the mathematical model. The theoretical results are compared with the experimental pressure distributions in Fig. 6 for the FLO-28 case, and in Fig. 7 for the FLO-30 case. The following observations are in order:

1) Like FLO-22, both FLO-28 and FLO-30 accurately predict the sectional lift coefficient values.

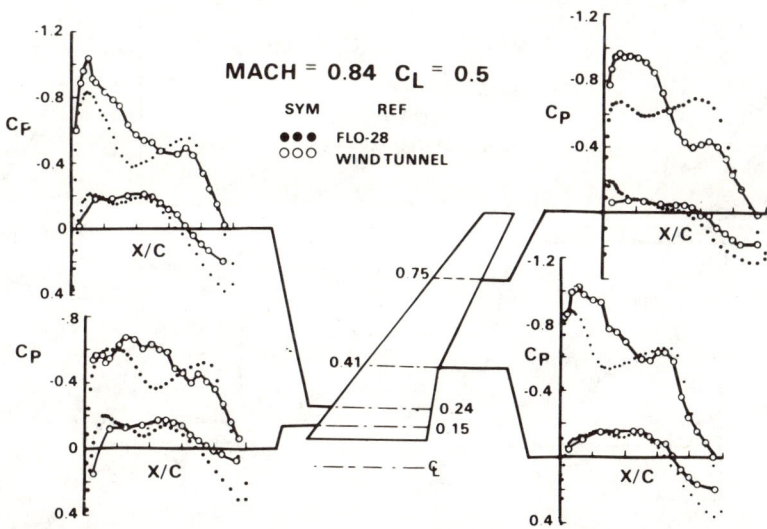

Fig. 6 Calculated vs experimental pressure distributions for wing 53, FLO-28.

2) The underestimation of the supercritical levels of the pressure coefficient is even more pronounced for FLO-28 and FLO-30 than it was for FLO-22.

3) Likewise, the overprediction of the aft loading is more pronounced in the FLO-28 and FLO-30 pressure distributions.

4) The predictions of the pressure distribution at the wing station next to the fuselage 15% semispan are substantially different between FLO-28 and FLO-30. The FLO-30 distribution is in closer agreement with the experimental data, at least in the general shape of the distribution if not in actual details and levels. It should be pointed out that qualitatively FLO-22 showed the same kind of agreement with the experimental results shown by FLO-30 at this next-to-the-body station, even though it did not incorporate any simulation of body effects.

From the preceding discussion it is obvious that no improvement in the agreement between theoretical and experimental pressure distributions accrued from the use of the more advanced

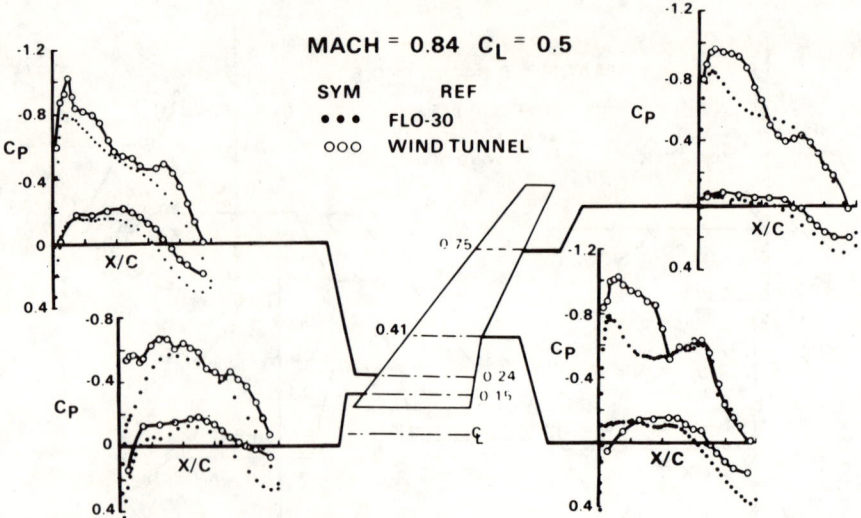

Fig. 7 Calculated vs experimental pressure distributions for wing 53, FLO-30.

codes FLO-28 and FLO-30. The major discrepancies between the theoretical and experimental results were not resolved by the capability provided by both FLO-28 and FLO-30 to model the fuselage. In the present case the viscous effects predominate, overshadowing any improvement in accuracy that might result from the inclusion of fuselage effects. On the other hand, the use of the FLO-22 code showed some significant practical advantages, namely, it was 6-10 times less expensive in computer cost and it was easier and more reliable to run, requiring far less manipulation of the input data to achieve a successful run. These practical advantages led to an effort to improve the predictive capabilities of FLO-22. The code modifications developed towards this objective are discussed next.

Modifications to FLO-22

Three major additions or modifications were made to FLO-22, Fig. 8, as follows:

Viscous Correction

This is simply done by adding the boundary-layer displacement thickness to the wing surface

- **VISCOUS CORRECTION**
  STRIPWISE 2-D INTEGRAL
  BOUNDARY LAYER

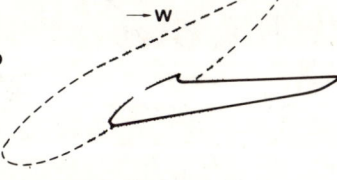

- **FUSELAGE SIMULATION**
  LINEAR THEORY BODY-INDUCED
  SPANWISE FLOW IMPOSED
  AT "PLANE-OF-SYMMETRY"

- **CONSISTENT CALCULATION OF VELOCITY**

Fig. 8 Modification to FLO-22.

geometry. A number of potential flow relaxation cycles, a boundary-layer displacement thickness calculation, and a modification of the surface geometry are iterated a specified number of times. This displacement thickness is calculated in an iterative fashion by stripwise application of Truckenbrodt's two-dimensional laminar/turbulent boundary-layer integral method.[3,4]

Fuselage Effect Simulation

This is accomplished by calculating the spanwise flow induced by the body at the vertical plane containing the wing-body intersection. This flow is computed using a linear theory singularity distribution along the fuselage axis, and it is imposed in FLO-22 as the boundary condition at the "plane-of-symmetry" instead of the usual zero spanwise flow condition.

Consistent Calculation of Velocity

In the standard version of the FLO-22 code, the calculation of the velocity components is done by finite differencing the velocity potential. In the subroutines that solve the velocity potential equations and apply the boundary conditions, central differences of the velocity potential are used throughout, except at the plane of symmetry. At this plane, for stability reasons, a mixture of upwind and downwind differences are used to calculate the velocity compo-

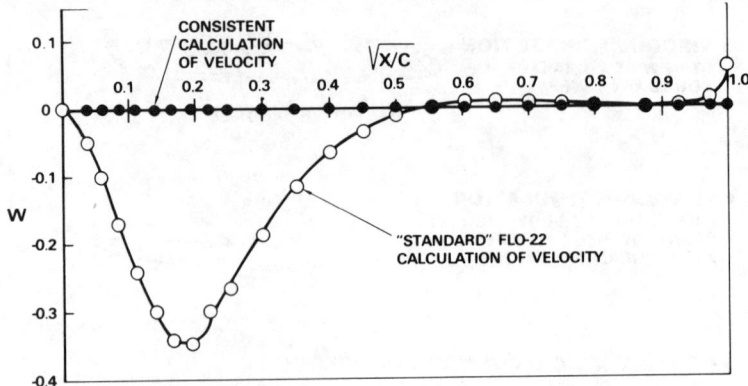

Fig. 9   Spanwise velocity at root station.

nent normal to the plane of symmetry. Later on, when the velocity components are recalculated for the computation of the pressure coefficients, the different scheme that has been used in the solution process at the root station is not taken into account. This inconsistency amounts to calculating the pressure distribution for a velocity field that flows into the wall, Fig. 9, which is physically unrealistic if the root station is a plane of symmetry or a wall. If consistency is used when calculating the pressure coefficients at the root station, i.e., mixed upwind/downwind differencing of the potential for computing the spanwise velocity component, then there is no flow into the wall and the pressure distribution corresponds to the flow that has been actually computed.

It should be pointed out that even after the introduction of the last modification, the FLO-22 code is not yet totally satisfactory for predicting the pressure distribution at the root station of swept wings. This problem is illustrated in Fig. 10 where comparisons are made between the modified FLO-22 results and those of a panel type method at a low Mach number. The accuracy of panel methods at low subsonic Mach numbers is well established, and it can be observed that the agreement between both theoretical results is excellent for the wall-mounted unswept wing case. When the comparison is made for the sweptback

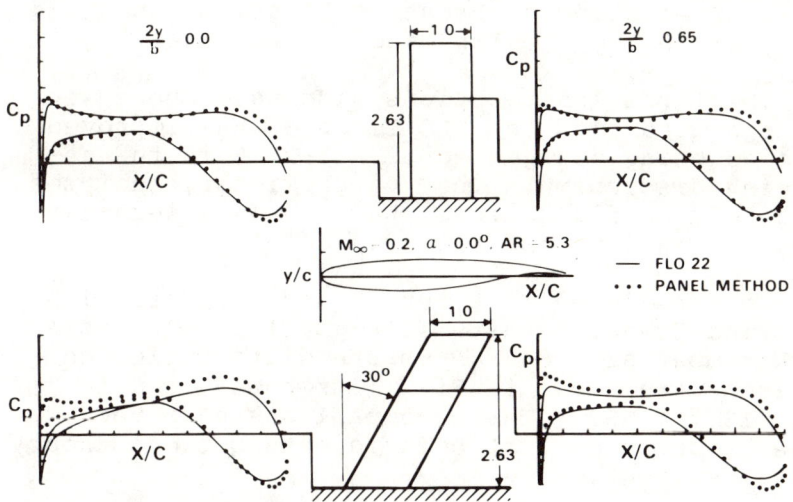

Fig. 10 Theoretical pressure distribution comparison.

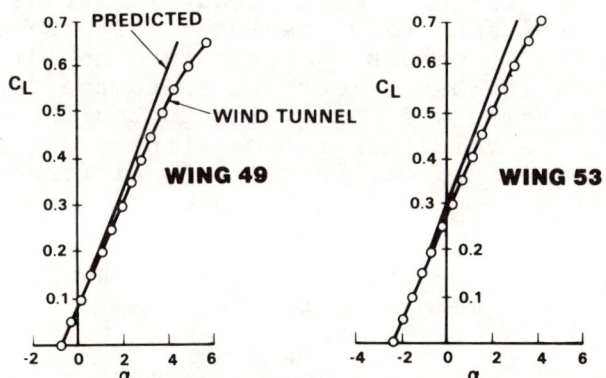

Fig. 11 Lift coefficient vs angle of attack - prediction vs experiment at Mach 0.84.

wing case, significant discrepancy exits between the FLO-22 and the panel method results at the first station next to the wall. Though not shown here, this discrepancy increases with increasing sweep angle. Neither the solution nor the nature of this problem are known at the present time. Even though this discussion has addressed the FLO-22 code specifically, similar problems have been encountered in the use of the more advanced codes, FLO-28 and FLO-30.

Prediction of Pressure Distribution Using the Improved FLO-22 Code

The three modifications discussed above were incorporated into the FLO-22 code, the improved version being designated as FLO-22.5 in the remaining discussion. The two previously analyzed wing-body configurations, W49 and W53, were again used to evaluate the FLO-22.5 code.

The predicted lift curves are compared with the wind-tunnel results in Fig. 11 at the design Mach number of 0.84. Pressure distribution comparisons are shown in Fig. 12 for W49, and in Fig. 13 for W53. These comparisons have been made by matching lift coefficients between theory and experiment.

The agreement between theory and experiment for the wing 49 pressure distribution is, in general, very satisfactory. Shock locations and strengths are accurately predicted. The somewhat more negative C_p values computed for the 47% semispan station may be attributed to the fact that such station was not a wing control station. Therefore, the thickness ratio of the model as actually fabricated may be slightly different

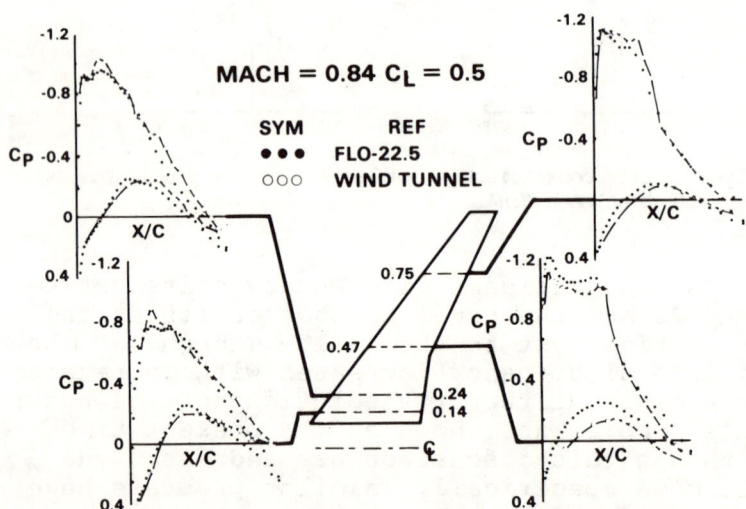

Fig. 12 Calculated vs experimental pressure distributions for wing 49, FLO-22.5.

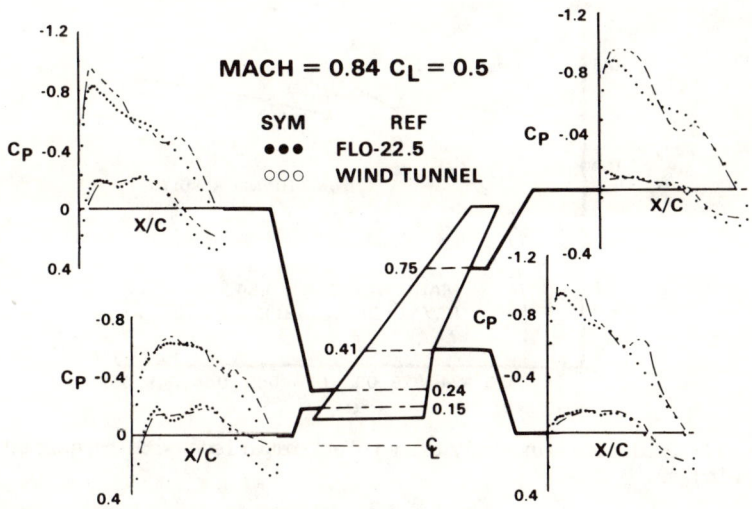

Fig. 13 Calculated vs experimental pressure distributions for wing 53, FLO-22.5.

than the one represented in FLO-22.5, due to the variations in lofting between control stations.

The improvement in the quality of the theory-experiment correlation is more significant for the more advanced supercritical-type wing W53, Fig. 13. This improvement is particularly evident for the inboard panel, 15% and 24% semispan stations. Apparently, the rather crude ways of simulating body and boundary-layer effects that have been incorporated in FLO-22.5 are quite effective. Notwithstanding the observed improvement in the accuracy of the pressure distribution predictions, it is quite evident that additional improvements are needed. The major shortcoming lies in the simulation of viscous effects, particularly in the trailing edge and shock/boundary-layer interaction regions.

Prediction of Drag

Unfortunately, the above improvements do not extend to the capability to compute drag, as shown in Figs. 14 and 15 for wings 49 and 53, respectively. Neither polar shape nor drag levels are acceptably predicted. The theoretical drag is the sum of the friction drag value, ob-

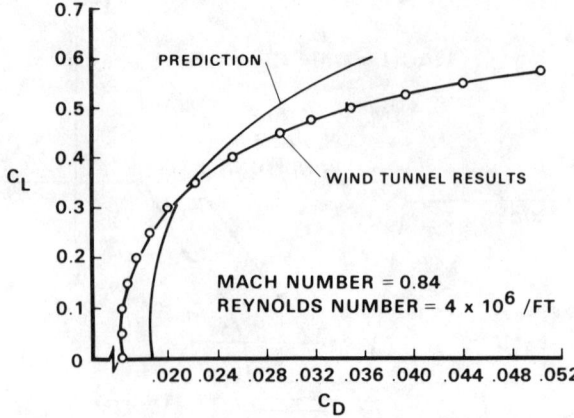

Fig. 14 Predicted wing-body drag polar comparison with experiment for wing 49.

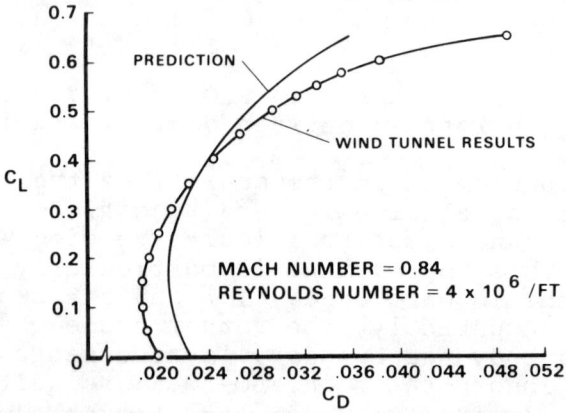

Fig. 15 Predicted wing-body drag polar comparison with experiment for wing 53.

tained by applying a modified Squire-Young formula to integral boundary-layer results, and the pressure drag (vortex + wave) calculated by the integration of the pressure times surface slope. The major source of error appears to lie in the pressure drag computation, as indicated by the drag rise comparison of Fig. 16. The theoretical predictions overestimate the measured drag rise by about 100%, all of the contribution to the theoretical drag rise coming from pressure drag, the theoretical skin friction drag slightly diminishing with increasing Mach number.

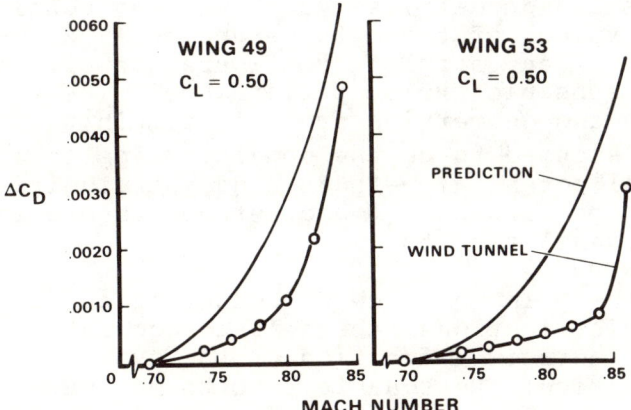

Fig. 16 Predicted drag rise comparison with experiment.

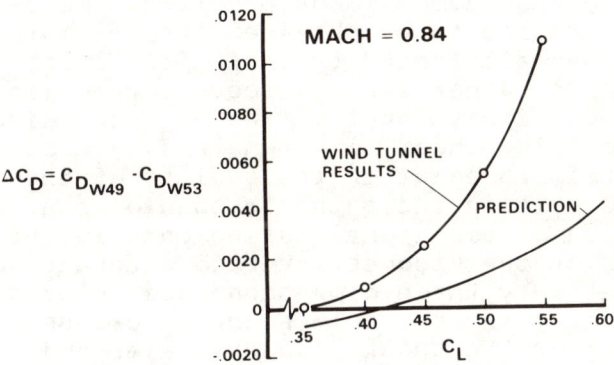

Fig. 17 Predicted drag reduction - comparison with experiment.

Even though drag levels may not be accurately predicted, the computational methodology may still be a useful tool for making design decisions if trends are predicted accurately and consistently. The capability of present transonic flow methods may be considered generally adequate for this task, but Fig. 17 clearly indicates that there is need for improvement.

Conclusions and Recommendations

The use of three-dimensional finite-difference codes that solve the full potential flow equation has been evaluated within the context of transonic transport wing analysis. The flow conditions and geometries that have been considered are typical of those to be encountered in the design of any airplane configuration optimized for subsonic cruise performance. These conditions and geometries are well within what is generally assumed to be the domain of applicability of full potential methods, and yet significant discrepancies were found between theoretical and wind-tunnel results.

From the point of view of computing surface pressure distributions, the need to account for viscous effects is a function of the type of airfoil section. Reasonable results may be obtained for the more conventional low aft-loading airfoils without having to include any boundary-layer displacement effect. But the inclusion of viscous effects becomes mandatory if any meaningful answers are to be obtained for the more advanced, more aft-loaded type of airfoil sections. Even a rather crude viscous correction, such as the stripwise application of a two-dimensional integral boundary-layer method, yields significant improvement in the quality of the predictions. It is the author's opinion that the incorporation of additional refinements in the calculation of the viscous/inviscid interaction, such as the truly three-dimensional calculation of the boundary-layer displacement thickness and the modeling of the shock/boundary-layer and trailing edge interactions is of crucial importance for enhancing the practical value of these codes.

The method of calculating drag by surface integration of the pressure distribution, the standard procedure built into the codes discussed in this paper, is totally inadequate. More accurate ways of computing drag must be used. A lot of detailed flowfield information is computed as part of the solution process of these analysis codes. This information should be sufficient to calculate drag with adequate accuracy, i.e., the problem is not the lack of information but the present manner of using it.

Finally, cost effectiveness and "robustness" or reliability are desirable qualities for computer codes that cannot be overemphasized. As a good example of this, the more advanced codes evaluated during this work, FLO-28 and FLO-30, though fundamentally more accurate and complete than FLO-22, were not fully used due to their high computer cost and the need to fine tune the input data in order to get a successful run. If these numerical methods are to be extended to more realistic and, therefore, more complex geometries, significantly faster and more reliable solution techniques must be developed. Otherwise, the potential benefits offered by computational aerodynamics will not be fully achieved.

Acknowledgments

The work that provided the experience and data presented in this paper was part of a collaborative program between the NASA/Ames Research Center and the Lockheed-California Company. The author wishes to acknowledge the guidance and support received from R. M. Hicks, of the NASA/Ames Research Center, who sponsored this program. Thanks are due to P. Raj and J. S. Reaser, of the Lockheed-California Company, who did most of the analysis and computer code modification work.

References

[1] Jameson, A. and Caughey, D. A., "Numerical Calculation of the Transonic Flow Past a Swept Wing," New York University ERDA Report COO-3077-140, June 1977.

[2] Jameson, A. and Caughey, D. A., "Progress in Finite-Volume Calculations for Wing-Fuselage Combinations," AIAA Journal, Vol. 18, Nov. 1980, pp. 1281-1288.

[3] Truckenbrodt, E., "A Method of Quadrature for Calculation of the Laminar and Turbulent Boundary Layer in Case of Plane and Rotationally Symmetrical Flow," NACA TM 1379, May 1955.

[4] Scholz, N., "Zur rationallen Berechnung laminarer und turbulenter kompressibler Grenzschichten mit Warmeubergang," Zeitschrift fur Flugwissenschaften, Vol. 7, 1959 pp. 33-39.

Part III
Alternative Prediction Methods

The preceding section has been concerned with finite-difference methods for predicting transonic flows, and while these techniques constitute the mainstream of ideas there are alternative viewpoints. The four articles in this section present a point of view not considered elsewhere in this book. Whether the techniques noted here will replace or complement finite-difference methods is open to question at present, but it is always illuminating to read an alternative argument.

Chapter XV.

Nonlinear Green's Function Method for Unsteady Transonic Flows

Kadin Tseng* and Luigi Morino†
Boston University, Boston, Mass.

Nomenclature

a_∞	=	speed of sound
AR	=	aspect ratio
c	=	chord
C_p	=	pressure coefficient
$E(\bar{P})$	=	see Eq. (4)
G	=	Green's function
K	=	nondimensional transonic parameter, $\kappa U_\infty / \beta^3$
ℓ	=	reference length
M	=	Mach number, U_∞ / a_∞
$\bar{n}$	=	normal to σ_B
$\bar{n}_W$	=	normal to σ_W
$\bar{N}$	=	normal to Σ_B
$\bar{N}_S$	=	normal to Σ_S
$\bar{N}_W$	=	normal to Σ_W
$\bar{p}$	=	point having coordinates x,y,z
$\bar{p}_*$	=	control point, (x_*, y_*, z_*)
$\bar{P}$	=	point having coordinates X,Y,Z

Presented at the Transonic Perspective Symposium, NASA/Ames Research Center, Moffett Field, Calif., Feb. 18-20, 1981. Copyright © 1981 by the American Institute of Aeronautics and Astronautics. All rights reserved.
*Research Associate, Department of Aerospace and Mechanical Engineering.
†Professor, Department of Aerospace and Mechanical Engineering, Director, Center for Computational and Applied Dynamics.

$\bar{P}_*$	=	control point, (X_*, Y_*, Z_*)
r	=	$|\bar{p}-\bar{p}_*|$
r_β	=	see Eq. (A.7)
R	=	$|\bar{P}-\bar{P}_*|$
t	=	time
T	=	nondimensional time, $U_\infty t/\ell$
U	=	$\partial\Phi/\partial X$
U_∞	=	velocity of undisturbed flow
x, y, z	=	space coordinates
X, Y, Z	=	nondimensional Prandtl-Glauert coordinates, $x/\beta\ell$, y/ℓ, z/ℓ
β	=	$(1-M^2)^{1/2}$
$\Delta\Phi$	=	discontinuity of Φ across the wake, $\Phi_1 - \Phi_2$
$\Delta\Psi$	=	discontinuity of Ψ across the shock, $\Psi_1 - \Psi_2$
θ	=	propagation time from $\bar{P}$ to $\bar{P}_*$, Eq. (A.10)
θ	=	see Eq. (A.14)
$\hat{\theta}$	=	see Eq. (A.15)
κ	=	transonic parameter, $(\gamma+1)U_\infty/a_\infty^2$
Ξ	=	nondimensional nonlinear terms, $\chi\ell/U_\infty$
Π	=	convection time of wake vortices, see Eq. (A.19)
σ	=	surface surrounding body and wake
σ_B	=	surface of body in x,y,z space
σ_W	=	surface of wake in x,y,z space
Σ_B	=	surface of body in X,Y,Z space
Σ_S	=	surface of shock in X,Y,Z space
Σ_W	=	surface of wake in X,Y,Z space
τ	=	thickness ratio
φ	=	perturbation velocity potential
ϕ	=	velocity potential, $U_\infty x + \varphi$
Φ	=	nondimensional perturbation velocity potential, $\varphi/U_\infty\ell$
χ	=	nonlinear terms
ψ	=	normal wash in x,y,z space
Ψ	=	normal wash in X,Y,Z space
$\bar{\nabla}$	=	del operator in x,y,z space
$\bar{\nabla}_0$	=	del operator in X,Y,Z space

$[\ ]^\theta$	=	evaluation at time $t_*-\theta$, see Eq. (A.10)
$[\ ]^\theta$	=	evaluation at time $T_*-\theta$, see Eq. (9)

Subscripts and Superscripts

B	= body
S	= shock wave
TE	= trailing edge
W	= wake
∞	= freestream condition

Introduction

The work presented in this paper differs from most of the other papers in this volume in two major respects. First, it deals with unsteady flows while most of the other papers deal with steady-state configurations. Second, this is the only work based on the Green's function method. As indicated by other works presented in this volume, most researchers in the field of transonic computational fluid dynamics have been concentrating on the finite-difference methods.

The objective of this paper is to present some fundamental results obtained by Tseng during work on his doctoral thesis[1] and to discuss their significance. Therefore only as few numerical results as necessary to demonstrate the basic points are presented here (extensive numerical results are included in Ref. 1). It is the purpose of this paper to clarify why we believe that the Green's function method is to be considered a very valid alternate computational method to the finite-difference method (at least in the specific field of unsteady three-dimensional transonic flows around complex configurations). Our recommendation is that a considerable amount of effort be devoted to this method, which at this moment seems to be at least as promising, in many respects, as the finite-difference method.

It should be noted that the method proposed here may be considered to be an extension, to three-dimensional unsteady flows, of the well-known integral equation methods (e.g.,

Refs. 2 - 8). Extensive reviews of these and other methods are presented in this volume. No attempt is made here to duplicate the effort of the experts in the field (who are much more knowledgeable than we are in the field of transonic aerodynamics). Therefore the review of the work relevant to this paper is limited to the Green's function method in subsonic and supersonic flows (see following section), since it is expected that many people in the transonic field are not familiar with the level of sophistication of the Green's function method in the linear range and therefore might not be able to appreciate the relevance of the new development of the method in the field of transonic aerodynamics.

The Green's function method for transonic flow is outlined in a section bearing that heading with mathematical details given in Appendices A and B. The issue of shock capturing is then discussed followed by an assessment of the method and concluding remarks, respectively.

Review of the Linear Green's Function Method

Before introducing the Green's function method for transonic aerodynamics, it is important to review the basic developments of the Green's function method for linear (subsonic and supersonic) unsteady aerodynamics, with emphasis on the works which are relevant to this paper.

Consider first the field of incompressible potential aerodynamics, which is governed by the Laplace equation

$$\nabla^2 \varphi = 0 \qquad (\bar{P} \varepsilon V) \qquad (1)$$

where V is the fluid volume outside the boundary surface σ. The boundary conditions are $\varphi = 0$ at infinity and

$$\psi = \frac{\partial \varphi}{\partial n} = (-U_\infty \bar{i} + \bar{v}) \cdot \bar{n} \qquad \text{on } \sigma \qquad (2)$$

where $U_\infty \bar{i}$ is the velocity of the undisturbed flow and $\bar{v}$ is the velocity of the point on the surface of the body.

Using Green's theorem, Eq.(1) yields

$$E_* \varphi_* = \oiint_\sigma \{ \frac{\partial \varphi}{\partial n} \frac{-1}{4\pi r} - \varphi \frac{\partial}{\partial n}(\frac{-1}{4\pi r}) \} d\sigma \qquad (3)$$

where $r = |\bar{P} - \bar{P}_*|$, $\bar{n}$ is the outward unit normal and

$$E_* = 1 \quad \bar{P}_* \text{ outside } \sigma \qquad (4)$$
$$ = 0 \quad \bar{P}_* \text{ inside } \sigma$$

This equation is well known (see for instance Ref. 9). The foundations of the potential theory are given by Kellogg[10] who is the major reference for most of the works in potential incompressible aerodynamics. Important mathematical results can also be found in Refs. 11 and 12. All of the so-called panel aerodynamic methods (e.g., Refs. 13-17) are based upon integral equations presented in Ref. 10 which can be obtained directly from Eq. (3) (see for instance Refs. 9 and 10).

However, strangely enough, this equation per se had not been used as 'the' equation to solve the problem until 1972, when Morino (see Refs. 18-20) noted that $E_* = 1/2$ if $\bar{P}_*$ is a smooth point of the surface σ. In this case Eq.(3) is an integral equation relating the unknown potential φ to its normal derivative $\psi = \partial \varphi / \partial n$ (normal component of the perturbation velocity, or normal wash), which is known from the boundary condition on the surface of the body, Eq.(2).

The unit source

$$G = -1/4\pi r \qquad (5)$$

is known as the Green's function for the infinite space. Hence strictly speaking the method should be called the 'infinite space Green's function method.'

Eq.(3) is still valid for lifting bodies if σ surrounds both body and wake. Isolating the contribution of the wake, one obtains

$$E_* \varphi_* = \oiint_{\sigma_B} \{ \frac{\partial \varphi}{\partial n} \frac{-1}{4\pi r} - \varphi \frac{\partial}{\partial n}(\frac{-1}{4\pi r}) \} d\sigma + \iint_{\sigma_W} \Delta\varphi \frac{\partial}{\partial n}(\frac{-1}{4\pi r}) d\sigma \qquad (6)$$

where σ_B is the closed surface of the body, σ_W is the open surface of the wake,

$$\Delta\varphi = \varphi_1 - \varphi_2 \qquad (7)$$

is the discontinuity across the wake(s), and $\bar{n}$ is pointing from side 2 to side 1 of the wake(s).

The fact that this equation had never been used in potential aerodynamics is puzzling to the authors. A possible explanation may be found in the remarks by Hess who notes that the integral equation given by Eq.(3) with $E_* = 1/2$

> is often referred to as the integral equation formulation of the Neumann problem [Ref. 11], but it is not the one that is so considered in the fundamental work by Kellog [Ref. 10].
> Despite the naturalness of the above formulation, the fact is that it has never been used as the basis of a general method of potential flow calculation. One possible reason for this neglect is the fact that the integral equation [3] has a non-unique solution, which can lead to numerical difficulties in applications. The non-uniqueness is evident from the fact that a constant value of φ on σ gives zero velocity in [volume] V.(Ref.14, pp. 9, 10)

The nonuniqueness is true only for internal flows, but not for external flows: in contrast to the interior problem, the exterior Neumann problem for Laplace equation has a unique solution (Ref. 12, pp. 607-610). It is unfortunate that this point has always been obscure in the aerodynamics literature and has delayed the development of a very powerful method in the field of potential aerodynamics.

The major strength of the method is that it may be easily extended to linear unsteady subsonic and supersonic flows (see Refs. 18-28). As shown by Morino in Refs. 18, 20, and 26 if the motion of the surface is infinitesimal, straightforward application of the Green's function method yields

(for $E_* = 1/2$)

$$\Phi(\overline{P}_*, T_*) = \oiint_{\Sigma_B} \{[\Psi]^\theta (\frac{-1}{2\pi R}) - [\Phi]^\theta \frac{\partial}{\partial N}(\frac{-1}{2\pi R}) + [\dot{\Phi}]^\theta (\frac{-1}{2\pi R})\frac{\partial \hat{\theta}}{\partial N}\} d\Sigma$$

$$- \iint_{\Sigma_W} \{[\Delta\Phi]^\theta \frac{\partial}{\partial N}(\frac{-1}{2\pi R}) - [\Delta\dot{\Phi}]^\theta (\frac{-1}{2\pi R})\frac{\partial \hat{\theta}}{\partial N}\} d\Sigma \qquad (8)$$

where all the expressions are in the Prandtl-Glauert space ($X=x/\beta\ell$, $Y=y/\ell$, $Z=z/\ell$) and θ and $\hat{\theta}$ are given by Eqs. (A.14) and (A.15). Also $\dot{\Phi} = \partial\Phi/\partial T$, whereas

$$[\ ]^\theta = [\ ]_{T=T_*-\theta} \qquad (9)$$

A similar expression is obtained for supersonic flows (see Ref. 26). Finally the general theory for a surface moving with respect to the frame of reference (e.g., a helicopter rotor) is given in Ref. 20 (see last section of Appendix A). It is important to emphasize that Eq. (9) is the only equation known to the authors which gives the solution for the three-dimensional fully unsteady, compressible flows, around arbitrarily complex aircraft configurations, explicitly in the time domain. Actual time domain results were obtained in 1978 by Morino and Tseng.[28]

In order to appreciate the relevance of this work a few remarks may be in order. A production version code SOUSSA-P 1.1 (steady, oscillatory, unsteady, subsonic, and supersonic aerodynamics, production version 1.1, Refs. 26 and 27) has been developed. Also the method has been adopted by several researchers, notably Suzuki and Washizu[29] and Summa[30] for steady flow and Dusto and Epton[31] and Maskew[32] for unsteady flows. In addition the formulation has been incorporated in the program PANAIR[17].

The Green's Function Method for Transonic Flows

As indicated in the previous section, the Green's function method is quite developed and well accepted in the linear unsteady aerodynamics community. On the other hand, there seems to be a considerable amount of skepticism about the method in the transonic aerodynamic community. Not only the usefulness and practicality of the method, but even the validity of the theoretical foundations are often being questioned: 'How can you use the Green's function method, which is based upon the theory of linear operators, to solve a nonlinear problem? How can you use the subsonic (i.e., elliptic) Green's function to study mixed type (i.e., elliptic and hyperbolic) flows?' The best answers to these questions is that the well-established integral equation methods[2-8] are also based upon the Green's function technique and also deal with nonlinear mixed type flows. Another issue that has been puzzling many of our colleagues in the transonic field is the question of the shock: 'How do you capture the shock? What conditions do you impose?'

In order to clarify these points, let us review the salient phases of the development of the Green's function method for the transonic problem. We hope that in this case such a review might be more helpful than just mathematics (for completeness the mathematical details are given in Appendices A and B). Back in 1977, we were just interested in extending the formulation to a range (high subsonic) where the nonlinear terms are not negligible (as in the linear subsonic flow), but still small compared to the linear terms. In this case (shock-free flow) the effect of the nonlinear terms χ can be evaluated step by step.

More precisely, the governing equation is now

$$\nabla^2 \varphi - \frac{1}{a_\infty^2}\left(\frac{\partial}{\partial t} + U_\infty \frac{\partial}{\partial x}\right)^2 \varphi = \chi \qquad (10)$$

where χ includes all the nonlinear terms. If χ were known then Eq. (10) is a nonhomogeneous wave equation and the use of Green's function method (see Appendix A) yields, in Prandtl-Glauert variables:

$E_* \Phi(\bar{P}_*, T_*)$

$$= \oiint_{\Sigma_B} \{[\Psi]^\theta (\frac{-1}{4\pi R}) - [\Phi]^\theta \frac{\partial}{\partial N}(\frac{-1}{4\pi R}) + [\dot{\Phi}]^\theta (\frac{-1}{4\pi R})\frac{\partial \hat{\theta}}{\partial N}\} d\Sigma$$

$$- \iint_{\Sigma_W} \{[\Delta\Phi]^\theta \frac{\partial}{\partial N}(\frac{-1}{4\pi R}) - [\Delta\dot{\Phi}]^\theta (\frac{-1}{4\pi R})\frac{\partial \hat{\theta}}{\partial N}\} d\Sigma$$

$$+ \iiint_V [\Xi]^\theta (\frac{-1}{4\pi R}) dV \qquad (11)$$

Equation (11) may be interpreted, from a physical point of view, as follows: the effect of the presence of the body may be simulated by replacing the body with a layer of sources, $-1/4\pi R$, on the surface of the body (with intensity equal to the normal wash, Ψ, at time $T_*-\theta$), plus a layer of doublets $\partial(1/4\pi R)/\partial N$ on the surfaces of the body and the wake (with intensity equal to the potential, Φ, on the body and the potential discontinuity, $\Delta\Phi$, on the wake, also at delayed time $T_*-\theta$), plus a layer of 'ratelets' (from rate doublets) $(-\partial\hat{\theta}/\partial N)/4\pi R$ on the surface of the body and the wake (with intensity equal to the time derivative of the potential, $\dot{\Phi}$, on the body and the time derivative of the potential discontinuity, $\Delta\dot{\Phi}$, on the wake, at time $T_*-\theta$). In addition, the effect of the nonlinear terms is equal to the one of a distribution of sources in the flowfield (with intensity equal to Ξ at time $T_*-\theta$). This combination yields $\Phi=0$ inside Σ_B, thereby preventing wave propagation inside the surface of the body. It should be noted that θ is equal to the time necessary for a disturbance to propagate from $\bar{P}$ to $\bar{P}_*$ at the speed of sound (Ref. 20, Sec. 2). In other words, θ is equal to the usual acoustic time delay (distance over speed of sound) where the distance is calculated from the location that $\bar{P}$ had at time $T=T_*-\theta$ (when the disturbance was generated) to the location that $\bar{P}_*$ has at time T_*. Therefore Eq. (11) is simply a mathematical representation of the familiar phenomenon of acoustical wave propagation, but with a frame of reference moving with respect to the medium.

It is apparent that Eq. (11) differs from Eq. (8) (where $E_*=1/2$) for the presence of the volume integral (which is equal to zero if $\Xi=0$). The modification is so minor that it was too tempting not to try it out. In essence, Eq. (11) after suitable space and time discretization (see Appendix B) can be used to evaluate the values of Φ at the nodal points on the surface of the body and in the field. Once the values of Φ are known, nonlinear terms $\Xi = K\Phi_X \Phi_{XX}$ can be evaluated by finite difference[*]. An integration by parts was used initially to avoid the evaluation of Φ_{XX}. However, the numerical formulations with and without the integration by parts are practically identical. This point is analyzed in detail later.

Results for steady-state flows were presented in 1978[28]. Encouraged by the initial results, we tried to see how far we could push the method, and found that if the supersonic region is large, the iteration could diverge. We then tried to use the time domain approach hoping to obtain the steady-state solution through a transient analysis. Linear results were also presented in 1978[28] but after adding the nonlinear terms the step-by-step integration turned out to be unstable. We were dealing with the same type of problems encountered a few years earlier by the people working on the finite-difference method. We therefore tried to 'borrow' some of their experience. After introducing a backward-central differencing scheme[33] of the type introduced by Murman and Cole the integration scheme became stable. Typical results are presented in Figs. 1 and 2. (As mentioned in the Introduction, only as few results as necessary to illustrate the basic points are presented here. Extensive results are given in Ref. 1.) The results were obtained through an actual time domain analysis (it took 40 time steps to reach convergence). The case under consideration is the one of a rectangular wing of aspect ratio AR = 4, thickness ratio τ = 0.06, and Mach number M = 0.908. Figure 1 presents a comparison with the finite-difference results of Ref. 34, whereas Fig. 2 presents an analysis of

---

[*] The earliest work using this mixed integral-equation/finite-difference technique appears to be that of Ogana[6]. The smearing of the shock, introduced by this technique, is analyzed in detail later.

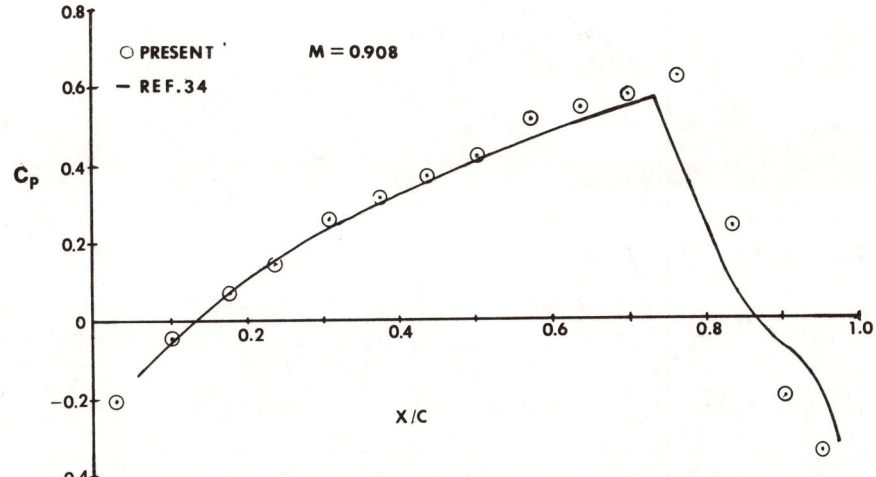

Fig. 1 Pressure distribution near the mid span of a 6% thick circular biconvex wing with rectangular planform, AR=4 and M =0.908.

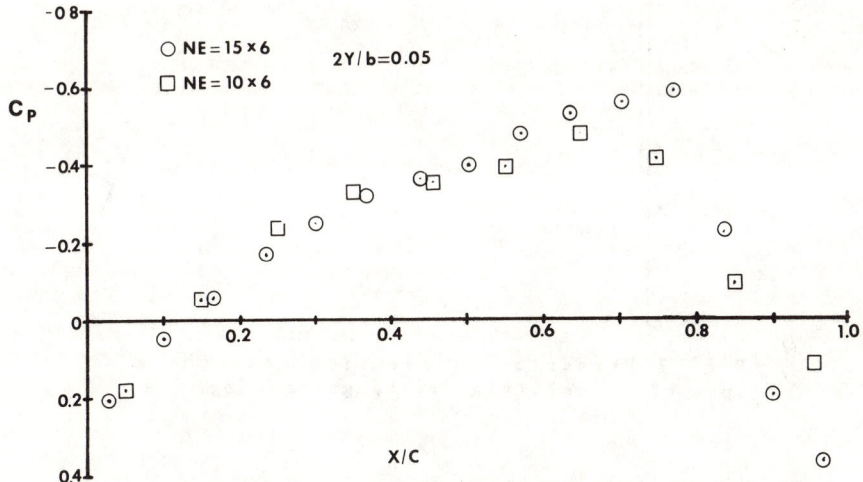

Fig. 2 Pressure distributions of a 6% thick biconvex wing with rectangular planform, M =0.908.

the sensitivity to the number of elements. It was clear, from the analysis of these results, that the method was capable of reproducing the experimental trend and furthermore, that the increase in number of elements yields more

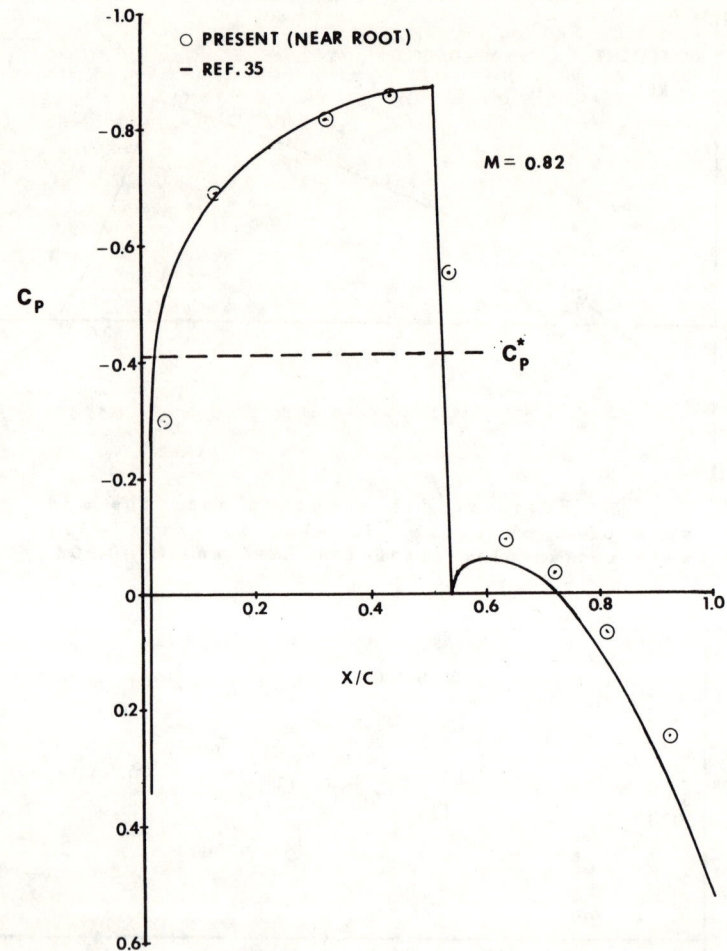

Fig. 3 Pressure distribution near the midspan of a nonlifting rectangular wing.

pronounced shock-like behavior. It became apparent from other similar results that the method was requiring too many elements to produce a shock as sharp as the one obtained with the finite-difference technique. Hence, a considerable period of time was dedicated to trying out different algorithms to see whether a sharper shock could be obtained without an excessive increase in the number of elements. Good results were obtained by evaluating Φ_X behind the shock from the normal shock condition

$$U_1 + U_2 = 2/K \tag{12}$$

where $U = \partial \Phi / \partial X$. Equation (12) is obtained from the shock condition, Eq. (25), with $N_X = 1$ (normal shock). The results for a rectangular wing of aspect ratio AR = 6 with NACA 0012 sections, and Mach number M = 0.82 are compared in Fig. 3 with the results of Ref. 35.

Recently, encouraged by the results, actual time domain (oscillatory) applications were attempted. Very preliminary results are presented here. More extensive results will be included in Ref. 1. In order to assess the accuracy and above all the stability of the method, a very small disturbance problem was analyzed first, since linear results obtained with the linear code SOUSSA were available for comparison. The results shown in Fig. 4 are for a rectangular wing of aspect ratio AR = 5, with NACA 64A006 sections, and Mach number M = 0.875 for a pitching motion of the type

$$z/c = A e^{ikT}(x - x_{LE})/c \tag{13}$$

with A = 0.01 and $k = \omega c / U_\infty = 0.12$. The results are in reasonable agreement with the linear ones and indicate that the method of integration (see Appendix B) is extremely stable (eight time steps per cycle were sufficient to obtain the results). They also indicate that the correct trend is obtained. However, additional results are needed to validate the accuracy of the results. They are included in Ref. 1.

Shock-Capturing Nature of Method

An a posteriori explanation of the discovery of the shock-capturing nature of the shock is given in this section. Again for the sake of simplicity and clarity, only the steady-state problem is considered here. The discussion includes both the theoretical (integral equation) and the numerical (space discretized equations) aspects. The theoretical part deals with

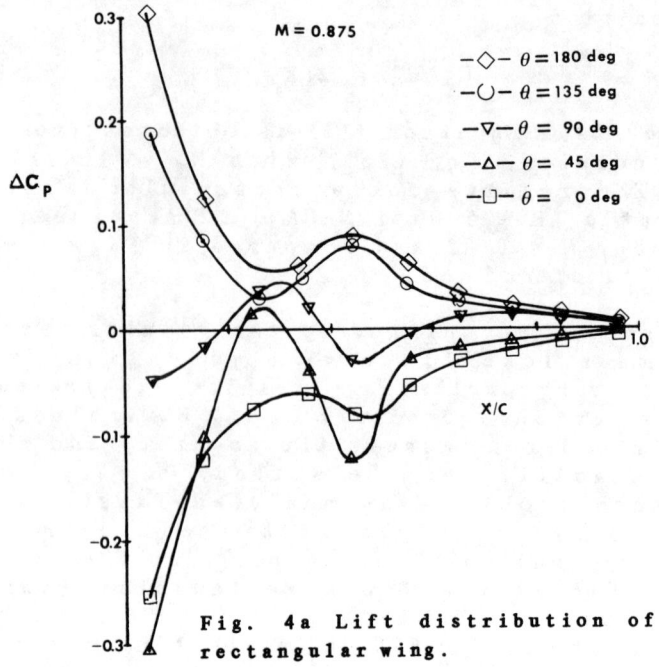

Fig. 4a Lift distribution of rectangular wing.

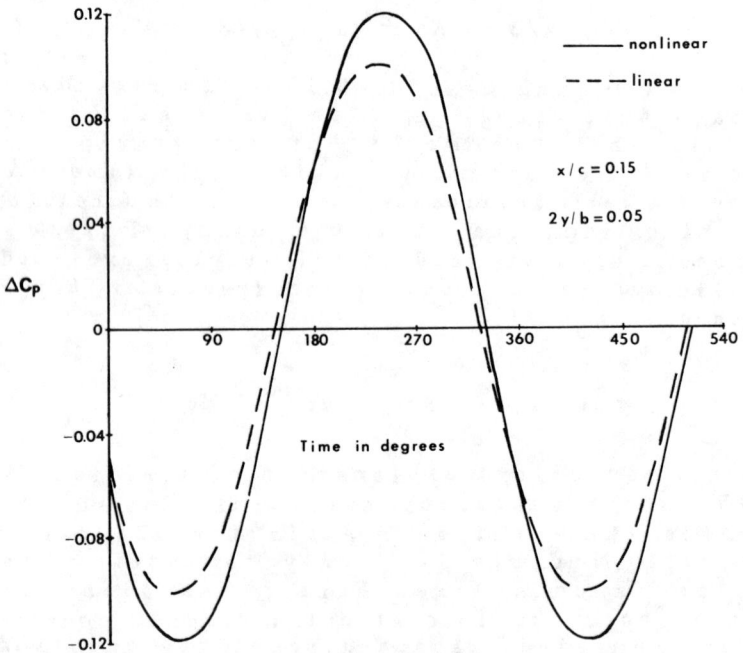

Fig. 4b Time history of lift on rectangular wing of Fig. 4a.

GREEN'S FUNCTION METHOD FOR TRANSONIC FLOWS

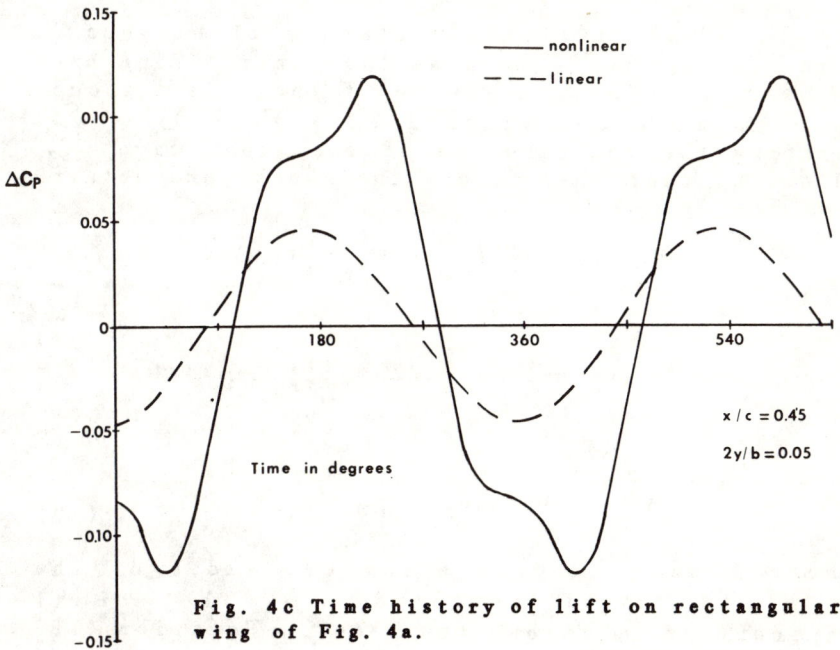

Fig. 4c Time history of lift on rectangular wing of Fig. 4a.

nonlinear terms of the type

$$\chi = \bar{\nabla} \cdot \bar{w} \tag{14}$$

Again for simplicity the discussion of the numerical formulation is limited to nonlinear terms of the type

$$\chi = \kappa\, \varphi_x \varphi_{xx} \tag{15}$$

For steady state, Eq. (10), in Prandtl-Glauert dimensionless variables, reduces to

$$\nabla_0^2 \Phi = \Xi = \bar{\nabla}_0 \cdot \bar{w} \tag{16}$$

where $\bar{\nabla}_0$ is the del operator in Prandtl Glauert variables and $\Phi = \varphi/U_\infty \ell$ and $\Xi = \chi \ell/U_\infty$. The Green's theorem for Poisson's equation (without shock waves) yields

$$4\pi E_* \Phi_* = -\oiint_{\Sigma_{BW}} [\Psi \frac{1}{R} - \Phi \frac{\partial}{\partial N}(\frac{1}{R})]\, d\Sigma \tag{17}$$

$$+ \iiint_V \Xi \frac{1}{R}\, dV$$

(see Eq. (A.11) for steady state without shock), where $\Sigma_{BW} = \Sigma_B + \Sigma_W$ is a surface surrounding body and wake. On the other hand if there is a shock, $\Sigma_B + \Sigma_W$ must be replaced with $\Sigma_B + \Sigma_W + \Sigma_S$. Isolating the contribution of the shock wave, Σ_S, and noting that $\Delta\Phi=0$ across the shock, one obtains

$$4\pi E_* \Phi_* = -\iint_{\Sigma_{BW}} [\Psi \frac{1}{R} - \Phi \frac{\partial}{\partial N}(\frac{1}{R})] d\Sigma$$

$$- \oiint_{\Sigma_S} \Delta\Psi \frac{1}{R} d\Sigma + \iiint_V \Xi \frac{1}{R} dV \quad (18)$$

where

$$\Delta\Psi = \partial\Phi_1/\partial N_1 - \partial\Phi_2/\partial N_1 \quad (19)$$

(where 1 and 2 indicate the two sides of the shock) is the discontinuity of the normal component of $\overline{\nabla}_0\Phi$ across the shock. Equation (18) is equal to Eq. (A.18) for steady state without explicit wake contribution. Noting that

$$(\overline{\nabla}_0 \cdot \overline{W})/R = \overline{\nabla}_0 \cdot (\overline{W}/R) - \overline{W} \cdot \overline{\nabla}_0(1/R) \quad (20)$$

and using the divergence theorem yields

$$4\pi E_* \Phi_* = -\oiint_{\Sigma_{BW}} [(\Psi + \overline{W} \cdot \overline{N})\frac{1}{R} - \Phi \frac{\partial}{\partial N}(\frac{1}{R})] d\Sigma$$

$$- \iiint_V \overline{W} \cdot \overline{\nabla}_0(\frac{1}{R}) dV \quad (21)$$

The contribution of the shock does not appear in Eq. (21), because of the shock boundary condition (see Refs. 33, 36, and 37)

$$\Delta\Psi + \Delta\overline{W} \cdot \overline{N}_1 = 0 \quad (22)$$

In particular, if [see Eq. (15)]

$$\Xi = K\Phi_X \Phi_{XX} \quad (23)$$

(where $K = \kappa U_\infty / \beta^3$), i.e.,

$$\overline{W} = (1/2) U^2 \bar{i} \qquad (24)$$

(with $U = \partial \Phi / \partial X$), Eq. (21) yields

$$4\pi E_* \Phi_* = -\oiint_{\Sigma_{BW}} [(\Psi + \frac{1}{2} K U^2 N_X) \frac{1}{R} - \Phi \frac{\partial}{\partial N}(\frac{1}{R})] \, d\Sigma$$

$$-\iiint_V \frac{1}{2} K U^2 \frac{\partial}{\partial X}(\frac{1}{R}) \, dV \qquad (25)$$

The shock boundary condition is (see Refs. 33, 36, and 37)

$$\Delta \Psi + (1/2) K \Delta(U^2) N_X = 0 \qquad (26)$$

Equation (25) was originally used in Tseng's doctoral work. Results indicating that the presence of the shock was captured were obtained. It was apparent that the 'captured' shock did not exhibit an actual discontinuity in the solution but simply a sharp change in pressure. Therefore, the solution being continuous, one could have inverted the integration by parts and obtain Eq. (17) (i.e., Eq. (18) without the contribution from the shock).

After a more careful analysis it became apparent that the shock contribution was now 'imbedded' into the volume integral of Eq. (17). In order to clarify this concept, it is convenient to make use of the Dirac delta function. If a function $f(x)$ is discontinuous at 0 and f' is its derivative, then

$$\int_{-\infty}^{\infty} f g' \, dx = -\int_{-\infty}^{\infty} f' g \, dx - [f(0^+) - f(0^-)] g(0) \qquad (27)$$

But if the discontinuity is 'smoothed out,' then, indicating with $\hat{f}$ the smoothed function,

$$\int_{-\infty}^{\infty} \hat{f} g' \, dx = -\int_{-\infty}^{\infty} \hat{f}' g \, dx \qquad (28)$$

Comparing Eqs. (27) and (28) it is apparent that the limit of $\hat{f}(x)$ is $f(x) + [f(0^+) - f(0^-)]\delta(x)$ (recall that the derivative of the Heaviside step function is the Dirac delta function). In other words Eq. (17) (with no contribution from the shock) is valid, even in the presence of the shock, if in evaluating Ξ suitable Dirac delta functions are introduced at the points of discontinuity of $\overline{W}$. A more detailed mathematical derivation is given in Ref. 1. Here it is sufficient to underline that as long as the derivatives are evaluated carefully, Eq. (17) is able to capture shocks (even actual discontinuities, if layers of Dirac delta functions are introduced).

It is even more remarkable that the space discretized formulations for Eqs. (17) and (25) are practically indistinguishable from each other. For simplicity consider the nonlinear terms of the type given by Eq. (15) and infinitesimally thin wing ($N_X=0$). Then it is sufficient to show that the numerical approximation of the volume integrals in the two equations are equivalent. Consider the volume V divided, for simplicity into cylinders (with rectangular cross sections) parallel to the x axis and indicate with V_i the volume, and with A_i the area of the cross section of the ith cylinder. Then the volume integral in Eq. (25) may be evaluated approximately as $I' = \sum_i I_i$ where

$$I'_i = -\iiint_{V_i} \frac{1}{2}KU^2 \frac{\partial}{\partial X}\left(\frac{1}{R}\right) dV \cong -\int_{-\infty}^{\infty} U^2 \frac{dJ}{dX} dX$$

$$= \lim_{N\to\infty} \sum_{h=-N}^{N-1} U^2_{h+\frac{1}{2}} \int_{X_h}^{X_{h+1}} \frac{dJ}{dX} dX$$

$$= \lim_{N\to\infty} \sum_{h=-N}^{N-1} U^2_{h+\frac{1}{2}} (J_{h+1} - J_h) \qquad (29)$$

where $U = \partial \Phi / \partial X$ and

$$J(X) = \frac{1}{2} K \iint_{A_i} \frac{1}{R} \, dA \qquad (30)$$

whereas the subscript h indicates evaluation at $X = X_h = h\Delta X$. On the other hand the integral in Eq. (17) may be evaluated as $I'' = \sum_i I''_i$, where

$$I''_i = \iiint_{V_i} \frac{1}{2} K \frac{\partial U^2}{\partial X} \frac{1}{R} \, dV \cong -\int_{-\infty}^{\infty} \frac{\partial U^2}{\partial X} J \, dx$$

$$= \lim_{N \to \infty} \sum_{h=-N}^{N} (U^2_{h+\frac{1}{2}} - U^2_{h-\frac{1}{2}}) \frac{1}{\Delta X} \int_{X_{h-\frac{1}{2}}}^{X_{h+\frac{1}{2}}} J \, dX$$

$$= \lim_{N \to \infty} \sum_{h=-N}^{N} (U^2_{h+\frac{1}{2}} - U^2_{h-\frac{1}{2}}) J_h \qquad (31)$$

The two results are identical since by 'summing by parts' one obtains

$$\sum_{h=-N}^{N-1} U^2_{h+\frac{1}{2}} (J_{h+1} - J_h) =$$

$$\sum_{h=-N}^{N} (U^2_{h+\frac{1}{2}} - U^2_{h-\frac{1}{2}}) J_h - U^2_{N+\frac{1}{2}} J_N - U^2_{-N-\frac{1}{2}} J_{-N} \qquad (32)$$

and

$$\lim_{N \to \infty} U_N = \lim_{N \to \infty} J_N = 0 \qquad (33)$$

Therefore, in the limit it is impossible to distinguish 'at the numerical-algorithm level,' which one of the two integral equations is being used. Therefore, since Eq. (25) can capture the shock,[1] Eq. (17) has the same capability. It should be emphasized that in order to yield accurate results it is necessary that the total 'amount of sources' in the field be conserved, i.e., that

$$\int_{X_1}^{X_2} \Xi \, dX = \frac{1}{2} K \Phi_X^2 \Big|_{X_1}^{X_2} = \frac{1}{2} K (U_2^2 - U_1^2) \quad (34)$$

not be affected by the representation used for between X_1 and X_2.

Assessment of Green's Function Method

From the discussion presented in the previous sections, the following remarks can be made. First, the Green's function method has been extended to unsteady three-dimensional potential transonic (nonlinear) flows. Therefore the method is now applicable to the complete range of the potential flow formulation (incompressible, subsonic, transonic, and supersonic). Second, the method is of shock-capturing type with its well-known advantages. Third, the input geometry consists of the aircraft surface geometry (similar to that of all the panel-aerodynamics codes), plus volume elements of the finite-element type (only in the region where nonlinear terms are important). This not only reduces considerably the number of unknowns (with consequent reduction on the central processing unit (CPU) time), but also simplifies considerably the use of the code: no cumbersome grid generation is required by the Green's function method. Fourth, the method seems to be exceptionally stable, although the addition of nonlinear terms of the type $\varphi_y \varphi_{xy}$ and $\varphi_x \varphi_{yy}$ (necessary to capture swept shocks) might affect the stability. Fifth, the method, per se, is valid for complex configurations (e.g., wings - body - tail) although the code is now limited to simple configurations, because of the finite-difference evaluation of the nonlinear terms.

It should be emphasized that the method also has certain disadvantages. First, the calculation of the coefficients [see Eq. (B.8)] requires a considerable amount of CPU time. For this reason, the method is more attractive for unsteady problems (in which the CPU time for the calculations of the coefficients is a small fraction of the CPU time for the step-by-step solution of the equations). Second, in contrast to the finite-difference method, the matrices are fully populated. However, the size of the matrix is considerably lower for two reasons: the unknowns in the volume are limited to the region of non-negligible nonlinear terms and the number of unknowns along the chord is much less than the finite-difference method. These two factors (the matrices being fully populated and the size of the matrices being smaller) seem to compensate each other and at the present time the CPU time/time step is of the same order of magnitude for both the finite-difference[38] and the Green's function method. However, because of favorable stability characteristics, the number of steps/cycle required by the Green's function method (eight steps/cycle, see Fig. 4) is much less than that required by the finite-difference method (a few hundred[38] to 1000 steps/cycle for three-dimensional flows and typically 300 steps/cycle for two-dimensional flows[39]). In other words the Green's function method appears to be at least one order of magnitude faster than the finite-difference method.

It should be emphasized that we are aware of the fact that these results are very preliminary and therefore it is premature to draw any conclusions. However, if these results are confirmed, the Green's function method appears to be a very powerful candidate for the solution of unsteady three-dimensional transonic potential flows.

Concluding Remarks

Summarizing, an extension of the Green's function method to nonlinear unsteady transonic flows has been presented. The method was originally intended for nonlinear subcritical (shock free) flows (see Ref. 28). Extensive results for this case are presented in Ref. 1. After introducing classical modifications [backward-central differencing of the type introduced by Murman and Cole[32] and use of the normal shock condition, Eq. (16)], it appears that the method is capable of reproducing the correct pressure discontinuities (Fig. 3). All the steady-state results were obtained through a time domain analysis which indicated that converged results were obtained with very few time steps (of order of 40), thereby indicating that the time domain analysis could be used as a powerful alternative to the iteration schemes. Also, application to actual unsteady (oscillatory) problems indicated that very few time steps (8/cycle) were sufficient for both stability and accuracy. Finally, the shock-capturing nature of the method has been analyzed in detail.

The method can be thought of as an extension of the classical integral equation method. It may be of interest to quote Nixon and Hancock,[40] who in their reappraisal of the integral equation methods, in 1976, concluded: 'Satisfactory progress is being made with the integral equation methods. Integral equation have a number of advantages. First, the solution is fully conservative. Second, it is not necessary to attempt to evaluate the full field accurately. Third, the total number of iteration for overall convergence is relatively low, of order 30.' Similar remarks are valid for the Green's function method as well.

Considerable additional work is needed, such as extension of the code to complex configurations and the use of the artificial viscosity method by Piers and Slooff,[8] instead of the backward-central differencing of Murman and Cole.[33] Also, shock fitting is a very important item, since it might improve considerably the rate of convergence. Finally, extensive numerical experimentation is necessary in order to verify and confirm the issues of stability and accuracy of the method.

In conclusion, the Green's function method provides a unified approach for the analysis of the complete range (incompressible, subsonic, transonic, and supersonic) of potential unsteady three-dimensional flows around complex configurations. In particular, for the transonic range, the method is shock-capturing and requires a very simple input geometry (no cumbersome grid generation is needed). Finally, the method appears to be extremely stable: if the preliminary results are confirmed, the method is at least one order of magnitude faster than the finite-difference method.

Appendix A: Transonic Integral Equation

Considered in this appendix is the integral equation for unsteady transonic potential aerodynamics for an aircraft having arbitrary shape. The motion of the aircraft is assumed to consist of small perturbations, starting at t = 0, with respect to the constant speed motion. The objective of this formulation is to describe the functional relationship between aerodynamic potential and its normal derivative (normal wash, $\psi = \partial\varphi/\partial n$) on the fluid boundary.

The analysis presented in this section is based upon an integral formulation presented in Refs. 19 and 20, which includes completely arbitrary motion. For the sake of clarity and simplicity, the unsteadiness of the aircraft is assumed to consist of small (infinitesimal) perturbations around the steady-state configuration so that the surface can be assumed to be fixed in deriving the integral equation. The general formulation, for arbitrarily moving surfaces, is given in Refs. 19 and 20 and is summarized in the last part of this appendix.

Green's Theorem for Transonic Potential Aerodynamics

The purpose of this analysis is to use Green's function method to obtain a representation of the potential in terms of its value (and the values of its derivatives) on the surface of the body and the wake (as well as nonlinear terms in the flowfield).

Note that the equation of the aerodynamic potential given by Eq. (10) is not valid on the wake and the shock wave, where discontinuities on φ and ψ exist. Thus, consider the volume V in which Eq. (10) is valid. At any instant of time, this volume is given by the whole physical space except the volume, V_B, occupied by the body and the infinitesimally thin layers, V_W and V_S, representing the wake and the shock wave respectively. As mentioned above, the volume V is assumed to be time-independent (time dependence is considered later). Define the function E

$$E(x,y,z,t) = 1 \quad \text{in } V$$
$$= 0 \quad \text{in } V_B + V_W + V_S \quad (A.1)$$

This function represents the domain of validity of the equation of the potential and will be called 'domain function.' Consider the surface of discontinuity of the function E, that is, the surface, σ, surrounding the volume $V_B + V_W + V_S$. Let $S(x,y,z) = 0$ be the equation of the surface σ.

Note that the surface σ is composed of three branches. The first, σ_B, is the surface of the body, $S_B(x,y,z) = 0$. The second is the surface, σ_W, of the wake, $S_W(x,y,z) = 0$. Note that this surface σ_W is considered twice, since σ is a closed surface. In other words, the two sides of the wake are considered to be two independent surfaces having the same equation (but opposite outwardly directed normals). The third one is the surface, σ_S, of the shock, for which similar considerations are valid.

As mentioned previously, the present formulation is based upon the Green's function method. The infinite space Green's function for the equation of potential is defined by

$$\nabla^2 G - (1/a_\infty^2)(d^2G/dt^2) = \delta(x-x_*, y-y_*, z-z_*, t-t_*) \quad (A.2)$$

(where $d/dt = \partial/\partial t + U_\infty \partial/\partial x$ and δ is the Dirac delta function) with $G = 0$ at infinity.

Multiplying the equation of the aerodynamic potential, Eq. (10), by the Green's function G and subtracting Eq. (A.2) multiplied by φ, yields

$$\overline{\nabla} \cdot (G\overline{\nabla}\varphi - \varphi\overline{\nabla}G) - (1/a_\infty^2) d/dt(G \, d\varphi/dt - \varphi \, dG/dt)$$
$$= G\chi - \varphi\delta \quad (A.3)$$

Multiplying Eq. (A.3) by the domain function E, defined by Eq. (A.1), integrating over the whole four-dimensional space time, and introducing the outwardly directed normal to the surface

$$\bar{n} = \bar{\nabla}S/|\bar{\nabla}S| \qquad (A.4)$$

(where 'outwardly' is understood as 'going from the body into the fluid,' that is, from the region E = 0 into the region E = 1), using divergence theorem and making use of the definition of the Dirac delta function, one obtains

$$E(\bar{p}_*) \, \varphi(\bar{p}_*, t_*) = \int_{-\infty}^{\infty} \iint_{\sigma} [(G \, \partial\varphi/\partial n - \varphi \, \partial G/\partial n)$$

$$- (1/a_\infty^2) \, U_\infty n_x (G \, d\varphi/dt - \varphi \, dG/dt)] \, d\sigma \, dt$$

$$+ \int_{-\infty}^{\infty} \iiint_{V} G \chi \, dV \, dt \qquad (A.5)$$

Equation (A.5) is the desired Green's theorem for potential transonic aerodynamics.

<u>Transonic Integral Equation</u>

In this paper, the undisturbed flow is assumed to be subsonic, i.e., $M = U_\infty/a_\infty < 1$. In this case the Green's function is given by (Ref. 20)

$$G = -(1/4\pi r_\beta) \, \delta_\theta \qquad (A.6)$$

where

$$r_\beta = \{(x-x_*)^2 + \beta^2 [(y-y_*)^2 + (z-z_*)^2]\}^{1/2} \qquad (A.7)$$

with

$$\beta = (1-M^2)^{1/2} \qquad (A.8)$$

whereas

$$\delta_\theta = \delta(t-t_*+\theta) \qquad (A.9)$$

with

$$\theta = [r_\beta + M(x-x_*)]/a_\infty \beta^2 \qquad (A.10)$$

Combine Eqs. (A.5) and (A.6), perform the integration with respect to time, and introduce the generalized Prandtl-Glauert transformation and nondimensionalization $\Phi = \varphi/U_\infty \ell$, $X = x/\beta \ell$, $Y = y/\ell$, $Z = z/\ell$, $T = U_\infty t/\ell$, $\Xi = \chi \ell / U_\infty$, where ℓ is a reference length. The resulting equation is (see Ref. 26 for details)

$$4\pi E(\overline{P}_*) \Phi(\overline{P}_*, T_*) = - \oiint_\Sigma [\Psi]^\theta \frac{1}{R} d\Sigma$$

$$+ \oiint_\Sigma [\Phi]^\theta \frac{\partial}{\partial N}\left(\frac{1}{R}\right) d\Sigma - \oiint_\Sigma [\dot{\Phi}]^\theta \frac{1}{R} \frac{\partial \widehat{\theta}}{\partial N} d\Sigma$$

$$+ \iiint_V [\Xi]^\theta \frac{1}{R} dV \qquad (A.11)$$

where

$$R = [(X-X_*)^2 + (Y-Y_*)^2 + (Z-Z_*)^2]^{1/2} \qquad (A.12)$$

and

$$[\ ]^\theta = [\ ]\Big|_{T=T_*-\theta} \qquad (A.13)$$

with

$$\theta = U_\infty \theta/\ell = [M(X-X_*)+R]M/\beta \qquad (A.14)$$

and

$$\widehat{\theta} = [M(X_*-X)+R]M/\beta \qquad (A.15)$$

In addition, $\partial/\partial N$ is the normal derivative in the Prandtl-Glauert space and

$$\Psi = \frac{\partial \Phi}{\partial N} \qquad (A.16)$$

indicates the component of the nondimensional velocity in the direction of the normal $\overline{N}$ to the surface Σ of the X,Y,Z space (not in the direction of the normal $\overline{n}$ to the surface σ of the physical space) and is known from the boundary conditions. The relationship between Ψ and ψ is given in Ref. 26 and for $n_x \simeq 0$ is approximated as

$$\Psi = \psi / U_\infty \qquad (A.17)$$

Contribution of Wake and Shock Wave

In order to understand the nature of the aerodynamic operator, Eq. (A.11), it is convenient to isolate the contribution of the wake and shock wave. Note that, as mentioned earlier, the surface σ is composed of three branches. The first is the (closed) surface of the body, Σ_B. The second is the (open) surface of the wake, Σ_W, and the third is the open surface of the shock wave, Σ_S. Note that the surface of the wake is considered twice since it is a closed surface. In other words, the two sides of the wake are considered to be two independent surfaces, having the same equation but opposite outwardly directed normals, $\bar{N}_1$ and $\bar{N}_2$, respectively. Similar considerations are valid for the shock. Note also that $\Delta\Psi = 0$ on the shock wave and $\Delta\Phi=0$ on the wake (see also Sec. 2.2.3 of Ref. 26 for further details). Hence Eq. (A.11) may be rewritten as

$$4\pi E(\bar{P}_*)\Phi(\bar{P}_*,T_*) = - \oiint_{\Sigma_B} [\Psi]^\Theta \frac{1}{R} d\Sigma$$

$$+ \oiint_{\Sigma_B} [\Phi]^\Theta \frac{\partial}{\partial N}(\frac{1}{R}) d\Sigma - \oiint_{\Sigma_B} [\dot{\Phi}]^\Theta \frac{1}{R} \frac{\partial \hat{\Theta}}{\partial N} d\Sigma$$

$$- \iint_{\Sigma_S} [\Delta\Psi]^\Theta \frac{1}{R} d\Sigma + \iint_{\Sigma_W} [\Delta\Phi]^\Theta \frac{\partial}{\partial N}(\frac{1}{R}) d\Sigma$$

$$- \iint_{\Sigma_W} [\Delta\dot{\Phi}]^\Theta \frac{1}{R} \frac{\partial \hat{\Theta}}{\partial N} d\Sigma + \iiint_V [\Xi]^\Theta \frac{1}{R} dV \quad (A.18)$$

The boundary condition on the wake is (see Ref. 26)

$$\Delta\Phi(\bar{P},T) = \Delta\Phi(\bar{P}_{TE},T-\Pi) \quad (A.19)$$

where Π is the nondimensional time necessary for a particle in the wake to travel (along a vortex line, within the steady flow) from the point $\bar{P}_{TE}$ (origin of the vortex line at the trailing edge), to the point $\bar{P}$.

It should be reiterated that, in Eq. (A.18), the surface Σ_B is assumed to be fixed with respect to the frame of reference. However, the effect of the motion of the surface is retained in the boundary conditions (see Appendix C of Ref. 26). Also note that (consistent with the hypothesis of small perturbation with respect to the steady-state configuration) the surface of the wake and the shock wave are assumed to be the one of the steady-state case. Moving surfaces are considered in the following section.

Moving Shock Waves

In many cases of practical interest, the shock waves cannot be appoximated as fixed in space. In order to avoid the impression that the formulation cannot be used in this case the general formulation of Ref. 20 (for surfaces having arbitrary motion) is briefly summarized here and applied in particular to the case of moving shock.

If the surface of the body is moving (along with its wake and the shock waves) then the volume V used to define the function E in Eq. (A.1) will be time dependent. Hence the function E will also be time dependent. Repeating the same procedure outlined in the first part of Appendix A, one obtains [Ref. 20, Eq. (3.32)]

$$4\pi E(\bar{p}_*, t_*) \varphi(\bar{p}_*)$$

$$= -\oiint_{\sigma^\theta} [\bar{\nabla}S \cdot \bar{\nabla}\varphi - \frac{1}{a_\infty^2} \frac{dS}{dt} \frac{d\varphi}{dt}]^\theta \frac{1}{r_\beta} \frac{1}{|\bar{\nabla}S^\theta|} d\sigma^\theta$$

$$+ \oiint_{\sigma^\theta} [\bar{\nabla}S \cdot \bar{\nabla}\frac{1}{r_\beta} - \frac{1}{a_\infty^2} \frac{dS}{dt} \frac{d}{dt}(\frac{1}{r_\beta})]^\theta \frac{[\varphi]^\theta}{|\bar{\nabla}S^\theta|} d\sigma^\theta$$

$$- \frac{\partial}{\partial t_*} \oiint_{\sigma^\theta} [\bar{\nabla}S \cdot \bar{\nabla}\theta - \frac{1}{a_\infty^2} \frac{dS}{dt}(1+U_\infty \frac{\partial \theta}{\partial x})]^\theta \frac{1}{r_\beta} \frac{[\varphi]^\theta}{|\bar{\nabla}S^\theta|} d\sigma^\theta$$

$$- \iiint_{-\infty}^{\infty} [EX]^\theta \frac{1}{r_\beta} dV \qquad (A.20)$$

where σ^θ is the 'retarded surface' defined by

$$S(x,y,z,t_*-\theta) = 0 \qquad (A.21)$$

whereas

$$\overline{\nabla}S^\theta = \overline{\nabla}S(x,y,z,t_*-\theta) \qquad (A.22)$$

Introducing the generalized Prandtl-Glauert transformations and assuming that the surface of the body and the wake are time independent but the shock wave is moving one obtains [see Ref. 20, Eq. (6.40)] an expression similar to Eq. (A.11) with the shock integral $I_S = \iint_{\Sigma_S} [\Delta\Psi]^\theta (1/R) d\Sigma$ replaced by

$$I_S = \iint_{\Sigma_S^\theta} [\Delta\Psi + \frac{M^2}{\beta} V_S \Delta \frac{\partial \Phi}{\partial X}]^\theta \frac{1}{R} \frac{|\overline{\nabla}_o S|^\theta}{|\overline{\nabla}_o S^\theta|} d\Sigma_S^\theta \qquad (A.23)$$

where $V_S = -(\partial S/\partial T)/|\overline{\nabla}_o S|$ is the velocity of the shock wave surface (in direction normal to the shock wave surface). A simple expression for evaluating $|\overline{\nabla}_o S|^\theta/|\overline{\nabla}_o S^\theta|$ is given in Ref. 26, Eq. (C.11).

Appendix B: Numerical Solution of the Integral Equation

Equation (A.18) fully describes the problem of nonlinear unsteady potential transonic aerodynamics around complex configurations. In order to solve this problem, it is necessary, in general, to obtain a numerical approximation for Eq. (A.18). The numerical formulation is presented in this Appendix. First, a general formulation for the space discretization using an arbitrary finite-element representation is presented, followed by the zeroth order formulation used in SUSAN. Then the time discretization is discussed.

Space Discretization

Consider the integral equation given by Eq. (A.18). Because of the shock-capturing nature of the method (considered in Shock-Capturing Nature of Method), the explicit treatment of the contribution of the shock is not needed here. In order to discretize the space integral operator over the surface Σ_B it is convenient to use a finite element representation for the normal wash Ψ and the potential Φ. [Finite-element representation is meant here in a very broad sense: actually any interpolation formula of the type given by Equations (B.1) and (B.2) is consistent with the formulation presented here. However, for arbitrarily complex configurations, only the finite-element interpolation, including splines over patches, is sufficiently general to be of interest here.] Using a general finite-element representation, it is possible to write

$$\Psi(\overline{P},T-\theta) = \sum_{h=1}^{N_B} \Psi_h(T-\theta_h)N_h(\overline{P})$$

$$\Phi(\overline{P},T-\theta) = \sum_{h=1}^{N_B} \Phi_h(T-\theta_h)N_h(\overline{P}) \quad (B.1)$$

where $\Psi_h(T-\theta_h)$ and $\Phi_h(T-\theta_h)$ are time-dependent values of Ψ and Φ at the point $\overline{P}_h$ on Σ_B at the time $T-\theta_h$ (where θ_h is the disturbance propagation time from $\overline{P}_h$ to $\overline{P}_*$); furthermore $N_h(\overline{P})$ are prescribed global shape functions, obtained by standard assembly of the element shape function. The points $\overline{P}_h$ will be referred to as nodes; N_B is the total number of nodes on the body. For simplicity the same shape functions are used for Ψ and Φ, although this is not essential to the method.

Next consider the integration over the wake. In order to facilitate the use of Eq. (A.19), it is convenient to divide the wake into strips defined by (steady-state) vortex lines emanating from the nodes on the trailing edge. The strips are then divided into elements with nodes along the vortex lines. The potential discontinuity can then be expressed as

$$\Delta\Phi(\overline{P},T-\theta) = \sum_{n=1}^{N_W} \Delta\Phi_n(T-\theta_n) L_n(\overline{P}) \quad (B.2)$$

where N_W is the number of nodes on the wake, $\Delta\Phi_n(T-\theta)$ is the value of $\Delta\Phi$ at the nth node $\overline{P}_n^{(W)}$

the wake at time $T-\theta_n$ (where θ_n is the propagation time from $\bar{P}_n^{(W)}$ to $\bar{P}_*$), and $L_n(\bar{P})$ is the global shape function relative to the nth node of the wake. Note that according to Eq. (A.19)

$$\Delta\Phi_n(T) = \Delta\Phi_{m(n)}^{(TE)}(T-\Pi_n) \qquad (B.3)$$

where $m=m(n)$ identifies the trailing edge node which is on the same vortex line as the nth node $\bar{P}_n^{(W)}$. Furthermore, Π_n is the time necessary for the vortex point to be convected from the trailing edge node $\bar{P}_{(n)}^{(TE)}$ to the wake node $\bar{P}_n^{(W)}$. It may be worth noting that $\Delta\Phi_m^{(TE)} = \Phi_{h_1} - \Phi_{h_2}$ where h_1 and h_2 identify the trailing edge nodes (e.g., upper and lower sides, respectively) on the body corresponding to the mth node on the trailing edge. Therefore, it is possible to write

$$\Delta\Phi_{m(n)}^{(TE)} = \sum_{h=1}^{N_B} S_{nh}\Phi_h \qquad (B.4)$$

where $S_{nh}=1$ ($S_{nh}=-1$), if h identifies the upper-side (lower-side) node $\bar{P}_h$ on the body corresponding to the nth node $\bar{P}_n^{(W)}$ on the wake (i.e., $\bar{P}_h$ coincides with the node $\bar{P}_{m(n)}^{(TE)}$ on the trailing edge), and $S_{nh} = 0$ otherwise. Thus,

$$\begin{aligned}
S_{nh} &= +1 && \text{if } \bar{P}_h = \bar{P}_{m(n)}^{(TE)} \text{ is on the upper side of } \Sigma_B \\
&= -1 && \text{if } \bar{P}_h = \bar{P}_{m(n)}^{(TE)} \text{ is on the lower side of } \Sigma_B \\
&= 0 && \text{otherwise} \qquad (B.5)
\end{aligned}$$

Finally consider the integration over the volume. The nonlinear terms can be expressed as

$$\Xi(\bar{P},T-\theta) = \sum_{q=1}^{N_V} \Xi_q(T-\theta_q)M_q(\bar{P}) \qquad (B.6)$$

where N_V is the number of nodes in the fluid volume and its boundary, $\Xi_q(T-\theta_q)$ is the value of Ξ at the qth volume node $\bar{P}_q^{(V)}$ at time $T-\theta_q$ (where θ_q is the propagation time from $\bar{P}_q$ to $\bar{P}_*$) and $M_q^q(\bar{P})$ is the global shape function relative to the qth volume node.

Combining Eqs. (B.1) and (B.2) one obtains (for notational simplicity T_* is replaced with T in this appendix)

$$\Phi_j(T) = \sum_{h=1}^{N_B} B_{jh}\Psi_h(T-\theta_{jh})$$

$$+ \sum_{h=1}^{N_B} C_{jh}\Phi_h(T-\theta_{jh}) + \sum_{h=1}^{N_B} D_{jh}\dot{\Phi}_h(T-\theta_{jh})$$

$$+ \sum_{n=1}^{N_W} F_{jn}\Delta\Phi_n(T-\theta_{jn}) + \sum_{n=1}^{N_W} G_{jn}\Delta\dot{\Phi}_n(T-\theta_{jn})$$

$$+ \sum_{q=1}^{N_V} H_{jq}\Xi_q(T-\theta_{jq}) \qquad (B.7)$$

with $\Phi_j(T) = \Phi(\bar{P}_j,T)$ where $\bar{P}_j$ is a control point (i.e., either a body node or a volume node), and

$$B_h = -\frac{1}{4\pi E_*} \oiint_{\Sigma_B} N_h(\bar{P}) \frac{1}{R} \, d\Sigma \bigg|_{\bar{P}_*=\bar{P}_j}$$

$$C_h = \frac{1}{4\pi E_*} \oiint_{\Sigma_B} N_h(\bar{P}) \frac{\partial}{\partial N}\left(\frac{1}{R}\right) d\Sigma \bigg|_{\bar{P}_*=\bar{P}_j}$$

$$D_h = -\frac{1}{4\pi E_*} \oiint_{\Sigma_B} N_h(\bar{P}) \frac{1}{R} \frac{\partial \hat{\theta}}{\partial N} \, d\Sigma \bigg|_{\bar{P}_*=\bar{P}_j}$$

$$F_n = \frac{1}{4\pi E_*} \iint_{\Sigma_W} L_n(\bar{P}) \frac{\partial}{\partial N}\left(\frac{1}{R}\right) d\Sigma \bigg|_{\bar{P}_*=\bar{P}_j}$$

$$G_n = -\frac{1}{4\pi E_*} \iint_{\Sigma_W} L_n(\bar{P}) \frac{1}{R} \frac{\partial \hat{\theta}}{\partial N} \, d\Sigma \bigg|_{\bar{P}_*=\bar{P}_j}$$

$$H_q = \frac{1}{4\pi E_*} \iiint_V M_q \frac{1}{R} dV \bigg|_{\overline{P}_* = \overline{P}_j} \quad (B.8)$$

and according to Eqs. (B.4) and (B.5)

$$\Delta\Phi_n(T) = \sum_{h=1}^{N_B} S_{nh}\Phi_h(T-\Pi_n) \quad (B.9)$$

The rest of this section deals with the transonic formulation used in SUSAN. This formulation is a particular case of the general formulation presented above, and is briefly illustrated here. Divide the surface of the aircraft Σ_B into small elements, Σ_h. Consider the shape function, N_h, equal to one inside Σ_h and equal to zero outside Σ_h, i.e.,

$$N_h(\overline{P}) = 1 \quad \text{if } \overline{P}\epsilon\Sigma_h$$
$$= 0 \quad \text{otherwise} \quad (B.10)$$

A point located on the element Σ_h and identified as the center of the element will be designated as the point at which Φ and Ψ are evaluated. Equation (B.1) may thus be interpreted as saying that, within the element Σ_h, the normal wash and the potential are approximated with the values Ψ_h and Φ_h at the center, $\overline{P}_h$, of the element Σ_h. [Note that the shape functions given by Eq. (B. 10) may be called zeroth-order shape functions. Therefore, the formulation presently used in SUSAN may be called zeroth-order finite-element formulation.]

Next, note that using Eq. (B.10), Eq. (B.8) yields for instance

$$B_{jh} = -\frac{1}{4\pi E_*} \iint_{\Sigma_h} \frac{1}{R} d\Sigma \bigg|_{\overline{P}_* = \overline{P}_j} \quad (B.11)$$

If Σ_h is a quadrilateral element, Σ_h is approximated with a hyperboloidal element and the integrals are calculated analytically (see Ref. 22). Zeroth-order shape functions are also used on the wake and on the volume.

Time Discretization

Equation (B.7) indicates the nature of the aerodynamic operator relating potential and normal wash as obtained by using finite-element representation to discretize the spatial problem. In this section Eq. (B.8) will be discretized in time. Set

$$t = (m+\alpha)\Delta t \qquad (0 \leq \alpha < 1) \qquad (B.12)$$

Consider a function $F(t)$. Assuming that $F(t)$ varies linearly in t and using Eq. (B.12), one obtains

$$F(t) = F[(n+\alpha)\Delta t] \qquad (B.13)$$
$$= (1-\alpha) F(n\Delta t) + \alpha F[(n+1)\Delta t]$$

On the other hand, using a backwards finite-difference scheme,

$$\frac{\partial F(t)}{\partial t} = \frac{F(t) - F(t-\Delta t')}{\Delta t'} \qquad (B.14)$$

where $\Delta t'$ is an arbitrary time increment not necessarily the same as Δt.

Setting $t_k = k\Delta t$ and using Eq. (B.1), Eq. (B.7) combined with Eqs. (B.2, B.3, and B.4) yields

$$\Phi_j(k) = \sum_h B_{jh}[(1-\alpha_{jh})\psi_h(k-m_{jh}) + \alpha_{jh}\psi_h(k-m_{jh}-1)]$$

$$+ \sum_h C_{jh}[(1-\alpha_{jh})\Phi_h(k-m_{jh}) + \alpha_{jh}\Phi_h(k-m_{jh}-1)]$$

$$+ \sum_h D_{jh}[(1-\alpha_{jh})\Phi_h(k-m_{jh}) + \alpha_{jh}\Phi_h(k-m_{jh}-1)$$

$$- (1-\alpha_{jh}')\Phi_h(k-m_{jh}') - \alpha_{jh}'\Phi_h(k-m_{jh}'-1)]/\Delta t'$$

$$+ \sum_n F_{jn}[(1-\alpha_{jn})\Delta\Phi_{m(n)}^{TE}(k-m_{jn}) + \alpha_{jn}\Delta\Phi_{m(n)}^{TE}(k-m_{jn}-1)]$$

$$+ \sum_n G_{jn}[(1-\alpha_{jn})\Delta\Phi_{m(n)}^{TE}(k-m_{jn}) + \alpha_{jn}\Delta\Phi_{m(n)}^{TE}(k-m_{jn}-1)$$

$$-(1-\alpha'_{jn})\Delta\Phi_{m(n)}^{TE}(k-m'_{jn}) - \alpha'_{jn}\Delta\Phi_{m(n)}^{TE}(k-m'_{jn}-1)]/\Delta t'$$

$$+\sum_q H_{jq}[(1-\alpha_{jq})\Xi_q(k-m_{jq}) + \alpha_{jq}\Xi_q(k-m_{jq})] \quad (B.15)$$

where Δt has been dropped for notational simplicity (i.e., $\Phi(k\Delta t) = \Phi(k)$) and

$$\begin{aligned}
\theta_{jh} &= (m_{jh} + \alpha_{jh})\Delta t \\
\theta_{jh} + \Delta t' &= (m'_{jh} + \alpha'_{jh})\Delta t \\
\theta_{jn} + \Pi_n &= (m_{jn} + \alpha_{jn})\Delta t \quad (B.16) \\
\theta_{jn} + \Pi_n + \Delta t' &= (m'_{jn} + \alpha'_{jn})\Delta t \\
\theta_{jq} &= (m_{jq} + \alpha_{jq})\Delta t
\end{aligned}$$

Note that in Eq. (B.15), the unknown $\Phi_h(k)$ appears on both sides of the equation, since, on the right-hand side,

$$\Phi_h(k-m_{jh}) = \Phi_n(k) \quad \text{for } m_{jh} = 0 \quad (B.17)$$

In any event, at each time step the only unknowns are $\Phi_h(k)$. Hence Eq. (B.15) can be used step by step to solve for $\Phi_h(k)$ at all time steps given appropriate initial conditions.

Acknowledgments

This work was supported by NASA Grant NGR 22-004-030 by NASA/Langley Research Center to Boston University. The authors wish to thank Dr. E. Carson Yates, Jr. for the invaluable help received in performing this work.

References

[1] Tseng, K., Doctoral Dissertation, Boston University, Department of Mathematics, in preparation.

2. Oswatitsch, K., 'Die Geschwindigkeitsverteilung ab Symmetrische Profilen beim auftreten lokaler Uberschallgebiete.' *Acta Physica Austriaca*, Vol. 4, 1950 pp. 228-271.

3. Spreiter, J. R. and Alksne, A. Y., 'Theoretical Prediction of Pressure Distributions on Non-Lifting Airfoils at High Subsonic Speeds,' NACA Rept. 1217, 1955.

4. Norstrud, H., 'High Speed Flow Past Wings,' NASA CR-2246, 1973.

5. Nixon, D., 'Transonic Flow Around Symmetric Aerofoils at Zero Incidence,' *Journal of Aircraft*, Vol. 11, No. 2 Feb. 1974, pp. 122-124.

6. Ogana, W., 'Solution of Transonic Flows by an Integro-Differential Equation Method,' NASA TM 78490, 1978.

7. Nixon, D., 'Calculation of Unsteady Transonic Flows Using the Integral Equation Method,' *AIAA Journal*, Vol. 16, No. 9, Sept. 1978, pp. 976-983.

8. Piers, W. J., and Slooff, J. W., 'Calculation of Transonic Flow by Means of a Shock-Capturing Field Panel Method,' AIAA Paper No. 79-1459, AIAA Computational Fluid Dynamics Conference, Williamsburg, Va., July 1979.

9. Lamb, H., *Hydrodynamics*, Cambridge University Press, London, 1932, pp. 58-61.

10. Kellogg, O. D., *Foundations of Potential Theory*, Frederick Ungar Publishing Co., New York, 1929, pp. 236-287.

11. Ikebe, Y., Lynn, M. S., and Timlake, W. P., 'The Numerical Solution of the Integral Equation Formulation of the Neumann Problem,' *Information Processing 68*, North Holland Publishing Co., Amsterdam, 1969, pp. 1-9.

12. Smirnov, V. I., *A Course of Higher Mathematics*, Vol. IV, *Integral Equation and Partial Differential Equations*, Pergamon Press, New York, 1964, pp. 617-627.

13. Smith, A. M. O. and Pierce, J., 'Exact Solution of the Neumann Problem. Calculation of Plane and Axially Symmetric Flows about or within Arbitrary Boundaries,' Douglas Aircraft Company Report No. 26988, April 1958.

14. Hess, J. L. and Smith, A.M.O., 'Calculation of Potential Flow about Arbitrary Bodies,' *Progress in Aeronautical Sciences*, Vol. 8, Pergamon Press, New York, 1966, pp. 1-138.

[15] Hess, J. L., 'Numerical Solution of the Integral Equation for the Neumann Problem with Application to Aircraft and Ships,' Symposium on Numerical Solution of Integral Equations with Physical Applications, SIAM, Fall Meeting 1971, Douglas Aircraft Company, Engineering Paper 5987, Oct. 1971.

[16] Hess, J., Johnson, F. T., and Rubbert, P. E., 'Panel Methods,' AIAA Professional Study Series, Seattle, Wash., July 1978.

[17] Magnus, A. E. and Epton, M. E., 'PAN AIR- A Computer Program for Predicting Subsonic or Supersonic Linear Potential Flows about Arbitrary Configurations Using a Higher Order Panel Method,' Vol. 1, Theory Document (Version 1.0), NASA-CR 3251, April 1980.

[18] Morino, L., 'Unsteady Compressible Potential Flow Around Lifting Bodies Having Arbitrary Shapes and Motions,' Boston University, Department of Aerospace Engineering, TR-72-01, June 1972.

[19] Morino, L., 'Unsteady Compressible Potential Flow Around Lifting Bodies: General Theory,' AIAA Paper No. 73-196, Jan. 1973.

[20] Morino, L., 'A General Theory of Unsteady Compressible Potential Aerodynamics', NASA CR-2464, 1974.

[21] Morino, L. and Kuo, C. C., 'Subsonic Potential Aerodynamics for Complex Configurations: A General Theory,' *AIAA Journal*, Vol. 12, Feb. 1974, pp. 191-197.

[22] Morino, L., Chen, L. T. and Suciu, E. O., 'Steady and Oscillatory Subsonic and Supersonic Aerodynamics Around Complex Configurations,' *AIAA Journal*, Vol. 13, March 1975, pp. 368-374.

[23] Morino, L. and Chen, L. T., 'Indicial Compressible Potential Aerodynamics Around Complex Aircraft Configuration,' NASA SP-347, 1975, pp. 1067-1110.

[24] Tseng, K. and Morino, L., 'A New Unified Approach for Analyzing Wing-Body-Tail Configurations with Control Surfaces,' AIAA Paper No. 76-418, San Diego, Calif., July 1976.

[25] Morino, L. and Tseng, K., 'Steady, Oscillatory and Unsteady, Subsonic and Supersonic Aerodynamics (SOUSSA) for Complex Aircraft Configurations,' in AGARD-CP-227, *Unsteady Aerodynamics*, Ottawa, Canada, Sept. 1977, pp. 3-1 to 3-14.

[26] Morino, L., 'Steady, Oscillatory, and Unsteady Subsonic and Supersonic Aerodynamics - Production

Version 1.1 (SOUSSA-P 1.1), Vol. 1, Theoretical Manual,' NASA CR-159130, 1980.

[27] Smolka, S. A., Preuss, R. D., Tseng, K., and Morino, L., 'Steady, Oscillatory, and Unsteady Subsonic and Supersonic Aerodynamics- Production Version 1.1 (SOUSSA-P 1.1), Vol. 2, User/Programmer Manual,' NASA CR-159131, 1980.

[28] Morino, L. and Tseng, K., 'Time-Domain Green's Function Method for Three-Dimensional Nonlinear Subsonic Flows,' AIAA Paper No. 78-1204, Seattle, Wash., 1978.

[29] Suzuki, S. and Washizu, K., 'Calculation of Wing-Body Pressures in Incompressible Flow Using Green's Function Method,' Journal of Aircraft, Vol. 17, No. 5, May 1980, pp. 326-331.

[30] Summa, I. M., 'Communication at the Round Table on Helicopter Wakes,' NASA/Ames Research Center, Nov. 1980.

[31] Dusto, A. R., and Epton, M. A., 'An Advanced Panel Method for Analysis of Arbitrary Configurations in Unsteady Subsonic Flow,' NASA CR-152323, Feb. 1980.

[32] Maskew, B., 'Unsteady Potential Flow Analysis of Rotor Blade Shapes,' Analytical Methods Rept. 8005, Apr. 1980.

[33] Murman, E. M. and Cole, I. D., 'Calculation of Plane Steady Transonic Flows,' AIAA Journal, Vol. 9, Jan. 1971, pp. 114-121.

[34] Bailey, F.R, and Steger, J.L., 'Relaxation Techniques for the Three-Dimensional Transonic Flow About Wings', AIAA Paper No. 72-189, San Diego, Calif., Jan. 1972.

[35] Lee, K. D., Dickson, L. J., Chen, A. W., and Rubbert, P. E., 'An Improved Matching Method for Transonic Computations,' AIAA Paper No. 78-1116, Seattle, Wash., July 1978.

[36] Yoshihara, H., 'A Survey of Computational Methods of 2D and 3D Transonic Flows with Shocks,' General Dynamics, Convair Aerospace Division, Report GDCA-ERR-1726, Dec. 1972.

[37] Freedman, M. I., Tseng, K., and Morino, L., 'Shock Conditions -a General Formulation,' in preparation.

[38] Borland, C. J., Rizzetta, D. P., and Yoshihara, H., 'Numerical Solution of Three-Dimensional Unsteady Transonic Flow Over Swept Wings,' AIAA Paper No. 80-1369, Seattle, Wash., July 1980.

[39] Ballhaus, W.F. and Goorjian, P.M., 'Implicit Finite Difference Computations of Unsteady Transonic Flow about Airfoils Including the Treatment of Irregular Shock Wave Motions,' AIAA Journal, Vol. 15, No. 12, 1977.

[40] Nixon, D. and Hancock, G. J., 'Integral Equation Methods - A Reappraisal,' in *Symposium Transonicum II*, edited by K. Oswatitsch and D. Rues, Springer Verlag, New York, 1976, pp. 174-182.

Chapter XVI.

Hybrid Approach to Transonic Inviscid Flow with Moderate to Strong Shock Wave

Tsze C. Tai*
*David Taylor Naval Ship Research and Development Center,
Bethesda, Md.*

Introduction

The finite-difference relaxation (FDR) technique for solving the transonic potential flow equations constitutes the main thrust of transonic flow research.[1-5] The method is well developed, and relatively easy to implement. However, it is based on the isentropic flow assumption and is, therefore, limited to the weak shock-wave solution. The shock wave has to be smeared out a finite length instead of the exact inviscid jump condition. The technique predicts the pressure distribution around airfoils very well near the design flow conditions.[6] As the freestream Mach number increases, the numerical results deviate from experiments considerably.[6] The deviation appears predominately in the shock-wave region where the potential flow equation becomes most inadequate in describing the physical phenomenon. In this region, the flow would be more properly represented by the Euler equation - if we restrict ourselves to the inviscid cases for the time being. Because of the extensive computation requirement and peculiar numerical problems, a steady-state Euler solution for the subject problem using the finite-difference technique seems to be some time in the future.

The method of integral relations (MIR), on the other hand, solves the Euler equations with a small computation requirement. Previous applications of the method to tran-

---

Presented at the Transonic Perspective Symposium, NASA/Ames Research Center, Moffett Field, Calif., Feb. 18-20, 1981. This paper is a work of the U.S. Government and therefore is in the public domain.
  *Research Aerospace Engineer, Aviation and Surface Effects Department.

sonic flows past airfoils include those by Holt and Masson,[7] Melnik and Ives,[8] Sato,[9] and Tai[10] for various flow conditions in the transonic regime. The main advantage in using MIR is its ability to solve the full inviscid flow equations directly; thus the assumption of isentropic flow is not necessary. It is valid, therefore, for transonic flows with moderate to strong shock waves. The exact Rankine-Hugoniot relations can be applied at the shock to give the correct pressure jump. However, because of the multiple iterative processes involved, the whole solution procedure cannot be automated without man-machine interactions.[10]

To take advantage of both methods and yet avoid their shortcomings, a combination of the FDR techniques with MIR - a hybrid method, appears to offer a promising capability for solving more general transonic flow problems. The idea was introduced by the author in 1980.[11] Since then the approach has been implemented in a CDC 6600/6700 computer. In the following sections, the procedure is described and results of test cases are discussed.

Hybrid Method

The basic idea for the hybrid method is illustrated in Fig. 1. The overall mixed flowfield, except for the shock region, is governed by the potential flow equation to be solved by a finite-difference relaxation technique. In the shock region and the region downstream of the shock, the flow is governed by Euler equations to be solved by the method of integral relations. For simplicity, the approach is easily illustrated with reference to a Cartesian coordinate system. Therefore, the potential flow-FDR code developed by Carlson[12] and the Euler-MIR code by the present author,[10] both using Cartesian coordinates, are summarized below.

Potential Flow Solution (Overall Flowfield)

If the entire flowfield is assumed isentropic, the flow is everywhere irrotational. The velocity is related to a potential function, Φ, as

$$U_i = \frac{\partial \Phi}{\partial x_i} = \text{grad } \Phi \qquad (1)$$

The exact equation for the potential function for two-dimensional compressible flow can be written in Cartesian coordinates as

$$(a^2 - \Phi_x^2) \Phi_{xx} - 2\Phi_x \Phi_y \Phi_{xy} + (a^2 - \Phi_y^2) \Phi_{yy} = 0 \quad (2)$$

where

$$\Phi_x = \frac{\partial \Phi}{\partial x}, \quad \Phi_{xx} = \frac{\partial^2 \Phi}{\partial x^2}, \quad \Phi_{xy} = \frac{\partial^2 \Phi}{\partial x \partial y}, \quad \text{etc.}$$

and a is the speed of sound. Equation (2) is elliptic if the local Mach number $M_\infty < 1$ and hyperbolic if $M_\infty > 1$. Once the solution is found, the velocities are calculated from Eq. (1). Equation (2) is valid only in cases with weak shock waves where the change of entropy can be neglected. By definition of the irrotational flow, there is no transonic wave drag.

The full potential flow equation, Eq. (2), can be recast into the form of a perturbation potential

$$(a^2 - U^2)\phi_{xx} - 2UV\phi_{xy} + (a^2 - V^2)\phi_{yy} = 0 \quad (3)$$

where U and V are given by

$$U = \Phi_x = q_\infty (\cos \alpha + \phi_x)$$

$$V = \Phi_y = q_\infty (\sin \alpha + \phi_y) \quad (4)$$

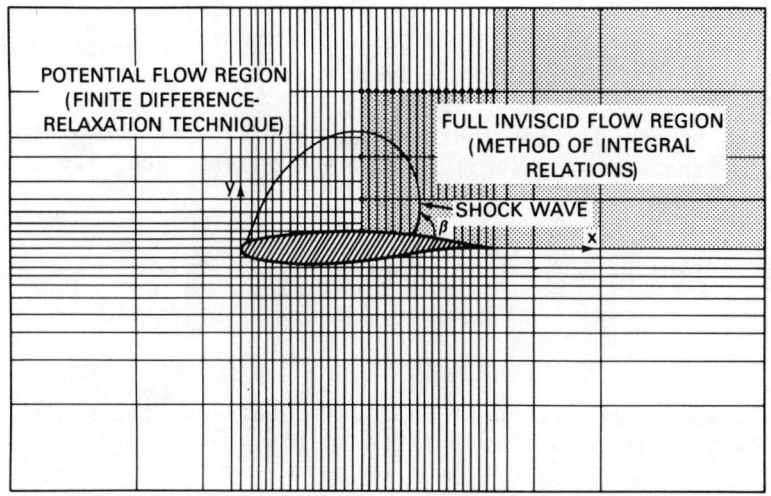

Fig. 1 The hybrid method in Cartesian coordinates.

along with the boundary conditions of the velocity normal to the airfoil surface being equal to zero and a velocity potential satisfying the field behavior at infinity.

To incorporate the coordinate stretching in the physical plane and to avoid computationl problems in the supersonic region, the potential flow equation is further arranged in rotated coordinates parallel and perpendicular to the local velocity for finite differencing. The numerical solution is obtained iteratively by using column relaxation. Detailed descriptions of the procedure can be found in Ref. 12. Similar schemes for determining the potential flow solution are available elsewhere,[3,4] which also can be used in the present approach.

Solution to Euler Equations (Shock-Wave Region)

The Euler equations that govern a full inviscid transonic flow can be written in nondimensional form normalized by freestream values:

Continuity
$$\frac{\partial(\rho U)}{\partial x} + \frac{\partial(\rho V)}{\partial y} = 0 \tag{5}$$

x Momentum
$$\frac{\partial}{\partial x}(KP + \rho U^2) + \frac{\partial}{\partial y}(\rho UV) = 0 \tag{6}$$

y Momentum
$$\frac{\partial}{\partial x}(\rho UV) + \frac{\partial}{\partial y}(KP + \rho V^2) = 0 \tag{7}$$

State
$$P = \rho^\gamma \exp\left(\frac{S_2 - S_1}{c_v}\right) \tag{8}$$

where
$$K = 1/\gamma M_\infty^2$$

The boundary conditions are as follows: at the airfoil surface, the normal velocity component equals zero, i.e.,

$$q_n = 0$$

and at infinity, the flow is undisturbed, i.e.,

$$P = \rho = U = 1$$
$$V = 0$$

To apply the method of integral relations, the system of flow equations must be written in divergence form

$$\frac{\partial}{\partial x} A(x,y,U,\ldots) + \frac{\partial}{\partial y} B(x,y,U,\ldots) = Q(x,y,U,\ldots) \tag{9}$$

The divergence form of Eqs. (5-7) may then be integrated outward from the airfoil surface (but not necessarily normal to the surface) to each strip boundary, successively at some x station. This procedure reduces the partial differential equations to ordinary ones. To perform the integration, the integrand is approximated by interpolation polynomials, for example A, by

$$A = \sum_{k=0}^{\bar{N}} a_k(x) (y - y_0)^k \qquad (10)$$

where $\bar{N}$ = the number of strips; $a_k(x)$ = constants evaluated at strip boundaries; and y_0 = the location of the base strip boundary. In principle, the actual flow variation may be represented by an increasing number of strips.

Using a basic second-order approximation for Eq. (10), the method can be implemented with three strips for desired accuracy. The process is illustrated in Fig. 2. The idea is to treat the whole integration domain as two different effective regions, which are designated by strip boundaries (0, 1, 2) and (a, b, c). In each effective region, the

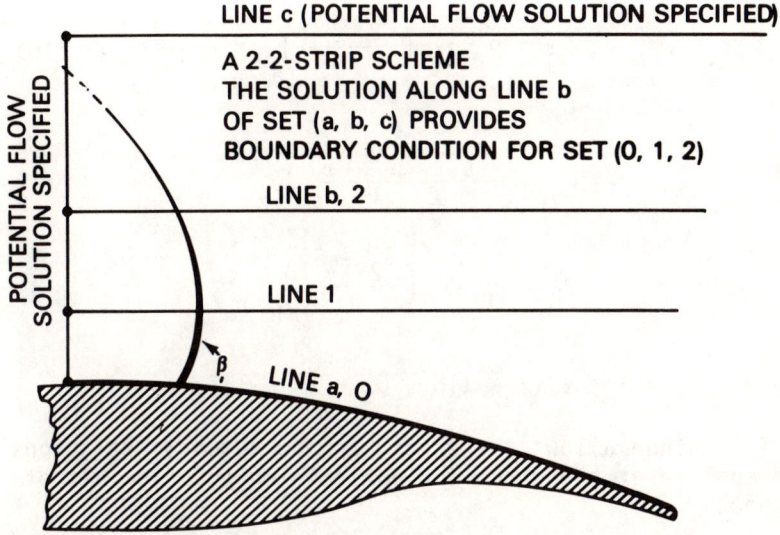

Fig. 2 Integration scheme for the method of integral relations in the shock region.

flowfield is approximated in the usual way by a second-order polynomial. The MIR is first applied to the boundaries a, b, c with the purpose of determining the flow conditions along line b. On the uppermost boundary c, flow conditions are specified by the potential flow solution. Finally, MIR is applied in the more disturbed part of the field along the boundaries 0, 1, and 2.

The resulting differential equations take on the form along boundaries a and b, for example,

$$\frac{dU_a}{dx} = f_1(x,y,U_a,V_a,\ldots) \tag{11}$$

$$\frac{dV_a}{dx} = f_2(x,y,U_a,V_a,\ldots) \tag{12}$$

$$\frac{dU_b}{dx} = g_1(x,y,U_b,V_b,\ldots) \tag{13}$$

$$\frac{dV_b}{dx} = g_2(x,y,U_b,V_b,\ldots) \tag{14}$$

$$\rho_a = \left(\frac{C_s - U_a^2 - V_a^2}{C - 1}\right)^{\frac{1}{\gamma - 1}} \tag{15}$$

$$P_a = \rho_a^\gamma \exp\left(\frac{S_2 - S_1}{c_v}\right)_a \tag{16}$$

$$\rho_b = \left(\frac{C_s - U_b^2 - V_b^2}{C - 1}\right)^{\frac{1}{\gamma - 1}} \tag{17}$$

$$P_b = \rho_b^\gamma \exp\left(\frac{S_2 - S_1}{c_v}\right)_b \tag{18}$$

where $C = 1 + 2/(\gamma - 1)M_\infty^2$; $C_s = (C - 1)P_2/\rho_2$

The method allows the exact Rankine-Hugoniot relations to be applied at the shock for determining the flow aft of the shock

$$\rho_2 = \rho_1 \left[\frac{(\gamma + 1) M_1^2 \sin^2 \beta}{(\gamma - 1) M_1^2 \sin^2 \beta + 2}\right] \tag{19}$$

$$P_2 = P_1 \left[1 + \frac{2\gamma(M_1^2 \sin^2\beta - 1)}{\gamma + 1} \right] \quad (20)$$

where β is the oblique shock angle and subscripts 1 and 2 designate flow properties before and after the shock. The increase in entropy is obtained in terms of P_2 and ρ_2:

$$S_2 - S_1 = c_v \ln(P_2/\rho_2^\gamma) \quad (21)$$

where $S_2 - S_1$ varies from one streamline to the other.

Solution Procedure

First, the potential flow solution is sought by the FDR scheme developed by Carlson.[12] For supercritical flow with a moderate or strong shock wave, the flow properties so calculated will be valid throughout the whole flowfield, except for the shock-wave region which is shaded in Fig. 1. The values at the boundaries of the shaded region are then taken as the initial condition (those along the vertical line) and the boundary condition (those along the horizontal line) for the numerical solution of the Euler equations by MIR. It is noted that other finite-difference codes such as those by Bauer et al[3] or Jameson,[4] should be equally applicable to the present procedure.

With the initial and boundary conditions provided, the MIR can be readily implemented similar to the procedure developed earlier.[10] In the present case, however, the overall procedure for using the MIR is substantially simplified by:

1) Elimination of iteration processes for determining the sonic point (a saddle-type singularity) and the upstream flow;

2) Reduction of the integration domain.

Thus, the essential difficulty in using the MIR (item 1) is removed and the numerical stability problem associated with a large number of strips (item 2) can be avoided. A complete solution procedure now consists of two (instead of five before) iteration processes, which are illustrated in Fig. 3 and described in the following two sections. Also, in treating the shock wave, a representation of the shock geometry is required. It is briefly discussed in the third section.

Determination of the Shock Location

The procedure for determining the shock-wave location was based on the hypothesis that the shock-wave location is determined by the condition whereby the flow returns to its undisturbed state sufficiently far downstream. During the iteration process, a number of shock foot locations are assumed and the Rankine-Hugoniot relations applied. After the shock wave, the integration resumes down to the trailing edge and finally through the downstream region. The results for each boundary condition is satisfied, i.e., whether the flow based on a particular shock location approaches a uniform state in the far downstream. The shock location that meets this criterion is considered "correct" and the others "incorrect."

In practical computations, however, a complete uniform flow cannot be obtained because of accumulated numerical errors. Under these circumstances, a general approach is to consider that the solution is satisfactory when the freestream value can be bracketed by two integral curves based on two shock locations. This is shown in Fig. 4 where typical pressure distributions along the y = 0 boundary are plotted for several assumed shock locations on a Garabedian-Korn airfoil at M_∞ = 0.752. Attention is given to the curves based on shock locations at x/c = 0.691 and 0.696. Although the difference in the shock locations is minimal, the difference in the resulting pressures is not. It is observed that the freestream value is bracketed in these two sets of curves. It is also interesting to note that the freestream pressure is bracketed not only along the y = 0 boundary but also along other intermediate strip

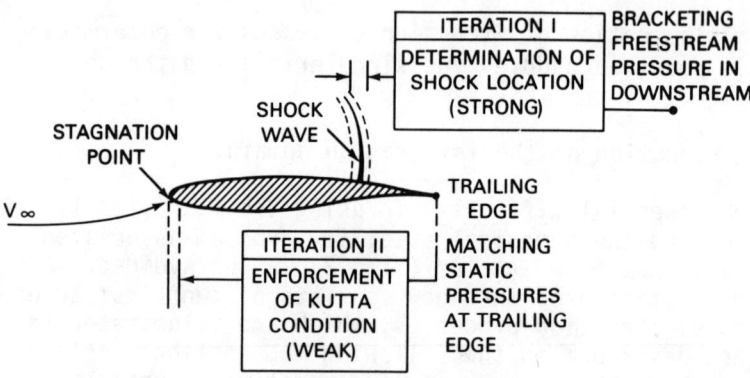

Fig. 3 Solution procedures for the hybrid method.

boundaries. The results imply that the downstream boundary condition is satisfied for all the strips. The actual shock location, therefore, lies somewhere between x/c = 0.691 and x/c = 0.696.

The principle behind the present iteration is that satisfaction of the downstream flow condition implies a downstream influence on the entire flow. The downstream influence propagates to the upstream throughout the whole flowfield in subcritical flows. In a supercritical flow, however, part of the influence propagates up to and stops at the shock wave because the embedded supersonic region in front prevents the propagation of any influence further upstream. In fact, because it is close to the surface, this part of the influence contributes to the major portion of the downstream feedback. That is why the downstream flow is so closely affected by the shock-wave location.

Enforcement of Kutta Condition

The Kutta condition is satisfied by matching the static pressures at the trailing edge. The surface pressure at the trailing edge, calculated by the MIR, generally differs from that of the potential flow solution of the FDR scheme. To match the altered upper surface pres-

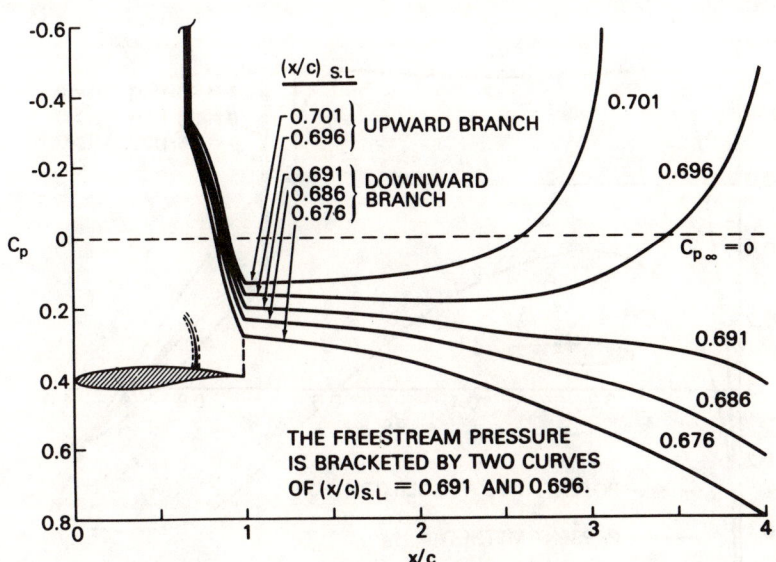

Fig. 4 Downstream pressure distribution along y = 0 for a Garabedian-Korn airfoil at M_∞ = 0.752 and α = 0.

sure with the original potential pressure on the lower surface at the trailing edge, the potential flow solution would have to be recalculated with a new circulation accounting for the pressure difference. For purposes of this matching, the pressure values should be those which satisfy the imposed and downstream boundary conditions. That is, they should be the converged values from the iteration for the determination of the shock location. Experience in using MIR indicates that this iteration is rather insensitive. Inasmuch as it involves the converged shock location, the present iteration, therefore, provides an overall downstream influence resulting from the change of the shock condition to the upstream region.

Shock Geometry

For a supercritical flow with moderate or strong shock waves, the embedded supersonic region may cross more than one strip boundary. It is, therefore, necessary to assume the shape of the shock wave in order to locate the proper station to apply the shock relations at the intermediate strip boundary.

Generally, the shock wave is curved in a nonlinear flowfield. To be consistent with MIR, a quadratic form for the shape of the shock wave is suggested

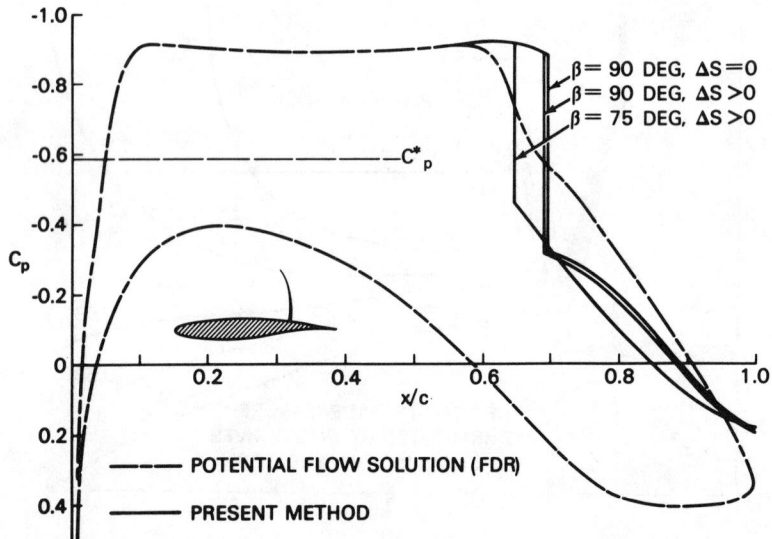

Fig. 5 Pressure distribution on a Garabedian-Korn airfoil at M_∞ = 0.752 and α = 0.

$$\frac{x}{c} = a_0 + a_1 Y + a_2 Y^2 \qquad (22)$$

where $Y = (y - y_0)/c$. Unknown coefficients a_0, a_1, and a_2, can be determined by two conditions at the shock foot (the location and slope of the shock foot) and one condition at the nearest strip boundary (the local slope of the shock).

For inviscid flows, the shock foot should be normal to the airfoil surface, i.e., $\beta = 90$ deg. However, previous experience shows that a better correlation between the theory and the experiment may be achieved for $\beta < 90$ deg. The lower the β value, the better the correlation. The result is not too surprising for the reason that the shock wave is always leaning to the wall experimentally, so that $\beta < 90$ deg may account for the influence of viscosity to a certain extent. Of course, any β less than 90 deg would be of an empirical nature and has no theoretical justification in a purely inviscid theory. Nevertheless, because of its simplicity in simulating the viscous effect, a variable β will be used to parametrically study the flow phenomenon.

Results and Discussion

Numerical results were obtained by embedding MIR for the shock region in Carlson's finite-difference grids using a Cartesian coordinate system. The flows over a Garabedian-Korn airfoil at $M_\infty = 0.752$ and $M_\infty = 0.78$ were considered. Three strips were employed for the MIR solution in the shock region. The MIR calculation started near the middle of the chord, sufficiently far enough to include the shock. At the shock, the Rankine-Hugoniot relations were applied, with or without entropy change. Also, with entropy change ($\Delta S > 0$), the effect of shock angle, β, was examined.

The results indicate that the surface pressures calculated by the MIR agree closely with the original FDR solution for the first few steps and then deviates after $x/c = 0.55$, as shown in Fig. 5. The flow then decelerates at a somewhat slower rate than the FDR solution before the shock. The case of $\beta = 90$ deg and $\Delta S = 0$ corresponds to a full potential flow with an isentropic jump at the shock. An increase in entropy moves the shock wave forward. It is of interest to note that the entropy increase has a similar effect as the nonconservative form of the finite-differencing scheme. Furthermore, in the case of an oblique shock, the smaller the shock angle, the more forward the shock location. As the oblique angle decreases, the MIR result tends to simulate the effect of viscosity of a physical flow.

The shock location was determined by the hypothesis discussed in the preceding section. To allow sufficient variation of the downstream flow for bracketing purposes, the FDR data were no longer used as the boundary condition beyond the trailing edge. Instead, the MIR integration domain was extended to the outer boundary of the FDR scheme and the freestream conditions was applied there. The downstream results, shown in Fig. 4, have been discussed in the section entitled Determination of Shock Location and, therefore, will not be repeated here.

The boundaries of the main MIR integration domains, i.e., the shock region, seem to be properly specified by the FDR data. The basic features of the Euler equations, i.e., the effect of entropy change can be easily evaluated by the present method. This is evidenced by comparing the present results with those obtained previously by solving the whole Euler flowfield with MIR.[10] Both indicate that if there is an increase in entropy across the shock wave, its location moves forward. Also, the qualitative trend of the effects of oblique shock angle in the present results is consistent with the earlier data.[10] A replot of the previous Euler-MIR results on the same airfoil at approximately the same Mach number (Fig. 6 in Ref. 10), in the present Cp scale is shown in Fig. 6. Both solutions indicate that as β decreases, the shock waves move forward. A smaller β value seems to have simulated the effect of viscosity more closely.

The advantage of using the Euler solution becomes more apparent as the freestream Mach number increases. This is indicated by considering the same airfoil at M_∞ = 0.78. Numerical results for this case, along with experimental data of Ref. 6, are shown in Fig. 7. It is observed that 1) the difference between the isentropic jump (β = 90 deg, ΔS = 0) and the Rankine-Hugoniot jump (β = 90 deg, ΔS > 0) is more distinctive, and 2) the agreement between the numerical results and experimental data is considerably improved by using the present method. As the freestream Mach number further increases, the FDR potential flow solution would become more erroneous and the use of the present method would be more justified and desirable.

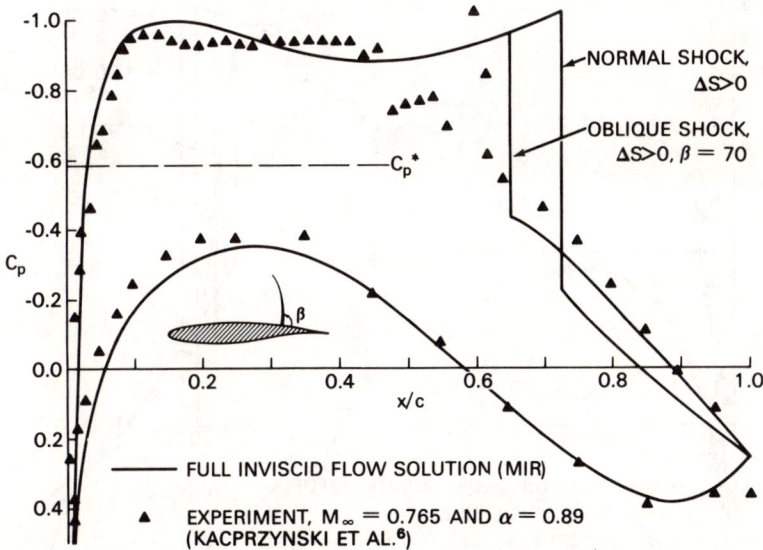

Fig. 6 Pressure distribution on a Garabedian-Korn airfoil at M_∞ = 0.75 (from Ref. 10).

In both cases, the static pressure at the trailing edge is considerably reduced. The recalculation of the potential flow solution by FDR has not been reflected in Figs. 6 and 7. If the updating of the trailing edge condition had been completed, the lower surface pressure near the trailing edge would have decreased accordingly. It would generally improve agreement between the theoretical results and the measurements because numerical predictions often yield higher pressure values than experimental data in this region.

Conclusions and Recommendations

A hybrid method which uses a finite-difference relaxation technique for solving the overall transonic potential flowfield and the method of integral relations for obtaining the Euler solution in the shock region is formulated. Based on results from two test cases, it is found that:

1) The essential feature of the Euler solution, i.e., the effect of entropy change across the shock wave, can be found by the present method. The integration domain for the Euler equations seems to be properly specified by the potential flow data along its boundaries.

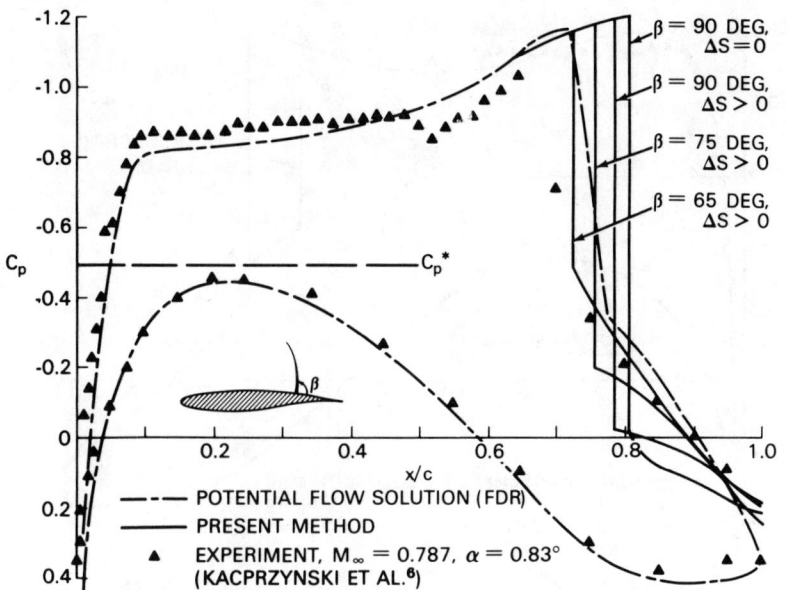

Fig. 7 Pressure distribution around a Garabedian-Korn airfoil at $M_\infty = 0.78$ and $\alpha = 0$.

2) For supercritical flows with moderate to strong shock waves, the difference between a full potential flow (isentropic) and full inviscid flow (nonisentropic) is rather distinctive and the use of the present method would be more justified and desirable.

The approach yields an approximate solution to the Euler equations in a relatively simple manner. Refinement and further development of the method are recommended:

1) Develop a scheme that will effectively adjust the static pressure of the lower surface at the trailing edge to match the new upper surface pressure resulting from the change of shock condition.

2) Integrate the total pressure around the shock surface to determine the theoretical wave drag.

3) Develop numerical logics that will help automate the complete solution procedure with the finite-difference potential flow programs into an integrated computer code. The increase in program size and computation time is estimated to be about 25% of the original FDR program.

Acknowledgment

This research was supported by the Naval Air Systems Command (AIR-320D, AIRTASK 9R023-02) under the cognizance of D. Kirkpatrick, and by the Independent Research program at the David Taylor Naval Ship Research and Development Center (Work Unit 1606-105).

References

[1] Emmons, H.W., "Flow of a Compressible Fluid Past a Symmetrical Airfoil in a Wind Tunnel and Free Air," NACA TN 1746, Nov. 1948.

[2] Murman, E.M., "Analysis of Embedded Shock Waves Calculated by Relaxation Methods," AIAA Journal, Vol. 12, No. 5, 1974, pp. 626-632.

[3] Bauer, F., Garabedian, P., Korn, D., and Jameson, A., Supercritical Wing Sections II, Springer-Verlag, New York, 1975.

[4] Jameson, A., "Transonic Potential Flow Calculations using Conservative Form," Proceedings of AIAA 2nd Computational Fluid Dynamics Conference, Hartford, Conn., 1975, pp. 148-161.

[5] Ballhaus, W.F. Jr., "Some Recent Progress in Transonic Flow Computations," presented at the Lecture Series on Computational Fluid Dynamics, von Karman Institute for Fluid Dynamics, Belgium, March 1976.

[6] Kacprzynski, J.J., Ohman, L.H., Garabedian, P.R., and Korn, D.G., "Analysis of the Flow Past a Shockless Lifting Airfoil in Design and Off-Design Conditions," National Research Council of Canada, Ottawa, Canada, Aero Rept. LR-544, Nov. 1971.

[7] Holt, M. and Masson, B.S., "The Calculation of High Subsonic Flow Past Bodies by the Method of Integral Relations," Proceedings of the Second International Conference on Numerical Methods in Fluid Dynamics, Vol. 8, Springer-Verlag, New York, 1971, pp. 207-214.

[8] Melnik, R.E. and Ives, D.C., "Subcritical Flows of Two-Dimensional Airfoils by a Multistrip Method of Integral Relations," Proceedings of the Second International Conference on Numerical Methods in Fluid Dynamics, Vol. 8, Springer-Verlag, New York, 1971, pp. 243-251.

[9] Sato, J., "Application of Dorodnitsyn's Technique to Compressible Two-Dimensional Airfoil Theories at Transonic Speeds," National Aerospace Lab., Tokyo, Japan, Rep. TR-220T, Oct. 1970.

[10] Tai, T.C., "Transonic Inviscid Flow over Lifting Airfoils by the Method of Integral Relations," AIAA Journal, Vol. 12, No. 6, June 1974, pp. 798-804; also NSRDC Rept. 3424, July 1971.

[11]Tai, T.C., "Theoretical Calculation of Viscous-Inviscid Transonic Flows," presented at the Lecture Series on Shock-Boundary Layer Interaction in Turbomachine von Karman Institute for Fluid Dynamics, Belgium, June 1980; also DTNSRDC Rept. 80/104, Aug. 1980.

[12]Carlson, L.A., "Transonic Airfoil Flowfield Analysis Using Cartesian Coordinates," NASA CR-2577, Aug. 1975; also in Journal of Aircraft, Vol. 13, No. 5, May 1976, pp. 349-356.

Chapter XVII.

Application of a Shock-Turbulent Boundary-Layer Interaction Theory in Transonic Flowfield Analysis

G. R. Inger[*]
University of Colorado, Boulder, Colo.

Introduction

Shock/boundary-layer interaction can significantly influence not only the local transonic flow on missiles, wings, and turbine blades but its influence can also extend downstream within the boundary layer and thereby alter the global aerodynamic properties of lift, drag, and pitching moment. It is therefore important that shock/boundary-layer interactions and their Reynolds and Mach number scaling be properly modeled in engineering flowfield prediction methods for supercritical aerodynamic bodies. This paper describes the application of a nonasymptotic triple deck theory of transonic shock/turbulent boundary-layer interaction which provides such a tool for nonseparating two-dimensional flows over a wide range of practical Reynolds numbers; contains a brief description of the essential features of the theoretical model; and describes how the results of a comprehensive parametric study of this theory may be used to develop a generalized "viscous wedge" model of the local interaction which embodies proper scaling behavior as well as an approximate account of incipient separation that is in good agreement with experimental trends. We then examine the application of this theory as an element in global viscous flowfield analyses of supercritical airfoils. In such problems it will be shown that even in nonseparating cases the changes across the interaction may significantly alter the subsequent turbulent boundary-layer

---

Presented at the Transonic Perspective Symposium, NASA/Ames Research Center, Moffett Field, Calif., Feb. 18-20, 1981. Copyright © 1981 by George R. Inger. Published by the American Institute of Aeronautics and Astronautics with permission.
[*]Professor and Chairman, Department of Aerospace Engineering Sciences.

behavior for appreciable distances, especially when larger downstream adverse pressure gradients are present.

Brief Outline of the Local Interaction Theory

Unlike significantly separated flow where the disturbance flow pattern associated with a nearly normal shock/boundary-layer interaction is a very complicated one involving a bifurcated shock pattern,[1] the unseparated case pertaining to turbulent boundary layers up to roughly $M_1 \simeq 1.3$ has instead a much simpler type of interaction pattern which is more amenable to analytical treatment. With some judicious simplifications, it is possible to construct a fundamentally based approximate theory of the problem in the latter case. Consider a known adiabatic boundary-layer profile $M_o(y)$ subjected to small transonic disturbances due to an impinging weak and nearly normal shock. In the practical Reynolds number range of interest here ($10^5 \leq Re_L \leq 10^8$), it has been established[2-4] that the local interaction disturbance field in the neighborhood of the impinging shock organizes itself into three basic layered regions or "decks" (Fig. 1): 1) an outer region of potential inviscid flow above the boundary layer, which contains the incident shock and interactive wave systems; 2) an intermediate deck of frozen shear stress rotational inviscid disturbance flow occupying the outer 90% or more of the incoming boundary-layer thickness; 3) an inner shear disturbance sublayer adjacent to the wall which accounts for the interactive skin friction perturbations (and hence any possible incipient separation) plus most of the upstream influence of the interaction. The "forcing function" of the problem here is thus impressed by the outer deck upon the boundary layer; the middle deck couples this to the response of the inner deck but in so doing can itself modify the disturbance field to some extent, while the slow viscous flow in the thin inner deck reacts very strongly to the pressure gradient disturbances imposed by these overlying decks. In treating this interactive field we employ a nonasymptotic method[5] that is an extension to turbulent flow of Lighthill's approach[6] because of its essential soundness and adaptability to practical engineering problems, similarity to related types of multiple deck approaches that have proven highly successful in treating turbulent boundary-layer response to strong adverse pressure gradients, and the large body of turbulent boundary-layer interaction data plus recent numerical studies with the full Navier-Stokes equations which support the predicted results in a variety of

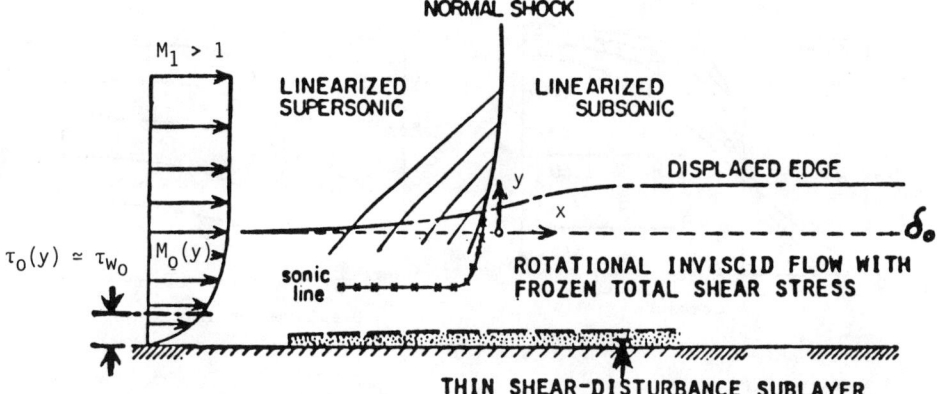

Fig. 1 Triple-deck structure of the local disturbance field in a transonic interaction.

problems (see the survey in Ref. 5). Moreover, this approach provides at realistic Reynolds numbers a treatment of the inner deck pressure gradient terms plus the middle deck $\partial p/\partial y$ and streamline divergence effects, along with simplifying approximations that render the resulting theory tractable from an engineering standpoint.

A very detailed description of the abovementioned nonasymptotic triple deck analysis can be found in Refs. 5 and 7 and hence will not be given here. The resulting predictions, such as typically illustrated in Fig. 2, describe all the essential global features of the mixed transonic character of the nonseparating normal shock/turbulent boundary-layer interaction problem including the interactive pressure distribution and upstream influence, displacement thickness and local shape factor, and interactive skin friction up to incipient separation. This interaction theory employs for the incoming turbulent boundary-layer velocity profile a very general Composite Law of the Wall-Law of the Wake profile model due to Walz,[8] which is characterized by three parameters (M_1, boundary-layer thickness Reynolds number and the incoming shape factor). The influence of both shock obliquity[9] and wall curvature[10] have also been examined in detail and incorporated into the theory. Very extensive parametric studies and detailed comparisons with experiment have shown that it gives a very good account of all the important engineering features of the interaction over a wide range of Mach and Reynolds number conditions. Moreover, the important but heretofore ignored influence of in-

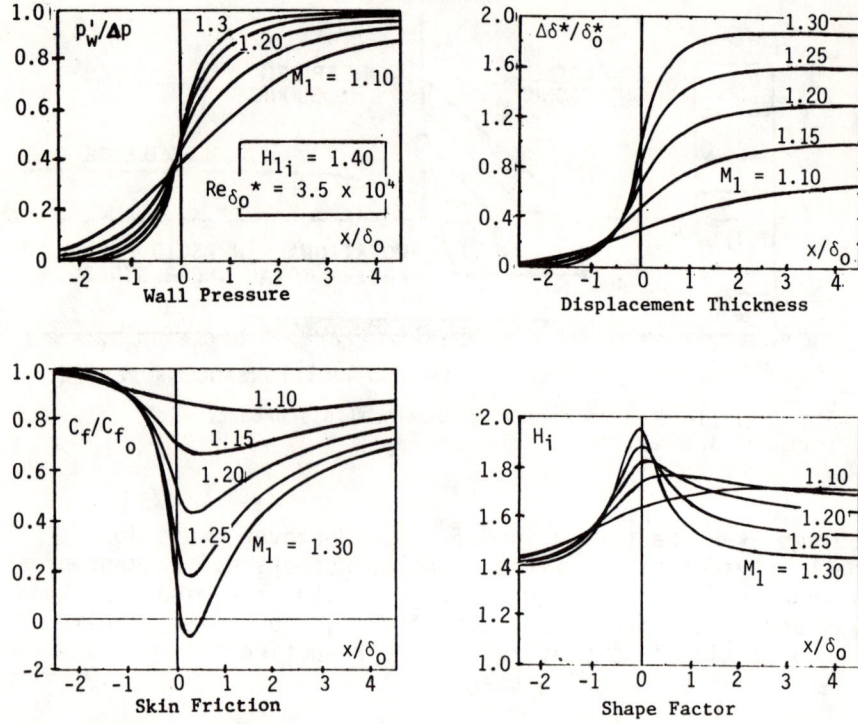

Fig. 2 Typical features of normal shock/turbulent boundary-layer interactions without separation.

coming boundary-layer shape factor H_{1_i} (and hence the upstream pressure gradient history) has been determined.[5,7]

Generalized Viscous Ramp Model of the Interaction

In certain engineering applications to global flow-field analysis computer programs for wings or turbine blades, it has proved expedient to model the interaction as a simple local inviscid δ^* - "bump" or "ramp." A serious deficiency of this approach is that it does not account for the dependence of the bump shape and size on Reynolds number, shock strength and boundary-layer shape factor, while the additional interaction effects on the downstream boundary layer (such as C_f reduction) are ignored altogether. With the aforementioned parametric study results in hand, however, the present theory provides a much improved "vis-

APPLICATION OF SHOCK BOUNDARY-LAYER INTERACTION 625

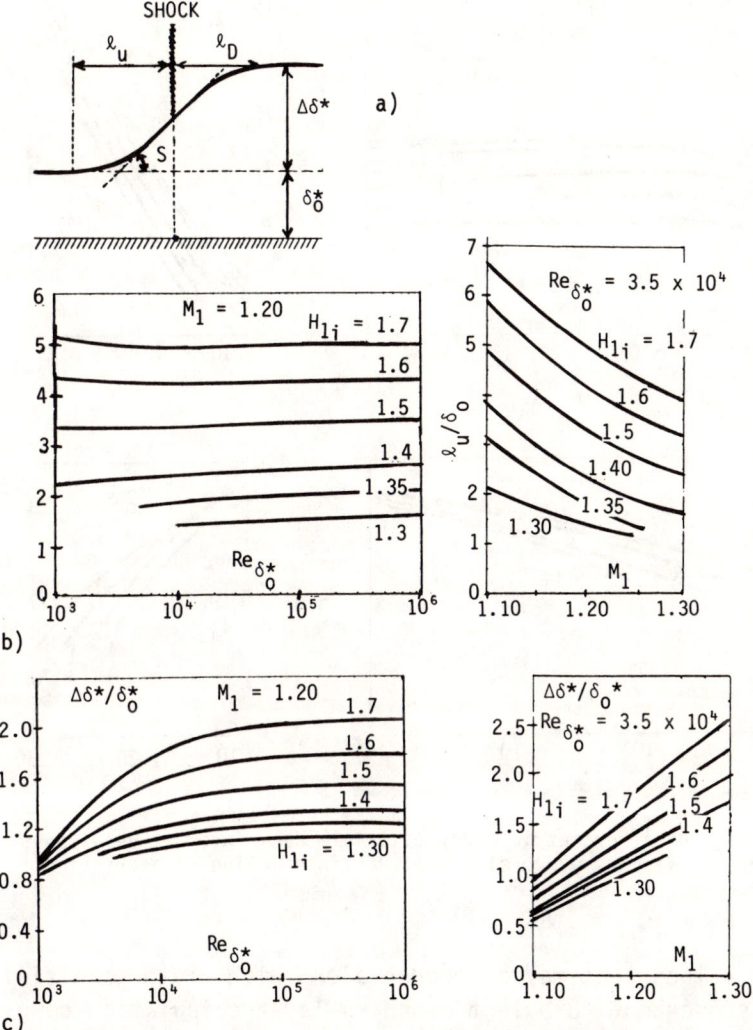

Fig. 3 Parametric study of triple deck interactive viscous wedge:
a) Parametric study of viscous wedge interaction model; b) Upstream influence distance; c) Overall displacement thickening.
(Figure 3 continued on next page)

cous ramp" representation of the interaction (see Fig. 3a) whose key physical features have the correct dependence on M_1, Re_{δ^*}, and H_{1i}.

Results for these viscous wedge properties taken from Ref. 11 are presented in Figs. 3b-3e, where the upstream and downstream influence distances, the slope and overall height of the δ^* - bump are plotted along with the downstream

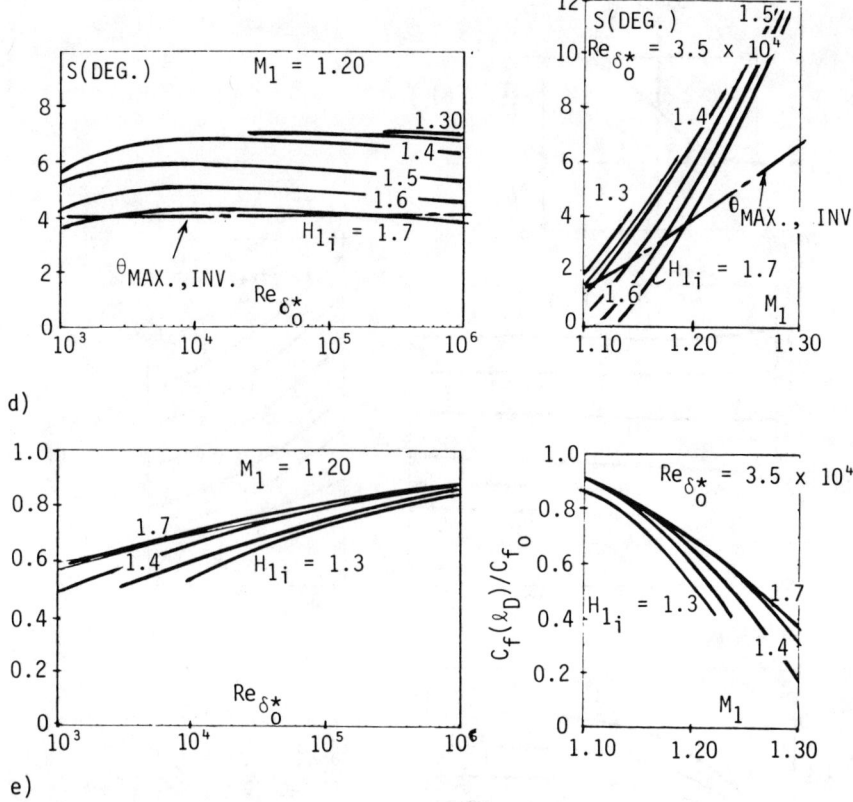

Fig. 3 (cont.) Parametric study of triple deck interactive viscous wedge: d) Viscous wedge slope; e) Postinteractive skin friction level.

C_f/C_{f_0} values that may be needed along with δ^* to reinitialize a subsequent turbulent boundary-layer calculation downstream. We note in general that the overall scale of the interaction (which can be a sensitive effect in both steady and unsteady applications where such viscous wedge models are employed) does not scale according to the undisturbed boundary-layer thickness even in the nonseparating case. It is further noted that the viscous wedge slopes are in rough agreement with the maximum attached shock deflection value observed empirically,[12] although here of course there is a dependence on Re_δ and H_{1_j} as well as Mach number. Finally, in all of these curves we see a significant dependence on the incoming boundary-layer shape factor that would appear to be an important consideration in practical applications.

Closed-form analytical fits to the various curves of Fig. 3 have been developed[11] which provide a very rapid yet complete modelling of the local interaction effects for incorporation as a local module in global inviscid boundary-layer analysis or design programs for supercritical airfoils.

Incipient Separation

The present theory, although it breaks down at separation, does yield a useful indication of incipient separation where $C_{f_{min}} \to 0$, owing to the particular attention paid to the treatment of the local interactive skin friction behavior.[5] Since this indication is of great practical interest, a parametric study of incipient separation conditions inherent in the present theory was carried out; the results for a normal shock on a flat surface are presented in Fig. 4a where the shock Mach number above which incipient separation occurs is plotted as a function of the Reynolds number with the shape factor as a parameter; also shown in the figure is the approximate experimental boundary determined by a careful examination[11] of a large number of transonic interaction tests, besides Nussdorfer's[13] $M \sim 1.30$ criterion for turbulent flow. It is seen that the theoretical prediction of a gradual increase in the incipient separation Mach number value with Reynolds number is in agreement with the trend of the data; moreover, the theoretical prediction of only a small influence of shape factor on the incipient separation conditions is also borne out by the lack of any consistent H effect for the same Re discernible in the data (Squire has observed a similar insensitivity to H_{1_i} in purely supersonic flow interactions[14]). The absolute values of the incipient separation Mach number predicted by the present interaction theory are seen to be consistently slightly lower than the average experimental value; this is attributable to the combined effects of the linearized inner deck theory (which overpredicts the pressure gradient effect on C_f) and the assumption of a normal shock when in fact most of the experiments likely entail some shock obliquity (which also delays separation to somewhat higher shock-strengths). It is interesting to note that Nussdorfer's[13] original incipient separation criterion, based as it was on a very limited base, does roughly go through the average of the data although it does not account for the proper Reynolds number effect.

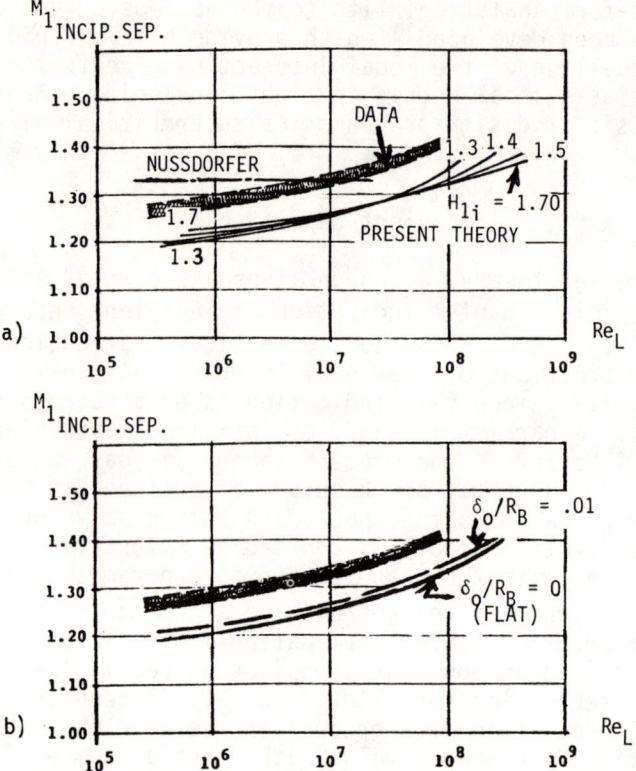

Fig. 4 Incipient separation condition: a) Flat surfaces; b) Curved-wall effect (H_{1_i} = 1.40).

Figure 4b shows the influence of wall curvature; it is seen to have only a small effect in delaying incipient separation to a slightly higher shock strength for a given Re_L and H_i, this being of the same order as the experimental data band.

Application to Global Flowfield Analysis

Downstream Effects from a Shock/Boundary-Layer Interaction

In addition to the increased displacement thickness, the foregoing discussion shows that the skin friction level following the interaction is significantly reduced; combined with the attendant sensitivity to the profile shape, this suggests that the subsequent downstream boundary-layer development may retain a memory of the interaction effects for a considerable distance (over and above a simple thickening), particularly as regards possible incipient separation in any adverse pressure gradients downstream. As indicated in Fig. 5 this downstream "interaction aftereffect" in the boundary layer influences the sensitive trailing edge region and thus may be important in the design and analysis of rear loaded airfoils, especially at higher lift coefficients with increasingly aft shock locations; it likewise may be important on three dimensional wing configurations where the shock interaction zones are well aft.

The aforementioned aftereffect question was therefore subjected to detailed study using the two-layer turbulent boundary-layer program of Moses[15] as a model of the downstream viscous flow; the program is coupled to the present interaction theory by initializing it with the postinteractive flow properties so as to account fully (both C_f and δ^*), partially (δ^* only) or not at all for the changes across the interaction. Calculations were then made of the subsequent downstream turbulent boundary-layer behavior (H, C_f, θ^*, δ^*) in various constant postshock adverse pressure gradients typical of airfoils for different assumed local interactive shock strengths and positions or Reynolds numbers. The results serve as a paradigm of the downstream sensitivity question in real flows.

A variety of cases were studied,[16] typical results of which are presented in Fig. 6, where we show the predicted behavior of the boundary-layer shape factor and skin friction in three increasingly strong adverse pressure gradients downstream of an interaction occurring at a typically rearward position; the consequences of fully, partially, or

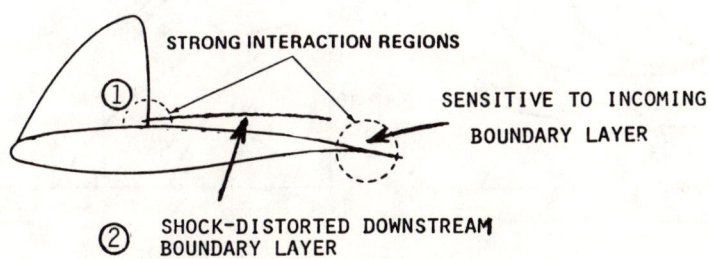

Fig. 5 Global viscous inviscid interaction problem for supercritical airfoils (schematic).

negligently treating the boundary-layer changes across the interaction are indicated. Generally, it is seen that the downstream behavior of the boundary layer is indeed sensitive to detailed modelling of the interactive effects and that this sensitivity increases with the strength of the downstream adverse pressure gradient. The adverse pressure gradient magnifies the subsequent influence of the skin friction (as well as the δ^* rise) across the interaction so that downstream separation tends to occur earlier than would be predicted by either neglecting or treating only the δ^* effect of the upstream interaction. As shown in Fig. 7, these predictions are supported by a comparison with boundary-layer measurements downstream of a nonseparating shock interaction zone on a supercritical airfoil; both the skin friction and shape factor data are poorly predicted when the interaction is neglected but are reasonably well predicted when the complete interaction effects are taken into account.

Examination of many such results leads to the further conclusion that such interactive aftereffects extend at least 20%-30% chord distances downstream on a typical air-

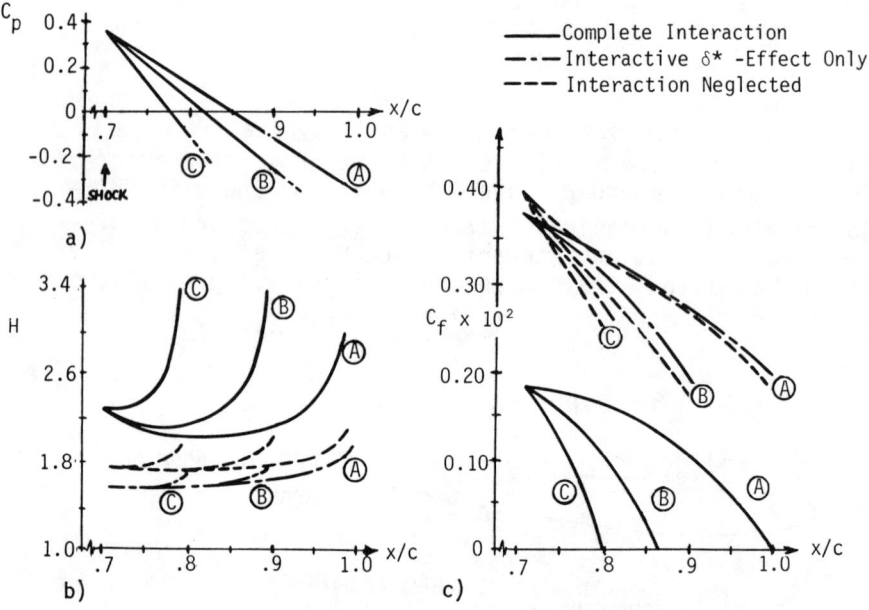

Fig. 6 Sensitivity study of interaction effects on downstream turbulent boundary-layer behavior: a) Postshock pressure gradients; b) Shape factor; c) Skin friction

foil or wing and increase (as expected) with either larger shock strength or decreasing Reynolds number. If the trailing edge region lies within this range of the shock, it is thus seen that a simple thickening effect alone is not sufficient to account for the interaction and may result in an inaccurate prediction of the rearward boundary-layer shape factor, skin friction, and incipient separation properties including their scaling. This is of practical importance for two major reasons: 1) in regions of sustained adverse pressure gradient that often follow the short scale interaction zone, the shape of the velocity profile and streamwise shear stress distribution (as well as thickness) are of considerable importance to the aerodynamic design of an airfoil or wing; 2) the altered boundary-layer properties (especially possible incipient separation) near the trailing edge and into the wake can further exert a powerful effect on the overall aerodynamics via their influence of the Kutta condition[17] and on possible buffet onset.

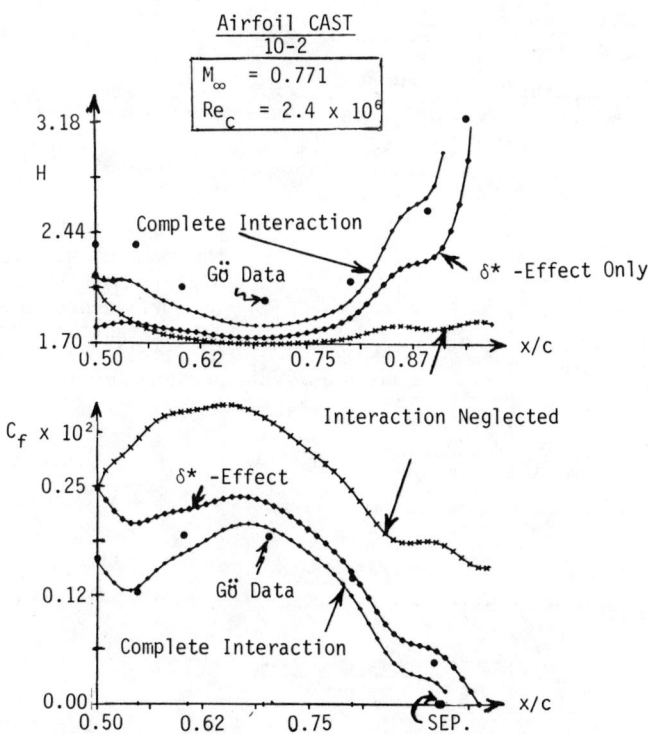

Fig. 7 Comparison of postinteraction turbulent boundary-layer predictions with experiment on a supercritical airfoil.

Supercritical Wing Section Flowfields

Nandanan et al.[12] have carried out an even more detailed study of interactions on actual supercritical airfoils including experimental comparisons. They developed a global computational method for transonic airfoil flow analysis which incorporates the present analytical solution for near normal shock/boundary-layer interaction into a state-of-the-art viscous inviscid computation code. Theoretical results obtained with this method were compared to representative data from boundary-layer and surface pressure measurements on three transonic airfoils in the DFVLR-AVA (Göttingen) Transonic Wind Tunnel; some examples of these comparisons are shown in Figs. 8a and 8b. The agreement between theory and experiment in both the boundary-layer dis-

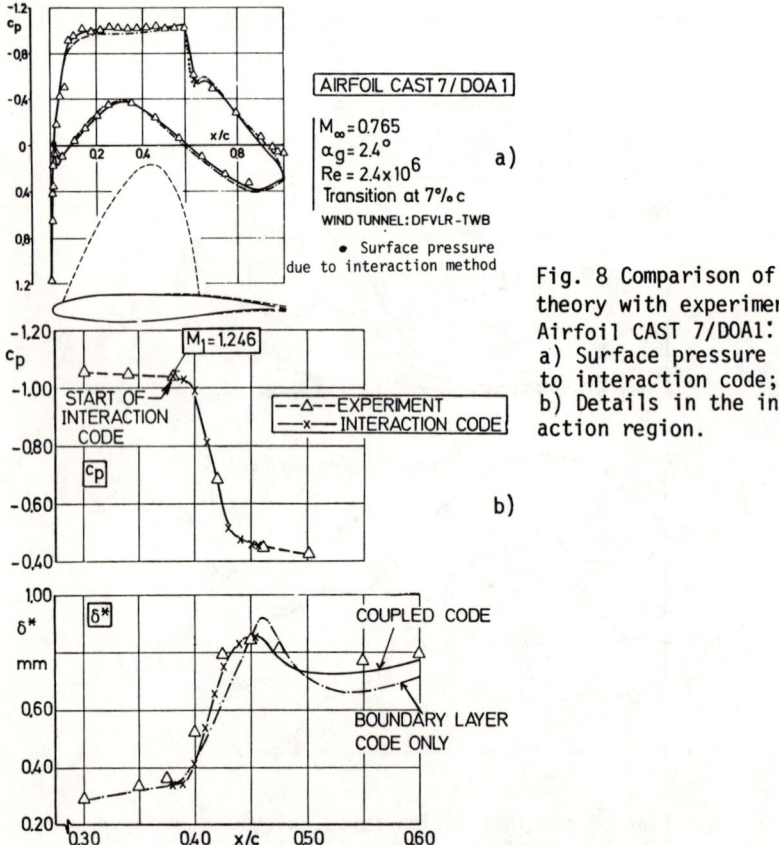

Fig. 8 Comparison of present theory with experiment. Airfoil CAST 7/DOA1:
a) Surface pressure due to interaction code;
b) Details in the interaction region.

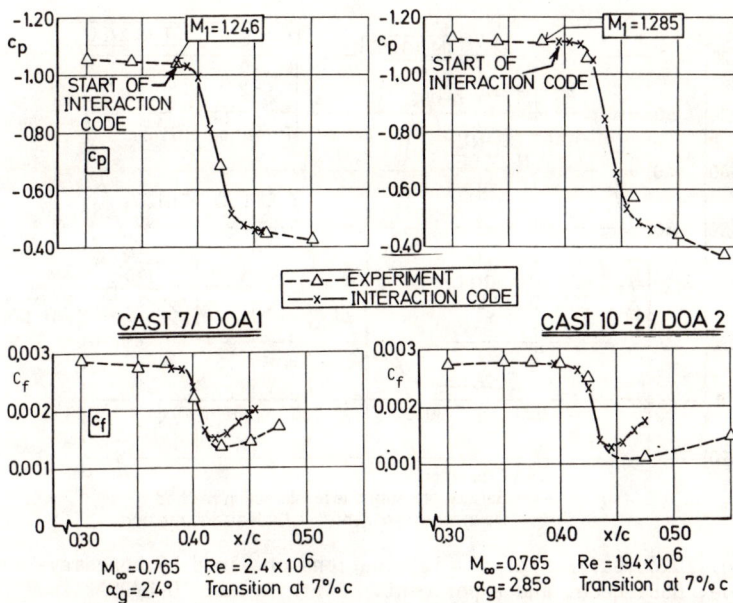

Fig. 9 Comparison of present interaction theory with experimental data of Stanewsky (DFVLR-Gö).

placement thickness and the surface pressure distributions was, for all test cases considered, quite good. The associated predictions of the local skin friction variation through the interaction zone also agree reasonably well with the values inferred from the experimental boundary-layer profiles via the Ludwig-Tillman relation (see, e.g., Fig. 9).

The results of this investigation indicated that treating the shock/boundary-layer interaction by conventional boundary-layer theory generally leads to a slight underprediction of the displacement thickness immediately downstream of the shock and, due to the amplifying effect of the sustained rear adverse pressure gradients, to an appreciable underestimation of the displacement thickness at the trailing edge (see Fig. 10). The latter is also clearly reflected in the pressure distribution and aerodynamic coefficients compared. Considering these results, one may conclude that it is generally necessary to include a physically correct treatment of shock wave/boundary-layer interaction in the analysis of transonic airfoil flow.

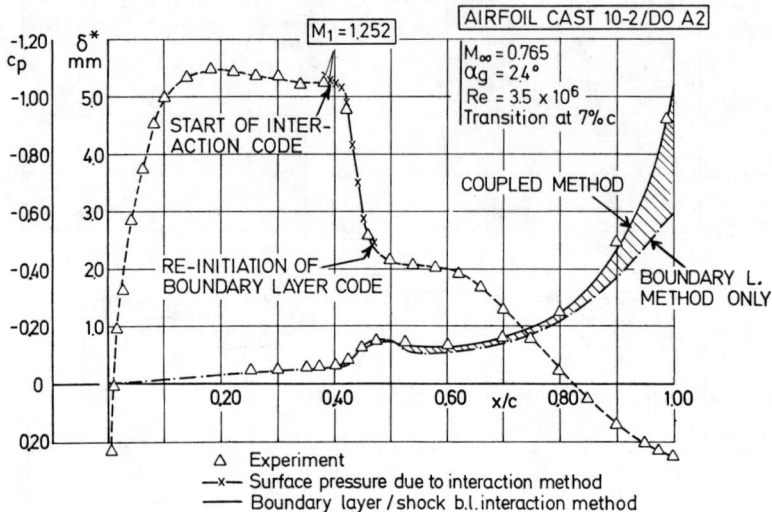

Fig. 10 Comparison of δ* distributions obtained by boundary-layer code, BSBI-code, and experiment. Airfoil CAST 10-2/DOA2.

Concluding Remarks

The results of this study have shown that it is now possible to incorporate as an interactive module within a global flowfield analysis the correctly modelled (and scaled) local shock/boundary-layer interaction effects for the nonseparating case. The nonasymptotic triple deck interaction theory involved covers a wide range of practical Reynolds numbers and turbulent boundary-layer profile shape factors including the effect of wall curvature; moreover, it gives an approximate indication of when incipient separation occurs. It was further shown that such theory is generally desirable when accurate predictions are desired in the important trailing edge region of rear loaded supercritical airfoils because the detailed changes across an upstream interaction can significantly alter the subsequent turbulent boundary-layer behavior for appreciable distances downstream.

References

[1] Ackeret, J., Feldman, F., and Rott, N., "Investigation of Compression Shocks and Boundary Layers in Gases Moving at High Speed," NACA TN-1113, 1947.

[2] Stewartson, K. and Williams, P. G., "Self-Induced Separation," Proceedings of the Royal Society A, Vol. 312, 1969, pp. 181-206.

[3] Inger, G. R. and Mason, W. H., "Analytical Theory of Transonic Normal Shock-Turbulent Boundary Layer Interaction," AIAA Journal, Vol. 14, Sept. 1976, pp. 1266-72; see also AIAA Paper 75-831, Washington, D.C., June 1975.

[4] Melnik, R. E. and Grossman, B., "Analysis of the Interaction of a Weak Normal Shock Wave with a Turbulent Layer," AIAA Paper 74-598, Palo Alto, Calif., June 1974.

[5] Inger, G. R., "Upstream Influence and Skin Friction in Non-Separating Shock Turbulent Boundary Layer Interactions," AIAA Paper 80-1411, Snowmass, Colo., July 1980.

[6] Lighthill, M. J., "On Boundary Layers and Upstream Influence, II. Supersonic Flow Without Separation," Proceedings of the Royal Society A, Vol. 217, 1953, pp. 478-507.

[7] Inger, G. R., "Some Features of a Shock-Turbulent Boundary Layer Interaction Theory in Transonic Flow Field," Proceedings of the AGARD Symposium on Computation of Viscous-Inviscid Interactions, Colorado Springs, Colo., Sept. 1980, pp. 18-1 to 18-66.

[8] Walz, A., Boundary Layers of Flow and Temperature, M.I.T. Press, Cambridge, Mass., 1969, p. 113.

[9] Inger, G. R. and Sobieczky, H., "Shock Obliquity Effect on Transonic Shock-Boundary Layer Interaction," Zeitschrift für Angewante Math und Mechanik, Vol. 58T, 1978, pp. 333-335.

[10] Inger, G. R. and Sobieczky, H., "Normal Shock Interaction with a Turbulent Boundary Layer on a Curved Wall, : VPI&SU Report Aero-088, Blacksburg, Va., Oct. 1978; see also AIAA Paper 81-1244, Palo Alto, Calif., June 1981.

[11] Deane, A., "Reynolds Number and Shape Factor Effects on Transonic Shock-Turbulent Boundary Layer Interactions Including Incipient Separation," M. S. Thesis, VPI&SU, Blacksburg, Va., Dec. 1980.

[12] Nandanan, M., Stanewsky, E. and Inger, G. R., "A Computational Procedure for Transonic Airfoil Flow Including a Special Solution for Shock-Boundary Layer Interaction," AIAA Paper 80-1389, Snowmass, Colo., July 1980.

[13] Nussdorfer, T. J., "Some Observations of Shock-Induced Turbulent Separation on Supersonic Diffusors," NACA RM E51L26, May 1956.

[14] Squire, L. C. and Smith, M. J., "Interaction of a Shock Wave with a Turbulent Boundary Layer Disturbed by Suction," Aeronautical Quarterly (to be published).

[15]Moses, H. L., "A Strip-Integral Method for Predicting the Behavior of Turbulent Boundary Layers," *Proceedings of the Computation of Turbulent Boundary Layers - 1968 AFOSR-IFP Stanford Conference*, Vol. 1, edited by S. Kline et al., Stanford U. Press, Calif., 1968.

[16]Inger, G. R. and Cantrell, J. C., "Application of Shock-Turbulent Boundary Layer Interaction Theory to Transonic Aerodynamics," *Proceedings of the 1979 USAF-Fed. Republic of Germany DEA Meeting*, April 1979; see also J. C. Cantrell, M. S. Thesis, VPI&SU, Blacksburg, Va., June 1979.

[17]Melnik, R. E., Chow, R. and Mead, H. R., "Theory of Viscous Transonic Flow Over Airfoils at High Reynolds Number," AIAA Paper 77-680, Albuquerque, N. Mex., June 1977.

Chapter XVIII.

Rapid Approximate Determination of Nonlinear Solutions: Application to Aerodynamic Flows and Design/Optimization Problems

Stephen S. Stahara*
Nielsen Engineering & Research, Inc., Mountain View, Calif.

Introduction

With the remarkable growth in capability of advanced computational methods to simulate a variety of complex fluid mechanic phenomena, it is apparent that a need exists for complimentary methods capable of alleviating, at least in part, the usage limitations imposed on these methods by their computational expense. The need becomes particularly compelling when large numbers of related cases are required, as in parametric or design studies. Techniques such as direct acceleration procedures provide an important means of reducing computer time by improving computational efficiency of the solution algorithm, but these and similar methods which enhance the solution algorithm itself represent only a partial answer. What is most desirable is a means to minimize the actual number of separate calculations required in a particular application by extending, over some parametric range, the usefulness of each individual solution determined by these computationally expensive procedures.

The classical approach of accomplishing this, involving the establishing and solving of a series of linear perturbation problems appears as an obvious choice. Recent studies,[1] however, have shown that for applications to sensitive flows such as typically occur in transonic situ-

---
Presented at the Transonic Perspective Symposium, NASA/Ames Research Center, Moffett Field, Calif., Feb. 18-20, 1981.
Copyright © American Institute of Aeronautics and Astronautics, Inc., 1981. All rights reserved.
*Manager, Theoretical Fluid Mechanics Department.

ations, the fundamental linear assumption of that technique is sufficiently restrictive that the useful range of parameter variation is so small to be of little practical use. An interesting alternative has recently been successfully examined[1-7] in which an approximation technique is used that employs two or more nonlinear base solutions determined by the full computational method to predict entire families of related nonlinear solutions. Here we provide results for several interesting applications of that method which demonstrate both its accuracy and its utility for engineering applications.

A crucial aspect of such a method is its ability to treat accurately regions where either discontinuities or high gradients exist. The concept of coordinate straining suggested by Nixon[2,3] is employed to account for the displacements of discontinuities due to parameter changes, and is extended to predict displacements of other high gradient locations such as stagnation points, maximum suction pressure points, etc. This is shown to result in highly accurate predictions in the vicinity of these regions.

Although the methodology developed is applicable to general nonlinear problems, the specific results reported are for aerodynamic applications. Single parameter and simultaneous multiple parameter perturbation results based on full potential solutions are presented for strongly supercritical transonic flows, which exhibit large surface shock movement over the parameter range studied, to demonstrate the accuracy of the method. Finally, and perhaps most importantly, results are provided for the preliminary application of the method coupled with an optimization procedure and applied to airfoil design problems. These results provide an evaluation of the method's ability to work accurately in a difficult design environment. Moreover, they demonstrate the potential of the method for reducing the computational work in such applications by an order of magnitude.

Analysis

Perturbation Concept and Methods

The basic hypothesis underlying the present procedure is that a range of solutions in the vicinity of a previously determined or base solution can be calculated to first-order accuracy in the incremental change of the varied parameter by determining a linearized unit perturbation solution Q_p defined according to the relation

$$Q = Q_o + \Delta \times \{Q_p\} \qquad (1)$$

where Q is the approximate solution for conditions differing from those of the base solution Q_o by an amount Δ of some arbitrary flow quantity. The effectiveness of such an approximation method depends upon the ability of the relationship defined by Eq. (1) to remain accurate over a range Δ of practical significance, and the fact that Q_p need be determined only once.

For the approximation method, Q_p is determined simply by differencing two nonlinear base flow solutions removed from one another by some nominal change of a particular flow or geometrical quantity and then dividing that result by the change in the perturbed quantity. Related solutions are determined by multiplying the unit perturbation by the desired parameter change and adding that result to the base flow solution. This simple procedure, however, only works directly for continuous flows for which the perturbation change does not alter the solution domain. For those perturbations which change the flow domain, coordinate stretching (usually obvious) is necessary to insure proper definition of the unit perturbation solution. Similarly, for discontinuous flows, coordinate straining is necessary to account additionally for movement of discontinuities due to the parameter change.

The attractiveness of the approximation method is that it is not restricted to a linear variation range but rather replaces the nonlinear variation between two base solutions with a linear fit. This de-emphasizes the sensitivity inherent in the classical linear perturbation equation approach. Furthermore, other than the approximation of a linear fit between two nonlinear base solutions, the method is not restricted by further approximations with respect to the governing differential equations and boundary conditions. Rather, it retains the full character of the original methods used to calculate the base flow solutions. Most importantly, no perturbation differential equations have to be posed and solved, only algebraic ones. In fact, it is not necessary to know the exact form of the perturbation equation, only that it can be obtained by some systematic procedure and the perturbations thus defined will behave in some "generally appropriate" fashion so as to permit a logical perturbation analysis. For situations involving perturbations of physical parameters, such as reported here, the governing perturbation equations are usually transparent, or at least readily derivable. In applying this method it is not necessary to work with primitive variables; rather the procedure can be applied directly to the final quantity desired. Moreover, because of the implicit nature of coordinate straining, the final approximation solution is nonlinear in the varied parameter.

The primary disadvantage of the method is that two base solutions are required for each parameter perturbation considered. Furthermore, both flows must be topologically similar, i.e., discontinuities or other characteristic features must be present in both base solutions used to establish the unit perturbation.

Coordinate Straining

The concept of employing coordinate straining to remove nonuniformities from perturbation solutions of nonlinear problems is well established and originally suggested by Lighthill[8] three decades ago. The basic idea of the technique is that a straightforward perturbation solution may possess the appropriate form, but not quite at the appropriate location. The procedure is to strain slightly the coordinates by expanding them as well as the dependent variables in an asymptotic series. It is often unnecessary to actually solve for the straining. It can generally be established by inspection. The final uniformly valid solution is then found in implicit form, with the strained coordinate appearing as a parameter.

In the original applications of the method,[9] it was applied in the "classical" sense; that is series expansions of the dependent and independent variables in ascending powers of some small parameter were inserted into the full governing equation and boundary conditions, and the individual terms of the series determined. An ingenious variation in the application of the method was made by Pritulo[10] who demonstrated that if a perturbation solution in unstrained coordinates has been determined and found to be nonuniform, the coordinate straining required to render that solution uniformly valid can be found by employing straining directly in the known nonuniform solution, and then solving algebraic rather than differential equations. The idea of introducing strained coordinates a posteriori has since been applied to a variety of different problems (see Ref. 9), and forms the basis of the current applications.

APPROXIMATE DETERMINATION OF NONLINEAR SOLUTIONS 641

The fundamental idea underlying coordinate straining as it relates to the present aerodynamic applications is illustrated graphically in Fig. 1. In the upper plot on the left, two typical transonic pressure distributions are shown for a highly supercritical flow about a nonlifting symmetric profile. The distributions can be regarded as related nonlinear flow solutions separated by a nominal change in some geometric or flow parameter. The shaded area between the solutions represents the perturbation result that would be obtained by directly differencing the two solutions. We observe that the perturbation so obtained is small everywhere except in the region between the two shock waves, where it is fully as large as the base solutions themselves. This clearly invalidates the perturbation technique in that region and most probably somewhat ahead and behind it as well. The key idea of a procedure for correcting this, pointed out by Nixon,[2,3] is first to strain the coordinates of one of the two solutions in such a fashion that the shock waves align, as shown in the upper plot on the right of Fig. 1, and then determine the unit perturbation. Equivalently, this can be considered as maintaining the shock wave location invariant during the perturbation process, and assures that the unit perturbation remains small both at and in the vicinity of the shock wave. Obviously, shock points are only one of a number of characteristic high gradient locations such as stagnation points, maximum suction pressure points, etc., in which the accuracy of the perturbation solution can degrade rapidly. The plots in the lower left part of the Fig. 1 indicate such a situation which contains multiple shocks and high gradient regions. Simultaneously straining at these locations, as indicated in the lower right plot, serves to <u>minimize</u> the unit perturbation over the entire domain considered and provides the basis of the high accuracy of the method.

Analytical Formulation

In order to provide the theoretical essentials of the method, consider the formulation of the procedure at the level of the full potential equation. We denote the operator L acting on the velocity potential equation for Φ as that which results in the two-dimensional full potential equation for Φ, i.e.

$$L[\Phi] = 0 \qquad (2)$$

If we now expand the potential in terms of zero- and higher-

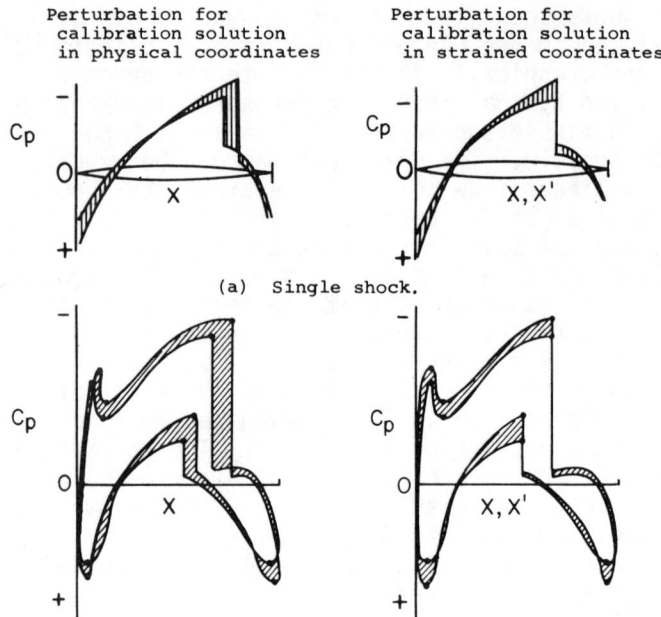

Fig. 1 Illustration of perturbation solution for calibration solution in physical and strained coordinates.

order components to account for the variation of some arbitrary geometrical or flow parameter q

$$\Phi = \Phi_0 + \varepsilon\Phi_1 + \ldots \quad ; \quad q = q_0 + \Delta q \qquad (3)$$

and then insert this into the governing Eq. (2), expand the result, order the equations into zero- and first-order components, and make the obvious choice of expansion parameter $\varepsilon = \Delta q$, we obtain the following governing equations for the zero- and first-order components

$$L[\Phi_0] = 0 \quad ; \quad L_1[\Phi_1] + \frac{\partial}{\partial q}L[\Phi_0] = 0 \qquad (4)$$

Here L_1 is a linear operator whose coefficients depend on zero-order quantities and $\partial L[\Phi_0]/\partial q$ represents a "forcing" term due to perturbation. Actual forms of L_1 and the "forcing" term are provided in Ref. 1 for a variety of flow and geometry parameter perturbations of a two-dimensional turbomachine, and in Ref. 4 for profile shape perturbations of an isolated airfoil. An important point regarding Eq. (4) for the first-order perturbation Φ_1 is that the equation represents a unit perturbation independent of the actual value of the perturbation quantity ε.

Appropriate account of the movement of discontinuities and maxima of high gradient regions due to the perturbation is now accomplished by the introduction of strained coordinates (s,t) in the form

$$x = s + \varepsilon x_1(x,t) \quad ; \quad y = t + \varepsilon y_1(s,t) \qquad (5)$$

where

$$x_1(s,t) = \sum_{i=1}^{N} \delta x_i x_{1_i}(s,t) \quad ; \quad y_1(x,t) = \sum_{i=1}^{N} \delta y_i y_{1_i}(s,t) \qquad (6)$$

and $\varepsilon \delta x_i$, $\varepsilon \delta y_i$ represent individual displacements of the N strained points, and $x_{1_i}(s,t)$, $y_{1_i}(s,t)$ are straining functions associated with each of the N strained points. Introducing the strained coordinate Eqs. (5) and (6) into the expansion formulation leaves the zero-order result in Eq. (4) unchanged, but results in a change of the following form for the first-order term

$$L_1[\Phi_1] + L_2[\Phi_0] + \frac{\partial}{\partial q} L[\Phi_0] = 0 \qquad (7)$$

Here the operators are understood to be expressed in terms of the strained (s,t) coordinates, and the additional operator L_2 arises specifically from displacement of the strained points. In Refs. 3 and 4, specific expressions for L_2 are provided for selected perturbations involving transonic small distrubance and full potential equation formulations. The primary point, however, with regard to perturbation Eq. (7) expressed in strained coordinates is that it remains valid as before for a unit perturbation and independent of ε.

In employing the correction method, Eq. (7) for the unit perturbation is solved by taking the difference between two solutions obtained by the full nonlinear procedure after appropriately straining the coordinates. If we designate the two solutions for some arbitrary flow quantity Q as base Q_0 and calibration Q_c, respectively, of the varied parameter, we have for the predicted flow at some new parameter value q (Ref. 5)

$$Q(x,y) = Q_0(s,t) + \frac{\varepsilon}{\varepsilon_0} [Q_c(\bar{x},\bar{y}) - Q_0(s,t)] \qquad (8)$$

where

$$\bar{x} = s + \varepsilon_0 x_1(s,t) \quad ; \quad \bar{y} = t + \varepsilon_0 y_1(s,t)$$

$$x = s + \frac{\varepsilon}{\varepsilon_0}[\bar{x} - s] \quad ; \quad y = t + \frac{\varepsilon}{\varepsilon_0}[\bar{y} - t] \qquad (9)$$

$$\varepsilon_0 = q_c - q_0 \quad ; \quad \varepsilon = q - q_0$$

Extension of this result to simultaneous multiple parameter perturbations is straightforward[3]; and that extension is provided in the following section where applications of the procedure are made to predict surface properties. Also provided are the particular forms of the straining functions Eq. (5) for those applications.

Application to Surface Properties

For the current applications, we have employed coordinate straining with the approximation method to predict surface pressure distributions for a variety of single parameter and multiple parameter geometrical and flow perturbations of isolated airfoils and cascades. In that instance where flow properties are required along some contour, the first-order solutions can be represented by

$$Q(x,\varepsilon_j) = Q_0(s) + \sum_{j=1}^{M} \varepsilon_j Q_{1_j}(s)$$

$$x = s + \sum_{j=1}^{M} \varepsilon_j x_1(s) \qquad (10)$$

where x is the independent variable measuring distance along the contour or a convenient projection of that distance, s is the strained coordinate, and ε_j a small parameter representing the change in one of M flow or geometrical variables which we wish to vary simultaneously.

In order to determine the first-order corrections $Q_{1_j}(s)$, we require one base and M calibration solutions in which the calibration solutions are determined by individually varying each of the M parameters by some nominal amount from the base flow value while keeping the others fixed at the base flow values.

In this way, the first-order corrections $Q_{1_j}(s)$ can be represented as

$$Q_{1_j}(s) = [Q_{c_j}(\bar{x}_j) - Q_0(s)]/\varepsilon_j^c \qquad (11)$$

where Q_{c_j} is the calibration solution corresponding to changing the jth parameter to a new value q_{c_j}, $\bar{x}_j$ is the strained coordinate pertaining to the Q_{c_j} calibration solution and ε_j^c represents the change $q_{c_j} - q_{0_j}$ in the jth parameter from its base flow value. If we now desire to keep invariant during the perturbation process a total of N points corresponding to discontinuities or high gradient maxima, the coordinates $\bar{x}_j$ and x can be given by

$$\bar{x}_j = s + \sum_{i=1}^{N} \varepsilon_j^c(\delta x_i^c)_j x_{1_i}(s) \qquad (12)$$

$$x = s + \sum_{j=1}^{M} \sum_{i=1}^{N} \varepsilon_j(\delta x_i^c)_j x_{1_i}(s) \qquad (13)$$

where

$$\varepsilon_j^c = q_{c_j} - q_{0_j} \quad ; \quad \varepsilon_j = q_j - q_{0_j} \qquad (14)$$

$$\varepsilon_j^c(\delta x_i^c)_j = (x_i^c - x_i^0)_j \quad ; \quad \varepsilon_j(\delta x_i^c)_j = \frac{\varepsilon_j}{\varepsilon_j^c}(x_i^c - x_i^0)_j \qquad (15)$$

Here $\varepsilon_j^c(\delta x_i^c)_j$ given in Eqs. (12) and (15) represents the displacement of the ith invariant point in the jth calibration solution from its base flow location due to the selected change ε_j^c in the q_j parameter given by Eq. (14), and $x_{1_i}(s)$ is a unit order straining function having the property that

$$x_{1_i}(x_k^0) = \begin{cases} 1 & k = i \\ 0 & k \neq i \end{cases} \qquad (16)$$

which assures alignment of the ith invariant point between the base and calibration solutions.

In addition to the single condition Eq. (16) on the straining function, it may be convenient or necessary to impose additional conditions at other locations along the contour. For example, it is usually necessary to hold invariant the end points along the contour, as well as to require that the straining vanish in a particular fashion in those locations. All of these conditions, however, do not serve to determine the straining uniquely. The nonuniqueness of the straining, nevertheless, can often be turned to advantage, either by selecting particularly simple classes of straining functions or by requiring the straining to satisfy further constraints convenient for a particular application.

In the present application, the problem posed for the first-order strained coordinate approximation is somewhat unusual in that the straining requirements are fixed a priori rather than determined as part of the solution. Consequently, it is unnecessary to examine a higher-order straining problem as is the standard procedure[9] for determining the straining. Rather, the straining is established on the basis of Eq. (16) regarding displacement of the invariant points. Since relatively simple classes of straining functions can accomplish this, the results obtained thus far with the approximation method have employed either various continuous polynomial straining functions or piecewise continuous linear straining functions.

The fact of nonuniqueness of straining functions, however, raises a further question of the dependence of the final approximation predicted result on choice of straining function. An initial example of the effect of employing two different straining functions for a strongly supercritical flow was provided in Ref. 3. Those results indicated no difference between the two approximation solutions. In Ref. 11, it was demonstrated that the final approximation result when employing strained coordinates in the present manner is formally independent of the particular straining function used, provided the function moves be invariant points to the proper locations. We have found this to be true for predictions at and in the vicinity of invariant points. However, we have additionally found that certain classes of straining functions can have the undesirable property of causing unwanted straining in regions removed from the invariant points. This has the effect of inducing spurious behavior in the approximation predictions in those regions. Examples of such results were provided in Refs. 6 and 7 and demonstrate in particular some of the limitations of various polynomial straining functions. A simple and direct means to avoid such spurious behavior, which has proven effective in all case studies undertaken thus far, is to employ linear piecewise continuous straining.

For linear piecewise continuous straining, the functional form of the straining can be compactly written. For

APPROXIMATE DETERMINATION OF NONLINEAR SOLUTIONS

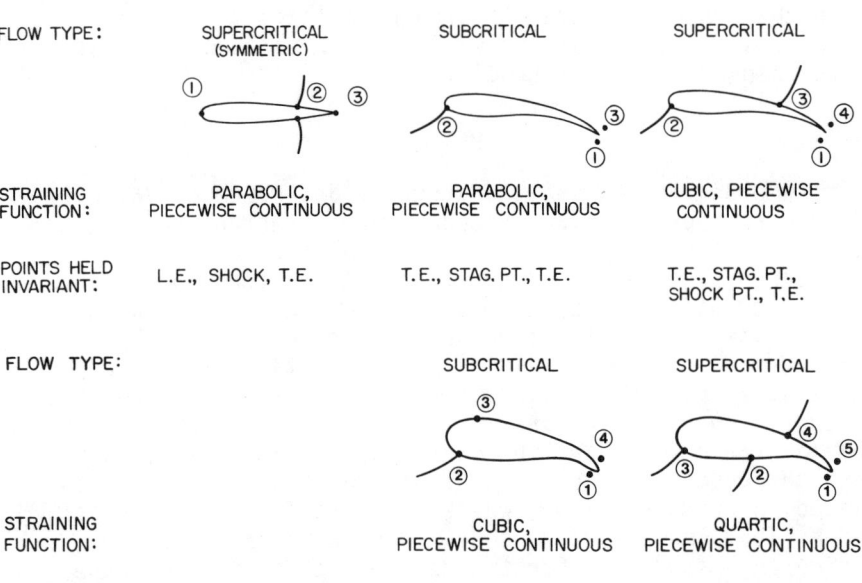

Fig. 2 Summary of various flows and straining functions considered.

example, Eq. (12) for $\bar{x}_j$ is given by

$$\bar{x}_j = x + \sum_{i=2}^{N-1} \left\{ \frac{x^0_{i+1} - s}{x^0_{i+1} - x^0_i} (x^c_i - x^0_i) \right.$$

$$\left. + \frac{x - x^0_i}{x^0_{i+1} - x^0_i} (x^c_{i+1} - x^0_{i+1}) \right\} H(x^0_{i+1} - s)H(s - x^0_i) \quad (17)$$

where H denotes the Heaviside step function. As discussed above, it is usually necessary to hold invariant both of the end points along the contour in addition to the points corresponding to discontinuities or high gradient maxima. Consequently, for the results reported here, the array of invariant points in the base and calibration solutions have been taken as

$$\begin{aligned} x^0_i &= \{0, x^0_1, x^0_2, \ldots, x^0_n, 1\} \\ x^c_{i_j} &= \{0, x^c_{1_j}, x^c_{2_j}, \ldots, x^c_{n_j}, 1\} \end{aligned} \quad (18)$$

where the contour length has been normalized to one. Figure 2 provides a summary of the various combinations of flows and straining functions studied.

Results

In order to examine the accuracy and range of validity of such an approximation procedure and to establish to what extent it is capable of providing results useful in an engineering analysis, we have tested the method in a variety of different perturbations, including applications in both isolated airfoils and compressor cascades. Results for a number of these preliminary applications are reported in Refs. 1, 6, and 7. Since the ability of the method to account accurately for the movement of discontinuities and maxima of high gradient but continuous regions is essential if such procedures are to be of general use, emphasis was placed on transonic flows which are strongly supercritical and exhibit large surface shock movement over the parametric range studied. For the results reported here, base flow theoretical solutions were determined at the transonic full potential level.[12,13,16] In the examples to follow, which were selected as typical from systematic calculations of a much larger number of cases, the choice of base and calibration solutions was often made at the limits of validity of the procedure to demonstrate how well the method performs under such conditions.

Single and Multiple Parameter Perturbations

In order to provide a more severe test of the approximation method than those considered in some of the preliminary examples previously studied, we have applied the method to a number of transonic flows that are characterized by surface pressure distributions having multiple shock and/or high gradient locations, such as those typified schematically in the lower plots of Fig. 1. Demonstration of the ability of the perturbation method to predict accurately such classes of flows is relevant to a variety of important practical flow situations, such as transonic turbomachinery applications, where typical flows are generally complex. To accomplish such a demonstration, we have investigated two separate classes of sensitive supercritical transonic flows; those with multiple shock waves, and those having a single shock together with multiple high gradient regions.

In Fig. 3, we present results for an angle of attack perturbation of supercritical lifting flows past a NACA 0012

profile at M_∞ = 0.80. These highly sensitive flows exhibit
two shocks, one on each upper and lower surface. The full
potential base and calibration flows employed to determine
the unit perturbation are at α = 0.50 deg and 0.20 deg,
respectively, and were obtained by solving the transonic
full potential equation using the code TAIR.[12] They are
illustrated as the dotted and dashed curves in the figure.
The open circles represent the approximation solutions at
several new angles of attack, and are shown for α = {0.0,
0.1, 0.4, 0.6} deg. Those approximation results are meant
to be compared with the solid lines which are the corresponding nonlinear solutions obtained by rerunning TAIR[12]
at the new angles of attack. Piecewise continuous linear
straining has been used with invariant points corresponding
to the lower trailing edge, lower surface shock point,
stagnation point, upper surface shock point, and upper

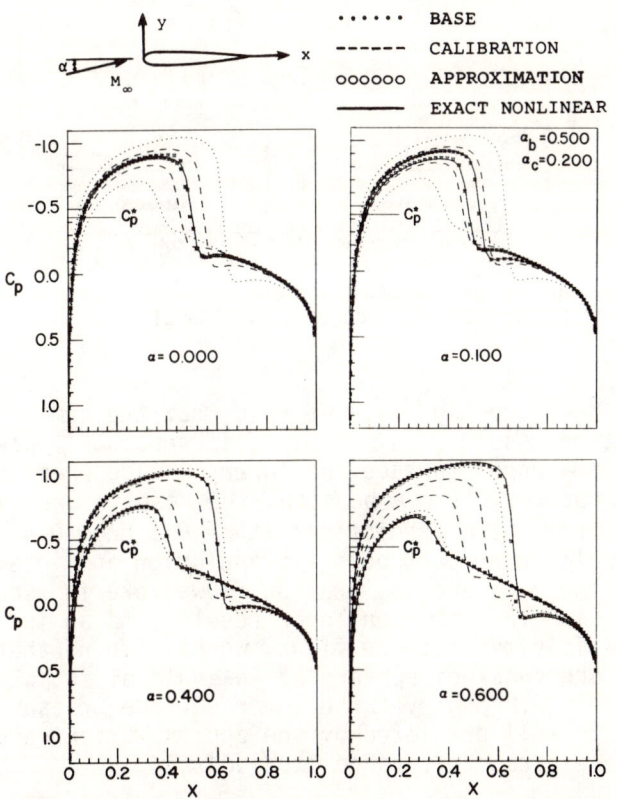

Fig. 3 Comparison of approximation and nonlinear surface pressures
for an angle of attack perturbation of an isolated NACA 0012
airfoil at M_∞ = 0.80.

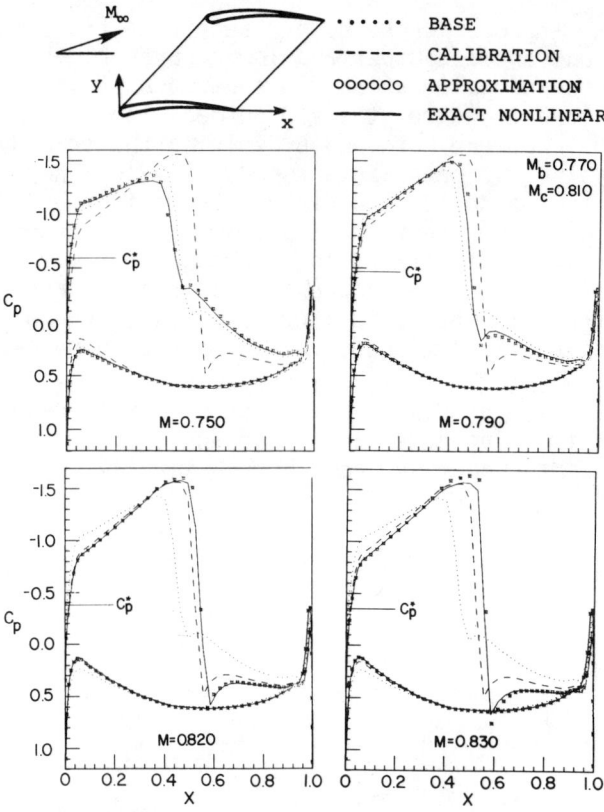

Fig. 4 Comparison of approximation and nonlinear surface pressures for a Mach number perturbation of a cascade of Jose Sanz blade profiles.

trailing edge (see Fig. 2). We note that the symmetrical extrapolation result at $\alpha = 0.0$ deg is separately predicted from both the upper surface and lower surface pressure distributions and, as can be seen, the results are quite good. The remaining predictions at $\alpha = 0.1$ deg, 0.4 deg, and 0.6 deg, which represent both extrapolation and interpolation from the base and calibration flows, are in excellent agreement with the full nonlinear result. As an indication of the sensitivity of these flows, we have found that the lower surface shock disappears at an angle of attack of approximately 0.8 deg; yet the lower surface pressure distribution is well predicted by the approximation result over the complete parametric range studied.

As an example of the ability of the method to predict a supercritical flow with multiple high gradient regions, in Fig. 4 we provide results for a Mach number perturbation of supercritical flow past a cascade composed of Jose Sanz[13] profiles. For these results, the oncoming and exit flow angles are 30.81 deg and 0.09 deg, respectively, the blade twist is 9.33 deg, while the gap to chord ratio is 1.028. The full potential base and calibration flow oncoming Mach numbers are M_∞ = 0.77 and 0.81, with comparisons of approximation and full nonlinear results shown at M_∞ = {0.75, 0.79, 0.82, and 0.83}. Piecewise continuous linear straining was employed with invariant points at the lower trailing edge, stagnation point, shock point, and upper trailing edge. As with the multiple shock example shown in Fig. 3, we note that the approximation predictions are in excellent agreement with the exact nonlinear results. In particular, we note that the method captures the variation of the plateau-like pressure distribution on the upper surface near the leading edge, the location and the strength of the shock, the postshock expansion region, the rapid expansion near the trailing edge, and the expansion on the lower surface near the stagnation point, indicating a capability for treating very general flow situations.

In Fig. 5, we provide corresponding results for an example involving the simultaneous perturbation of several parameters. Those results represent a four parameter perturbation of strongly supercritical full potential flows past a cascade of compressor blades having NACA four digit profiles. The base flow is for an oncoming Mach number of M_∞ = 0.78, thickness ratio, τ = 0.110, gap to chord ratio t = 3.2, and oncoming inflow angle α = 0.3 deg. The four calibration solutions to account for changes in these parameters involved individually varying each parameter from its base value to its calibration matrix value given by $\{M_\infty, \tau, t, \alpha\}$ = {0.790, 0.120, 3.0, 0.5 deg} while keeping the others at their base values. Comparison of predicted and exact nonlinear results are for parameter values of $\{M_\infty, \tau, t, \alpha\}$ = {0.785, 0.115, 3.1, 0.4 deg}. The base solution is indicated as the dashed line and provides some idea of the solution displacement. The four calibration solutions are not shown. This particular set of flows was selected because of the presence of multiple shocks and high sensitivity to parameter change. Linear piecewise continuous straining was employed with the invariant points being the lower surface trailing edge, lower surface shock, stagnation point, upper surface shock, and upper surface trailing edge. The comparisons between approximation and exact nonlinear results is again remarkably accurate. The predictions of the locations of both shocks on the upper and lower surface are given very well, as are the pressure distributions in the regions immediately ahead and behind those shocks.

Combination of Approximation Method with Optimization Procedures

The ultimate utility of the approximation methods developed and evaluated here lies in their application to problems involving the high frequency use of computational codes to determine a large number of related nonlinear flow solutions. In order to test the capability of the method to work effectively in such practical applications, we have combined the method with an optimization procedure[14] and have then made several preliminary case studies of the combination applied to airfoil design/optimization problems. The objectives of these initial applications were to examine the feasibility and potential computational savings of the combined approximation/optimization procedure, and to determine the accuracy of the approximation-predicted optimization results.

The particular airfoil design optimization problems selected for study involved the alteration of a baseline profile shape by adding to the baseline profile a set of shape functions according to the relation

$$Z(x) = Z_{base}(x) + \sum_{i}^{M} a_i F_i(x) \qquad (19)$$

where Z_{base} are the ordinates of the baseline airfoil, F_i are the shape functions, and the coefficients a_i are the design variables whose values are determined by the optimization program in its search to achieve a desired design improvement. The general class of geometric shape functions employed, and which have been used successfully in the past for optimizing supercritical airfoils,[15] consist of exponential decay functions and sine functions. These are of the general form $(1-x)x^p/e^{qx}$ and $\sin(\pi x^r)^n$, where the exponents p, q, r and n are selected to provide a desired maximum at a particular chordwise location. The exponential functions are employed to provide adjustments near the leading edge, while the sine functions are used to provide maximum ordinate changes at particular chordwise stations. Illustrations of the chordwise variation of typical members of these shape functions are provided in Ref. 15.

APPROXIMATE DETERMINATION OF NONLINEAR SOLUTIONS 653

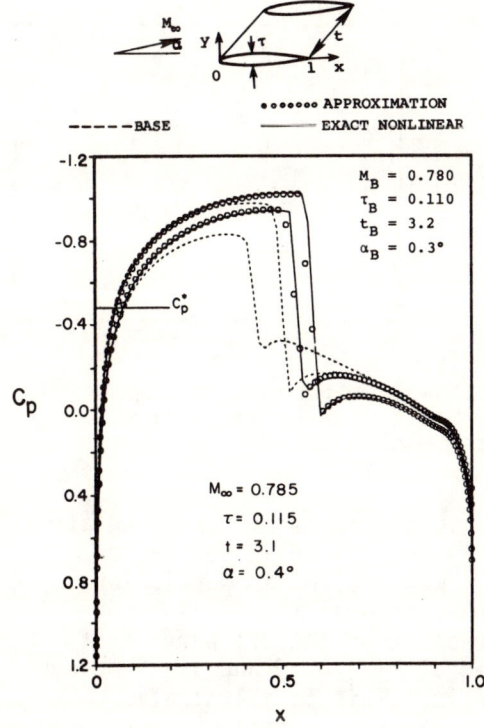

Fig. 5 Comparison of approximation and nonlinear surface pressures for the simultaneous four parameter perturbation of $(M_\infty, \tau, t, \alpha_\infty)$ of a cascade of NACA 00XX blade profiles.

For the initial application of the combined approximation/optimization method, we have examined subcritical flow at $M_\infty = 0.10$ and $\alpha = 5$ deg past a modified NACA 64A007 profile involving the nine profile shape functions.

$$F_i = 6(1-x)x^{p_i}/e^{q_i x} \qquad i = 1,2 \qquad (20)$$

$$F_i = \sin(\pi x^{r_i})^2 \qquad i = 3,9$$

where $(p_1, q_1) = (0.5, 15)$, $(p_2, q_2) = (0.25, 10)$, and $r_i = \{0.37, 0.50, 0.66, 0.87, 1.16, 2.41\}$. The exponential functions achieve their maxima within 5% of chord, while those for the sine functions are at {15%, 25%, 45%, 55%, 65%, 75%} of chord.

A strategy that has proven convenient for optimizing aerodynamic performance parameters[14] has been to recontour the profile shape so as to tailor the surface pressure distribution to conform to a desired distribution. This enables local control over the basic aerodynamic flow

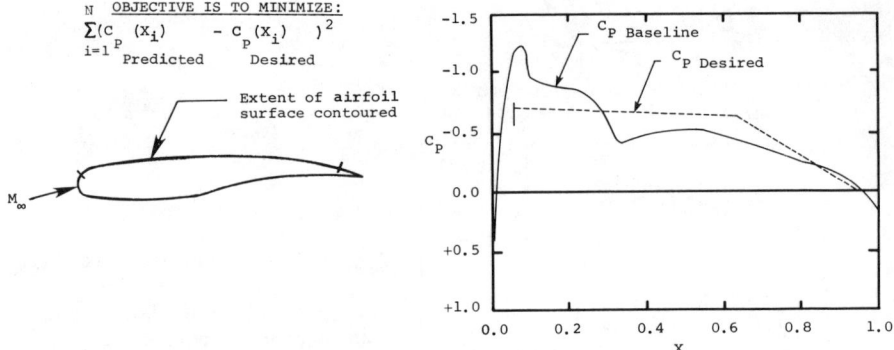

Fig. 6 Illustration of physical basis of optimization problem involving airfoil surface contouring to tailor the surface pressure distribution to a desired distribution.

property of interest, and provides a means of achieving aft pressure gradients sufficiently weak to avoid separation. In this initial case study, it was desired in particular to minimize both the peaky behavior near the leading edge and the strong compressive gradient on the aft portion of the upper surface that existed at $M_\infty = 0.10$ and $\alpha = 5$ deg on the NACA 64A007 baseline profile. This is illustrated schematically in Fig. 6. The objective function was taken as the minimization of the mean squared error between the predicted and desired surface pressure distributions, i.e.

$$\text{OBJ} = \sum_{i=1}^{N} [C_{p_D}(x_i) - C_{p_C}(x_i)]^2 \qquad (21)$$

where N represents the number of chordwise locations x_i where desired and calculated pressures are compared.

Recall that in order to initiate the approximation procedure in situations involving the simultaneous variation of M individual parameters from a baseline point, a matrix of M calibration solutions is required, each representing the solution change for a separate variation of each of the M parameters from its baseline value. Because optimum, or sometimes even typical, stepsizes for a particular optimization problem would not generally be known a priori, one of the goals of these initial studies was the demonstration that the approximation method was capable of working effectively even under severe conditions imposed by a poorly selected calibration solution matrix.

This was accomplished by examining the sensitivity and accuracy of the approximation predicted optimization results as a function of the initial design variable stepsizes of the calibration solution matrix.

Figure 7 shows the results of such a sensitivity study for this example, and indicates that even under extreme test case conditions caused by deliberately selected poor choices of design variable stepsize, the approximation method performs exceptionally well, never breaking down or yielding spurious results. Indicated on the plots are the final optimized design variable values after five search cycles as predicted by the approximation method ($\bullet$) for four different choices of initial design variables, i.e., δDV_i = {0.05, 0.02, 0.01, 0.001}. Also shown are the corresponding final design variable values predicted when employing the nonlinear aerodynamic solver[16] throughout ($\circ$). For the extreme interpolation case δDV_i = 0.05, except for design variables 3 and 5, the agreement is very good. For δDV_i = 0.02, a more reasonable stepsize choice, the approximation result is quite good for all the design variables, while for δDV_i = 0.01, the approximation prediction is essentially identical to the full nonlinear aerodynamic result. As a final illustration of the behavior of the approximation method under extreme extrapolation conditions, the lower right hand plot displays the results for δDV_i = 0.001. We note that for several of the design variables, the extrapolation range is of the order of 25 times the initial stepsize; yet the approximation predictions are quite reasonable. This is remarkable and indicative of the robustness of the procedure.

Finally, we point out that all four of the approximation predicted results illustrated in Fig. 7 are satisfactory in terms of the final objective function value obtained. These values are illustrated on the right of each of the plots. Provided for comparison are the initial ($\circ$ Initial) and final ($\circ$ Final) values obtained when using the nonlinear aerodynamic solver throughout. The value of the objective function evaluated at the final design variable point when using the approximation method is indicated by the solid circle ($\bullet$). However, the objective function value of interest is the result indicated by the solid square ($\blacksquare$) which represents that obtained by running the nonlinear aerodynamic solver at the approximation predicted final design point, and then using that solution to evaluate the objective function. This provides the overall ultimate check of the approximation predicted result. As can be

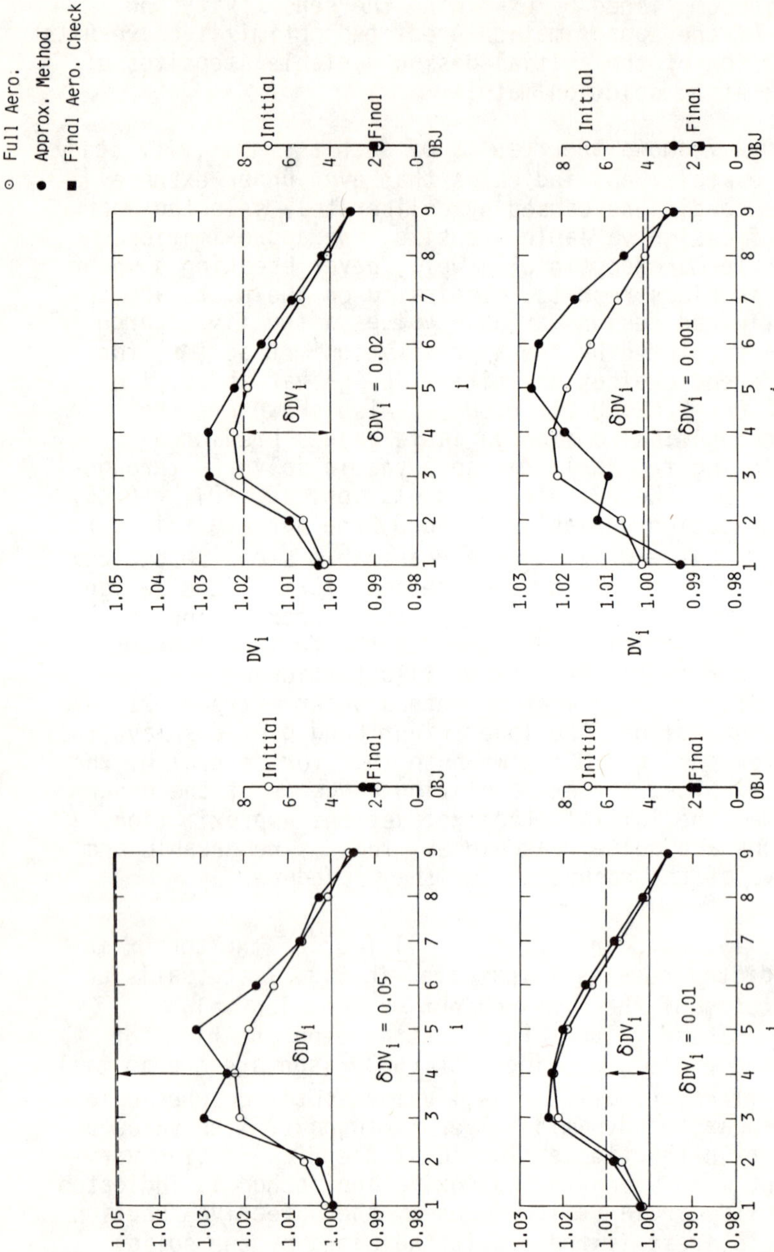

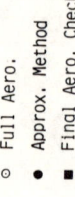

Fig. 7 Comparison of approximation predicted final design variables and objective function with full nonlinear result for various choices of initial design variable stepsize for flow past a 64A007 airfoil at $M_\infty = 0.10$ and $\alpha = 5$ deg.

APPROXIMATE DETERMINATION OF NONLINEAR SOLUTIONS   657

seen from Fig. 7, those results lie essentially on top of the final objective function result ($\bigcirc_{Final}$) obtained when using the nonlinear aerodynamic solver throughout.

The computational savings attained for this application are shown in Fig. 8. There, a comparison of the computational work versus reduction in objective function per optimization cycle is provided when using the approximation procedure (●) and when not using it but employing the nonlinear aerodynamic code (○) for each flow solution required by the optimizer. As can be seen, the computational time required for both the approximation method and when using the full nonlinear aerodynamic solver throughout are the same for the first cycle, since both require a matrix of M+1 nonlinear aerodynamic solutions. After that, the approximation predicted results required essentially no computational time for cycles 2 through 5, and then a slight amount for the one additional call to the

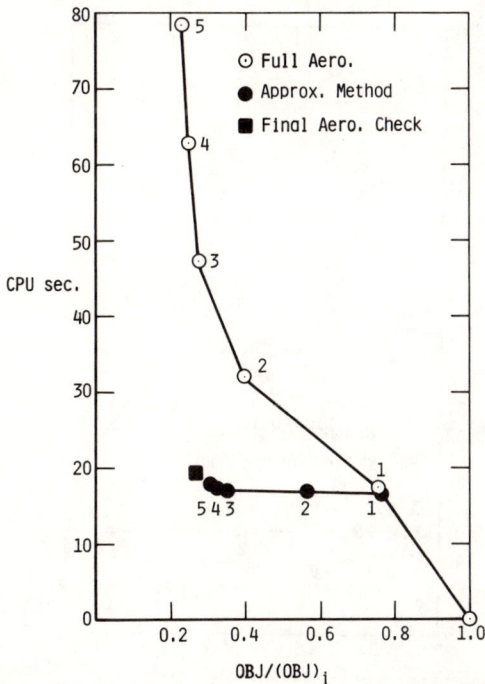

Fig. 8 Comparison of computational work and objective function reduction per optimization cycle between approximation procedure and full nonlinear aerodynamic result for nine design variable subcritical surface pressure tailoring optimization study.

aerodynamic solver for the final check calculation (■). The reduction in the ratio of final to initial objective function is $OBJ/(OBJ)_i = 0.22$ and required approximately 20 CPU seconds on the CDC 7600. In comparison, the result when not employing the approximation method required approximately 80 CPU seconds for the same reduction in objective function, indicating the approximation method is able to save 75% of the computational work in this example.

Similar studies for supercritical situations have demonstrated a corresponding capability and even greater potential computational savings. The final figure illustrates the computational savings achieved for one such supercritical case study involving upper surface pressure distribution tailoring of a NACA 0015 profile employing four design variables related to the shape functions

$$F_i = \sin(\pi x^{q_i})^3 \quad i = 1,4$$

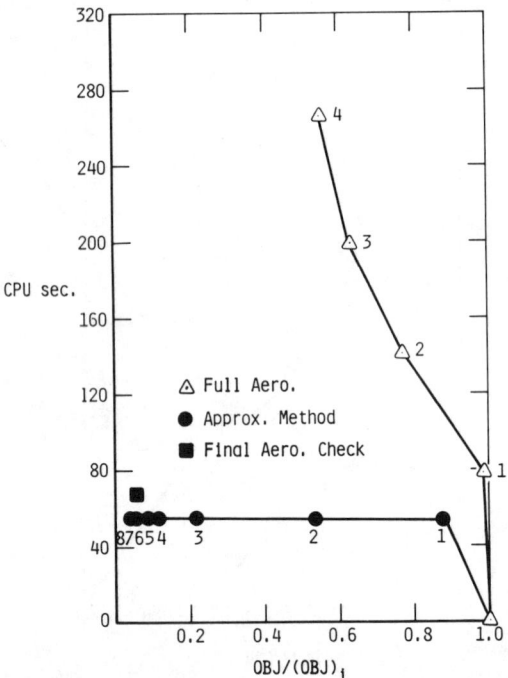

Fig. 9 Comparison of computational work and objective function reduction per optimization cycle between approximation procedure and full nonlinear aerodynamic result for four design variable supercritical surface pressure tailoring optimization case study.

with $q_i = \{0.301, 0.431, 0.576, 0.756\}$ and which have maxima at {10%, 20%, 30%, 40%} of chord. Shown in Fig. 9 is a comparison of computational work and objective function reduction per optimization cycle when using the approximation procedure (●) as opposed to not using it (△). Here we see clearly that the approximation method is able to drive the objective function to essentially zero, while the procedure using the nonlinear aerodynamic procedure throughout becomes fixed in a local minimum and is only able to reduce the objective function to 50% of its initial value. If we had carried the full aerodynamic result for eight optimization cycles, as was done with the approximation result, the time savings would have been over an order of magnitude.

This result demonstrates that it is possible in certain instances for the approximation method to provide not only a large savings in computational cost but also an improved optimization result. The latter is undoubtedly accomplished by the selection of a reasonable initial calibration solution matrix which permits the optimization method an enhanced rather than local view of the design solution space, thereby avoiding shallow local minima's in favor of a more global minima.

Concluding Remarks

An evaluation has been made of a procedure for determining highly accurate approximations to families of nonlinear solutions which are either continuous or discontinuous, and which represent variations in some arbitrary parameter. The procedure employs unit perturbations, determined from two or more nonlinear solutions which differ from one another by a nominal change in some geometric or flow parameter, to predict a family of related nonlinear solutions. Coordinate straining is used in determining the unit perturbation in order to account properly for the motion of discontinuities and maxima of high gradient regions. Calculations based on full potential nonlinear solutions and spanning a variety of flow and geometry perturbations of flows past airfoils and compressor cascades were carried out. Particular emphasis was placed on supercritical transonic flows which exhibit large surface shock movements over the parameter range studied. Approximation results, for both single parameter and multiple parameter perturbations, characterized by both extreme solution interpolation and extrapolation, were obtained in order to demonstrate the accuracy and range of validity of the method. Preliminary application of the approximation method coupled with an optimization procedure was made to typical airfoil design problems. These results provide an evaluation of the capability of the method to work in a design atmosphere and demonstrate its potential for savings in computational work.

Comparisons of the approximation results with the corresponding "exact" nonlinear solutions indicate a remarkable accuracy and range of validity of the method across the spectrum of examples reported. Solution interpolation and extrapolation are feasible and results at and in the vicinity of discontinuities and other high gradient regions are accurately predicted. Initial results from the combined approximation/optimization procedure demonstrate a high degree of accuracy of the method and the potential for reduction in computational work of an order of magnitude.

Acknowledgments

The results reported are based on research supported by NASA/Lewis Research Center under Contract NAS3-20836 with Mr. Aaron Synder as Technical Monitor. Special thanks are due to Dr. Terry L. Holst for making available the isolated airfoil full potential solver,[12] to Dr. Djordge S. Dulikravich for the cascade full potential theory.[13]

References

[1]Stahara, S. S., Chaussee, D. S., and Spreiter, J. R., "Perturbation Solutions for Transonic Flow on the Blade-to-Blade Surface of Compressor Blade Rows," NASA CR-2941, Jan. 1978.

[2]Nixon, D., "Perturbation of a Discontinuous Transonic Flow," AIAA Journal, Vol. 16, Jan. 1978, pp. 47-52.

[3]Nixon, D., "Perturbations in Two and Three-Dimensional Transonic Flows," AIAA Journal, Vol. 16, July 1978, pp. 669-709.

[4]Nixon, D., "Design of Transonic Airfoil Sections Using a Similarity Theory," NASA TN 78521, Oct. 1978.

[5]Nixon, D., "Perturbation Methods in Transonic Flow, " AIAA Paper No. 80-1367, Fluid and Plasma Dynamics Conference, Snowmass, Colo., July 14-16, 1980.

[6]Stahara, S. S., Elliott, J. P., and Spreiter, J. R., "A Rapid Method for the Approximate Determination of Nonlinear Solutions: Application of Aerodynamic Flows," ICAS Proceedings 1980, No. 80-75, Oct. 1980, pp. 324-337.

[7]Stahara, S. S., Elliott, J. P, and Spreiter, J. R., "A Rapid Perturbation Procedure for Determining Nonlinear Flow Solutions: Application to Transonic Turbomachinery Flows," NASA CR-3425, May 1981.

[8]Lighthill, M. J., "A Technique for Rendering Approximate Solutions to Physical Problems Uniformly Valid," Philosophy Magazine, Vol. 40, 1949, pp. 1197-1201.

[9]Van Dyke, M., Perturbation Methods in Fluid Mechanics, The Parabolic Press, Stanford, Calif., 1975.

[10]Pritulo, M. R., "On the Determination of the Uniformly Accurate Solutions of Differential Equations by the Method of Perturbation Coordinates," Journal of Applied Mathematics and Mechanics, Vol. 26, 1962, pp. 661-667.

[11]Nixon, D. and McIntosh, Jr., S. C., "Further Observations on the Strained Coordinate Method for Transonic Flows," AIAA Journal, Vol. 18, Dec. 1980, pp. 1540-1541.

[12]Holst, T. L. and Ballhaus, W. F., "Fast, Conservative Schemes for the Full Potential Equation Applied to Transonic Flows," AIAA Journal, Vol. 17, Feb. 1979, pp. 145-152.

[13]Dulikravich, D. S., "CAS2D-FORTRAN Program for Nonrotating Blade-to-Blade Steady, Potential Cascade Flows," NASA TP 1706, July 1980.

[14]Hicks, R. M., Murman, W. M., and Vanderplaats, G. N., "An Assessment of Airfoil Design by Numerical Optimization," NASA TMX-3092, July 1974.

[15]Hicks, R. M., and Vanderplaats, G. N., "Application of Numerical Optimization to the Design of Supercritical Airfoils without Drag-Creep," SAE Paper 770440, Business Aircraft Meeting, Wichita, Kans., March 29-April 1, 1977.

[16]Jameson, A., "Transonic Calculations for Airfoils and Bodies of Revolution," Grumman Aero. Rept. 390-71, Dec. 1971.

Epilogue

This book describes the state-of-the-art of transonic flow research from both an industry viewpoint and a research viewpoint. The main thrust of present research is in the field of computational fluid dynamics where a set of partial differential equations that describe the flow are numerically discretized and the resulting algebraic equations solved on a computer. However, not all of the research into transonic flow phenomena are numerical; for instance, recent advances in wind-tunnel testing techniques have led to more accurate test data for airplanes. Most of the recent advances in the subject have been considered in some detail by the authors of the preceeding chapters and it is not the intention to reiterate the various points here. Rather it is intended to outline the present status of research in the light of industry needs, and to point out areas that are neglected or those areas that require more basic work.

In the following discussion a fairly loose categorization is used in that the topics of numerical prediction methods and experimental work will be treated separately.

At present the main transonic prediction method is the finite-difference solution of the full potential equation or its small disturbance approximation. Solutions of the full potential equation are presently limited to the flow around simple wing-body configurations, although the inclusion of nacelles or a horizontal tail into these analyses is likely in the very near future. The less accurate small disturbance methods have been applied to fairly complex bodies but, of course, there is a trade between geometric complexity and accuracy. Both these computational methods have been coupled with a boundary-layer theory of either a strip application of a two-dimensional method or a three-dimensional method. Moving up the complexity scale of equations, there are very few Euler equation methods available, although there is considerable research effort in this area at the present time. The main advantage of solving the Euler equations is the removal of the irrotationality assumption which then allows treatment of flows with strong shock waves. Finally, there are some solutions of the Navier-Stokes equations available for very simple three-dimensional applications.

The potential equation computer codes seem to work reasonably well for simple geometries, such as those on a wing-body configuration typical of a transport. However, the accuracy for a wing-body configuration typical of a fighter is frequently less than adequate.

One problem that arises in assessing the accuracy of these computer codes is the unavailability of good experimental data for comparison. For example, it appears that there is not a good set of experimental data to validate the accuracy of a three-dimensional transonic potential equation/boundary-layer theory calculation. Recent work by Melnik[1] and his co-workers indicate that in two-dimensional boundary-layer theory the trailing edge behavior must be accurately modeled. A three-dimensional development of this theory would be difficult to assess without a good experimental data base.

There are three general uses of the full potential equation codes, namely analysis, optimization techniques, and inverse applications. An analysis code is one which solves the conventional problem, that is, given the body, what is the pressure distribution? The accuracy of these methods has been discussed above. The optimization technique couples an analysis code with a numerical optimization technique and seeks to minimize some "objective function," such as drag coefficient, by altering the shape of the wing. There are several problems with this technique at present. For instance, the drag cannot be accurately calculated by the numerical method, so that a minimization of a predicted drag may not necessarily lead to a reduction in actual drag. Most investigators now appear to minimize the difference between a specified and an actual pressure distribution over the wing; this is, perhaps, a more realistic objective function. There are also problems with the numerical optimization techniques and, to circumvent these, a considerable amount of ingenuity is sometimes necessary. This is costly in both time and money. The inverse methods solve the basic equations for a wing geometry having specified the surface pressure distribution. The problem here is to determine in advance what constitutes a good pressure distribution and whether this specified distribution will lead to a practical wing. A considerable degree of skill seems necessary to use this method. Finally, it should be noted that there is some doubt as to whether a pressure distribution can be specified at a specified freestream Mach number, since it appears[2] that the freestream Mach number must be found as part of the solution. It is not clear whether this restriction significantly reduces the practical power of the inverse method.

Solution techniques for the Navier-Stokes equations are available, but the computation time is long, and a considerable amount of research effort is being expended in trying to develop faster and more reliable algorithms. A major deficiency of numerically solving the Navier-Stokes equations is the large number of grid points needed to resolve the viscous phenomena. This contributes substantially to the required computation time. A second major problem concerns the turbulence model that is a crucial part of the technique. Since a large number of grid points are necessary for an adequate calculation, a simple algebraic eddy viscosity turbulence model is usually the only viable method because of the limitations of computer memory and speeds. The question then arises as to whether such a simple turbulence model can lead to a quantitatively adequate description of the flow. It should be noted that this type of turbulence model will give results that qualitatively describe the flow. It is possible that an integral boundary-layer theory will model the viscous effects for attached flow more accurately, because the empiricism in the method effectively contains a higher order turbulence model. If so, an alternative way to proceed would be to couple the Euler equations with a boundary-layer model.

In all of these finite-difference methods a critical problem is that of grid generation. The development of a satisfactory three-dimensional grid that conforms to a complex shape, such as an airplane, on one coordinate surface is a far from trivial task. This is a burgeoning field of research and hopefully this problem will be resolved in the intermediate future.

The main goal of the present family of computational methods is to model attached flows where both the viscous effects and shock waves are weak. This is a fairly satisfactory objective for transport configurations, but for fighter configurations it is woefully inadequate. The general feeling among those who design highly maneuverable military aircraft is that massively separated flow, whether from a strake or a canard, is something that is necessary for future design and that it must be controlled so as to interact favorably with the attached flow on the wing. There does not seem to be much work on the basic research necessary to develop a practical prediction method for this problem. Other "Cinderella" problems concern the prediction of roll, pitch, and yaw characteristics of airplanes in transonic maneuver. Also, the prediction of buffet, which is critical for the certification of a civil transport, is not in a very satisfactory state.

One interesting problem that arises in finite-difference solutions is that the equation that the numerical model actually solves may not be the differential equation that is discretized. For example, a numerical solution on a coarse grid of the Navier-Stokes equations resolves only the large rotational effects and not the viscous effects. Essentially the Euler equations are being solved numerically, and if this is so, then it is better to forget about the viscous terms at the start. In other words, the effects that are resolved by the numerical scheme should be clearly examined before any reliance is placed in the results.

Finally, in the realm of prediction methods, the calculation of unsteady flow phenomena are not as well developed as those for steady flow. Furthermore, these methods should be developed in such a way that structural effects can be easily incorporated, since it is probably necessary that aeroelastic effects should be included in the overall design optimization process.

One of the topics that was raised earlier is the validation of computer codes. The problem here is to develop an experiment that can be modeled numerically to a satisfactory accuracy. For example, the commonly used porous wind-tunnel walls cannot be modeled accurately at transonic speeds. A second point is that a real validation of a method requires a large number of comparative studies and there is some question as to the best way of achieving this. Aircraft companies usually have a large quantity of data, most of which are proprietary, which could be used for validation. However, often these companies do not have access to the computer power necessary to run the test cases. One suggestion is that NASA should provide the necessary computer power—perhaps the National Aerodynamic Simulator (NAS) will provide this capability. Another viewpoint is that industry should invest in its own computer resources and validate each technique "in-house."

Most of the articles in this book are concerned with prediction methods and there is a school of thought which holds that wind tunnels will be unnecessary in the future. However, wind-tunnel experiment is necessary for testing the complex geometries that prediction methods cannot treat, for example, a fighter configuration at high angle of attack. There is also the problem of validation of computational methods as discussed above. Some of the problems associated with wind-tunnel testing concern wall effects; others concern the ever-present mismatch of Reynolds number between wind-tunnel and flight conditions. Work is in progress on the first of these problems by several investigators and perhaps the National Transonic Facility will solve some of the scale effect problems. Finally, it should be noted that the day an airplane is flown without having been tested in a wind-tunnel is far off; no one has *that* much confidence in prediction methods.

The articles in this book do not purport to cover every aspect of transonic flow research. Some interesting additional questions that arise are listed below.

What kind of engineer should the universities be training to make best use of the rapidly changing prediction techniques and the more static wind-tunnel testing methods? It seems that in spite of the recent technical advances the engineer should have a good "feel" for what is right for an airplane design and what is wrong. This requires a combination of experience in traditional aerodynamics and an understanding of numerical algorithms.

Is there too much emphasis on developing techniques and not enough research into new concepts that would radically improve aircraft design? Also, is there too much emphasis on the sometimes complex numerical techniques at the expense of simpler, more approximate, techniques that would be useful in preliminary design?

Many of the problems associated with transonic flow have been resolved in the last decade, but there are still many more to be considered. It would be interesting to see a similar state-of-the-art review in 10 years' time.

David Nixon
August 1981

References

[1] Melnik, R.E., Chow, R., and Mead, H.R., "Theory of Viscous Transonic Flow Over Airfoils at High Reynolds Number," AIAA Paper 77-680, 1977.

[2] Volpe, G. and Melnik, R.E., "The Role of Constraints in the Inverse Design Problem for Transonic Airfoils," AIAA Paper 81-1233, 1981.

Author Index for Volume 81

Bhateley, I.C. 405
Blackwell, James A., Jr. 189
Bonner, E. 451
Bradley, Richard G. 149

Cox, R.A. 405

Dahlin, J.A. 511

Gingrich, P.B. 451

Haney, H.P. 431
Henne, P.A. 511
Hinson, B.L. 377
Hinson, M.L. 489

Inger, G.R. 621

Kerlick, G. David 239

Lomax, Harvard 297
Lores, M.E. 377
Lynch, F.T. 81

Mehta, Unmeel 297
Miranda, Luis R. 545
Morino, Luigi 565

Nixon, David 239

O'Neil, P.J. 467

Peavey, C.C. 511

Spreiter, John R. 3
Stahara, Stephen S. 637

Tai, Tsze C. 605
Tseng, Kadin 565

Verhoff, A. 467